Industrial Waste Treatment

Contemporary Practice and Vision for the Future

By

Nelson Leonard Nemerow

Industrial Waste Treatment

Editor

Nelson Leonard Nemerow

AMSTERDAM • BOSTON • HEIDELBERG • LONDON
NEW YORK • OXFORD • PARIS • SAN DIEGO
SAN FRANCISCO • SINGAPORE • SYDNEY • TOKYO
Butterworth-Heinemann is an imprint of Elsevier

Butterworth–Heinemann is an imprint of Elsevier
30 Corporate Drive, Suite 400, Burlington, MA 01803, USA
Linacre House, Jordan Hill, Oxford OX2 8DP, UK

Copyright © 2007, Elsevier Inc. All rights reserved.

No part of this publication may be reproduced, stored in a retrieval system, or transmitted in any form or by any means, electronic, mechanical, photocopying, recording, or otherwise, without the prior written permission of the publisher.

Permissions may be sought directly from Elsevier's Science & Technology Rights Department in Oxford, UK: phone: (+44) 1865 843830, fax: (+44) 1865 853333, E-mail: permissions@elsevier.com. You may also complete your request online via the Elsevier homepage (http://elsevier.com), by selecting "Support & Contact" then "Copyright and Permission" and then "Obtaining Permissions."

∞ Recognizing the importance of preserving what has been written, Elsevier prints its books on acid-free paper whenever possible.

Library of Congress Cataloging-in-Publication Data
Nemerow, Nelson Leonard.
  Industrial waste treatment : contemporary practice and vision for the future / Nelson Leonard Nemerow.
    p. cm.
  Includes index.
  ISBN-13: 978-0-12-372493-9 (alk. paper)
  ISBN-10: 0-12-372493-7 (alk. paper)
  1. Factory and trade waste—Purification. 2. Sewage disposal plants. I. Title.
  TD897.6.N46 2006
  628.4—dc22

2006021434

British Library Cataloguing-in-Publication Data
A catalogue record for this book is available from the British Library.

ISBN 13: 978-0-12-372493-9
ISBN 10: 0-12-372493-7

For information on all Butterworth–Heinemann publications
visit our Web site at www.books.elsevier.com

Printed in the United States of America
07 08 09 10   10 9 8 7 6 5 4 3 2 1

Working together to grow
libraries in developing countries
www.elsevier.com | www.bookaid.org | www.sabre.org

ELSEVIER   BOOK AID International   Sabre Foundation

BRIAN
(GRANDSON,
UNIVERSITY OF
CALIFORNIA
ENVIRONMENTAL STUDENT)

GLEN
(SON, SCRIPPS
RESEARCH INSTITUTE
SCIENTIST)

NELSON
(FATHER,
PHD, ELSEVIER
ENVIRONMENTAL BOOK
AUTHOR)

Contents

Prologue ix

List of Tables xi

List of Figures xv

Industrial Environmental History 1

Introduction to Industrial Waste Treatment 9

Part A—Twentieth Century **11**

1. Theories and Practices 13

2. Contaminant Concentration Reduction 25

3. Neutralization 35

4. Equalization and Proportioning 45

5. Removal of Suspended Solids 53

6. Removal of Colloidal Solids 79

7. Removal of Inorganic Dissolved Solids 89

8. Removal of Organic Dissolved Solids 105

9. Treatment and Disposal of Sludge Solids 149

10. Joint Treatment of Raw Industrial Waste with Domestic Sewage 175

| 11. | Hazardous Wastes | 245 |
| 12. | Removal of Industrial Air Contaminants | 355 |

Part B—Twenty-First Century   **361**

Foreword to the Twenty-First Century   i

Preface to the Twenty-First Century   iii

13.	Prologue to the Twenty-First Century	363
14.	Rationale of Environmentally Balanced Industrial Complexes	369
15.	Procedure for Industry in Attaining Zero Pollution	373
16.	Economic Justification for Industrial Complexes	379
17.	Realistic Industrial Complexes	405
18.	Potential Industrial Complexes	443
19.	Potential Municipal–Industrial Complexes	515
20.	Naturally Evolving Industrial Complexes	527
21.	Benefit-Related Expenditures for Industrial Waste Treatment	531
22.	Summary	545

Index   549

Prologue

In 1963, one year after Rachel Carson published her last book, *Silent Spring*, which warned of the interdependence of industrial waste pollution and human development, I published my first textbook on the subject of industrial waste treatment, *Theories and Practices of Industrial Waste Treatment*. Most of my current readers were not actively participating in this field at that time. Therefore, I feel impelled to republish—in modified and somewhat updated form—the considerable amount of historical theories and practices of twentieth-century industrial waste treatment. It is remarkable how much of this technical information has remained the same since the publishing of that original book, but it is difficult to locate a book from that long ago. I followed that book with updated books on the subject—*Liquid Wastes of Industry* in 1971, *Industrial Water Pollution* in 1978, *Industrial and Hazardous Waste Treatment* with Dr. Dasgupta in 1991, *Zero Pollution* in 1995, and *Strategies of Industrial and Hazardous Waste Management* with Dr. Agardy in 1998. But they too (except for the last two) would be hard to locate in most libraries. In fact, many of the publishers themselves have changed or gone completely out of the business. For an updated and excellent version of current conventional industrial waste treatment equipment and practices (including costs), I urge you to consult another Elsevier text, *Industrial Waste Treatment Handbook*. It was prepared by the collective effort of the firm Woodard and Curran of Portland, Maine. Many of the figures in this book are replicated from the above titles with permission from John Wiley and Sons.

I have made a special effort in this book to reference publications that are still in print and available for purchase. In that way readers may opt to select works that serve their specific interests.

I intend for this book to be an overview of the subject of industrial waste treatment and disposal as used in the twentieth century and how it is evolving into a new conceptual field as we enter the twenty-first century. Further, I have attempted to provide some historical data of people and concepts of industrial waste treatment for generations to come to look back on for a more complete understanding of its significance in industrial production and how we naturally evolved the solutions to which we must resort in the twenty-first century.

It is critical to note that this book not only recounts the past theories and practices, but even more importantly, confronts the present dilemma with innovative solutions to industrial wastes for the future.

Nelson Leonard Nemerow

List of Tables

1	Primary Personnel and Fields of Study in U.S. Universities	7
1.1	Typical Analyses of Sewage Effluents After Conventional Primary and Secondary Treatment	19
1.2	Composition of Secondary Treated Municipal Wastewater Effluents and Irrigation Water	20
2.1	Wastes from a Textile Mill	28
3.1	Cost Comparison of Various Alkaline Agents[a]	38
5.1	Rectangular Primary Settling-Tank Data	60
5.2	Circular Primary Tanks: Long-Term Performance Data[a]	62
5.3	Typical Efficiencies of Dissolved-Gas Flotation Treatment of Wastes	69
5.4	Results Obtained on Humus Tank Effluent at Eastern Sewage Works, London, England, December 30, 1966–January 13, 1967	75
6.1	Types and Characteristics of Colloidal Solids	80
6.2	Valence and Coagulant Dosage	84
6.3	Properties of Coal-Derived Granular Carbon for Waste Treatment	86
7.1	Typical Overall Coefficients in Evaporators	91
7.2	Elemental Composition of Green Algae	95
7.3	Occurrence of *Cyanophyceae* and *Chlorophyceae* in Massachusetts Lakes and Reservoirs	96
7.4	Refractory Containment Removal Techniques	100
8.1	Biological Degradation of Organic Constituents in Sewage	107
8.2	Materials Potentially Useful for Liners of Ponds Containing Hazardous Wastes	110
8.3	Summary of Disposal Systems	118
8.4	Summary of Disposal Systems	129
10.1	Industrial Contaminants and General Limiting Values for Discharge into Municipal Sewerage Systems[a]	187
10.2	Sewer Use Ordinance City of Palo Alto Regional Water Quality Control Plant (September 1995)	191
10.3	Maximum Allowable Discharge Limits for Wastewater	192
10.4	Allocation of Fixed Costs	194

10.5	Allocation of Operation and Maintenance Costs	195
10.6	Summary of Allocation of Fixed and Operating Cost	196
10.7	Calculation of Users' Charges Based on Three Factors	198
10.8	Cayadutta Creek Analyses in October 1964	201
10.9	Time of Flows from Station 5 Downstream Obtained 1 Week Before Stream-Sampling Program in October 1964	203
10.10	Summary of 7-Day Sampling of Cayadutta Creek in Dry Period, from 10/8/64 to 10/18/64	203
10.11	Minimum Flow Data of Measured Creek Compared with Cayadutta Creek	205
10.12	Normal Probability Distribution Analysis of Data (1927–1960)	206
10.13	Summary of Data Required from Cayadutta Creek Analyses in October 1964 for Churchill Method of Analysis	207
10.14	Churchill Analysis Applied to Cayadutta Creek Data	208
10.15	Summary of 24-Hour Sampling Results	209
10.16	Summary of 24-Hour Sampling Results	215
10.17	Composite Analysis (24 Hours) of Gloversville and Johnstown Wastewater	216
10.18	Summary of Total Loads for Treatment	218
10.19	Industrial Production During Sampling Days	220
10.20	Water Consumption Related to Production Percentage[a]	222
10.21	Industrial Waste Flow	224
10.22	Sludge Digestion (Laboratory Study)[a]	225
10.23	Activated-Sludge Pilot Laboratory Studies	226
10.24	Prototype Operating Data	230
10.25	Prototype Operating Results and Design Parameters	231
10.26	Inconsistencies Between Theory of Design and Actual Practice in Design	242
11.1	Cost of Hazardous Waste Disposal Practices	250
11.2	Hazardous Waste Industries	251
11.3	List of Hazardous Characteristics	252
11.4	RCRA-Regulated Hazardous Wastes	254
11.5	Typical Automotive Oil Waste Composition	255
11.6	Sources of Asbestos Wastes	269
11.7	Hazardous Waste Disposal in United States, circa 1980	275
11.8	Organic Chemical Treatment	276
11.9	Commercial Hazardous Waste Disposal Methods	276
11.10	Emerging Alternative Technologies, circa 1985	278
11.11	Incineration	285
11.12	Typical Industrial Laundry Wastewater Constituent Concentration	304
11.13	Some New Treatments for Hazardous Waste, Many Still on the Drawing Board	318
11.14	Water Quality Limits for Toxic Pollutants for Three Uses	328
11.15	Allowable Concentrations for Air Contaminants Resulting from Hazardous Waste Treatment, Storage, and Disposal Emissions (EPA)	334

List of Tables xiii

16.1	Comparison of Real Production Costs (1995 Dollars) of Free Standing Fertilizer and Cement Plants with EBIC Costs (1995 Dollars) per Ton of Fertilizer	382
16.2	Real Production Costs^a of Fertilizer and Cement Plants	391
16.3	Comparison of Classical Production Costs^a of Free-Standing Fertilizer and Cement Plants with Real Production Costs^a per Ton of Product	391
16.4	EBIC Costs in a Fertilizer^a-Cement Industrial Complex	392
16.5	Comparison of Classical Production Costs^a of Free-Standing Fertilizer and Cement Plants with EBIC Costs^a per Ton of Product	392
16.6	Comparison of Real Production Costs^a of Free Standing Fertilizer and Cement Plants with EBIC Costs^a per Ton of Product	392
16.7	Comparison of Production Cost^a of a Ton of EBIC Product (0.22 Ton Fertilizer + 0.78 Ton of Cement) and Equivalent Masses of Free-Standing Plant Products^a	393
17.1	Raw Materials Consumed for Portland Cement in United States 3 (Thousands of Short Tons)	407
17.2	External Raw Materials and Manufactured Products in Three-Industry Complex (Stage 3)	417
17.3	World Production	421
17.4	Sugarcane Composition	421
17.5	Chemical and Physical Composition of Bagasse and Mudcake (4)	423
17.6	Composition of Ash from Cachaza and Bagasse [Dasgupta 1983]	425
17.7	Raw Material Balance as Part of Total Production Cost	433
18.1	Air Emissions from Model Coke Plants	447
18.2	Wood-Preserving Processes	451
18.3	Amount of Raw Materials Consumed for Production of Portland Cement	457
19.1	Annual Cost and Environmental Benefits of Successful Synergies	525
21.1	Summary of Cost Data of Four Plants	539
21.2	Pulp and Paper Industry Treatment Costs, Sales Revenues (1970), and Sales Indices (Computed)	542

List of Figures

2.1.	Time Elapsed After Start of Equalization	29
3.1.	Neutralization Accomplished by Mixing of Wastes	36
3.2.	Nomograph for Treatment of Acid Wastes: A Chart for Determining the Amount of Alkaline Agent Needed	39
3.3.	Submerged-Combustion Pilot Unit as Used by Remy and Lauria (1958)	41
3.4.	Acid Required to Neutralize Industrial Wastes in Sewer	42
4.1.	Effect of Equalization	46
4.2.	Top View of an Equalizing Basin, with Perforated Inlet Pipe and Over-and-Under Baffles	46
4.3.	Side View of an Equalizing Basin, with Mechanical Agitators Instead of Baffles	47
4.4.	Waste-Metering System	49
4.5.	Effect of Proportioning	50
5.1.	Effect of (A) Doubling the Floor Area and (B) Halving the Depth of a Settling Basin	55
5.2.	Effect of Overflow Rate on BOD Removal	56
5.3.	Flocculation Increases Settling Rate	56
5.4.	Effect of Turbulence on Particle Path	57
5.5.	(A) Fast and Good Settling Characteristics Typical of Heavy Suspended Solids. (B) Medium and Normal Settling Characteristics Typical of Homogeneous Mixture of Solids. (C) Slow and Poor Settling Characteristics Typical of Highly Colloidal and Finely Divided Solids	57
5.6.	Inlet Zone of a Circular Tank (A) Occupies 20–40% of Tank Area. Inlet Zone of a Rectangular Tank (B) Occupies only 10–15% of Tank Area	57
5.7.	(A) Circulator Tank. (B) Square Tank	58
5.8.	Typical Dispersion Curves for Various Tanks (see text for explanation). The Vertical Axis Shows the Ratio of the Actual Concentration of Contaminant (C) to the Concentration of Contaminant Mixed with the Entire Tank Volume (C_o); the Horizontal Axis Shows the Ratio of the Actual Time (t) that a Concentration Takes to Reach the End of the Tank to (T), the Total Detention Period (vol/rate) (Adapted from Camp 1953.)	59

List of Figures

5.9.	Graphical Analysis of Settling Test Results	63
5.10.	Rate of Rise of Air Bubbles in Tap Water (Calculated by Means of Stoke's Law) as a Function of Bubble Size	68
5.11.	Solubility of Air in Distilled Water at Various Temperatures	70
5.12.	Methods of Dissolved-Air Flotation. (A) Adhesion of a Gas Bubble to a Suspended Liquid or Solid Phase. (B) The Trapping of Gas Bubbles in a Floc Structure as the Gas Bubbles Rise. (C) The Absorption and Adsorption of Gas Bubbles in a Floc Structure as the Floc Structure is Formed (Adapted from Vrablik 1960)	71
5.13.	Schematic Drawing of Pressure Flotation System	71
5.14.	North Water Filter	72
5.15.	The 48-inch-Diameter Sweco Separator Shown Is Screening Lint from Wastewater at the Eastern Overall Company, in Baltimore, Maryland. The Wastewater Is Fed onto the 60-Mesh Market-Grade Screen at a Rate of 300 gpm. The Screened Wastewater is Discharged to the Sewer (Photograph Courtesy Sweco, Inc., Los Angeles, California)	73
5.16.	Cutaway View of a 7½-Foot-Diameter Microstrainer	74
6.1.	Effect of Colloidal Type on Viscosity	80
6.2.	Stable Colloid	83
6.3.	(A) Granula-Carbon Reactivation Cycle. (B) Adsorber Configuration for Granular Carbon Waste Treatment	85
7.1.	Typical Dialysis Flow Diagram	92
7.2.	The Desalination Process Using Energy Recovery and Artificial Intelligence Control	98
8.1.	The Role of Algae in Stabilization Ponds	108
8.2.	Accelerated-Oxidation Pilot-Plant Basins	109
8.3.	Schematic Diagram of Step-Aeration Treatment. Step I, High Sludge Seed (4,000 ppm); Step II, 2,000 ppm; Step III, 1,000 ppm; Step IV, 800 ppm	112
8.4.	*Sphaerotilus*-Like Organism, Sheathed and Unsheathed (×620)	114
8.5.	Round-Ended Rods in a Capsule of Slime (×620)	115
8.6.	Effect of Temperature on Average BOD Reduction of a Synthetic Protein–Glucose Waste, Using a Dispersed-growth Aeration System, After 24 Hours of Aeration and No Settling	115
8.7.	Schematic Arrangement of Contact-Stabilization Process	116
8.8.	Complete-Mixing Activated-Sludge System	117
8.9.	Changes in Nitrogen Occurring in Filter	120
8.10.	Diagram of Experimental Controlled-Filtration System	122
8.11.	Single-Stage Trickling Filter (with Recirculation)	123
8.12.	Two-Stage Series-Parallel Biofiltration Process	123
8.13.	Schematic Arrangement of Wet-Combustion–Process Units	124
8.14.	Typical Cavitator System	126
8.15.	Typical Surface Equipment for Deep-Well Waste Injection from Waste Sump Underground	127

List of Figures xvii

8.16.	Typical Injection Well	128
8.17.	A Column Foam Fractionator	131
8.18.	Typical Layout of an Oxidation Ditch Treatment Plant	132
8.19.	A Comparison of Unit Disposal Costs	133
8.20.	Rotating Biological Contractor (RBC) Disc System	134
9.1.	Reduction of Volatile Matter in Raw Sludge by Digestion	153
9.2.	(A) The Coilfilter, a Aatented Machine for the Vacuum Filtration of Sludge. This Particular Machine, in Use Since 1953 at the Sewage-Treatment Plant at St. Charles, Illinois, has Filtering Media Made up of Two Layers of Alloy Steel Coiled Springs, Each Spring Made Endless by Joining Its Two Ends with a Threaded Plug. These Springs Discharge the Filter Cake After Each Revolution of the Cylinder and Are Then Washed Before They Reenter the Vat for Another Cycle. The Material at the Left, Which Looks Like a Length of Corduroy, Is Actually a Layer of Sludge (adapted from Komline-Sanderson Engineering Corp.) (B) Schematic Drawing of the Coilfilter Shown in (a) (Courtesy Komline-Sanderson Engineering Corp.)	155
9.3.	Effect of Elutriation on $FeCl_3$ Required for Conditioning of Sludge	156
9.4.	Schematic Diagram of the Zimpro Process for Sewage-Sludge Oxidation	160
9.5.	Apparatus for the Atomized-Suspension Technique	162
9.6.	Effect of Stack Temperature on Thermal Efficiency	163
9.7.	Flow Diagram for Heat Balance for a Flash-Drying and Incineration System	164
9.8.	Centrifuge Bowl Schematic	166
9.9.	Schematic Plan of Thickener Mechanism and Section of Tank	170
10.1.	Twelve Alternatives of Industrial-Waste Treatment Systems	176
10.2.	Allocation of Fixed Charges on the Intercepting Sewers	196
10.3.	Allocation of Fixed Costs for Treatment Plant	197
10.4.	Cayadutta Creek	199
10.5.	Population Growth of Gloversville and Johnstown	200
10.6.	Laboratory BODs for Stations 5 and 6 at 20°C. Each Point Represents an Average of Seven Samples Collected from the Creek on 7 Different Days at Different Times of Day, All During a Drought Flow Period (October 8–18, 1964)	204
10.7.	Minimum Flow of Cayadutta Creek and Expected Recurrence of This Level	207
10.8.	BOD Reduction Required at Station 5, at 12.4°C and 2 ppm DO Remaining and Based on a Waste Discharge of 23,750 lb/day	214
10.9.	Special Design curve for Computing Treatment Plant Requirements at 12.4°C and 2 ppm DO at Station 5	214
10.10.	Activated-Sludge Treatment: BOD Reduction Related to BOD Loading	227
10.11.	Field Prototype of the Gloversville–Johnstown Joint Treatment Plant	229
10.12.	Line Diagram and Hydraulic Profile of the Gloversville–Johnstown Joint Waste–Water Treatment Plant	234
10.13.	General View of the Gloversville–Johnstown Joint Treatment Plant	235

11.1.	Definition of a Solid Waste	246
11.2.	Definition of a Hazardous Waste	247
11.3.	Special Provisions for Certain Hazardous Waste	248
11.4.	Regulations for Hazardous Waste Not Covered in Diagram 3	249
11.5.	Wood-Preserving Processes	263
11.6.	Mobile PCBX unit	272
11.7.	Supersorbon Solvent Recovery	274
11.8.	Thermal Decontamination Process flow Diagram	282
11.9.	Flameless Thermal Oxidizer	288
11.10.	Typical System	289
11.11.	Typical System	290
11.12.	An Installation	291
11.13.	Soil and Groundwater Detoxification System	309
11.14.	Detoxifier System	310
11.15.	Waste Management Services Listing Form	313
11.16.	Material Available/Wanted Listing Form	314
11.17.	Separation of Hazardous Wastes Using Membrane Filtration	316
11.18.	RCRA TSD Part B Permit Application	324
11.19.	U.S. EPA Office Locations	325
11.20.	EPA Process Permit Flowsheet	326
12.1	Treatment of Waste Gases	356
16.1.	Mass Flow Diagram of a Free-Standing Fertilizer Plant	381
16.2.	Clean Environment Surrounding the Plants	383
16.3.	Free-Standing Cement Plant	385
16.4.	Mass Flow Diagram of a Free-Standing Cement Plant	386
16.5.	Mass Flow Diagram of a Fertilizer-Cement Complex	386
16.6.	Production of Hydrogen by Electric Disassociation of Water	397
17.1A.	Dry and Wet Processing	408
17.1B.	Dry and Wet Processing	409
17.2A.	Cement–Fertilizer–Municipal Complex	410
17.2B.	Environmentally Balanced Fertilizer-Cement Plant Complex Phase	411
17.3.	Schematic Diagram Environmentally Balanced Phosphate–Fertilizer–Cement Industrial Complex	413
17.4A.	Three-Industry Complex: Tannery–Slaughterhouse–Rendering	415
17.4B.	Three-Industry Complex: Tannery–Slaughterhouse–Rendering	416
17.5.	Raw Sugar Manufacturer-Flow Diagram	419
17.6.	Sugarcane Refinery-Based EBIC (Sugarcane–Power–Alcohol Complex)	420
17.7.	Sugarcane Production	422
17.8.	Current Sugar–Mill Situation	424
17.9.	Environmentally Balanced Sugarcane Complex	428
17.10.	Diagram of the Integrated Five-Plant Industrial Complex	431
17.11.	Pulp and Paper Mill Complex	435
17.12.	Paper Mill Complex	439
18.1.	Wood–Paper Mill Complex	445

List of Figures xix

18.2.	Steel Mill–Coke and Gas Plant Complex	447
18.3.	Metal Parts–Plastic Plant Complex	448
18.4.	Metal Finishing–Plastic Manufacturing Complex	449
18.5.	Organic Chemical–Wood Preserving Plants Complex	450
18.6.	Organic Chemical–Wood Preserving Plants Complex	451
18.7.	Biomass Power Plant–Municipal–Forestry–Agriculture Complex	452
18.8.	Steel Mill–Fertilizer–Cement Complex	454
18.9.	Schematic Diagram of a Typical Rotary Steam Kiln Boiler	457
18.10A.	Schematic Flow Diagram of the EBIC for Power Plant Industry	458
18.10B.	Simple Coal Power Plant Complexes	460
18.11.	Flowchart for Polystyrene Production	463
18.12.	Tubular Reactor Process for Low-Density Polyethylene Production	464
18.13.	Philips Process for High-Density Polyethylene Production	465
18.14.	Plastic Manufacturing Industrial Complex	467
18.15.	Fluosolids System (Dorr-Oliver, Inc.)	468
18.16.	Cement, Power, and Lime Flowchart	469
18.17.	Lumber Mill Complex—Four Products	471
18.18.	Power Plant-Agriculture Complex	472
18.19.	Cannery-Agriculture Complex	474
18.20.	Nuclear Power–Glass Block Complex	475
18.21.	Feedlot–Food Production Complex	478
18.22.	Gas-Producing Plants	480
18.23.	Flow Chart of Dilute Alcohol Production	481
18.24.	Water, Electricity, Chlorine, and Lye Plants Complex	483
18.25.	Power-Aluminium-Red Brick Plants Complex	484
18.26.	Corn Food–Alcohol Production Complex	486
18.27.	Restaurant-Paint Manufacturing Complex	488
18.28.	Oil Drilling Offshore-Seashore Recreation Complex	490
18.29.	Metal Plant-Dry Cleaning-Coffee Plant Complex	491
18.30.	Electrical Storing/Converting Voltage–Wax Plant Complex	493
18.31.	Nuclear Power–Waste Recovery–Cannery Complex	495
18.32.	Coal Power Plant-Desalination Water Plant Complex	497
18.33.	Vegetable Pickling Cannery–Inorganic Chemical Complex	498
18.34.	Sugar–Ethanol–Gasoline Complex	500
18.35.	Reclaimed Wireless Phones–Cement Plant–Concrete Products Plant	502
18.36.	Sugarcane-Fuel Briquette Complex	504
18.37.	Hog Production–Animal Feed Energy-Environmental Complex	505
18.38.	Seawater Desalination Plant–Borax Plant Complex	506
18.39.	Cow Feed lot–Power Plant–Fertilizer Complex	507
18.40.	Reused Plastic Waste–Consumer Products Complex	508
18.41.	Lumbar–Textile–Corngrowing–Alcohol Producing Industrial Complex	510
19.1.	Schematic Diagram of One Type of Municipal-Industrial Complex	516
19.2.	Municipal Waste Water–Agriculture (Food) Complex	517
19.3A.	Lake Industry–Village Complex	519

19.3B.	Food-Electricity and Water Production Plant	521
20.1	The Industrial Ecosystem at Kalundborg, Denmark	528
21.1.	Unit Price Calculation	536
21.2.	Sales Values, Treatment Costs, and Sales Indices of the Pulp and Paper Industry	541

Industrial Environmental History

Introduction to History of Environmental Wastes

The pollution of ground and surface waters in the United States began as soon as industry began producing manufactured goods and wasting liquids and solid matter simultaneously. One needs only to trace the Industrial Revolution period in the United States.

Industrial liquid effluents can be traced as far back as the nineteenth century. If we define pollution as that amount of industrial contamination that causes interference with the best usage of the receiving water, we can probably agree that this type of pollution did not begin until the turn of the twentieth century. Basic industries such as coal, power production, dairy, textile, cannery, tannery, and paper, which produce goods necessary for the sustenance of life, were the first to face the pollution problem.

Chemical industries, mainly inorganic, such as salt and salt degradation products including chlorine, lye, and soda ash, were next chronologically and in importance. These were followed by basic organic chemical plants, such as sugar, starch, and cellulosic wastes.

In the 1930s, all these industries began to be aware of the eventual danger of their wastes when sent untreated into waterways. It was natural for industry at that time to follow the lead of municipalities in using similar treatments to attempt to resolve their pollution problems. Then came World War II and its accelerated industrial production activity. New products and new wastes evolved and the quantity of older ones, such as metal plating, oil refining, and textiles increased tremendously.

Radioactivity, petrochemical, and synthetic organic chemicals were largely developed and surfaced in the environment in the 1940s and 1950s. During this period, major environmental problems surfaced with rapid and serious consequences.

Heretofore industrial waste treatment followed the examples of municipal waste treatment with certain variations in efficiency. The reason for success up to this point was the relative compatibility of these industrial and municipal wastes.

After the 1940s, however, new industrial products produced new wastes that exhibited toxicity as well as nonbiodegradability. These wastes did not respond to normal municipal (sewage) waste treatment. Industry found itself in a dilemma that persisted for several decades. Many decided to use antiquated municipal treatment methods despite knowing that the wastes would not respond appropriately and would find their way into the environment with disastrous results. At the same time industry and governmental agencies conducted serious studies to find appropriate treatment of these new wastes. Unfortunately, even today we find ourselves in much the same position as 50 years ago, although some progress has been made by recycling and changing production methods and materials.

The move to pollution prevention has been a slow one. It wasn't until the 1930s that the nation acknowledged that solving its pollution problems would require time, study, and most of all, money. Many of the early answers to these problems were stop-gap measures intended to prevent overloading the environment. All too often, "solutions" were selected based on economics. Eventually company managers began to realize that recovering and reusing or selling their wastes was more practical and economical than treating and disposing of them.

Reuse began in the 1950s, when companies started to recover metals from plating wastes, fatty hides and hair from tannery wastes, and fine fibers from papermill wastes. By the 1980s companies also were reusing water and burning waste oil to produce energy. Recovery and reuse or sale of waste products grew but has been hampered by transportation hazards and costs, the difficulty of locating buyers, frequent disparities between supply and demand, and industry's discomfort with fully disclosing waste characteristics to potential buyers.

In the late 1990s pollution prevention became the buzzword described as a "win-win" effort. The environment wins because fewer pollutants are created, and industry wins because preventing waste minimizes or eliminates liability and disposal costs.

One of the more promising recent innovations is the creation of emissions and effluent trading programs in which a company spends money to treat or eliminate waste and receives "credits" from an environmental control agency. These credits can be sold to other companies that, for whatever reason, cannot reduce their wastes. The company that sold the credits could use the money to offset part of the cost of reducing or eliminating its waste streams.

This practice too has been slow to gain universal acceptance, because companies are still reluctant to accept the idea of paying for something they have been getting for years at little or no cost. They also recognize that whether they buy credits or pay to reduce or treat their wastes, production costs will still rise. The concept, however, is a good one and, in my opinion, eventually will prevail.

Even pollution trading, however, will not bring us to zero pollution. Achieving this goal will require a more radical approach: the creation of environmentally balanced industrial complexes (EBICs). Under this concept, "compatible" plants would locate in the same complex so that one plant's wastes would be another's raw material. Again, this would reduce production costs while eliminating environmental damages.

Despite the potential advantages of EBICs, which I have been advocating since 1977, several constraints have prevented their widespread acceptance: Relocating plants can be costly, initiating negotiations between compatible plants is difficult, and matching production objectives and quantities is not easy. Even if these obstacles can be overcome, company managers will still require proof that EBICs would reduce "real" production costs, which include direct costs of waste treatment and the indirect costs of environmental damage in addition to the cost of labor, materials, and utilities.

In 1995 and 1996, several colleagues and I studied and reported on the economics of co-locating fertilizer and cement plants in a single complex. The primary objective was to prove that these industries could manufacture their products less expensively if they were located in the same complex. These results and recommendations will be described in detail in Chapter 4. Once real costs are calculated, a comparison can be made between these costs at an EBIC and those at facilities in different locations.

The concept of EBICs was originally proposed for the pulp and paper industry in 1977 (Nemerow 1977). Over the next 22 years we have published many papers containing potential industrial complexes for many industries. Most of these are described in the author's book, *Zero Pollution for Industry* (Nemerow 1995).

The field of industrial waste treatment from the 1940s to the 1980s evolved into one of industrial waste utilization in the 1990s. Society is demanding lower manufacturing costs as well as less environmental degradation. Outside of ceasing production, the use of EBICs is not only the logical answer, but the only response to society's urgent need. The EBIC system must include compatible industrial plants. Such a system completely changes our concept of industrial manufacturing. No longer can we locate industrial plants based solely on the economic marketability of a product, but now we must consider the usefulness of its waste as a raw material for an ancillary plant. To discharge waste untreated into the environment is simply not an alternative. And to discharge the same waste when treated is too costly for both the industry and society. Simple logic dictates that this waste be utilized directly by another manufacturer to save money for the plants and the quality of the receiving environment for society.

Progress of Industrial Revolution

The Industrial Revolution as described by most historians did not happen overnight. It was a gradual, slow-moving process. Its progress was greatly affected by certain manufacturers and certain means of production. Most historians concur that it began in the late 1700s, largely in European countries. Novel machinery introduced in and around 1820 (such as the steam engine and machine-driven textile looms) produced remarkable increases and changes in productivity. From the 1860s, moreover, these same industrial developments intensified to such a degree that the United States already possessed about one-quarter of global world's industrial production. Productivity was enhanced greatly by the advent and increased use of steam power, railways, electricity, and other instruments of modernization.

The German socialist author Friedrich Engels is credited with coining the term, "industrial revolution" in 1844. He also attributed this period to the transformation of an agrarian to an industrial society occurring in England from the mid-eighteenth to the mid-nineteenth centuries.

Progress of the Industrial Revolution was enhanced largely by developments in three major industrial sectors: (1) textile, (2) iron and steel, and (3) power. Each industry, growth was energized by specific patented inventions.

Prior to the rapid development of industry, all factories located on rivers for enhanced transportation of raw materials and products and were powered by this same water. Its elevation was such that it supplied water pressure to the factories to turn water wheels that furnished power to the machinery for production processes.

Textiles, previously manufactured by hand in a multitude of separate homes, began to be mass-produced in factories prior to the mid-eighteenth century. The patented invention in the textile industry was the "flying shuttle" in 1733. Weaving of cloth sped up dramatically when compared to hand-loom weaving. Many other inventions followed in this industry, culminating in Eli Whitney's cotton gin in the late 1800s. Other spin-offs occurred to society as a whole. When Elias Howe demonstrated in the mid-nineteenth century that a garment could be produced faster with his technology than by up to five women by hand, the textile industry became an American leader in productivity. Interestingly, this industry began in Lowell, Massachusetts, which also became the site of the first industrial pollution research laboratory in the United States.

Iron and steel, previously manufactured by subjecting iron ore to heat treatment with charcoal from trees, now could be reduced in England with coal. Coal proved much more efficient in converting iron ore to pig iron and then to steel. More power, however, was required to force air through the densely packed steel-making furnaces. Fortunately, James Watt's invention of the steam engine in the 1760s aided in providing this power, which was also used to pump water out of the coal mines. In the United States, like England before it, iron and steel plants located near the coal mines to facilitate transporting coal to the plant. As a result both coal mining and steel manufacturing were located primarily in the eastern United States such as Pennsylvania and Ohio.

Power, originated by Watt's steam engine, was further developed and enhanced by other inventors to make its equipment less bulky and more mobile. Henceforth, people as well as raw materials and products could be moved easily and more rapidly than ever before.

It was not until the late eighteenth or the early nineteenth century that steam power began to supplement hydropower. Although the latter was less expensive, steam power allowed industry to locate their factories at some distance from waterways and closer to their raw materials and product markets. By the mid-nineteenth century, railroads became available to industry to transport their goods more efficiently than ships.

Ironically, the U.S. government, spearheaded by Secretary of the Treasury Alexander Hamilton, encouraged the growth of the industrial revolution. Hamilton wanted to compete economically with the industrial nations of Europe despite Thomas Jefferson's belief in keeping the United States a nation of small farms.

Free usage of new patents before 1790 and establishment of a separate bank solely for encouraging industrialization enhanced rapid industrial development. Innovative industrial machinery, when patented after 1790, became a source of revenue for inventors to further encourage industrialization. Mass production of machine parts, stimulated by Eli Whitney's cotton gin and later his firearms parts around the end of the eighteenth century, signaled the real beginning of the industrial era as we know it today.

Working conditions in and around the nineteenth-century factories deteriorated once the scarce labor demand had been satisfied. This brought on labor unions and strikes to attempt to ameliorate working conditions for laborers. However, pollution of the environment outside the plants received little or no attention during this period. In all fairness, the reason for this probably was the sparseness of industrial plants and the abundance of receiving waters.

During the nineteenth century, raw materials and finished products were transported to and from markets by steamboats. Even these caused water pollution and environmental damages due to the many boiler explosions in the early boats. As steam engines became more reliable, steam-powered trains replaced boats as a cheaper form of transportation over land areas. The growth of rail transportation was once again stimulated and supported by financing by the government.

Communication by telegraph developed simultaneously with the railroads and their lines often followed the same path as the tracks. Industrialization was accelerated both by rapid communication and transportation. By the beginning of the twentieth century, communication was further enhanced by Bell's new telephone system and later enhanced by all of Thomas Edison's inventions.

In the late nineteenth century (around 1880), electricity for homes, offices, and factories completely changed life in the United States. Electricity production utilized water and produced wastewater contamination even if it was only in the form of increased temperature. But this was a start in electrical energy water pollution that continued to grow as sophistication in electricity output increased.

Industries such as meat packing and steel learned during the late nineteenth century how to keep prices low by controlling all the raw material as well as the production markets. Usually these industries were led by great innovators such as Swift or Carnegie in the above-mentioned industries or John D. Rockefeller in the oil industry.

Between 1880 and 1910, the United States experienced the greatest rate of industrial growth and rise in air and water pollution. People learned also how to live with polluted air and water especially in or near the cities.

As Corrick (1998) concludes, "the industrial revolution is thus far from over." Most historians agree that the age of industrial growth is an always continuing phenomenon. Corrick suggests that "its next location may not even be on earth, but in space, where manufacturing may be possible using raw materials mined on the moon or from asteroids." He laments that despite the benefits of the industrial revolution, "it has also given us a world whose water and air are polluted with industrial waste, and an urban lifestyle that is rushed and stressful." The sad truth is that we chose to tolerate this pollution rather than make the necessary effort to avoid it.

6   *Industrial Environmental History*

Developing countries, where industrialization occurs later in time, often use economics as a legitimate excuse for avoiding pollution control expenditures. They often follow the bad example previously set by their developed country brethren of trying to use dilution as the solution to pollution. In this way they maintain that they keep their production costs down and are more competitive with the rest of the world.

As a consultant to many of these countries I often heard the comment that "you escaped pollution costs for years, why shouldn't we do the same"?

With clothing, steel, and steam power leading the way, American industry rapidly developed into a world-dominating power. In America, the necessary ingredients of natural resources, increasing wealth, and product demand existed. Other early ancillary industries also flourished such as food, dairies, paper, transportation, and metals.

The environmentalist may note that coincident with the development of these industries were the use of large volumes of water and the simultaneous wastage of this same water with added contaminants. Hence was born the advent of what was to become the water pollution problems of the twentieth century.

Academic and Scientific Development of Environmental Knowledge in the United States

Environmental knowledge became established in and through various U.S. universities. Each of these universities contributed special environmental knowledge in specific areas. These specialties were dependent upon and selected by the leading professor(s) at these institutions. In Table 1, I have attempted to recall and list the primary universities involved, the major environmental subjects, and the leading professor(s). These should serve as a reminder of their contributions made to enhance the knowledge of water pollution abatement in the United States. Sadly, some of this knowledge was not put into practice until "after the fact" of pollution. Indeed, other knowledge was not made fully-available to the universities and general public by industries for a variety of self-serving reasons.

Universities were supplemented by research carried out by the U.S. Public Health Service, which also had its major areas of concentration and originators. It was and still is located in Cincinnati, Ohio, and focused on chemistry and microbiology with C. C. Rucchoft, Stream pollution with H. Streeter, toxicity and analysis with M. Ettinger and C. Tarzwell, and industrial wastes with H. Black.

Other research organizations such as river basin administrations and industrial research groups were involved in presenting and suggesting solutions to industrial waste problems. They include the following:

Ohio River Valley Sanitation Commission	River contaminants	E. Cleary
Delaware Water Resources Commission	River contaminants	J. Wright
National Council for Stream Improvement of the Pulp and Paper Industry	Papermill Waste Treatment	H. Gehm

National Dairymen Association	Milk waste treatment	H. Trebbler
Oil and Petrochemical	Oil refinery wastes	R. Weston
Textile Institute of Research	Textile waste treatment	S. Coburn
Mellon Research Institute	Coal waste research	W. Hodge

Early consulting engineers and scientists active in the environmental area include:

Consulting engineer	All wastes	S. Powell
Consulting chemist	All chemical wastes	R. Hess

The New Dilemma of the 1990s

After paying for the costs and consequences of avoiding pollution prevention over the nineteenth and twentieth centuries, in the last decade of the twentieth century and in the new millennium, industry faces the dilemma of accounting for the real costs of

TABLE 1
Primary Personnel and Fields of Study in U.S. Universities

Origin Timeline	University	Major Environmental Subject	Professor(s)
1920–1930s	Harvard	Public Health Engineering	G. Fair
			E. Moore
	Johns Hopkins	Health and Water Resources	A. Wolman
			J. Geyer
	Rutgers	Industrial Wastes	W. Rudolfs
		Wastes	H. Heukelekian
1930–1940s	Florida	Water	A.P. Black
			E. Phelps
			J. Kiker.
Mass. Inst. Tech.	San. Eng.	H.P. Eddy	L. Metcalf
U. California			R. EliassenE. Pearson
	Waste Treat.	P. McGaughey	W. Oswald
	U. Illinois	Water	W.R. Steele
	Rennselear Poly	Environment	E. Kilcawley
	U. Michigan	Pub. Health	G. Rideneaur
	U. Wisconsin	Env. Eng.	G. Rohlick
	New York Univ.	Env. Eng.	W. Ingram
			W. Dobbins
1940s	U. Minnesota	Public Health	G. Schropfer
	Penn State	Sewage Treat.	R. Stiemke
	Georgia Tech.	Sewage Treat.	R. Stiemke
			R. Ingols
	Univ. Texas	Waste Treat.	E. Gloyna
	Cal Tech.	Env. Science	J. McKee
	Purdue U.	Ind. Wastes	D. Bloodgood
	Univ. Illinois	Digestion	A. Buswell

pollution. These include external damage costs to society as well as internal cost responsibilities.

Industry finally realizes that it must include these external costs in their production costs. With this startling, sudden realization comes the understanding that it may be less costly to internalize all waste costs into its manufacturing costs. Just exactly how to do this is the dilemma!

One obvious method of accomplishing such internalization of waste problems and associated costs is to reuse all wastes. To some extent this has worked. But it soon became apparent that it was nearly impossible to find reasonable and economical use for all wastes.

Next, industry tried to export unreusable wastes to external markets. This method also failed to operate completely satisfactorily—mainly because of matching markets of supply and demand. Quality and quantity control of wastes also were difficult to maintain.

Sometimes the use of these last two methods led to a rather haphazard location of industrial waste reusers in the vicinity of the primary industrial waste producers. Transportation of wastes and matching them with the quality and quantity desired by reusers still remain a deterrent with this procedure.

But from these tentative and vague attempts the era of waste utilization rather than waste treatment is slowly emerging. This book is an attempt to describe how the waste utilization era will work.

There were those who speculated that the era of industrial waste treatment would end with the closing of the twentieth century. But your author believes that it will not only continue to be an important aspect of industrial operations, but will change to more innovative solutions to more complicated waste problems, especially in the developed nations of the world.

Clare Ansberry reports (2005) that "many experts believe that the pattern of past years will continue—that low skilled jobs making lower value mass produced items will keep migrating to countries where labor is plentiful and cheap, while manufacturing in industrial nations, such as the U.S., Japan, and Western Europe, will contain complex, value-added products and systems." Some of these products suggested include medical instruments (complex), kitchen cabinets (costly to ship), and frozen foods (perishable), bearings for automobiles, X-ray machines, washing machines, cars, and telephones.

Ansberry believes, as I do, that "what will ensure U.S. manufacturing's future is innovation, just as it has in the past" (Ansberry 2005). Ansberry adds clothing, computers, automation equipment, robotics, toys, sporting goods, drugs, garden machinery, motor vehicles, metal coating and screw machine products, and bolts and rivets industries to the list requiring innovative solutions.

References

Ansberry, C. 2003. Why U.S. Manufacturing Won't Die. *Wall Street Journal,* July 3.
Corrick, J. A. 1998. *The Industrial Revolution.* San Diego, CA: Lucent Books, Inc.
Nemerow, N. L. 1995. *Zero Pollution for Industry.* New York: John Wiley Publishing Company.
Nemerow, N. L., S. Farooq, S. Sengupta. 1977. *Industrial Complexes and Their Relevance for Pulp and Papermills,* vol. 3, no. 1, p. 133, Calcutta, India.

Introduction to Industrial Waste Treatment

Serious treatment of industrial wastes did not really begin in the United States until the 1930s—and then only in a few industries with easy and obvious remedies—such as screening feathers from poultry processing effluents and even the lagooning of certain papermill wastes. The extent of treatment grew during the next 70 years of the twentieth century. It broadened during this period to include many types of industries. The complexity of treatment also increased as environmental restrictions became greater. Growth was enhanced also by progress in industrial waste treatment research findings.

The goal of effective industrial waste treatment is directed towards the removal of all contaminants (see complete definition in Nemerow and Agardy 1998) that adversely impact the water as well as air and land environments. One should keep this goal in mind at all times regardless of the presence or absence of governmental or economic constraints.

Waste treatment findings and practices are presented largely in Part A of this book. Many of these carry over into the twenty-first century. I do not intend to maintain that the state of the art of waste treatment can be clearly divided into and by the two centuries. In fact, one can discover twentieth-century industries that were practicing the twenty-first century advanced treatment methods. Treatment decisions have always been affected by regulatory and economic constraints regardless of scientific knowledge. And they probably always will until we are able to include the costs of these constraints in industrial production decisions.

Twenty-first century thinking, if not actual practice, is presented in Part B. Keeping in mind that many older techniques are still being carried over into this century. In addition, as expected, the twenty-first century environmental engineer is benefiting from the massive industrial waste research and treatments utilized during the twentieth century. But I am certain that the reader realizes that the resources of our receiving environments (air, water, and land) are finite. And as such these resources must be protected with greater intensity and utilized much more economically than previously.

With the above in mind, I encourage the reader—student, consulting engineer, governmental regulatory agent, industrial producer, municipal planner, scientific investigator, or just plain resource-minded environmentalist—to review and absorb as much of the knowledge presented in Part A (supplemented with material in Nemerow and Agardy 1998) and proceed to utilize it in studying and evaluating the innovative twenty-first century ideals of Part B.

Reference

Nemerow, N. L., F. J. Agardy. 1998. *Strategies of Industrial and Hazardous Waste Management*. New York: John Wiley Publishing Company.

Part A

Twentieth Century

CHAPTER 1

Theories and Practices

Volume Reduction

In general, the first step in minimizing the effects of industrial wastes on receiving streams and treatment plants is to reduce the volume of such wastes. This may be accomplished by: (1) classifying wastes; (2) conserving wastewater; (3) changing production to decrease wastes; (4) reusing both industrial and municipal effluents as raw water supplies; or (5) eliminating batch or slug discharges of process wastes.

Classification of Wastes

If wastes are classified so that manufacturing-process waters are separated from cooling waters, the volume of water requiring intensive treatment may be reduced considerably. Sometimes it is possible to classify and separate the process waters themselves so that only the most polluted ones are treated and the relatively uncontaminated ones are discharged without treatment. The three main classes of wastes are as follows:

1. Wastes from manufacturing processes: These include waters used in forming paper on traveling wire machines, those expended from plating solutions in metal fabrication, and those discharged from washing of milk cans in dairy plants, dyeing and washing of textile fabrics, and washing of picked fruits from canneries.
2. Waters used as cooling agents in industrial processes: The volume of these wastes varies from one industry to another, depending on the total Btu's to be removed from the process waters. A single large refinery discharges 150 million gallons per day (mgd), of which only 5 mgd is process waste; the remainder is only slightly contaminated cooling-water waste. Cooling waters have been found to be contaminated by small leaks, corrosion products, or the effect of heat; however, these wastes usually contain little, if any, process matter and are generally considered nonpollutional. Power plants, however, represent an industry in which cooling waters are segregated and account for a high percentage of total volume of plant wastes, and may contain hazardous contaminants under infrequent malfunctioning conditions.

3. Wastes from sanitary uses: These will normally range from 25 to 50 gallons per employee per day. The volume depends on many factors, including size of the plant, amount of waste-product materials washed from floors, and the degree of cleanliness required of workers in the process operation.

Unfortunately, in most older plants, process, cooling, and sanitary wastewaters are mixed in a single pipeline; before 1930, industry paid little attention to segregating wastes to avoid stream pollution. Awareness of the problems and the differences in these three wastes led industry to practice separation when constructing new plants during World War II and thereafter.

Conservation of Wastewater

Water conservation is waste saved. Conservation begins when an industry changes from an "open" to a "closed" system. For example, a paper mill that recycles white water (i.e., water passing through a wire screen upon which paper is formed) and thus reduces the volume of wash waters it uses is practicing water conservation. Concentrated recycled wastewaters are often treated at the end of their period of usefulness, because usually it is impractical and uneconomical to treat the wastewaters as they complete each cycle. The savings are twofold: Water costs and waste-treatment costs are lower. However, many changes to effect conservation are quite costly and their benefits must be balanced against the costs. If the net result is deemed economical, then new conservation practices can be installed with assurance.

A paperboard mill may discharge 10,000 gallons of wastewater per ton of product, although there are many variations from one mill to the next. Paper mills may release as much as 100,000 gallons or as little as 1,000 gallons of wastewater per ton of product. The latter figure is usually the result of a scarcity of water and/or an awareness of the stream-pollution problem, and demonstrates what can be accomplished by effective waste elimination and conservation of water. One large textile mill reduced its water consumption by 50% during a municipal water shortage, without any decrease in production. The author observed that despite the savings to the mill, water usage returned to its original level once the shortage was over. This incident further illustrates the relative "cheapness" of water to the typical industrial plant manager of the mid-twentieth century.

Steel mills reuse cooling waters to quench ingots, and coal processors reuse water to remove dirt and other noncombustible materials from coal. Many industries installed countercurrent washing to reduce water consumption. By the use of multiple vats, the plating industry utilized makeup water so that only the most exhausted waters were released as waste. Automation, in such forms as water-regulating devices, also aided conservation of water. The introduction of conservation practices requires a complete engineering survey of existing water use and an inventory of all plant operations using water and producing wastes, to develop an accurate balance for peak and average operating conditions. For example, in 1950 Rudolfs and Nemerow found that recirculation of paperboard white water was objectionable because of slime formation that subsequently fouled fibers or even slowed down paper machines. Low pH and high temperature of these recirculated waters were found to abate the situation.

Changing Production to Decrease Wastes

Changing production to decrease wastes is an effective method of controlling the volume of wastes but is difficult to put into practice. It is hard to persuade plant managers to change their operations just to eliminate wastes. Normally, the operational phase of engineering is planned by the chemical, mechanical, or industrial engineer whose primary objective is cost savings. The main considerations of the environmental engineer, on the other hand, include the protection of public health and the conservation of a natural resource. Yet, there is no reason that both objectives cannot be achieved.

Waste treatment at the source should be considered an integral part of production. If the chemical engineer argues that it would cost the company money to change its methods of manufacturing to reduce pollution at the source, the environmental engineer can do more than simply enter a plea for the improvement of the environment. The environmental engineer can point out, for instance, that reduction in the amount of sodium sulfite used in dyeing, that of sodium cyanide used in plating, and that of other chemicals used directly in production has resulted in both reducing wastes and saving money. The engineer can also mention that balancing the quantities of acids and alkalis used in a plant often results in a neutral waste, along with saving chemicals, money, and time spent in waste treatment. Rocheleau and Taylor (1964) point out several measures that can be used to reduce wastes: improved process control, improved equipment design, use of different or higher quality raw materials, good housekeeping, and preventative maintenance.

Reusing Both Industrial and Municipal Effluents for Raw Water Supplies

Practiced mainly in areas where water is scarce or expensive, reusing industrial and municipal effluents for raw water supplies is proving a popular and economical method of conservation; of all sources of water available to industry, sewage plant effluent is the most reliable at all seasons of the year and the only one that is actually increasing in quantity and improving in quality. Although there are many problems involved in reusing effluents for raw water supply, it must be remembered that *any* water supply poses problems to cities and industries. Because the problems of reusing sewage effluents are similar to those of reusing industrial effluents, they are discussed here jointly.

Many industries and cities hesitate to reuse effluents for raw water supply. The reasons given (Keating and Calise 1954) include lack of adequate information on the part of industrial managers, difficulty negotiating contracts satisfactorily for both municipalities and industrial users, certain technical problems such as hardness, color, and so forth, and an aesthetic reluctance to accept effluents as a potential source of water for any purpose. Also, treatment plants are subject to shutdown and slug (sudden) discharges, both of which may make the supply undependable or of variable quality. In either case, industry may need an alternate source of water for these emergency situations. In addition, the "resistance to change in practice" factor cannot be overlooked as a major obstacle. However, as the cost of importing raw water supply increases, it would seem logical to reuse waste-treatment plant effluents to increase the present water supply by replenishing the groundwater. It cannot be denied that the ever-available treatment-plant effluent can produce a low-cost, steady water source through groundwater recharge. If any portion of a final industrial effluent can be reused, there will be less waste to treat and dispose. Similarly, reuse of sewage effluent will reduce the quantity of pollution discharged by the municipality. Still, in 1966 one of the deterrents to reusing treated municipal (or industrial) wastewater was the competition with lower-priced freshwater. For example, in San Diego, California, although supporters for using reclaimed waters from treated sewage have proclaimed that "it doesn't make sense to use water only once" and that "pushing for expensive reclaimed water doesn't make economic sense" when the county can buy a large supply of less costly water from Imperial County (Applebaum 1966).

The greatest manufacturing use of water is for cooling purposes. Because the volume of this water requirement is usually great, industries located in areas where water is expensive should consider reusing effluents. Even if the industry is fortunate enough to have a treated municipal water supply, the cost will usually be excessive in comparison, which may have a generally beneficial effect.

As far back as 1955, Smallwood and Nemerow presented a few examples of the water required to produce typical products we purchase:

0.25 gallons for every pound of rayon
0.11–0.25 gallons for every pound of butter
15–30 gallons for every pound of paper produced
5 gallons for every pound of paperboard produced
2.5–45.0 gallons for every pound of paper pulp (newspaper) produced
7.5–250 gallons for every can of food
3 gallons for every 3 pounds of chicken slaughtered
127 gallons for every hog slaughtered
0.75 gallons for every quart of milk that is bottled
8 gallons for every pound of hides tanned into leather
90 gallons for every pound of wool scoured
400 gallons for every guest room per day for hotels

Reusing municipal and industrial effluents saves water and brings revenue into the city. The design of wastewater-treatment plants will be greatly influenced because the effluent must satisfy not only conventional stream requirements but also those of industry.

Many cases are cited in the literature of industrial reuse of intermediate untreated effluents, such as white waters from paper machines as spray and wash waters. The practice of reusing treated industrial effluents, however, was still in its infancy; there are more instances of industrial reuse of municipal effluents. For example, Wolman (1948) described the design and performance of a sewage-effluent treatment plant producing treated water at a rate of about 65 mgd for use in steel-mill processing operations. The plant employed a conventional coagulation treatment, using alum combined with chlorination; final water averaged 5–10 ppm turbidity, with little or no coliform bacterial contamination. The most serious problem encountered was the presence of a high concentration of chlorides. Operating costs, exclusive of interest and amortization but including pumping costs, were $1.75 per million gallons (although this figure does not include the cost of raw-sewage treatment). It is interesting to compare this with the usual municipal cost of collecting, treating, and distributing raw water of $50–250 per million gallons, excluding fixed charges. Even when one adds $15–50 per million gallons for treating the raw sewage, the reusable effluent is much more economical than water obtained by developing a separate source of raw water. Treatment-plant reuse facilities at the Sun Oil Toledo refinery were evaluated (Mohler et al. 1964) for use as makeup water in the cooling towers. The cost savings resulting from elimination of municipal freshwater makeup were found to be $100,000 per year.

Keating and Calise (1954) list five main differences between most sewage-plant effluents and typical surface- or well-water supplies: (1) higher color, (2) higher nitrogenous content, (3) higher biochemical oxygen demand (BOD) content, (4) higher total dissolved solids, and (5) the presence of phosphates due to detergents. Industrial effluents may also possess these characteristic differences, as well as others such as higher temperature. Despite these contaminants, in many parts of the United States the effluent from properly operated secondary sewage plants is actually superior to available surface- or well-water supplies.

Dan Okun, a renowned twentieth-century environmentalist, was—and still is—recommending and strongly urging the reuse of treated sewage effluents for use as "secondary" water supplies for municipalities. The reader can refer to his many published papers on the subject. The debate goes on even in the twenty-first century on the acceptability of such a dual water-supply system.

The number and variety of return-flow and on-site reuse systems have increased. The overall reuse rate increased from 106% (of water reused) to 136% between 1954 and 1959 alone. Reuse in all industries other than steam-electric generation increased from 82% to 139% during the same period. In 1959, the primary metal, chemical, paper, oil, and food industries were particularly large reusers of water.

In 1957, El Paso Products Company founded a petrochemical complex near Odessa, Texas, designed to use sewage-plant effluent for cooling and boiler water. After pretreatment, the only problem encountered was foaming (largely eliminated by the switch to "soft" detergents in domestic use). Reusing sewage effluents often frees municipal or surface water for other valuable purposes. For example, reutilization of sewage for agricultural purposes in Israel could add 10% to its total water supply. It was found that the soil structure is improved by the organics in sewage, but where industrial wastes (particularly heavy metals) are present, treatment beyond oxidation ponds is needed.

"Dry" cleaning of processing equipment, instead of washing with water, can greatly reduce the volume of wastewater. However, this will still leave a solid waste for disposal rather than a liquid one. Hoak (1964) presents a set of conservation techniques largely adapted from his experience in steel mills:

1. Install meters in each department to make operators cost and quantity conscious
2. Regulate pressure to prevent needless waste
3. Use thermostatic controls to save water and increase efficiency
4. Install automatic valves to prevent loss through failure to close valves when water is no longer needed
5. Use spring-closing sanitary fixtures to prevent constant or intermittent flow of unused water

6. De-scale heat exchangers to prevent loss of heat transfer and subsequent inefficient and excessive use of cooling water
7. Insulate pipes so that water is not left running to get it either cold or hot
8. Instigate leak surveys as a routine measure
9. Use centralized control to prevent wastages from improper connections
10. Recirculate cooling water, thereby saving up to 95% of the water used in this process
11. Reuse, for example, blast-furnace cooling water for gas washing and clarified scale-pit water on blooming mills
12. Use high-pressure, low-volume rinse sprays for more efficiency and use a small amount of detergent, wetting agent, or acid to improve the rinsing operation
13. Recondition wastewater (often some minor in-plant treatment will provide water suitable for process use)

Eden and Truesdale (1968) give typical analyses of effluents from three towns in the south of England (Table 1.1). They found that the total solids content appeared to increase by about 340 mg/liter between the water supply and the sewage effluent derived from it. The total solids concentration is one of the chief limiting factors in reusing any wastewater; the number of times sewage can be reused for industrial water supply is controlled by the pickup of dissolved solids that can be removed only by expensive treatment methods. Some discussion of the contaminants listed in Table 1.2 is relevant to potential reuse of sewage effluents for industrial water. Many industrial purposes demand concentrations of suspended solids of less than 2 mg/liter, but sewage effluents contain considerably more than this and, even after tertiary treatment, often contain at least 7 mg/liter. The organic constituents of sewage effluents are still largely unknown. Absorption has been suggested as a method for reducing most of the organic matter. At Lake Tahoe, for example, it has been possible to reduce the organic matter (as measured by chemical oxygen demand [COD]) to less than 16 mg/liter by a combination of coagulation, filtration, and absorption. Detergents can also be removed in this manner to a theoretical minimum level of about 0.2 mg/liter. Additional removal of ammonia, nitrite, and nitrate is relatively expensive and difficult. Ammonia, which can be air-stripped at high pH values, is objectionable in concentrations of more than 0.1 mg/liter for drinking-water supplies that are to be chlorinated. Removal of phosphates is important whenever the water used by industry will be subjected to algae growth conditions. The Tahoe method will reduce the level of phosphate to less than 1.0 mg/liter; controlled activated-sludge and lime-precipitation methods are also effective. At high chlorine levels, it is possible to even remove many viruses. Because sewage effluents contain many types of microorganisms, they should be sterilized even for industrial-process use. In addition, color and hardness in sewage effluents may be harmful to certain industries.

Dowdy et al. (1976) derived a "typical" chemical composition of treated wastewater effluent from a selected number of cities. This composition was compared to that of water from the Colorado River—a source for crop production in several western states—for many of the water quality criteria important in irrigation.

TABLE 1.1
Typical Analyses of Sewage Effluents After Conventional Primary and Secondary Treatment

Constituent[a]	Source		
	Stevenage	Letchworth	Redbridge
Total solids	728	640	931
Suspended solids	15		51
Permanganate value	13	8.6	16
BOD	9	2	21
COD (chemical oxygen demand)	63	31	78
Organic carbon	20	13	
Surface-active matter			
Anionic (as Manoxol OT)	2.5	0.75	1.4
Nonionic (as Lissapol NX)			0.4
Ammonia (as N)	4.1	1.9	7.1
Nitrate (as N)	38	21	26
Nitrite (as N)	1.8	0.2	0.4
Chloride	69	69	98
Sulfate	85	61	212
Total phosphate (as P)	9.6	6.2	8.2
Total phenol			3.4
Sodium	144	124	
Potassium	26	21	
Total hardness	249	295	468
pH value	7.6	7.2	7.4
Turbidity (A.T.U.)[b]			66
Color (Hazen units)	50	43	36
Coliform bacteria (#/ml)	1,300		3,500

[a]Results are given in milligrams per liter, unless otherwise indicated.
[b]Absorptiometric turbidity units.
Adapted from Eden and Truesdale (1968).

Examination of Batch or Slug Discharges of Process Wastes

In "wet" manufacturing of a product, one or more steps are sometimes repeated, which results in production of a significantly higher volume and strength of waste during that period. If this waste is discharged in a short period, it is usually referred to as a *slug discharge*. This type of waste, because of its concentrated contaminants and/or surge in volume, can be troublesome to both treatment plants and receiving streams. There are at least two methods of reducing the effects of these discharges: (1) the manufacturing firm can alter its practice to increase the frequency and lessen the magnitude of batch dischargers; and (2) slug waste can be retained in holding basins from which they are allowed to flow continuously and uniformly over an extended (usually 24-hour) period. These are called *proportioning* and *equalization* (of slug wastes) and are described more fully in Chapter 4 of this book.

TABLE 1.2
Composition of Secondary Treated Municipal Wastewater Effluents and Irrigation Water

Parameter	Secondary Effluent[a]		Colorado River[b]	Irrigation Water Quality Criteria[c]
	Range	Typical		
Total solids	U	425	U	NA
Total dissolved solids	200–1,300	400	668.0	<2,000
pH	6.8–7.7	7.0	7.9	6.5–8.4
Biochemical oxygen demand	2–50	25	U	NA
Chemical oxygen demand	25–100	70	U	NA
Total nitrogen	10–30	20	U	<30
Ammonia nitrogen	0.1–25.0	10	U	NA
Nitrate nitrogen	1–20	8	0.1–1.2	NA
Total phosphorus	5–40	10	<0.02	NA
Chloride	50–500	75	55–77	<350
Sodium	50–400	100	71–97	<70
Potassium	10–30	15	4–6	NA
Calcium	25–100	50	66–163	NA
Magnesium	10–50	20	23–28	NA
Boron	0.3–2.5	0.5	0.10–0.54	<3.0
Cadmium (μg/liter)	<5–220	<5	<1–69	10
Copper (μg/liter)	5–50	20	<10–10	200
Nickel (μg/liter)	5–500	10	<1–4	200
Lead (μg/liter)	1–200	5	<5	5,000
Zinc (μg/liter)	10–400	40	<3–12	2,000
Chromium (μg/liter)	<1–100	1	<1	100
Mercury (μg/liter)	<2–10	2	<0.1–0.1	NA
Molybdenum (μg/liter)	1–20	5	2–8	10
Arsenic (μg/liter)	<5–20	<5	4–16	100

Note: All units in milligrams per liter unless otherwise noted as micrograms per liter (μg/liter). U, unavailable; NA, not applicable.
[a]Adapted from Asano et al. (1984) and Treweek (1985).
[b]Radtke et al (1988).
[c]From Westcot and Ayers (1985) and National Academy of Sciences (1973).

Example of Twentieth-Century Practice of Volume Reduction

An unusual example of reducing the volume of wastewater was described by Zimmerman et al. (1995). They determined the optimal liquid storage volume at minimum cost of a biosolid settling tank. The liquid volume arriving at different settling rates determined the amount of liquid to be discharged to waste. They concluded that obviously the thickening rate was the most significant factor affecting the required tank volume (and, hence, the liquid amount to be decanted and wasted). Although these authors were primarily interested in designing the proper-sized tank, they also revealed a connection between the thickening rate and the increase or decrease in liquid volume to be wasted. Therefore, they showed that this volume could be reduced by controlling the settling rate of biosolids.

Review Questions

1. What are the three major classifications of industrial wastes at an industrial plant?
2. What are the implications of these three types of wastes?
3. What do we mean by "industrial water conservation?"
4. What is another method of reducing volume? Give examples.
5. What advantage do we get by reducing waste volume?
6. What is another method of reducing volume?
7. What is the greatest factor influencing an industry to reuse its wastewater?
8. What is usually the greatest deterrent to industrial reuse of wastewaters?
9. How can we encourage water conservation in an industrial plant?
10. What is yet another method of reducing the volume of wastewaters?

References

Applebaum, S. B. 1966. Industry does benefit from pollution control. *Water Wastes Eng.* 3:46.
An expensive waste [editorial]. *San Diego Union Tribune*, June 25, 1966, pp. 6–12.
Clarke, F. E. 1962. Industrial re-use of water. *Ind. Eng. Chem.* 54:18.
Eden, G. E., G. A. Truesdale. 1968. Reclamation of water from sewage effluents. Paper presented at Symposium on Conservation and Reclamation of Water, 28 November 1967, London. London: Water Pollution Research Laboratory. Reprint no. 519.
Flourishing on sewage-plant effluent. 1966. *Chem. Process.* 29:30.
Hershkovitz, S. Z., F. Feinmesser. 1967. Utilization of sewage for agricultural purposes. *Water Sewage Works* 113:181.
Hoak, R. D. 1964. Water resources and the steel industry. *Iron Steel Engr.* May:87.
Keating, R. J., V. J. Calise. 1954. *The Treatment of Sewage Plant Effluent for Water Reuse in Process and Boiler Feed*. Union, NJ: Graver Water Conditioning Company. Technical Reprint T-129.
Marks, R. H. 1967. Waste water treatment. *Power* 111:S32.
Mohler, E. F., Jr., H. F. Elkin, L. R. Kumnick. 1964. Experience with reuse and biooxidation of refinery wastewater in cooling tower systems. *J. Water Pollution Control Fed.* 36:1380.
Morris, A. L. 1967. Water renovation. *Ind. Water Eng.* 4:18.
National Association of Manufacturers and Chamber of Commerce of the United States, in cooperation with Nation Task Committee on Industrial Wastes. 1967. *Water in Industry*. Washington, DC.
National Research Council. 1996. *Use of Reclaimed Water and Sludge in Food Production*. pp. 19–20. Washington, DC: National Research Council.

Rawn, A. M., F. R. Bowerman. 1957. Planned water reclamation. *J. Sewage Ind. Wastes* 29:1134.

Renn, C. E. 1967. Serendipity at Hempstead—a study in water management. *Ind. Water Eng.* 4:25.

Rice, J. K. 1966. Water management to reduce wastes and recover water in plant effluents. *Chem. Eng.* 73:125.

Rocheleau, R. F., E. F. Taylor. 1964. An industry approach to pollution abatement. *J. Water Pollution Control Fed.* 36:1185.

Rudolfs, W., N. L. Nemerow. 1950. Some factors affecting slime formation and freeness in boardmill stock. *TAPPI* July:33, 7, 321.

Unwin, H. D. 1967. In plant wastewater management. *Ind. Water Eng.* 4:18.

Smallwood, C., Jr., L. Nemerow. 1955. *Water for Industry*. Facts for Industry Series Bulletin No. 2. School of Engineering, North Carolina State College Raleigh, North Carolina, October 1955, p. 1.

Wolman, A. 1948. Industrial water supply from processed sewage treatment plant effluent at Baltimore, Maryland. *Sewage Works J.* 20:15.

Zimmerman, R., D. Richard, and R. Mann. 1995. In the thick of things. *Water Environ. Technol.* May:7, 5, 42.

Suggested Reading

Alexander, D. E. 1959. Wastewater transformation at Amarillo. II. Industrial phase. *Sewage Ind. Wastes* 31:1107.

Berg, E. J. 1959. Considerations in promoting the sale of sewage treatment plant effluent. *Sewage Ind. Wastes* 30:96.

Besselievre, E. B. 1959. Industries recover valuable water and by-products from their wastes. *Wastes Eng.* 30:760.

Besselievre, E. B. 1960. Industry must reuse effluents. *Wastes Eng.* 31:734.

Black, A. P. 1964. Statement by Dr. A. P. Black. *J. Sanit. Eng. Div. Am. Soc. Civil Engrs.* 90(SA4):11.

Burrell, R. 1964. Users of effluent water in sewage treatment plants. *Sewage Works J.* 18:104.

California State Water Pollution Control Board. 1955. *Direct Utilization of Waste Waters*. Sacramento, CA.

California State Water Pollution Control Board. 1957. Industry utilizes sewage and waste effluent for processing operations. *Waste Eng.* 28:444.

Cecil, L. K. 1964. Sewage treatment plant effluent for water reuse. *Water Sewage Works* 111:421.

Clarke, F. E. 1962. Industrial re-use of water. *Ind. Eng. Chem.* 54:18.

Cohn, M. M. 1956. A million tons of steel with sewage. *Wastes Eng.* 27:309.

Connell, C. H. 1957. Utilization of waste waters. *Ind. Wastes* 2:148.

Connell, C. H., E. J. Berg. 1959. Industrial utilization of municipal wastewater. *Sewage Ind. Wastes* 31:212.

Connell, C. H., M. C. Forbes. 1964. Once used municipal water as industrial supply. *Water Sewage Works* 111:397.

1957. Copper mining plant squeezes water dry. *Public Works* 88:125.

Eliezer, R., R. Everett, J. Weinstock. 1963. *Contaminant Removal from Sewage Plant Effluents by Foaming.* Cincinnati, OH: U.S. Public Health Service. Advanced Waste Treatment Research Publication no. 5.

Elkin, H. F. 1955. Successful initial operation of water re-use at refinery. *Ind. Wastes* 1:75.

Gerster, J. A. 1963. *Cost of Purifying Municipal Waste Waters by Distillation.* Cincinnati, OH: U.S. Public Health Service. Advanced Waste Treatment Research Publication no. 6.

Geyer, J. C. 1957. Reuse of sewage effluents for industrial water supply. In: *Proceedings of the Sixth Southern Municipal and Industrial Waste Conference, April 1957, at North Carolina State College, Raleigh.*

Gloyna, E, J. Wolff, J. Geyer, and A. Wolman. A report upon present and prospective means for improved re-use of water. Unpublished observations.

Hoak, R. D. 1961. Industrial water conservation and re-use. *TAPPI* 44:40.

Hoppe, T. C. 1960. Industry will reuse effluent in future waste economy drive. *Wastes Eng.* 31:596.

Hoot, R. A. 1948. Plant effluent use at Fort Wayne. *Sewage Works J.* 20:908.

Howell, G. A. 1963. Re-use of water in industry. *Public Works* 94:114.

Jenkins, S. H. 1964. Re-use of water in industry. II. The composition of sewage and its potential use as a source of industrial water. *Water Sewage Works* 111:411.

Kabler, P. W. 1963. Bacteria can be a nuisance. *Chem. Eng. Progr.* 59:23.

Keating, R. J., V. J. Calise. 1955. Treatment of sewage plant effluent for industrial re-use. *Sewage Ind. Wastes* 27:773.

Keefer, C. E. 1956. Bethlehem makes steel with sewage. *Wastes Eng.* 27:310.

Middleton, F. M. 1964. Advance treatment of waste waters for re-use. *Water Sewage Works* 111:401.

Morris, J. C., W. J Weber. 1964. *Preliminary Appraisal of Advanced Wastes Treatment Process.* Washington, DC: U.S. Department of Health, Education and Welfare. Advanced Waste Treatment Research Publication no. 2.

Morris, J. C., W. J. Weber. 1964. *Adsorption of Biochemically Resistant Materials from Solution.* Washington, DC: U.S. Department of Health, Education and Welfare. Advanced Waste Treatment Research Publication no. 9.

Powell, S. T. 1948. Some aspects of the requirements for the quality of water for industrial uses. *Sewage Works J.* 20:36.

Powell, S. T. Adaptation of treated sewage for industrial use. *Ind. Eng. Chem.* 48:2168.

Randall, D. J. 1964. Reclamation of process water. *Water Sewage Works* 111:414.

Scherer, C. H. 1959. Sewage plant effluent is cheaper than city water. *Wastes Eng.* 30:124.

Scherer, C. H. 1959. Wastewater transformation at Amarillo. *Sewage Ind. Wastes* 31:1103.

Sessler, R. E. 1955. Waste water use in soap and edible-oil plant. *Sewage Ind. Wastes* 27:1178.

Silman, H. 1962. Re-use of water industry. I. The re-use of water in the electroplating industry. *Chem. Ind.* 49:2046.

Stanbridge, H. H. 1964. From pollution prevention to effluent re-use. Part I. *Water Sewage Works* 111:446.

Stanbridge, H. H. 1964. From pollution prevention to effluent re-use. Part II. *Water Sewage Works* 111:494.

Stephan, D. G. 1963. Water renovation, what it means to you. *Chem. Eng. Progr.* 59:19.

Stone, R., J. C. Merrell, Jr. 1958. Significance of minerals in waste-water. *Sewage Ind. Wastes* 30:928.

Tolman, S. L. 1959. Reclaiming valuable water and bark. *Wastes Eng.* 30:21.

Veatch, N. T. 1948. Industrial uses of reclaimed sewage effluents. *Sewage Works J.* 20:3.

Williamson, J. N., A. M. Heit, C. Calmon. 1964. *Evaluation of Various Adsorbents and Coagulants for Waste Water Renovation.* Washington, DC: U.S. Department of Health, Education and Welfare. Advanced Waste Treatment Research Program no. 12.

CHAPTER 2

Contaminant Concentration Reduction

Waste strength reduction is the second major objective for an industrial plant concerned with waste treatment. Any effort to find means of reducing the total pounds of polluting matter in industrial wastes will be well rewarded by the savings earned by reduced requirements for waste treatment. The strength of wastes may be reduced by: (1) process changes; (2) equipment modifications; (3) segregation of wastes; (4) equalization of wastes; (5) by-product recovery; (6) proportioning wastes; and (7) monitoring waste streams.

Process Changes

In reducing the strength of wastes through process changes, sanitary engineers are concerned with wastes that are most troublesome from a pollutional standpoint. Their problems and therefore their approach differ from those of plant engineers or superintendents. Sometimes tremendous resistance by a plant superintendent must be overcome in order to effect a change in process. Superintendents possess considerable security because they can do a familiar job well. Why should they jeopardize their position merely to prevent stream pollution? The answer is obvious. Industry dies when its progress stops. No manufacturer can meet present-day market competition without continually, and critically, reviewing and analyzing its production techniques. In addition, pollution abatement can no longer be considered by industry as a "optional" act; on the contrary, it must be regarded as a vital step in preserving water resources for all users. Many industries have resolved waste problems through process changes. Two such examples of progressive management are the textile and metal-fabricating industries. On the other hand, the leather industry still generally uses lime and sulfides (major contaminants of tannery wastes), although it is known that amines and enzymes could be substituted. The lag between research and actual application is often extensive, and is caused by many operational difficulties.

In a detailed study of the iron and steel industry, your author (Nemerow 1976) revealed three major process changes that resulted in reduced environmental pollution load:

1. Dry quenching of coke (instead of wet quenching)
2. Hydrochloric acid pickling (instead of sulfuric acid pickling)
3. Direct reduction of iron ore (instead of coking and blast furnace plants)

Textile-finishing mills were faced with the disposal of highly pollutional wastes from sizing, kiering, de-sizing, and dyeing processes. Starch had been traditionally used as a sizing agent before weaving, and this starch, after hydrolysis and removal from the finished cloth, was the source of 30–50% of the mill's total oxygen-demanding matter. The industry began to express an interest in cellulosic sizing agents, which would exhibit little or no biochemical oxygen demand (BOD) or toxic effect in streams. Several highly substituted cellulosic compounds, such as carboxymethyl cellulose, were developed and used in certain mills, with the result that BOD contributed by de-sizing wastes was reduced almost in direct relation to the amount of cellulosic sizing compound used.

In the metal-plating industries (Davis 1957), seven changes of process or materials were suggested to eliminate or reduce cyanide strengths: (1) change from copper-cyanide plating solutions to acid-copper solutions; (2) replace the $CuCN_2$ strike before the copper-plating bath with a nickel strike; (3) substitute a carbo-nitriding furnace, which uses a carburizing atmosphere and ammonia gas, for the usual molten cyanide bath; (4) use "shot blasts" or other abrasive treatment on nonintricate parts instead of H_2SO_4, in pickling of steel; (5) substitute H_3PO_4 for H_2SO_4 in pickling; (6) use alkaline de-rusters instead of acid solutions to remove light rust, which occurs during storage (the overall pH will be raised closer to neutrality by this procedure, which will also alleviate corrosive effects on piping and sewer lines); and (7) replace soluble oils and other short-term rust-preventive oils applied to parts after cleaning with "cold" cleaners. These cleaners can be used in both the wash and rinse solutions. They inhibit rust chemically rather than by a film of oil or grease.

A Pennsylvania coal-mining company modified its process to wash raw coal with acid mine waste rather than a public or private water supply. In this way, the mine drainage waste is neutralized while the coal is washed clean of impurities. In one analysis, for example, the initial mine water had a pH of 3, an acidity of 4,340 ppm as $CaCO_3$, and an iron content of 551 ppm. The wastewater finally discharged from the process had a pH of 6.7–7.1 and an iron content of less than 1 ppm.

Equipment Modifications

Changes in equipment can reduce the strength of the waste, usually by reducing the amount of contaminants entering the waste stream. Often quite small changes can be made in present equipment to reduce waste. For instance, in pickle factories, screens placed over drain lines in cucumber tanks prevent the escape of seeds and pieces of cucumber, which adds to the strength and density of the waste. Similarly, traps on the discharge pipeline in poultry plants prevent emission of feathers and pieces of fat. Your author (Nemerow 1977) recommended another procedure for accomplishing pollution reduction by equipment modification. The method is to change the production procedure to "dry-collect" as much waste material as possible from manufacturing machines and operating floors rather than "hosing down" the same matter into drains.

An outstanding example of waste strength reduction (with a more extensive modification of equipment) occurred in the dairy industry. Trebler (1944) redesigned the large milk cans used to collect farmers' milk. The new cans were constructed with smooth necks so that they could be drained faster and more completely. This prevented a large amount of milk waste from entering streams and sewage plants. Dairy farmers also installed drip pans in assembly lines to collect milk that drains from the cans after they have been emptied into the sterilizers. The drip-pan contents are returned to the milk tanks daily.

In the chemical industry, Hyde (1965) described a chemical plant that achieved a 23% decrease in average BOD through the installation of calandrias on open-bottom steam stills and by using refrigerated condensers ahead of vacuum jets, among other process modifications.

Segregation of Wastes

Segregation of wastes reduces the strength or the difficulty of treating the final waste from an industrial plant. It usually results in two wastes: one strong and small in volume and the other weaker, with almost the same volume as the original unsegregated waste. The small-volume strong waste can then be handled with methods specific to the problem it presents. In terms of volume reduction alone, segregation of cooling waters and storm waters from process waste will mean a saving in the size of the final treatment plant. Many dye wastes, for example, can be more economically and effectively treated in concentrated solutions. Although this type of segregation may increase the strength of the waste being treated, it will typically produce a final effluent containing less polluting matter.

Another type of segregation is the removal of one particular process waste from the other process wastes of an industrial plant, which renders the major part of the waste more amenable to treatment, as illustrated in the following examples.

A textile mill manufacturing finished cloth produced the wastes listed in Table 2.1. The combined waste was quite strong, difficult and expensive to treat, and very similar to laundry waste. However, when the liquid kiering waste was segregated from the other wastes, chemically neutralized, precipitated, and settled, the supernatant (the part that remained on the surface) could be treated chemically and biologically along with the other three wastes, because the strength of the resulting mixture was considerably less than that of the original combined waste. This type of segregation is also practiced in metal-finishing plants, which produce wastes containing both chromium and cyanide, as well as other metals. In almost all cases, it is necessary to segregate the cyanide-bearing wastes, make them alkaline, and oxidize them. The chromium wastes, on the other hand, have to be acidified and reduced. The two effluents can then be combined and precipitated in an alkaline solution to remove the metals. Without segregation, poisonous hydrogen cyanide gas would develop as a result of acidification. A method was patented (Koelsh-Folzer-Werke 1966) that allows the separation of

TABLE 2.1
Wastes from a Textile Mill

	Gray Water	White Water	Dye Waste	Kier Waste	Combined Waste
pH	4.0	7.3	11.0	11.8	9.4
Total solids, ppm	2,680	420	2,880	18,880	1,560
Suspended solids, ppm	224	67	148	218	156
Oxygen consumed, ppm	1,560	31	556	4,900	460

paint from wastewater by precipitation with ferric chloride and/or ferric sulfate, along with calcium hydroxide.

In treating the waste of a large poultry plant, the blood from the killing room floor was interfering with the treatment of the remainder of the chicken waste. Nemerow and Dasgupta (1991) recommended that the blood be scraped, swept, and disposed of with the screenings. In this way, a high BOD waste was segregated from the remaining plant process waste and treated separately.

Segregation of certain wastes is of great advantage in all industries. It is dangerous, however, to arrive at a blanket conclusion that segregation of strong or dangerous wastes is always desirable. Just the reverse technique—complete equalization—may be necessary in certain circumstances.

Equalization of Wastes

Plants that have many products from a diversity of processes prefer to equalize their wastes. This requires holding wastes for a certain period, depending on the time taken for the repetitive processes in the plant. For example, if a manufactured item requires a series of operations that take 8 hours, the plant needs an equalization basin designed to hold the wastes for that 8-hour period. The effluent from an equalization basin is much more consistent in its characteristics than each separate influent to that same basin. Stabilization of pH and BOD and settling of solids and heavy metals are among the objectives of equalization. Stable effluents are treated more easily and efficiently than unstable ones by industrial and municipal treatment plants. Sometimes equalization may produce an effluent that warrants no further treatment. The graph in Figure 2.1 illustrates one of the beneficial effects of equalization.

A large chemical corporation producing a predominantly acid waste found it advantageous to equalize its wastes for a 24-hour period in an earthen holding basin. After this equalization, a nearby plant, producing a highly alkaline waste, pumped its waste into the acid-waste effluent for neutralization. Considerably greater neutralizing power would be required for the acid waste were it not equalized to iron out the peaks before neutralization.

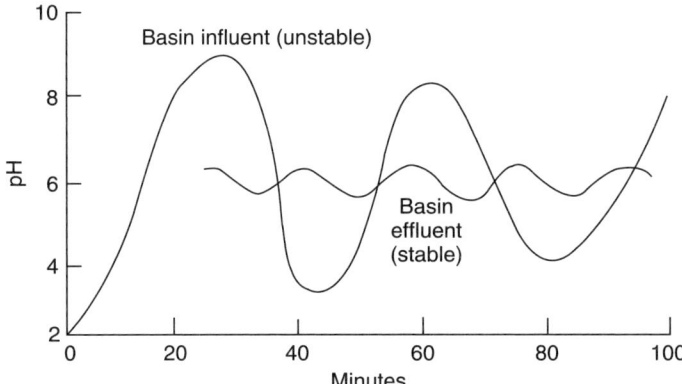

FIGURE 2.1. Time elapsed after start of equalization.

Salt in large quantities is used to "cure" cucumbers into pickles. The salt brine waste represents a large volume percentage of the total processing waste and can cause disproportionate objectionable effects when discharged into a receiving water. In at least one case in which your author was involved, I recommended holding the brine wastes in large vats for slow discharge with the remaining plant waste over long periods. In this way, the total plant waste was equalized with the brine wastes to minimize the salt effect.

A textile-finishing mill that discharged its waste into a domestic secondary sewage-treatment plant upset the efficiency of the plant. Although this waste represented only about 10% of the total being treated, it caused fluctuations, primarily in pH and BOD, which were responsible for the plant's difficulties. The solution was to build an equalization basin capable of detaining the waste long enough to reduce the fluctuations in pH and BOD. In addition, the mill decided to deliver the equalized waste to the city treatment plant at three different rates of flow: the highest flow rate corresponded to the time when the greatest amount of sewage was reaching the plant, and vice versa. This gave a more constant dilution of the mill's waste with domestic sewage.

By-Product Recovery

"By-product recovery" is the utopian aspect of industrial-waste treatment, the one phase of the entire problem that may lead to economic gain. Yet many consultants deprecate this approach to the solution of waste problems. Their attitude is based mainly on statistics concerning the low percentage of successful by-products developed from waste salvage. However, any use of waste materials obviously eliminates at least some of the waste that eventually must be disposed of, and the search for by-products should be encouraged if only because it provides management with a clearer insight into processing and waste problems. All waste contains by-products, the exhausted materials used in the process. Because some wastes are very difficult to treat at low cost, it is advisable for the industrial managers concerned to consider the possibility of building a recovery plant that will produce a marketable by-product while solving a troublesome waste problem. There are many examples of positive results from adapting waste-treatment procedures to by-product recovery.

Metal-plating industries use ion exchangers to recover phosphoric acid, copper, nickel, and chromium from plating solutions. The de-ionized water, without any further treatment, is ideal for boiler-feed requirements. For final recovery of valuable chromium, copper, and nickel, companies use vacuum evaporation of the concentrated plating solutions. A nickel-wire plating plant, faced with a nickel shortage, made the plating waste alkaline with soda ash and precipitated nickel as the carbonate, and then dried the sludge and treated it to recover the nickel. A silver-plating plant spent about $120,000 a year on waste treatment, of which $60,000 was returned as credit for silver recovered from the waste. The electrical industry recovered silver, gold, and (as by-products) water, valuable metals, and acids. Plants such as Scotscraft, Inc. reported the recovery and reuse of by-product cyanide from plating wastes. A system of evaporation is used here to result in an overall cost saving.

Specialty paper mills, with the aid of multiple-effect evaporators, recover caustic soda from cooking liquors. Chemical plants spray dilute waste acids into hot, lead-lined, brick-faced towers to concentrate the acids for reuse. Pharmaceutical houses recover molds by drying the cake from vacuum filters or evaporating spent broth in multiple-effect evaporators. Distilleries screen the "slop" and thicken it for by-product use. Yeast factories evaporate a portion of their waste and sell the residue for cattle feed.

Even sewage plants have entered the by-product business. Methane gas from sewage digesters is commonly used for heat and power, and some cities make fertilizers and vitamin constituents from digested and dried sewage sludges. The sewage plant in Bradford, England, recovers grease by cracking with sulfuric acid and precipitating with alum iron salts.

Classic examples of multiple usages of waste are the sulfite waste-liquor by-products from paper mills. They are used in fuel, road binder, cattle fodder, fertilizer, insulating compounds, boiler-water additives, and flotation agents, and in the production of alcohol and artificial vanillin. There are some 2,000 U.S. patents for products made from waste sulfite liquor.

Packinghouses and slaughterhouses recover waste blood, which is used as a binder in laminated wood products and in the manufacturing of glue; they also sell waste greases to rendering plants.

The dairy industry treats skim milk with dilute acid to manufacture casein. Casein manufacturers in turn use their waste to precipitate albumin. The resulting albumin waste is used in the crystallization of milk sugar, and the residue from this process is used as poultry feed. Calcium and sodium lactate are also produced from skim milk, and dried and evaporated buttermilk is used for chicken feed. It is even rumored that chocolate ice cream originated as a by-product of the dairy industry.

Some companies, such as rendering plants, are in business primarily to develop by-products from other plants' waste products. Many rendering plants make feeds and fertilizers from chicken feet and feathers and recover grease, which is used to make soap.

Once a by-product is developed and put into production, it is difficult to identify the new product with a waste-treatment process. For example, when sugar is extracted from sugar cane, a thick syrupy liquid known as "blackstrap molasses" is left. This molasses used to be so cheap that it was almost given away. Today, it has many uses, with one of the best-known being in the production of commercial alcohol. People have even found a use for the cane stalks; an insulating wallboard, called Celotex, is made from it.

These are only a few of the many ways in which industry can turn waste into usable products. Although the problem of waste disposal usually persists, it is greatly lessened by the utilization of waste for by-products. In the final analysis, both economic considerations and compliance with the requirements of pollution abatement play a major role in any decisions involving by-product recovery. For a more complete treatise of by-product utilization, the reader is referred elsewhere (Nemerow 1995).

As described earlier in this chapter, the salt brine waste from the pickle processing plant offers great potential for by-product recovery. The brine waste can be electrolyzed to produce chlorine at the anode (after conversion by steam) and sodium at the cathode. The sodium can then be passed through a limestone bed and converted to soda ash. Both the chlorine and the soda ash can then be sold and reused as by-products. I have shown this more fully in Chapter 16 of this book.

Proportioning Wastes

By proportioning its discharge of concentrated wastes into the main sewer, a plant can often reduce the strength of its total waste to the point at which it will need a minimum of final treatment or will cause the least damage to the stream or treatment plant. It may prove less costly to proportion one small but concentrated waste into the main flow, according to the rate of the main flow, than to equalize the entire waste of the plant in order to reduce the strength.

Monitoring Waste Streams

Sophistication in plant control should include that of wastewater controls. Remote sensing devices that enable the operator to stop, reduce, or redirect the flow from any process when its concentration of contaminants exceeds certain limits are an excellent method of reducing waste strengths. In fact, accidental spills are often the sole cause of stream pollution or malfunctioning of treatment plants and these can be controlled, and often eliminated completely, if all significant sources of waste are monitored.

Accidental Spills

Accidental discharges of significant process solutions represent one of the most severe pollution hazards. Because many accidental discharges go unobserved and are usually small in volume, they should be given special attention by the waste engineer. However, it is almost impossible to prevent every potential accident from occurring.

I would like to point out that all so-called *accidental spills* are actually not completely preventable and, therefore, cannot (in truth) really be called *accidents*. At some time, and under some circumstances, the so-called accidental spill will become a reality. Therefore, the environmental engineer must design for such spills with "backup" prevention plans.

There are some measures that can be taken to reduce the likelihood of accidents and severity when and if they occur. Some suggestions for general use include the following:

1. Make certain that all pipelines and valves in the plant are clearly identified.
2. Allow only certain designated and knowledgeable persons to operate these valves.
3. Install indicators and warning systems for leaks and spills.
4. Install double-wall tanks.
5. Provide for detention of spilled wastewater in holding basins or lagoons until proper waste treatment can be accomplished.
6. Monitor all effluents—quantity and quality—to provide a positive public record, if necessary.
7. Establish a regular maintenance program of all pollution-abatement equipment and all production equipment that may result in a liquid discharge to the sewer.

Example of Twentieth-Century Practice of Contaminant Concentration Reduction

Goldfield (1980) found and reported that although there was little difference obtained between general plant ventilation and local exhaust systems, five procedures could be used to lessen the contaminant concentration of asbestos to the Environmental Protection Agency–required standards of 0.5–0.1 fibers/cm^3:

1. Improve enclosure of dust sources
2. Increase the amount and effectiveness of local exhaust ventilation
3. Improve housekeeping
4. Improve work practices
5. Change processes and equipment

Although Goldfield's recommendations apply primarily to inside air contaminant concentration reduction, they could also be easily applied to liquid waste concentration reduction.

Review Questions

1. What do we mean by strength reduction? Why should we use it?
2. Give an example of how a process change can reduce the strength of wastewater.

3. Give a classic example of an equipment modification to reduce the strength of wastewater.
4. How does segregation reduce the strength of the wastewater? Give an example.
5. What do we mean by equalization to reduce the strength of wastewater?
6. When is it profitable to install by-product recovery? How does this help in strength reduction?
7. Give two examples of by-product recovery by industry.
8. How can proportioning industrial waste reduce its strength?
9. What advantage is gained by installing modern methods of monitoring waste contaminants as far as strength reduction is concerned?

References

Davis, L. 1957. Industrial wastes control in the General Motors Corporation. *Sewage Ind. Wastes* 29:1024.

Dillon, K. E. 1967. Waste disposal made profitable. *Chem. Eng.* 74:146.

Factory recovers cyanides from plating wastes. 1965. *Water Works Wastes Eng.* 2:65.

Goldfield, J. 1980. Contaminant concentration reduction: general ventilation versus local exhaust ventilation. *Am. Ind. Hygiene Assoc. J.* 41(Nov):812.

Hyde, A. C. 1965. Chemical plant waste treatment by ten methods. *J. Water Pollution Control Fed.* 37:1486.

Koelsh-Folzer-Werke, A. G. 1966. Separating paint from waste or circulating water containing paint. British Patent 1,016,673. *Chem. Abstr.* 64:649423.

Nemerow, N. L. 1977. *General Environmental Protection Manual for the Industrial Sector*, p. 49. Washington, DC:

Nemerow, N. L. 1976. *Iron and Steel Environment Management.* A report prepared for and published by UNIDO Vienna, Austria, July.

Nemerow, N. L. 1995. *Zero Pollution of Industry*. New York: John Wiley.

Nemerow, N. L., A. Dasgupta Van Nostrand. 1991. *Industrial and Hazardous Waste Treatment,* p. 338. New York: Reinhold Publishing Co.

Rosengarten, G. M. 1967. Union Carbide Corporation's water pollution control program. *Water Sewage Works* 114:R181.

Sanders, M. E. 1967. Implementation to meet the new water quality criteria. *Water Sewage Works* 114:R-5.

Trebler, H. A. 1944. Waste saving by improvements in milk plant equipment. In: *Proceedings of 1st Industrial Waste Conference, Purdue University, Layfayette, Indiana, November 1944,* pp. 6–21.

CHAPTER 3

Neutralization

"Neutralization" can be defined for usage in this book as the treatment of industrial waste so that it is neither too acidic nor too alkaline for safe discharge. There are several possible reasons that an industry neutralized its wastewater during the twentieth century and will continue to do so:

1. To render it compatible with the treatment of municipal sewage when joint treatment is practiced
2. To continue reason no. 1; more specifically, to make certain that its pH does not kill or otherwise inactivate the microorganisms that are being used to biologically oxidize the organic matter content
3. To prevent corrosion of pipelines and equipment leading from the industry to its ultimate destination
4. To comply with municipal or other governmental ordinances of excessive acid or alkaline conditions in sewers or receiving waters.
5. To continue reason no. 4; more specifically, to make certain that the waste discharge pH does not kill fish or otherwise affect other organisms in receiving waters

Excessively acid or alkaline wastes should not be discharged into a receiving stream without treatment. A stream even in the lowest classification—that is, one classified for waste disposal and/or navigation—is adversely affected by low or very high pH values. This adverse condition is even more critical when sudden slugs of acids or alkalis are imposed upon the stream.

At a pH of only 6.5, trout of three species (brook, brown, and rainbow) have shown significantly reduced hatching of eggs and growth. When the pH is lowered to 5.5, bass, walleyed pike, and rainbow trout have been reported to be eliminated (Boyle 1981), as have declines in trout and salmon populations. Below pH 5, most fish are unable to survive. This low pH causes female fish to deter laying of their eggs, and if laid, the fish are very sensitive in the egg, larval, and fish frog stages. Low pH can interfere with the salt balance that freshwater species of fish need to maintain in their body tissues and blood plasma. Acid ioniges or other entities activate many metals already present, such as aluminum, which can be toxic to the fish even at pH values normally considered safe.

There are many acceptable methods for neutralizing overacidity or overalkalinity of wastewater, such as: (1) mixing wastes so that the net effect is a near-neutral pH; (2) passing acid wastes through beds of limestone; (3) mixing acid wastes with lime slurries or dolomitic lime slurries; (4) adding the proper proportions of concentrated solutions of caustic soda (NaOH) or soda ash (NA_2CO_3) to acid wastes; (5) blowing waste boiler-flue to alkaline wastes; (6) producing CO_2 in alkaline wastes; and (7) adding sulfuric acid to alkaline wastes.

The material and method used should be selected on the basis of the overall cost, as material costs vary widely and equipment for using various agents will differ with the method selected. The volume, kind, and quantity of acid or alkali to be neutralized are also factors in deciding which neutralizing agent to use.

In any lime neutralization treatment, the waste engineer should establish a minimum acceptable effluent pH and allow adequate reaction time for an acid effluent to reach this minimum pH. This will usually save considerable unnecessary expense (Lewis and Yost 1950). In many cases, a mill can cut down on neutralization costs by providing sufficient detention time and sacrificing some efficiency in subsequent biological treatment (if used). During storage of alkaline wastes in contact with air, CO_2 will slowly dissolve in the waste and lower the pH. However, detention time alone, within feasible limits, will not effect as low a final pH as can be obtained by the use of neutralizing chemicals. Because biological treatment is more efficient at pH values nearer neutrality, prior neutralization by chemicals renders such treatment more effective.

Mixing Wastes

Mixing of wastes can be accomplished within a single plant operation or between neighboring industrial plants. Acid and alkaline wastes may be produced individually within one plant and proper mixing of these wastes at appropriate times can accomplish neutralization (Figure 3.1), although this usually requires some storage of each waste to avoid slugs of either acid or alkali.

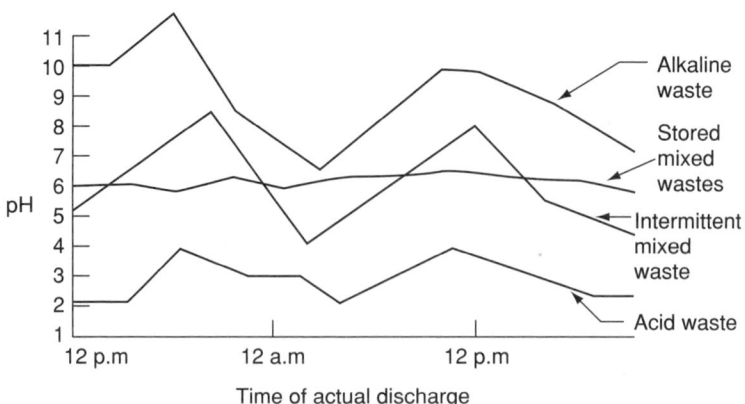

FIGURE 3.1. Neutralization accomplished by mixing of wastes.

If one plant produces an alkaline waste that can be pumped conveniently to an area adjacent to a plant discharging an acid waste, an economical and feasible system of neutralization results for each plant. For example, a building-materials plant producing an alkaline (lime and magnesia) waste pumps the slurry, after some equalization, about one-half mile to mix with the effluent from a chemical plant, producing an acid waste. The neutralized waste resulting from this combination is more readily treatable for final disposal, and both plants thereby solve problems in economics, politics, and engineering. In another instance, Hyde (1965) reports the use of a 500,000-gallon reservoir ahead of an anaerobic digestion pond to mix various types of plant wastes before treatment. The resulting pH of the reservoir effluent ranged from 6.5 to 8.5.

Limestone Treatment for Acid Wastes

Passing acid wastes through beds of limestone was one of the original methods of neutralizing them (Gehm 1944; Reidl 1947). The wastes can be pumped up or down through the bed, depending on the head available and the cost involved, at a rate of about 1 gallon/min (gpm) per square foot (ft^2) or less. Neutralization proceeds chemically according to the following typical reaction:

$$CaCO_3 + H_2SO_4 \rightarrow CaSO_4 + H_2CO_3$$

The reaction will continue as long as excess limestone is available and in an active state. The first condition can be met simply by providing a sufficient quantity of limestone; the second condition is sometimes more difficult to maintain. A sulfuric acid solution must be diluted to an upper limit of about 5% and applied at a rate less than 5 gpm/ft^2 to avoid fouling the bed. According to Jacobs (1951), no attempt should be made to neutralize sulfuric acid above 0.3% concentration or at a rate of feed less than 1 gpm/ft^2 because of the low solubility of calcium sulfate. Excessive acid will precipitate the calcium sulfate and cause subsequent coating and inactivation of the limestone.

Disposing of the used limestone beds can be a serious drawback to this method of neutralization, because the used limestone must be replaced by fresh limestone at periodic intervals, with the frequency of replacement depending on the quantity and quality of acid wastes being passed through a bed. When there are extremely high acid loads, foaming may occur, especially when organic matter is also present in the waste.

Lime-Slurry Treatment for Acid Wastes

Mixing acid wastes with lime slurries is an effective procedure for neutralization (Hoak 1944; Rudolfs 1943a,b; Smith 1943). The reaction is similar to that obtained with limestone beds. In this case, however, lime is used up continuously because it is converted to calcium sulfate and carried out in the waste. Though slow acting, lime possesses a high neutralizing power and its action can be hastened by heating or oxygenating the mixture. It is relatively inexpensive, but in large quantities, cost can be an important factor.

Hydrated lime is sometimes difficult to handle, because it tends to arch, or bridge, over the outlet in storage bins and possesses poor flow properties, but it is particularly adaptable to neutralization problems involving small quantities of acid waste, as it can be stored in bags without special storage facilities.

In an actual case (Dickerson and Brooks 1950), neutralization of nitric and sulfuric acid wastes in concentrations up to about 1.5% (in the case of sulfuric acid) was accomplished satisfactorily by using a burned dolomitic stone containing 47.5% CaO, 34.3% MgO, and 1.8% $CaCO_3$. The concentration of acid was limited to the stated 1.5%, at least in part, because of the absence of dilution water to vary the percentage. This stone provided the additional advantage of holding residual sulfation to a minimum, an impossibility with any of the high-calcium limes (Jacobs 1947).

Caustic-Soda Treatment for Acid Wastes

Adding concentrated solutions of caustic soda or sodium carbonate to acid wastes in the proper proportions results in faster, but more costly, neutralization. Smaller volumes of the agent are required, because these neutralizers are more powerful than lime or limestone. Another advantage is that the reaction products are soluble and do not increase the hardness of receiving waters. Caustic soda is normally bled into the suction side of a pump discharging acid wastes. This method is suitable for small volumes, but for neutralizing large volumes of acid wastewater, special proportioning equipment (see Chapter 2) should be provided, as well as a suitably sized storage tank for the caustic soda with a multiple-speed pump for direct addition of the alkali to the flow of acid wastes.

We have now discussed four methods of neutralizing acid wastes. Before moving on to alkaline wastes, compare the "basicity" and costs of the acid-neutralizing methods and agents (Table 3.1).

TABLE 3.1
Cost Comparison of Various Alkaline Agents[a]

Chemical	Cost, $/Ton (approx.)	Basicity Factor[b]	Cost, $/Ton of Basicity
NaOH (78% Na_2O)	106	0.687	154
Na_2CO_3 (58% Na_2O)	57	0.507	112
MgO	83	1.306	64
High-calcium hydrated lime	14	0.710	20
Dolomitic hydrated lime	14	0.912	15
High-calcium quicklime	11	0.941	12
Dolomitic quicklime	11	1.110	10
High-calcium limestone	4	0.489	8
Dolomitic limestone	4	0.564	7

[a]Based on 1954 cost quotations.
[b]A measure of the alkali available for neutralization (grams of equivalent CaO/per/gram).
Adapted from Hoak (1950).

Because the basicity factor, as shown in Table 3.1, is one of the vital factors in selecting a neutralizing agent, Hoak (1958) provides not only a method for computing this factor but also a nomograph for calculating the pounds of neutralizing agent required per gallon of waste (Figure 3.2). He determines the acid value by titrating a 5-ml sample of sulfuric acid waste with an excess amount of 0.5 N NaOH and back-titrating with 0.5 N HCl to a phenolphthalein endpoint. The basicity factor of the lime (or neutralizing agent) is determined by titrating a 1-g sample of alkaline agent with an excess of 0.5 N HCl, boiling the sample for 15 minutes, and back-titrating with 0.5 N NaOH to the phenolphthalein endpoint. The acid value (line B in Figure 3.2) and basicity factor (line A in Figure 3.2) are then connected in Hoak's nomograph to find the pounds of alkaline agent required per gallon of acid waste (line C in Figure 3.2).

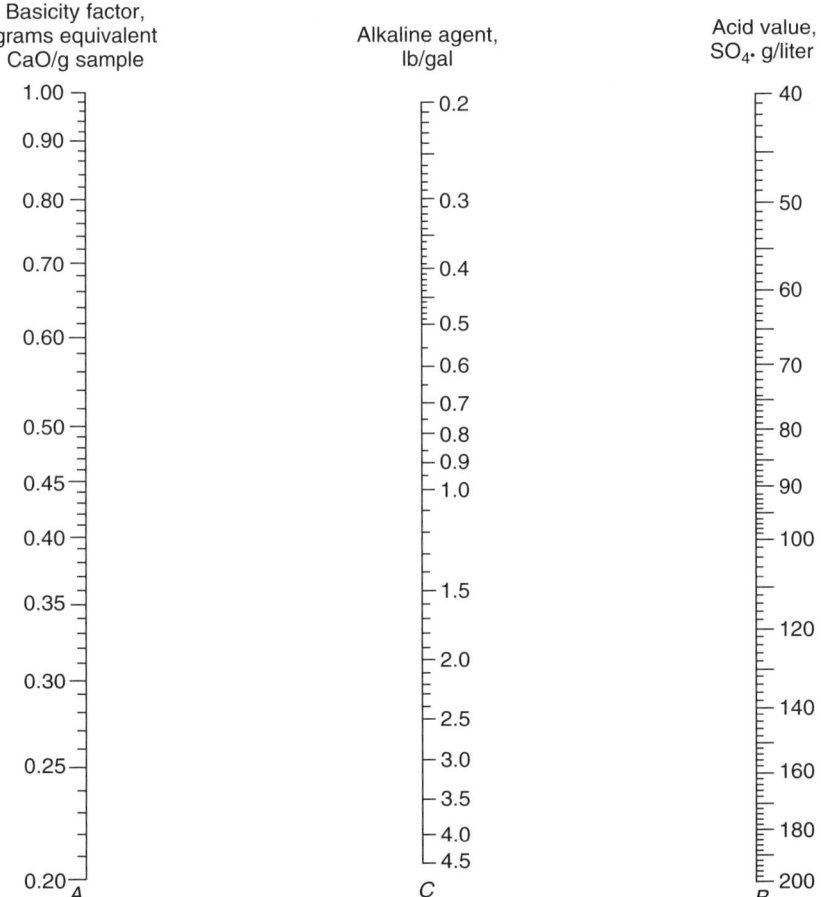

Connect scales A and B with a straight edge and read the result on scale C

FIGURE 3.2. Nomograph for treatment of acid wastes: a chart for determining the amount of alkaline agent needed (adapted from Hoak 1958).

When sodium hydroxide is used as a neutralizing agent for carbonic and sulfuric acid wastes, the following reactions take place:

$$Na_2CO_3 + CO_2 + H_2O \rightarrow 2NaHCO_3,$$

carbonic acid waste

$$2NaOH + CO_2 \rightarrow Na_2CO_3 + H_2O;$$

$$NaOH + H_2SO_4 \rightarrow NaHSO_4 + HOH,$$

sulfuric acid waste

$$NaHSO_4 + NaOH \rightarrow Na_2SO_4 + HOH.$$

Both these neutralizations take place in two steps and the end-products depend on the final pH desired. For example, one treatment may require a final pH of only 6, and thus, $NaHSO_4$ would make up the greater part of the products; another treatment may require a pH of 8, with most of the product being Na_2SO_4.

We shall now take up the subject of neutralization of alkaline wastes.

Using Waste Boiler-Flue Gas

Blowing waste boiler-flue gas through alkaline wastes is a relatively new and economical method for neutralizing them. Most of the experimental work has been carried out on textile wastes (Beach and Beach 1956; Nemerow 1956; Rudolf 1943; Steele 1954; *Treatment of Alkaline Sulfur Dye Waste with Flue Gas* 1956). Well-burned stack gases contain approximately 14% carbon dioxide. CO_2 dissolved in wastewater will form carbonic acid (a weak acid), which in turn reacts with caustic wastes to neutralize the excess alkalinity as follows:

$$CO_2 + H_2O \rightarrow H_2CO_3,$$

flue gas   wastewater   carbonic acid

$$H_2CO_3 + 2NaOH \rightarrow Na_2CO_3 + 2H2O,$$

carbonic acid   caustic soda in wastewater   soda ash

$$H_2O$$

$$H_2CO_3 + Na_2CO_3 \rightarrow 2NaHCO_3 + H_2O.$$

Excess carbonic acid   soda ash in waste   sodium bicarbonate in waste

The equipment required usually consists of a blower placed in the stack, a gas pipeline to carry the gases to the waste-treatment site, a filter to remove sulfur and unburned carbon particles from gases, and a gas diffuser to disperse the stack gases in the wastewater. Stack gases evolve hydrogen sulfide from wastewater that contains any appreciable quantity of sulfur, and this H_2S must be burned, absorbed, or vented positively to the upper atmosphere to prevent nuisance conditions.

Carbon Dioxide Treatment for Alkaline Wastes

Bottled CO_2 is applied to wastewater in much the same way as compressed air is applied to activated-sludge basins. It neutralizes alkaline wastes on the same principle as boiler-feed gases (i.e., it forms a weak acid [carbonic acid] when dissolved in water) but with much less operating difficulty. The cost may be prohibitive, however, when the quantity of alkaline wastes is large. A textile mill (Nemerow 1956) producing about 6 million gallons/day of alkaline waste studied the practical aspects of this method and found that installation of the equipment necessary to provide bottled CO_2 would cost about $150,000 and the power and fuel to generate it about $275/day—a considerable expense, even for so large a plant.

Producing Carbon Dioxide in Alkaline Wastes

Another way to produce carbon dioxide is to burn gas underwater. This process is called *submerged combustion* and has been used in the disposal of nylon wastes (Remy and Lauria 1958) to neutralize the waste before biological treatment. In pilot-plant studies, the researchers (Remy and Lauria 1958) investigated submerged combustion on a continuous basis, using an evaporation vessel, a burner with flame jets submerged below the waste surface in the vessel, a bustle in which air and natural gas were mixed to form a combustible mixture, and other equipment to measure air, gas, and waste flows and the weight of waste volatilized during each run. A schematic drawing of the submerged-combustion plant used is shown in Figure 3.3. Researchers concluded that submerged combustion, rather than aeration, should be used to treat part of the plant waste, for

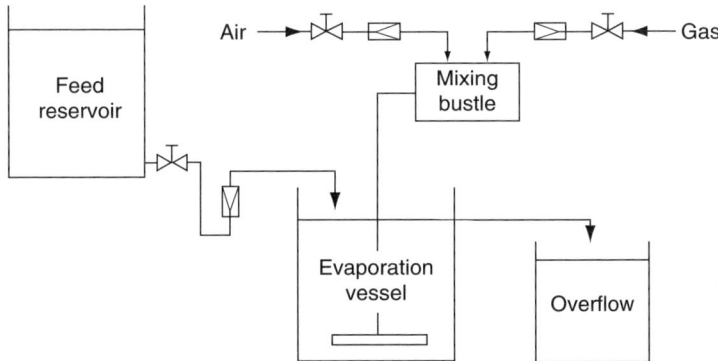

FIGURE 3.3. Submerged-combustion pilot unit as used by Remy and Lauria (1958).

economic reasons. (The researchers in this case, however, were primarily concerned with stripping toxic materials from the waste, rather than with neutralizing it.) Krofchak (1962) describes this method of neutralization and suggests its use for spent pickle liquors and spent electrolytes from nickel refining. CO_2 may also be produced by fermentation of an alkaline, organic waste; the resulting pH is thus lowered. EbaraInfilco, Ltd. in 1965 patented such a process for fermenting alkaline beet-sugar wastes with yeast; the CO_2 produced can be used for neutralization and the excess yeast as forage.

Sulfuric Acid Treatment for Alkaline Wastes

The addition of sulfuric acid to alkaline wastes is a fairly common, but expensive, means of neutralization. Sulfuric acid can cost as much as 2 or 3 cents per pound, although it may be as low as 1 cent per pound in large quantities. Storage and feeding equipment requirements are low as result of its great acidity, but it is difficult to handle because of its corrosiveness. The neutralization reaction that occurs when it is added to wastewater is as follows:

$$2NaOH + H_2SO_4 \rightarrow Na_2SO_4 + 2H_2O$$

wastewater   sulfuric acid as neutralizer   resulting neutral salt

A titration curve of the alkaline waste neutralized with various amounts of H_2SO_4 is helpful to ascertain the quantities of acid required for neutralization to definite pH values and the relevant costs. Figure 3.4 represents the titration curve of an actual mixed alkaline waste in Niagara Falls, New York.

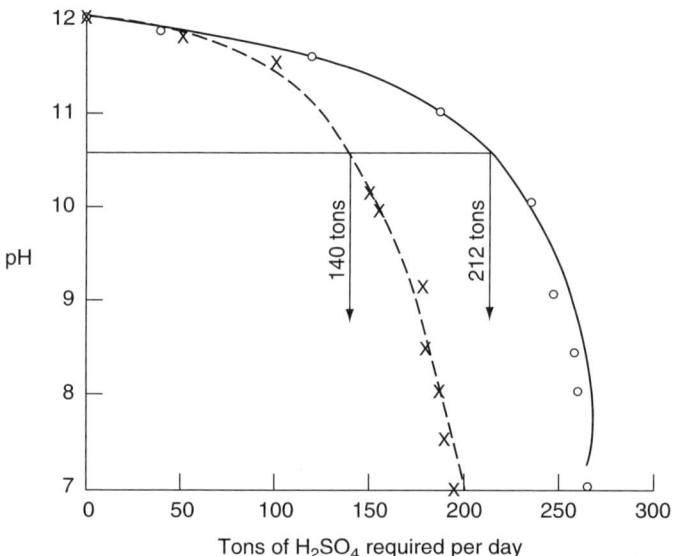

FIGURE 3.4. Acid required to neutralize industrial wastes in sewer.

During most of the twentieth century, a widely held opinion was that a high pH—say, 10 or higher—would so adversely affect biological oxidation that acid neutralization would be required. In some cases this may be true, but an industrial waste with a pH of 10 could be effectively degraded aerobically (biologically) without neutralization (Emerson and Nemerow 1969). The reader can readily realize the great potential savings in capital and operating costs of omitting the neutralization step here and in other cases of high alkalinity.

Acid-Waste Utilization in Industrial Processes

In some situations, it may be possible to use acid wastes to get the desired result in industrial processing: wash, cool, or neutralize products. For example, Dillon (1967) reports the use of acid-mine drainage water for cleaning raw coal. Mine wastewater occurs in large quantities in the coal industry. These waters are usually acidic and contain sulfates of iron and aluminum; if they are used to wash raw coal, neutralization results, because coal contains calcium and magnesium carbonates. Dillon (1967) describes the treatment of 600 tons of raw coal per hour, with an average of 225 gpm of mine wastewater. The pH of the mine water is thereby raised from 3.0 to neutrality.

Example of Twentieth-Century Practice of Neutralization

Drury (1999) found that the unusual use for pH neutralization of anaerobic digestion of acid mine drainage wastewater could raise the pH of the wastewater, as well as having other contaminant removal effects. He reported that adding whey to the anaerobic substrate reactor improved the long-term treatment efficiency for pH neutralization, alkalinity production, and sulfate, iron, zinc, and manganese removal.

Effluent pH—according to Drury (1999)—from the reactor with whey addition was relatively constant at 6.5.

Review Questions

1. Name seven major methods of neutralizing both acid and alkaline wastes.
2. What are the four major factors to be considered when neutralizing wastes?
3. What effect can the storage of alkaline wastes have on pH and resulting neutralization?
4. Under what conditions would you select mixing of wastes as a solution to neutralization?
5. When would you use limestone beds or filters for acid wastes?
6. What is the advantage of using lime slurry rather than limestone beds for neutralization?
7. When would you use NaOH or Na_2CO_3 for neutralization?

References

Beach, C. J., M. G. Beach. 1956. Treatment of alkaline dye wastes with flue gas. In: *Proceedings of 5th Southern Municipal and Industrial Waste Conference, April 1956,* p. 162.

Boyle, R. H. 1981. An American tragedy. *Sports Illustrated* 55(Sep 21):13, 75.

Dickerson, B. W., R. M. Brooks. 1950. Neutralization of acid wastes. *Ind. Eng. Chem.* 42:599.

Dillon, K. E. 1967. Waste disposal made profitable. *Chem. Eng.* 13(March):146.

Drury, W. 1999. Treatment of acid mine drainage with anaerobic solid substrate reactors. *Water Environ. Res.* 71(Sep/Oct):6, 1244.

Emerson, D. B., Nemerow, N. L. 1969. High solids, biological aeration of unneutralized, unsettled tannery wastes. In: *Proceedings of the 24th Purdue Industrial Waste Conference, May 7, 1969, Lafayette, Indiana,* p. 867.

Gehm, H. W. 1944. Neutralization with up-flow expanded limestone bed. *Sewage Works J.* 16:104.

Hoak, R. D. 1944. Neutralization studies on basicity of limestone and lime. *Sewage Works J.* 16:855.

Hoak, R. D. 1950. Acid iron wastes neutralization. *Sewage Ind. Wastes* 22:212.

Hoak, R. D. 1958. A neutralization nomograph. *Ind. Wastes* 3:D-48.

Hyde, A. C. 1965. Chemical plant waste treatment by ten methods. *J. Water Pollution Control Fed.* 37:1486.

Jacobs, H. L. 1947. Acid neutralization. *Chem. Eng. Progr.* 43:247.

Jacobs, H. L. 1951. Neutralization of acid wastes. *Sewage Ind. Wastes* 23:900.

Krofchak, O. 1962. Submerged combustion evaporation of acid wastes. *Ind. Water Wastes* 7:63.

Leidner, R. N. 1966. Burns Harbor—waste treatment planning for a new steel plant. *J. Water Pollution Control Fed.* 38:1767.

Lewis, C. J., L. J. Yost. 1950. Lime in waste acid treatment. *Sewage Ind. Wastes* 22:893.

Nemerow, N. L. 1956. Holding and aeration of cotton mill finishing wastes. In: *Proceedings of 5th Southern Municipal and Industrial Waste Conference, April 1956,* p. 149.

Reidl, A. L. 1947. Neutralization with up-flow limestone bed. *Sewage Works J.* 19:1093.

Remy, E. D., D. T. Lauria. 1958. Disposal of nylon wastes. In: *Proceedings of 13th Industrial Waste Conference, May 1958, Purdue University Engineering Extension,* Series no. 96, p. 596.

Rudolfs, W. 1943a. Pretreatment of acid wastes. *Sewage Works J.* 15:48.

Rudolfs, W. 1943b. Neutralization with lime. *Sewage Works J.* 15:590.

Smith, F. 1943. Neutralization of pickle liquor. *Sewage Works J.* 15:157.

Steele, W. R. 1954. Application of flue gas to the disposal of caustic textile wastes. In: *Proceedings of 3rd Southern Municipal and Industrial Conference, March 1954,* p. 190.

Treatment of alkaline sulfur dye waste with flue gas [research report no. 8]. 1956. *Proc. Am. Soc. Civil Engrs.* 82(SA-5):1078.

CHAPTER 4

Equalization and Proportioning

Equalization

Industry will always realize benefits from delivering a constant quantity and quality of waste from its plant. It will further benefit from delivering this waste in proportion to that occurring in its discharge environment—be it river or sewer line. The reason for these facts is that receiving environments can always cope better with a constant load of contaminant and in proportion to its own existing load. In fact, an industrial plant can sometimes forgo any further expensive treatment of its wastes after equalization and proportionment.

Equalization is a method of retaining waste in a basin so that the effluent discharged is fairly uniform in its water quality characteristics (pH, color, turbidity, alkalinity, biochemical oxygen demand [BOD], and so forth). A secondary but significant effect is that of lowering the concentration of effluent contaminants. This is accomplished not only by ironing out the slugs of a high concentration of contaminants but also by physical, chemical, and biological reactions that may occur during retention in equalization basins. For example, the increases in industrial waste reported by Fall (1965) at Peoria have greatly varied the organic loading at the treatment plant. A retention pond serves to level out the effects of peak loadings on the plant while substantially lowering the BOD and suspended-solids load to the aeration unit. Air is sometimes injected into these basins to provide: (1) better mixing; (2) chemical oxidation of reduced compounds; (3) some degree of biological oxidation; and (4) agitation to prevent suspended solids from settling.

The size and shape of the basins vary with the quantity of waste and the pattern of its discharge from the factory. Most basins are rectangular or square, although Metzger (1967) found that triangular tanks produce satisfactory flow distribution. The capacity should be adequate to hold, and render homogeneous, all the waste from the plant. Almost all industrial plants operate on a cycle basis; thus, if the cycle of operations is repeated every 2 hours, an equalization tank that can hold a 2-hour flow will usually be sufficient. If the cycle is repeated only every 24 hours, the equalization basin must be big enough to hold a 24-hour flow of waste. Herion and Roughhead (1964) reported the use of 72-hour equalization for a pharmaceutical waste to ensure ample mixing. This period (three times the 24-hour cycle of operations) was selected as the proper detention time to

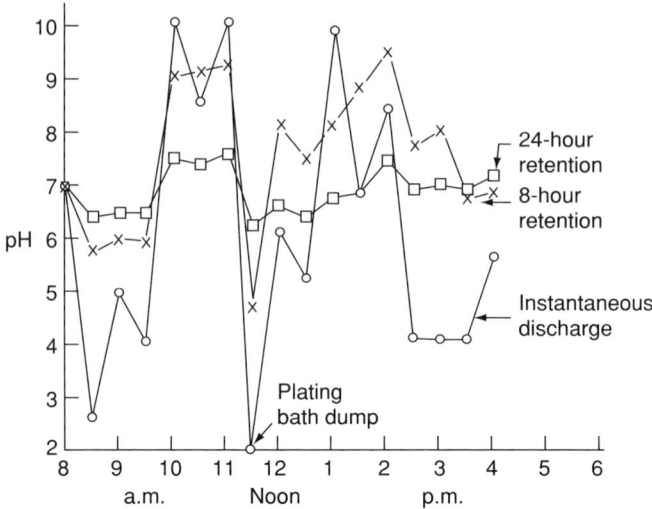

FIGURE 4.1. Effect of equalization.

not disrupt the biota of the activated-sludge units. In a wool-finishing-mill waste containing dieldrin (a mothproofing insecticide), an equalization period of 44 days was necessary to yield a receiving stream concentration of less than 0.0005 mg/liter. Figure 4.1 compares the effects of 8-hour and 24-hour detention periods on the final pH of metal-plating waste.

The mere holding of waste, however, is not sufficient to equalize it. Each unit volume of waste discharged must be adequately mixed with other unit volumes of waste discharged many hours previously. This mixing may be brought about in the following ways: (1) proper distribution and baffling; (2) mechanical agitation; (3) aeration; and (4) combinations of all three.

Proper distribution and baffling is the most economical, though usually the least efficient, method of mixing. Still, this method may suffice for many plants. Horizontal distribution of the waste is achieved by using either several inlet pipes, spaced at regular intervals across the width of the tank, or a perforated pipe across the entire width. Over-and-under baffles are advisable when the tank is wide because they provide more efficient horizontal and vertical distribution (Figure 4.2). Baffling is especially

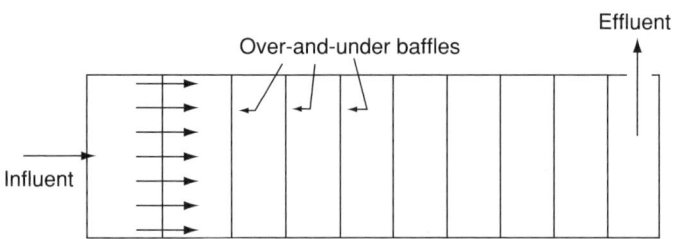

FIGURE 4.2. Top view of an equalizing basin, with perforated inlet pipe and over-and-under baffles.

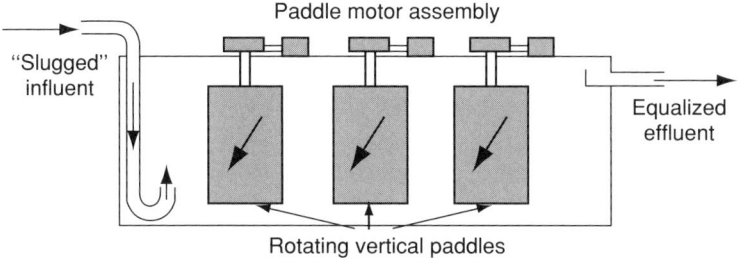

FIGURE 4.3. Side view of an equalizing basin, with mechanical agitators instead of baffles.

important when several types of waste enter the basin at various locations across the width. The influent should be forced to the bottom of the basin so that the entrance velocity prevents suspended particles from sinking and remaining on the bottom.

Mechanical agitation eliminates most of the need for baffles and generally provides better mixing than baffles alone. One typical arrangement (Rudolfs and Millar 1946), shown in Figure 4.3, uses three wooden gate–type agitators spaced equidistantly along the center line of the length of the tank. Agitators operated at a speed of 15 rotations/min (rpm) by a 3-horsepower (hp) motor are usually adequate.

The design in Figure 4.3 approximates the theoretically ideal tank, because of its relatively high efficiency at similar detention times, as a result of mechanical mixing, and because it prepares various types of chemical waste for direct disposal or final treatment. If subsequent treatment is necessary, the process is made easier because the problem of waste with rapidly changing characteristics varying from one extreme to the other is eliminated. Rudolfs and Millar (1946) recommended this method of equalization when: (1) limited space is available; (2) removal of suspended solids is not desired; (3) there are rapid fluctuations in the characteristics of the waste; and (4) facility of subsequent treatment is a goal.

This type of equipment is good not only for equalization but also for dilution, oxidation, reduction, or any other function in which one wants chemical compounds discharged to react with compounds discharged before or after them, to produce a desired effect.

Aeration of equalizing basins is the most efficient way to mix types of waste, but it is also the most expensive. To aerate an equalizing basin takes about half a cubic foot of air per gallon of waste. Aeration facilitates mixing and equalization of waste, prevents or decreases accumulation of settled material in the tank, and provides preliminary chemical oxidation of reduced compounds, such as sulfur compounds. It is of special benefit in situations in which wastes have varying character and quantity, excess of reduced compounds, and some settleable suspended solids.

Proportioning

Proportioning means the discharge of industrial wastes in proportion to the flow of municipal sewage in the sewers or to the stream flow in the receiving river. In most cases, it is possible to combine equalization and proportioning in the same basin. The effluent from the equalization basin is metered into the sewer or stream according to a predetermined schedule. The objective of proportioning in sewers is to keep constant the percentage of industrial wastes to domestic sewage flow entering the municipal sewage plant. This procedure has several purposes: (1) to protect municipal sewage treatment using chemicals from being impaired by a sudden overdose of chemicals contained in the industrial waste; (2) to protect biological-treatment devices from shock loads of industrial wastes that may inactivate the bacteria; and (3) to minimize fluctuations of sanitary standards in the treated effluent.

The rate of flow of industrial waste varies from instant to instant, as does the flow of domestic sewage, and both empty into the same sewage system. Therefore, the industrial waste must be equalized and retained, and then proportioned to the sewer or stream according to the volume of domestic sewage or stream flow. To facilitate proportioning, a holding tank should be constructed with a variable-speed pump to control the effluent discharge. Because the domestic sewage treatment plant is usually located some distance from an industry, signaling the time and amount of flow is difficult and sometimes quite expensive. For this reason, many industries have separate pipelines through which they pump their wastes to the municipal treatment plant. The wastes are equalized separately at the site of the municipal plant and proportioned to the flow of incoming municipal wastewater. Separate lines are not, of course, always possible or even necessary. One textile mill found that it could effectively proportion its waste to the variable domestic sewage flow by adjusting the valve on the holding-tank effluent pump at 8:00 A.M., 12:00 noon, and 7:00 P.M.

There are two general methods of discharging industrial waste in proportion to the flow of domestic sewage at the municipal plant: manual control related to a well-defined domestic sewage flow pattern, and automatic control by electronics.

Manual control is lower in initial cost but less accurate. It involves determining the flow pattern of domestic sewage for each day of the week over a period of months. Usually one does this by examining the flow records of the sewage plant or by studying the hourly water-consumption figures for the city. It is better to spend time on a careful investigation of the actual sewage flow than to make predictions based on miscellaneous nonpertinent records. Actual investigative data should be used to support those records that are applicable to the case.

Automatic control of waste discharge according to sewage flow involves placing a metering device that registers the amount of flow at the most convenient main sewer connection. This device translates the rate of flow in the sewer to a recorder located near the plant's holding tank. The pen on the recorder actuates either a mechanical (gear) or a pneumatic (air) control system for opening or closing the diaphragm of the pump. There are, of course, many variations of automatic flow-control systems. Although their initial cost is higher than that of manual control, they will usually return the investment many times by the savings in labor costs.

Some industrial and municipal sewage plant superintendents think that the best time to release a high proportion of industrial waste to the sewer is at night, when the domestic sewage flow is low. Whether night release is a good idea depends on the type of treatment used and the character of the industrial waste. If the treatment is primarily biological and the industrial wastes contain readily decomposable organic matter and no toxic elements, discharging the largest part of the industrial waste to the treatment plant at night is indeed advisable, because this ensures a relatively constant organic load delivered to the plant day and night.

One equipment manufacturing company recommends a three-component system for automatic proportioning of wastes into sewers (Figure 4.4): (1) a kinematic manometer with integral pneumatic transmitter; (2) a remotely located indicator program controller, which receives air signals and has a precut time pattern cam for continuously adjusting the set point of the pneumatic controller to give a waste-flow rate in accordance with the desired pattern; and (3) a diaphragm-actuated, motor-controlled valve that is actuated by the air signal from the program controller. Practically speaking, the length of the pneumatic capillary tubing limits the physical separation between the sensing components, but this difficulty can be overcome with an electrical system.

The typical waste-flow proportioning system (Bubbler System) shown in Figure 4.4, as supplied by Fischer and Porter Co., consists essentially of the three separate devices described in the previously. Item 1 (Figure 4.4), with a linear air-pressure output of 3–15 pounds/in^2, has a flow range of 17–170 gallons/min of an industrial waste (specific gravity assumed, 1.1). Item 2 (Figure 4.4) is a remotely located indicator for receiving air signals from Item 1, as explained in the text. Item 3 (Figure 4.4) is an automatic valve capable of operating at a maximum pressure drop of 10 pounds at maximum flow rate. This valve is actuated by air signals from the program controller, Item 2 (Figure 4.4).

Another arrangement for proportioning industrial wastes in a situation in which pipelines are flowing only partly full or waste flows in open channels, is the use of a weir, flume, or Kennison nozzle in the main flow line to measure the flow. A float-operated transmitter (either electrical or pneumatic) is connected to this measuring device and the electrical or pneumatic signals are used to actuate a flow splitter in a proportioning weir tank (such as is provided by Proportioneers, Inc.).

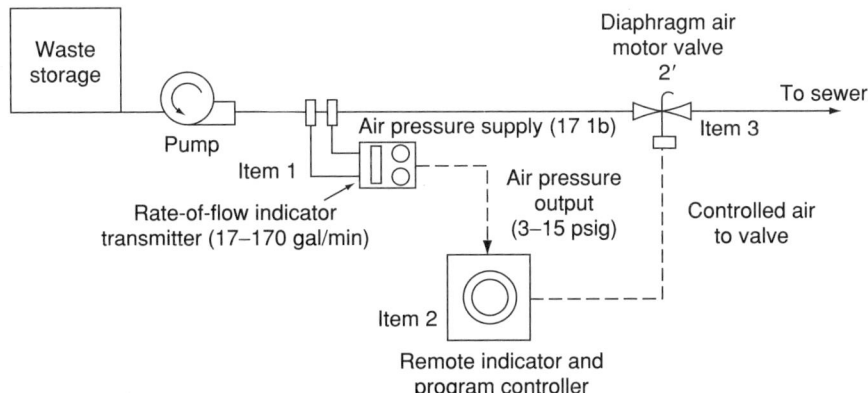

FIGURE 4.4. Waste-metering system (courtesy Fischer and Porter Company).

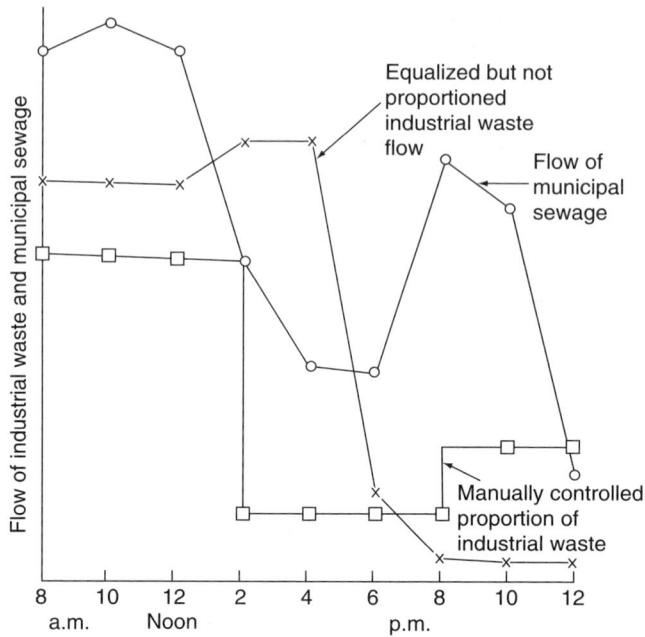

FIGURE 4.5. Effect of proportioning.

The Belle, West Virginia, works of the DuPont Nemours Company has been impounding its waste in two 2.5-million-gallon tanks and releasing it to the Kanawha River in proportion to the river flow for more than 10 years (Hyde 1965). This has been necessary owning to the flashiness of the river flows. Figure 4.5 compares the effects of both equalization and proportioning on the flow at a municipal treatment plant.

By equalizing a tannery mill waste while aerating it for 24 hours and proportioning the effluent for further treatment, it was possible to obtain high BOD removals previously found unattainable (Nemerow et al. 1978).

Example of Twentieth-Century Practice of Equalization and Proportioning

Crumb and West (2000) proposed a better and more cost-effective way to meet permit requirements during wet weather. Their pilot plant results show that removal efficiency of all processes were considerably better than conventional clarification. Using an enhanced high-rate clarification (EHRC) process, Crumb and West (2000) showed that treatment of peak wet weather flow produced effluent suitable for discharge into the Trinity River. In their presentation, Crumb and West (2000) provide a schematic diagram of the blending of storm flow with domestic wastewater to yield efficient removal of contaminants, thus using a modification of the equalization and proportioning process.

Review Questions

1. Define *equalization,* including its purpose.
2. What are four methods of mixing to effect equalization?
3. What are the objectives of proportioning of industrial wastes?
4. What are the problems associated with proportioning industrial wastes into municipal sewers?

References

Crumb, S. F., R. West. 2000. Blended flow process alleviates wet weather woes cost-effectively for Fort Worth, Texas. *Water Environ. Technol.* April:43.
Fall, E. B. 1965. Retention pond improves activated sludge effluent quality. *J. Water Pollution Control Fed.* 37:1194.
Gibbs, C. V., R. H. Bothel. 1965. Potential of large metropolitan sewers for disposal of industrial wastes. *J. Water Pollution Control Fed.* 37:1417.
Herion, R. W., H. O. Roughhead. 1964. Two treatment installations for pharmaceutical wastes. In: *Proceedings of 18th Industrial Waste Conference, 1964, Purdue University Engineering Extension Series,* p. 218. Bulletin no. 115. Lafayette, Indiana.
Hyde, A. C. 1965. Chemical plant waste treatment by ten methods. *J. Water Pollution Control Fed.* 37:1486.
Metzger, I. 1967. Triangular tank for equalizing liquid wastes. *Water Sewage Works* 114:9.
Nemerow, N. L., D. Warne, F. Falk. 1978. A new and effective solution for treatment of tannery wastewater. In: *Proceedings of the 33rd Annual Purdue Industrial Waste Conference, May 9–11, 1978.* Lafayette, Indiana.
Rudolfs, W., J. N. Millar. 1946. A method for accelerated equalization of industrial wastes. *Sewage Works J.* 18:686.
Texas Water and Sewage Works Association. 1955. *Manual for Sewage Plant Operators,* pp. 342–345. Austin: Texas Water and Sewage Works Association.
Wilroy, R. D. 1964. Industrial wastes from scouring rug wool and the removal of dieldrin. In: *Proceedings of 18th Industrial Wastes Conference, 1964, Purdue University Engineering Extension Series,* p. 413. Bulletin no. 115. Lafayette, Indiana.

CHAPTER 5

Removal of Suspended Solids

The removal of suspended solids from any industry's waste plays an important part in its overall waste treatment program for the following reasons:

1. Elimination of a major portion of the contaminant (often as high as 20–30%)
2. Segregation of one type of contaminant so that it can be further treated more easily and economically
3. Removal of this type of constituent so that the remaining contaminants are more homogeneous and likewise can be treated more efficiently—usually chemically or biologically
4. When their removal is the sole method of waste treatment, it reduces the unsightliness or visibility from the public viewers and even prevents damage to the habitats of riverbeds, which are not generally visible to the receiving water user

Sedimentation Theory

Although sedimentation is a method of treatment used in almost all domestic-sewage treatment plants, it should be considered for industrial waste treatment only when the industrial waste is combined with domestic sewage or contains a high percentage of "settleable" suspended solids, such as those found in cannery, paper, sand and gravel, coal-washery, and certain other wastes. The efficiency of sedimentation tanks depends, in general, on the following factors:

- Detention period
- Wastewater characteristics
- Tank depth
- Floor surface area and overflow rate
- Operation (cleanliness)
- Temperature
- Particle size
- Inlet and outlet design

- Velocity of particles
- Density of particles
- Container-wall effect
- Number of basins (baffles)
- Sludge removal
- Pretreatment (grit removal)
- Flow fluctuations
- Wind velocity

Although settling tanks have been used for other purposes, such as grease flotation, equalization, and biochemical oxygen demand (BOD) reduction, they are primarily used for removing settleable suspended matter. Theoretically, a suspended particle in a wastewater solution will continue to settle at a fixed velocity relative to the solution, as long as the particle remains discrete; when it coalesces with other particles, its size, shape, and resulting density will change, as will its settling velocity. Coagulation, or self-flocculation, of particles causes an increase in velocity. In liquid wastes containing high percentages of suspended solids, greater reductions in the suspended solids will occur primarily because of increased flocculation. The fixed settling velocity will also be altered by changes in the temperature and density of the liquid solvent through which the particle is moving. Rising layers of warmer liquid can cause eddying and a disturbance in the settling of particles; an increased density in the lower layers of liquid can deter the particle from settling to the bottom. These factors can interfere with settling to such an extent that particles may be carried out of the tank with the effluent.

Tank depth is also important. The deeper the tank (all other factors being equal), the better the chance of preventing the deposited solids from being re-suspended—for example, by sudden scouring due to turbulence caused by unequal flow distribution or by exposure to wind or temperature effects—and thus being carried out with the effluent. This is especially important when sludge is stored in sedimentation basins for lengthy periods before pumping. If the solids are continuously removed from the bottom of settling tanks as soon as they land, shallower tanks can be built.

Surface area is another factor affecting tank efficiency, and engineers agree that floor area must be adequate to receive all the particles to be removed from the wastewater. However, many state health departments, when establishing acceptable dimensions for settling basins, do so on the basis of standard detention periods. In certain designs, this method may not provide adequate floor area and complete settling is not achieved.

Figure 5.1 illustrates the effect of doubling the floor area and halving the depth of a settling basin, with volume and detention time remaining constant. Theoretically, the basin in Figure 5.1B will remove twice as many discrete particles as the basin in Figure 5.1A. Therefore, the engineer should strive to design settling basins that are as shallow as possible and contain ample floor area. However, tanks less than 6 feet deep have been found impractical from an operational standpoint, because they are subject to upsetting by scouring or velocity of currents. The floor area is increased most satisfactorily by extending the length of the basin.

Because the percentage of particles reaching the bottom of the settling basin also depends on the rate of waste flow, an expression correlating horizontal flow with the floor or surface area has been devised. It is commonly referred to as the *overflow rate*

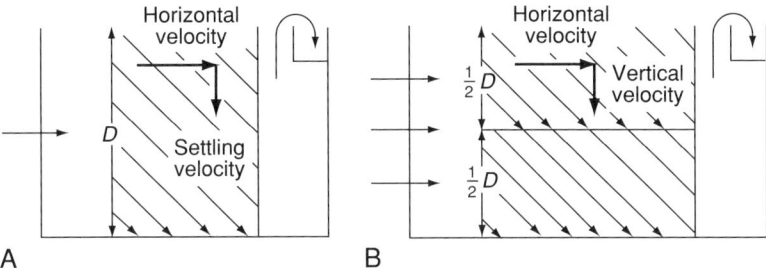

FIGURE 5.1. Effect of (**A**) doubling the floor area and (**B**) halving the depth of a settling basin.

and is expressed as gallons per square foot per day (g/ft^2/day). Typical overflow rates vary from 200 to 800 g/ft^2/day for primary sedimentation basins and from 1,000 to 3,000 g/ft^2/day for final tanks, because particles in the final tanks usually settle more rapidly than those in primary basins. Exceptions include grit particles, which settle faster than the average particle in primary basins, and activated sludge floc, which tends to slow down the settling rate in secondary basins. Because of these discrepancies, both primary and secondary settling basins are often designed for the same overflow rates. Lower overflow rates for domestic-type wastes generally result in the removal of more suspended solids and BOD, as shown in Figure 5.2 (Great Lakes–Upper Mississippi River Board of State Sanitary Engineers 1960). For further details on theories of sedimentation, the reader is referred to Eckenfelder (1966).

Unfortunately, actual settling velocities may vary from theoretical formulations. Turbulence and flocculation are the main causes of variation. Another factor is that velocities do not remain constant throughout a cross-sectional area of a tank. The settling velocity of discrete particles of diameter d in a quiescent viscous fluid is given by

$$V = 4/3 \bullet gd/C_d\,(p_s - 1/p),$$

where C_d is the drag coefficient between the fluid and the particle, g is the acceleration due to gravity, and p_s and p are the densities of the particle and the fluid. The drag coefficient does not remain constant but varies with the Reynolds number, R, which equals pdV/μ. The correlation between C_d and R has been plotted in various textbooks, but a trial-and-error procedure is still required to obtain V.

Turbulence in sedimentation basins has both a positive and a negative effect on the settling velocity of a particle. It causes eddies that carry some particles down and some up (as shown in Figure 5.1), so it both helps flocculation and hinders sedimentation. Settling or rising velocities can be unequal, depending on the local circumstances causing the turbulence, such as increased horizontal velocity of water at the inlet.

Other factors that induce eddying include wind, unequal distribution of flow, changes in temperature, and changes in density of the liquid at various depths. Eddying generally decreases the settling velocity and efficiency of operation, while flocculation generally increases the overall total of solids removed. The influence of flocculation is illustrated in Figure 5.3, where θ is the angle of vertical settling of the average particle.

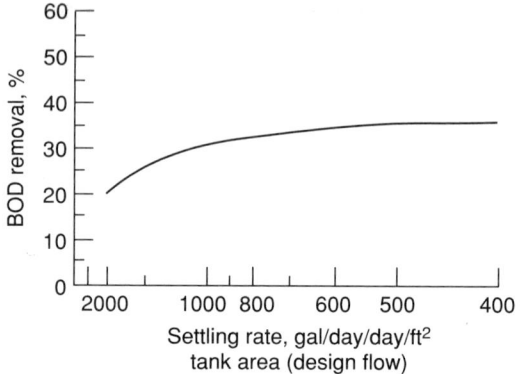

FIGURE 5.2. Effect of overflow rate on BOD removal.

Because shallow tanks appear to induce more flocculation, and because of other reasons (such as reaching the bottom sooner), they are preferred over deep tanks, provided that scouring of settled particles is prevented. The average settling velocities of particles in industrial wastes vary appreciably (Figure 5.4).

The percentage of suspended solids removed depends on the tank design, which in turn depends on the demands of the particular situation. Design engineers have begun to use either circular or square tanks instead of the conventional rectangular basins for reasons of space or economics. Circular tanks require less form work, materials, and land space than rectangular basins for large flows and for any size of tank. However, they are less efficient, because of: (1) reduced length of effective settling zone; and (2) short-circuiting (wastewater leaving the tank before theoretical detention time) (Figure 5.5). The efficiency of circular tanks has been increased somewhat by the introduction of peripheral feed with center draw-off. This system eliminates turbulence at the inlet.

The relative percentages of total transverse distance occupied by the inlet zones of circular and rectangular tanks are shown in Figure 5.6. Because the inlet zone of a circular tank occupies such a large portion of the horizontal particle path, special care

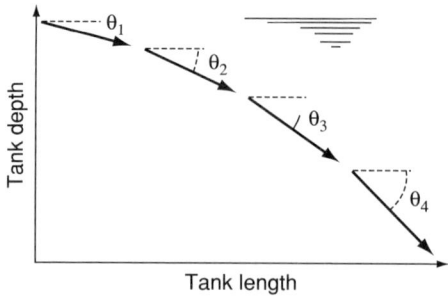

FIGURE 5.3. Flocculation increases settling rate.

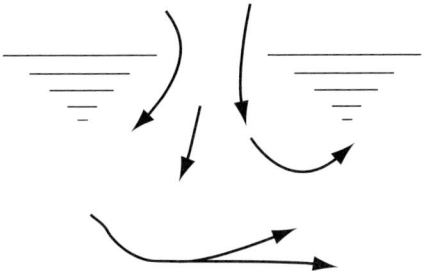

FIGURE 5.4. Effect of turbulence on particle path.

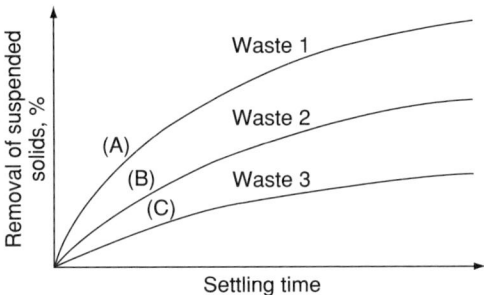

FIGURE 5.5. (A) Fast and good settling characteristics typical of heavy suspended solids. (B) Medium and normal settling characteristics typical of homogeneous mixture of solids. (C) Slow and poor settling characteristics typical of highly colloidal and finely divided solids.

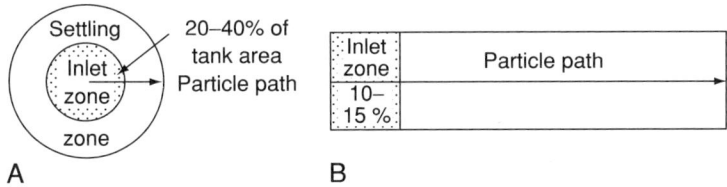

FIGURE 5.6. Inlet zone of a circular tank (A) occupies 20–40% of tank area. Inlet zone of a rectangular tank (B) occupies only 10–15% of tank area.

must be used in designing inlet and outlet devices. The slightest disturbance in flow conditions will tend to disrupt the operation of a circular tank, but with long, narrow, rectangular tanks, the design of the inlet and outlet zones becomes less important.

"Short-circuiting" means that effective sedimentation is not taking place in the entire volume of the settling tank; that is, a given entering volume of waste is hindered from spreading uniformly throughout the tank in a quiescent manner, so it reaches the effluent weir before the theoretical detention time has been used. This is essentially true in all tanks, regardless of shape, but it seems to occur most readily in circular and square tanks, as illustrated in Figure 5.7. To avoid short-circuiting, some state regulatory agencies specify a minimum distance between the inlet and exit of the tank. It has also been demonstrated graphically by Camp (1953) (Figure 5.8) that different shapes of sedimentation tanks cause different degrees of short-circuiting. Villemonte et al. (1966) showed that hydraulic efficiencies, predicted by basin dispersion curves, are related directly to the basin performance, measured by suspended-solids reduction.

In Figure 5.8, the higher peaks occurring over shorter ranges of t/T indicate the absence of short-circuiting. Curve A is a theoretical one for an ideal instantaneous dispersion of a slug with entire tank contents. (Short-circuiting approximates this.) Curve B is for a circular tank and indicates that some suspended contaminant reaches the outlet after about 15% of the detention period, and the greatest concentration of matter reaches the outlet after about 50% of the detention period. Curve C shows the situation in a wide rectangular tank, which approximates a square one. Curve D refers to a long, narrow rectangular tank and indicates that no contaminant reaches the end of the tank until after 50% of the detention period and most reaches the outlet after about 80% of the detention period. Curve E is the dispersion curve of a round-the-end, long, baffled rectangular chamber, with great length compared to width and depth. This type of tank gives a theoretical maximum contaminant content in the effluent after 100% of the detention period, but little or none before this time. The reader can readily appreciate from a study of Figure 5.8 the importance of proper design of sedimentation tanks. Preference should be given, wherever possible, to long rectangular tanks with proper baffling.

A major objective of sedimentation is to produce sludge with the highest possible solids concentration. As will be seen in Chapter 9, the volume and weight of sludge requiring final disposal is a major factor in waste treatment. A relatively new piece of equipment to achieve this objective is the Clarithickener, which combines sludge separation in circular settling tanks and thickening by means of slowly rotating picket-fence arms. Other methods of decreasing short-circuiting include effective inlet and outlet

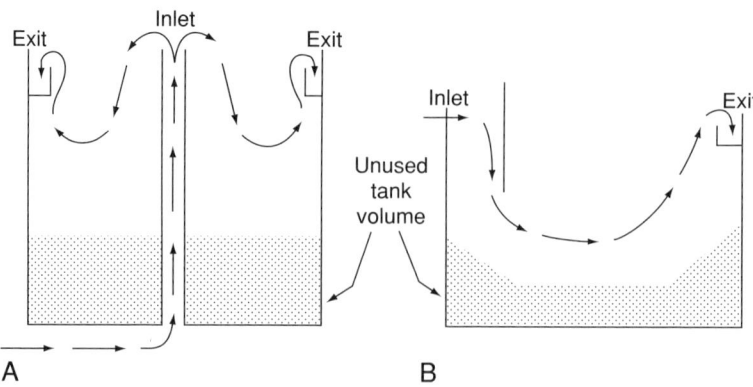

FIGURE 5.7. (**A**) Circulator tank. (**B**) Square tank.

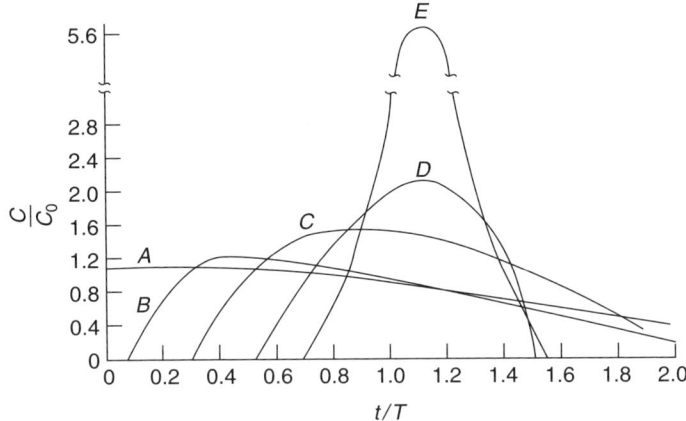

FIGURE 5.8. Typical dispersion curves for various tanks (see text for explanation). The vertical axis shows the ratio of the actual concentration of contaminant (C) to the concentration of contaminant mixed with the entire tank volume (C_o); the horizontal axis shows the ratio of the actual time (t) that a concentration takes to reach the end of the tank to (T), the total detention period (vol/rate) (adapted from Camp 1953.)

design, properly located baffling, inboard weirs, and modification of existing sedimentation tanks to obtain better flow distribution.

Although the differences between domestic sewage and industrial wastes are often quite significant, some general statements made for domestic sewage hold true for all wastes. Normally, with detention periods of 2 hours, primary sedimentation basins remove 50–70% of the suspended solids in the influent. Data collected from wastewater treatment plant superintendents (Bramer and Hoak 1966) are presented in Tables 5.1 and 5.2 to show design criteria and efficiencies of removal for rectangular and circular tanks.

In an unpublished study by Nemerow and McWeeny in 1972, the settling basin was separated into two zones by placing a series of vertical baffles perpendicular to the flow, with the top of the baffles below the surface of the sewage. The waste flows across the tops of the baffles following the path of least resistance, while the particles settle through the baffles into the sludge collection zone. The baffles dampen out the turbulence caused by sludge removal along the bottom of the basin. A removal efficiency of 85% of suspended solids at 691 gpd/ft² of suspended solids was obtained, a 20% increase over the then-current design expectation. A removal efficiency of 65% of suspended solids was obtained at a loading rate of 4,600 gallons/ft² with a baffle spacing of 0.25 feet.

Design of Sedimentation Process Units

Sedimentation processes are very effective in removing suspended solids in industrial wastewater. Clarifiers, either rectangular or circular, are most commonly used in

text continues on page 63

TABLE 5.1
Rectangular Primary Settling-Tank Data[a]

Plant Location	No. of Tanks	Length, ft	Width, ft	Depth, ft	Length Width	Length Depth	Flow, mgd	Detention, hr	Over-Flow, gpd/ft²	Weir Rate, gpd/ft	Raw Suspended Solids, mg/liter	Removal Suspended Solids, %	Raw BOD, mg/liter	Removal BOD, %
Hartford, Conn.	8	100	68	8.8	1.5	11.4	24.30	3.53	450	56,800	173	61	240	42
Detroit, Mich.	8	270	117	13	2.3	20.8	418.00	1.41	1,650	408,000	184	44	153	39
Racine, Wis.	4	140	40	10.5	3.5	13.3	17.03	2.48	760	106,500	149	67	133	48
New York City, Bowery Bay	3	124	50	12	2.5	10.3	41.00	0.98	2,210	284,000	152	39	169	22
New York City, Tallmans Island	3	124	50	11.6	2.5	10.7	31.00	1.25	1,670	215,000	137	55	128	39
Fort Wayne, Ind.	3	100	33	13	3.3	7.7	18.70	1.25	1,890	94,500	409	61	231	34
Rochester, N.Y.	2	37	12	8	3.1	4.6	0.81	1.56	914	41,000	233	21	260	21
Marshalltown, Iowa	3	80	16	8	5.0	10.0	1.22	1.51	950	13,550	436	58	414	42
Kenosha, Wis.	4	132	32	10.4	4.1	12.7	12.77	2.49	755	100,000	138	48	102	48
Jackson, Mich.	3	67.3	31	10	2.2	6.7	0.17	1.22	1,470	118,000	193	16.1	134	22
Hammond, Ind.	6	120	16	13.25	7.5	9.0	20.70	1.32	1,800	24,000	273	30	206	25
New York City, 26th Ward	4	162	67	12	2.4	13.5	41.00	2.16	930	35,500	139	31	127	28
New York City, Hunts Point	4	168	108.9	12	1.5	14.0	95.00	1.70	1,300	97,000	140	48	113	30
Abington, Pa.	2	50	14	10	3.6	5.0	1.24	2.02	855	44,400	237	39	198	29
Portsmouth, Va.	4	100	15.25	10	6.5	10.0	7.36	1.49	1,200	46,000	153	63	185	45
Canton, Ohio	3	124	32	10.6	3.9	11.7	17.00	1.33	1,430	214,000	577	40	253	33
Niles, Mich.	6	75	14	9	5.4	8.3	2.30	1.86	362	27,200	250	69.2	106	57
Dallas, Tex.	2	180	50	12	3.6	15.0	19.40	2.00	1,080	24,000	358	66	256	41

Location														
Richmond, Ind.	4	95	16	14.5	5.9	6.5	6.10	2.64	990	25,000	159	40	133	23
Lansing, Mich.	16	87.5	16	10	5.5	8.7	16.45	2.45	735	23,700	445	76	201	68
Winsted, Conn.	2	65	12	9	5.5	7.2	0.50	5.00	320	20,800	130	75	170	51
Waterbury, Conn.	3	212.5	33	10	6.4	21.2	13.94	2.71	660	14,500	144	54	166	33
Oklahoma City, Okla.	3	85	33	10	2.5	8.5	5.19	2.91	619	20,400	242	50	228	31
Tampa, Fla.	4	170	40	13	4.2	13.1	12.30	5.12	455	17,300	215	69	183	37
Roanoke, Va.	2	120	32	10.5	3.8	11.4	7.76	1.87	1,010	120,000	230	67	190	51
Blackstone Valley, R.I.	2	230	68	10.8	3.4	21.1	12.21	4.97	390	62,000	212	62	333	12
East Hartford, Conn.	2	125	32	7.5	3.9	16.7	1.50	7.18	187	12,500	212	54	242	50
Milford, Conn.	2	55	16	9.75	3.5	5.1	9.70	4.40	400	21,800	150	79	130	72
Springfield, Mass.	4	115	50	14.5	2.3	7.9	17.5	3.36	761		160	49	145	26
Orrville, Ohio	2	43.8	16	10.4	2.7	4.2	0.73	3.65	515		342	64	415	18
New Haven, Conn.	3	145	31	11.5	4.7	12.6	14.7	1.90	1,090		176	49		
Cleveland, Ohio (Easterly)	8	115	50	15	2.3	7.7	97.7	1.27	2,120		240	37	149	35

[a] Data from plant superintendents. See Federation of Sewage and Industrial Wastes association (1959).

TABLE 5.2
Circular Primary Tanks: Long-Term Performance Data[a]

Location	Data Period		Average Flow, mgd	No. of Tanks	Diameter ft	Sidewater Depth, ft	Detention, hr	Overflow, gpd/ft²	Suspended Solids			BOD			Sludge	
	Years	No.							Raw, mg/liter	Effluent, mg/liter	Removal, %	Raw, mg/liter	Effluent, mg/liter	Removal, %	Solids, %	Volatile matter, %
Washington, D.C.	1944–45	2	136.3	12	106	14	1.88	1,350	163	83	49	173	120	30.5	8.05	67.5
Winnipeg, Man.	1943–44	2	22.8	2	115	12	1.98	1,100	348	159	55	310	231	25.5	9.0	70.5
Battle Creek, Mich.	1938–42	5	4.92	2	80	10	3.66	490	282	85	70	264	174	34.1	5.5	82.5
Buffalo, N.Y.	1939–41	3	135	4	160	15	1.6	1,690	209	114	46	138	107	22.5	5.8	59
Albuquerque, N. Mex.	1939–46	7	5	1	80	12.2	2.21	995	254	91	61	282	150	44.5	3.9	81
Yakima, Wash.	1942	1	9.5	4	90	9	4.32	373	110	23	74	175	92	50	7.0	74.4
Appleton, Wis.	1938–45	7	4.8	2	70	10	2.90	623	276	63	77	284	141	50	5.6	58
Baltimore, Md.	1939–44	4	89.5	3	170	12	1.64	1,360	214	83	61	281	204	27.5	3.9	82.7
Springfield, Ohio	1937–40	4	14.8	2	90	10	1.55	1,160	166	63	62	90	43	52		
Mansfield, Ohio	1944–45	2	3	1	65	12	2.38	905	208	87	58	227	139	38.8	4.2	76
Cedar Rapids, Iowa	1936–44	9	4.21	1	70	11.5	1.95	1,060	354	132	63	383	291	24	5.5	81.2
Austin, Tex.	1944–45	2	5.64	1	75	12	1.69	1,275	263	95	64	285	152	46.3	4.0	83
Denver, Colo.	1939–43	5	46	4	140	9.7	2.34	750	187	44	77	212	108	49	5.4	76
Ypsilanti, Mich.	1943–45	3	1.66	2	40	9	2.5	660	226	87	62	141	95	33	8.2	71.4
Monroe, Mich.	1938–46	8	4.3	2	85	7.5	3.55	378	329	75	77	135	73	46	5.2	67.7

[a] Data from plant superintendents and/or annual reports. See Federation of Sewage and Industrial Wastes Association (1959).

the application of sedimentation in wastewater treatment facilities. The design of the clarifiers is based on several factors, as follows:

- Influent total suspended solids (TSS) concentration
- Effluent TSS concentration
- Surface loading
- Detention time
- Sludge generation

Clarification is used as a process to remove suspended solids at different stages of industrial wastewater treatment. It is often used in the primary treatment stage to remove TSS or colloidal solids before treatment for removal of dissolved inorganic or organic materials. A typical example of this application is in the treatment of metal-finishing industry wastewater, in which the suspended solids are removed by primary clarification before other physicochemical processes are used to remove the dissolved heavy metals. Sedimentation is also commonly used in the secondary treatment stage, usually after biological treatment. An example of this application is the treatment of pulp and paper-mill wastewater in which primary clarification is followed by biological treatment, and then the biological solids are removed by secondary clarification. In this case, the clarifier is used not only to remove the TSS, but also to act as a thickener for the sludge generated. The design of the clarifier is based on different considerations depending on the stage of the treatment in which clarification is used.

Another consideration in the design of sedimentation processes is the characteristics of the suspended solids. In some cases, the suspended solids could be discrete particles, as in the case of grit, sand, or suspended metal scales or particles. These types of solids settle easily following the principles of discrete settling (Figure 5.9). In other

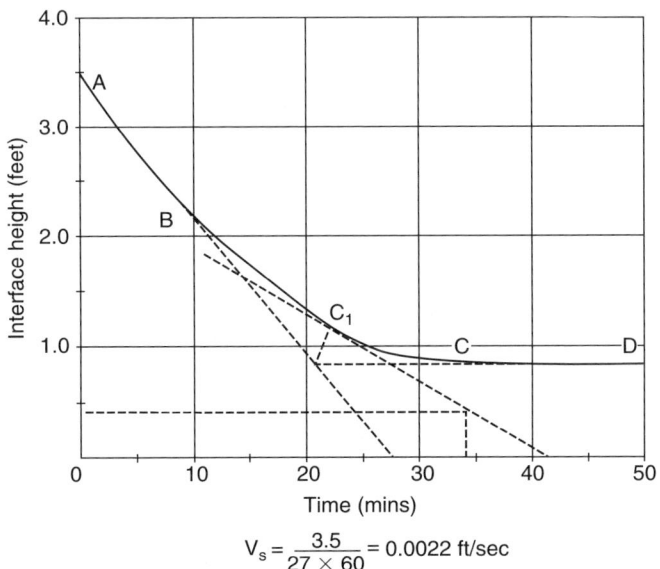

$$V_s = \frac{3.5}{27 \times 60} = 0.0022 \text{ ft/sec}$$

FIGURE 5.9. Graphical analysis of settling test results.

cases, the TSS could be composed of floc-type particles, and the settling characteristics of these suspended solids are different from those of discrete-type solids (Figure 5.9). Examples of the floc-type particles are biological flocs or chemically coagulated and flocculated particles. Because of the different settling characteristics of the suspended solids, it is important that batch-settling tests be conducted using settling columns before designing the sedimentation unit processes. In the settling test (Figure 5.9), the height of the interface between the clear supernatant liquid and the layer of suspended solids is noted with time as settling occurs in the column. (The data are plotted as shown in Figure 5.9.) The settling of the solids takes place in essentially two phases. The initial rate of settling (AB), known as *hindered settling,* is used to compute the area required for clarification. The second phase of settling (CD) represents the thickening of the sludge. A graphical method of combination of the rates of settling in the AB and CD portions of the curve is used to compute the area required for thickening. The larger of the two areas for clarification and thickening is used for sizing the clarifier:

$$V_s = Q/A_c$$

or

$$A_c = Q/V_s,$$

where

V_s = settling velocity in the hindered settling zone,
Q = hydraulic flow,
A_c = surface area required for clarification, and
V_s is computed as the slope of the line AB.
The area required for thickening is computed from the following equation:

$$A_c = Qt_u/H_o,$$

where

A_t = surface area required for thickening to desired solids concentration in sludge,
Q = hydraulic flow into the clarifier,
H_o = initial interface height of the settling column, and
t_u = time required to reach desired solids concentration in sludge.

t_u is computed by the following steps: (1) draw tangents to each of the portions AB and CD of the curve; (2) draw the bisector of the angle formed by the intersection of the two tangents (C_1 represents the critical concentration in the transition between the hindered and compression settling phases); (3) draw a tangent at C_1; and (4) draw a line parallel to the time axis at the interface height (H_u) corresponding to the desired solids concentration in the sludge (C_u). The time-scale intercept of tangent with the sludge concentration line is the required time t_u. H_u is computed as follows:

$$H_u = C_o H_o / C_u,$$

where
C_o = solids concentration in the influent,
C_u = desired solids concentration in the sludge, and
H_u = interface height at desired solids concentration in sludge.

Example

The results of a batch-settling column test for an industrial wastewater is given below:

Time (min)	Interface Height (ft)
0	3.5
5	2.8
10	2.2
15	1.5
20	1.2
25	0.9
30	0.7
35	0.6
40	0.5
45	0.5

Using the test data, determine the size of the clarifier. Flow = 0.8 mgd, influent solids concentration = 2,000 mg/liter, and sludge solids concentration = 1.5%.

Solution

The data from the settling test are plotted as shown in Figure 5.9. From the figure, V_s is computed as follows:

$$V_s = 3.5/27 \times 60 = 0.0022 \text{ ft/sec}$$

Area required for clarification:

$$A_c = Q/V_s$$
$$= (0.8 \times 1.547 \text{ ft}^3/\text{sec}) / (0.0022 \text{ ft/sec})$$
$$= 563 \text{ ft}^2$$

H_u is computed as follows:

$$H_u = C_o H_o / C_u$$
$$C_o = 2{,}000 \text{ mg/liter}$$
$$= 2{,}000/16{,}020 = 0.125 \text{ lb/ft}^3$$
$$C_u = (1.5 \times 10{,}000) / (16{,}020) = 0.94 \text{ lb/ft}^3$$
$$H_u = C_o H_o / C_u$$
$$= (0.125 \times 3.5) / (0.94) = 0.47 \text{ ft}$$

From the figure, $t_u = 33$ min for $H_u = 0.47$ ft.

Area required for thickening:

$$A_t = Ot_u / H_o$$
$$= (0.8 \times 1.547 \times 33 \times 60) / (3.5)$$
$$= 700 \text{ ft}^2.$$

Therefore, the size of the clarifier selected is 700 ft².

$$l{:}w = 4{:}1$$

Flotation

"Flotation" is the process of converting suspended substances and some colloidal, emulsified, and dissolved substances to floating matter (Hess et al. 1953). The term *flotation* includes both violently agitated froth flotation, as used in the separation of ores in the mining industry, and quiescent flotation, which is now becoming popular as an efficient method for the removal of most suspensions from wastewaters.

Small and difficult-to-settle particles in suspension can be flocculated and buoyed to the liquid surface by the lifting power of the many minute air bubbles that attach themselves to the suspended particles. Floated agglomerated sludge can be readily and continuously removed from the surface of the liquid by skimming. These skimmings are usually collected as a concentrated sludge and normally drain quite readily. A convenient practice is to detain the sludge float in a receiving tank for a few hours before draining the subnatant liquor from the bottom. The solids content of the float can be more than doubled by this concentration method; water is actually squeezed out of the float while the particles compact. Such a sludge float is usually quite stable and free from odors. Because the flotation process brings partially reduced chemical compounds into contact with oxygen in the form of tiny air bubbles, satisfaction of any immediate oxygen demand of the wastewater is thereby aided.

Typical vacuum flotation units first aerate the waste with air diffusers or mechanical beaters. Aeration periods are brief, some as short as 30 seconds, and require only about 0.025–0.05 ft^3 of air per gallon of wastewater. A brief de-aeration period is then provided at atmospheric pressure to remove large bubbles. The waste, at this point nearly saturated with dissolved air, passes to an evacuation tank that is enclosed and maintained under a vacuum of about 9 inches of mercury. This vacuum gives rise to bubbles, which cause flotation.

Pressure flotation differs from vacuum flotation in that air is injected into the waste under pressure, and bubbles of air are then formed when the waste is exposed to atmospheric pressure. Wastes are normally pressurized to about 30–40 lb/in.^{2} and retained at this pressure for approximately 1 minute. Some coagulant aids (alum and/or silica) and a small volume of air can be bled into the system at the suction end of the pump, where wastewater enters the tank. Passage through the pump usually suffices to provide good mixing of the chemicals and air with the waste. When released to the atmosphere in the flotation tank, the tiny rising bubbles trap suspended, colloidal, and (some) emulsified particles. The floated sludge is usually continuously skimmed and removed from the tank by sludge pumps.

Vrablik (1960) makes a distinction between two methods of flotation: dissolved air and dispersed air. Dispersed-air flotation generates gas bubbles by the mechanical shear of propellers, diffusion of gas through porous media, or by homogenizing a gas and liquid stream. Dissolved-air flotation generates gas bubbles by precipitation from a solution supersaturated with the gas. These bubbles are much smaller than dispersed-air bubbles, generally not exceeding 80 microns[1] in diameter, while dispersed-air bubbles often reach 1,000 microns in diameter.

To understand the theory of dissolved-air flotation, the student must investigate the gas, liquid, and solid phases, as they are brought into intimate contact with each other. Henry's Law indicates the relationship between the solubility of gas (in this case, dissolved air) and the total pressure:

$$C = kp,$$

where C is the concentration of gas in solution, k is the Henry's Law constant, and p is the absolute pressure above the solution at equilibrium.

By attachment to, or inclusion in, a suspended-solids structure or liquid phase, the bulk density of the paired system may be less than the density of the parent system, causing the agglomeration to be floated to the top. The gas bubbles, therefore, render a buoyancy to the original suspended particle in accordance with Archimedes' principle: The resultant pressure of a fluid on an immersed body acts vertically upward through the center of gravity of the displaced fluid and is equal to the weight of the fluid displaced. The resultant upward force exerted by the fluid on the body is called *buoyancy,* and this force is responsible for the floating of solids, which were originally somewhat heavier than the surrounding fluid.

---

[1] 1 micron = 0.0001 cm = 0.0000394 in.; 1 in. = 2.54 cm.

Because we are usually dealing with large volumes of water in waste treatment, detention time in flotation chambers becomes a critical factor. Detention time, in turn, is dependent primarily on the rate of rise of air bubbles in the water. This can best be expressed by Stokes' Law, which holds true for particles with a diameter of less than 130 microns:

$$V = kD^2,$$

where V is the rate of bubble rise (ft/min), k is Stokes' conversion factor (this includes all the factors that affect the rise or fall of bubbles, such as density or viscosity of the liquid, excluding the density of the bubble), and D is the diameter of the air bubble. The Stokes relationship is shown quantitatively in Figure 5.10.

Typical results obtained from samples of several industrial wastes (Hess et al. 1953) treated by dissolved-air flotation show suspended solids and BOD reductions of 69.0–97.5 and 60.0–91.8%, respectively (Table 5.3).

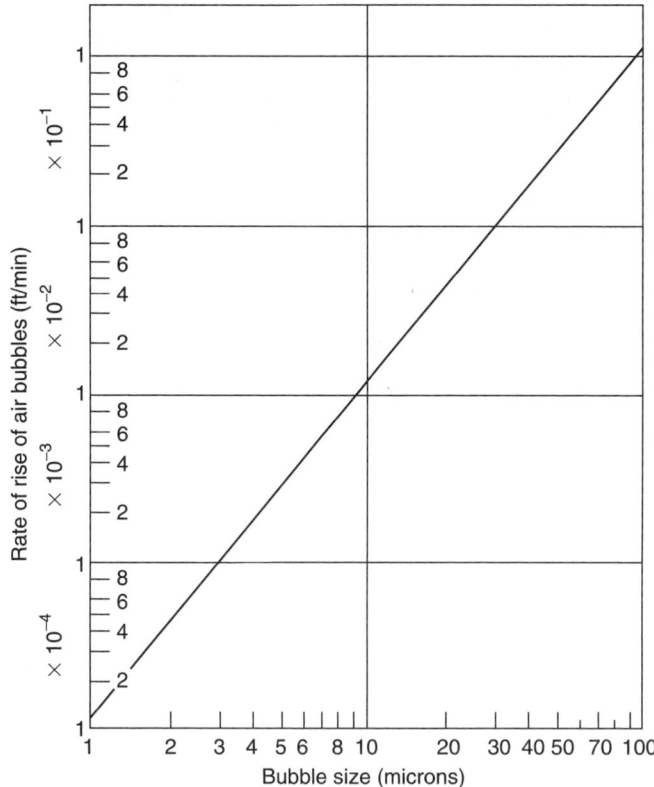

FIGURE 5.10. Rate of rise of air bubbles in tap water (calculated by means of Stoke's law) as a function of bubble size (after Vrablik 1960).

TABLE 5.3
Typical Efficiencies of Dissolved-Gas Flotation Treatment of Wastes

Waste Source	Suspended Solids in Influent, ppm	Reduction Obtained, %	BOD in Influent, ppm	Reduction Obtained, %
Petroleum production	441	95.0		
Railroad maintenance	500	95.0		
Meatpacking	1,400	85.6	1,225	67.3
Paper manufacturing	1,180	97.5	210	62.6
Vegetable-oil processing	890	94.8	3,048	91.6
Fruit-and-vegetable canning	1,350	80.0	790	60.0
Soap manufacture	392	91.5	309	91.6
Cesspool pumpings	6,448	96.2	3,399	87.0
Primary sewage treatment	252	69.0	325	49.2
Glue manufacture	542	94.3	1,822	91.8

Adapted from Quigley and Hoffman (1966).

Because almost twice as much air can be dissolved in water, all other factors being equal, at 0°C than at 30°C, the temperature of wastewater is a significant factor in the effectiveness of the flotation process. This relationship is shown in Figure 5.11.

Generally, air bubbles are negatively charged, the anions collecting mainly on the gas side of the interface, while the cations spread themselves out thinly on the water side of the interface. Because suspended particles or colloids may have a significant electrical charge, either attraction or repulsion will occur between these and air bubbles.

Vrablik (1960) made an extensive study of the three processes by which flotation may be caused: (1) adhesion of a gas bubble to a suspended liquor or solid phase; (2) the trapping of gas bubbles in a floc structure as the gas bubble rises; (3) the absorption of a gas bubble in a floc structure as the floc structure is formed. These three phenomena are illustrated in Figure 5.12. An illustration of pressure flotation is shown in Figure 5.13.

Finally, the engineer should be aware of both the advantages and the disadvantages of flotation as a waste-treatment process (Federation of Sewage and Industrial Wastes Association 1959). The advantages are as follows:

1. Grease and light solids rising to the top and grit and heavy solids settling to the bottom are all removed in one unit.
2. High overflow rates and short detention periods mean smaller tank sizes, resulting in decreased space requirements and possible savings in construction costs.
3. Odor nuisances are minimized because of the short detention periods, as well as in pressure and aeration-type units, because of the presence of dissolved oxygen in the effluent.
4. Thicker scum and sludge are obtained, in many cases, from a flotation unit than from gravity settling and skimming.

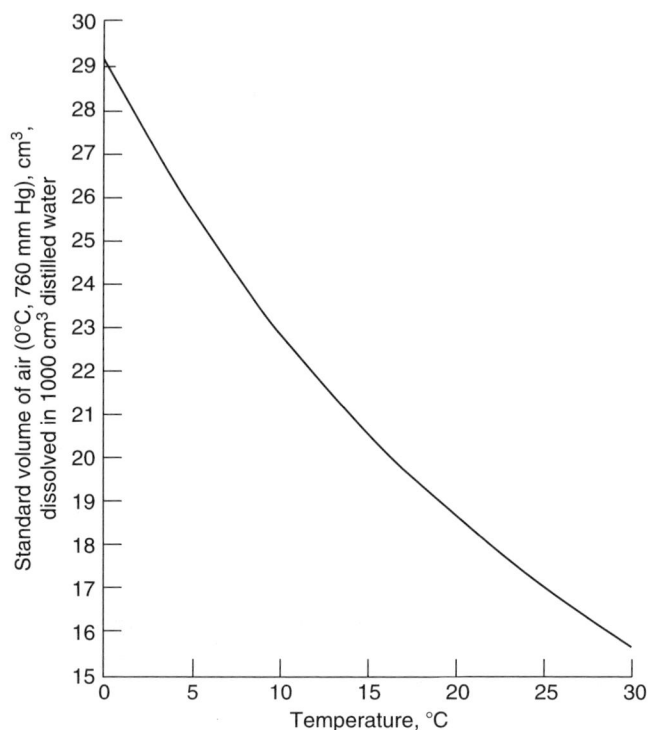

FIGURE 5.11. Solubility of air in distilled water at various temperatures (from *Handbook of Chemistry and Physics,* 1955).

The disadvantages are as follows:

1. The additional equipment required results in higher operating costs.
2. Flotation units generally do not provide treatment as effective as gravity-settling units, although efficiency varies with the waste.
3. The pressure type has high power requirements, which increases operating cost.
4. The vacuum type requires a relatively expensive airtight structure capable of withstanding a pressure of 9 inches of mercury; any leakage to the atmosphere will adversely affect performance.
5. More skilled maintenance is required for a flotation unit than for a gravity-settling unit.

Quigley and Hoffman (1966) deserve credit for daring to refer to flocculation and dissolved-air flotation as "secondary treatment" when it follows sedimentation. They describe an effective dissolved-air flotation system for treating oil-refinery wastes. By recycling treated effluent and using lime as a coagulant, they were able to obtain oil removals of 68–96%.

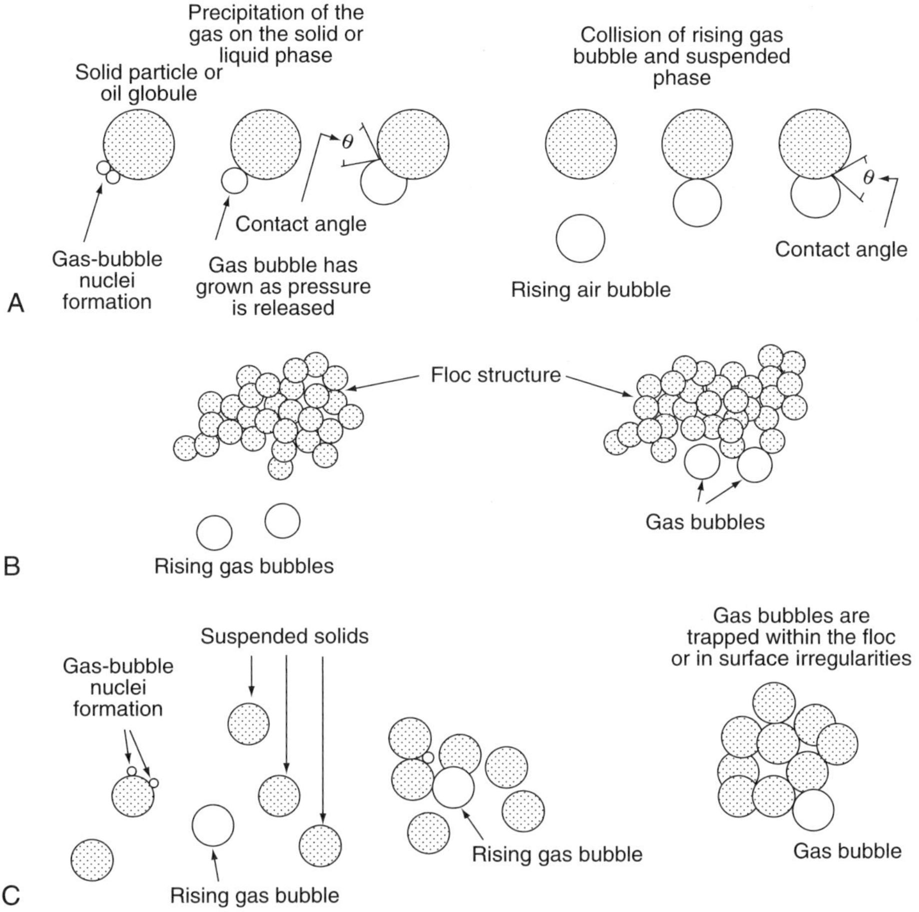

FIGURE 5.12. Methods of dissolved-air flotation. (**A**) Adhesion of a gas bubble to a suspended liquid or solid phase. (**B**) The trapping of gas bubbles in a floc structure as the gas bubbles rise. (**C**) The absorption and adsorption of gas bubbles in a floc structure as the floc structure is formed (adapted from Vrablik 1960).

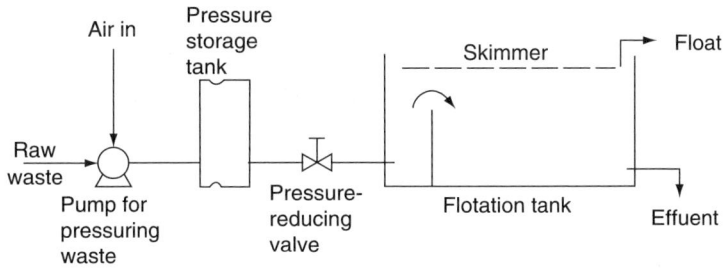

FIGURE 5.13. Schematic drawing of pressure flotation system.

Screening

Screening of industrial wastes is generally practiced on wastes containing larger suspended solids of variable sizes, such as from canneries, pulp and paper mills, or poultry processing plants. It is an economical and effective means of rapid separation of these larger suspended solids from the remaining waste material. In many cases, screening alone will reduce the suspended solids to a low enough concentration to be acceptable for discharge into a municipal sewer or a nearby stream. Often, considerable BOD is also removed by the screening process, the percentage removed varying almost directly with the size of the screen and the amount of BOD associated with the "screenable" solids. Screens are available in sizes ranging from coarse (10 or 20 mesh) to fine (120–320 mesh).

The North and Sweco screens are two major types used in industry today. The former are generally rotary, self-cleaning, gravity-type units (Figure 5.14). The latter are mostly circular, overhead-fed, vibratory units (Figure 5.15).

The rotary, gravity-type, waste-disposal screens are manufactured in several sizes to handle almost any volume of waste liquid. In general, they vary from 3 to 5 feet in diameter and from 4 to 12 feet in length and weigh between 1 and 5 tons. These screens separate solid and liquid constituents from waste materials at a location where they gravitationally flow or can be pumped into the screened cylinder. The machine's large drum rotates at 4 rpm. The lift paddles within the drum pick up the solid material from the water and deposit it into a stationary perforated hopper within the cylinder. The hopper holds a spiral screw conveyor that moves the solids to the

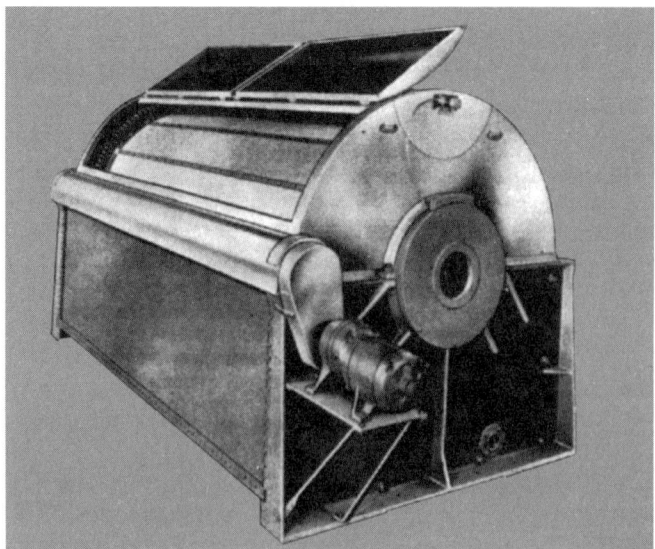

FIGURE 5.14. North water filter (photograph courtesy Green Bay Foundry and Machine Works, Green Bay, Wisconsin). Used with permission from John Wiley and Sons.

Removal of Suspended Solids 73

FIGURE 5.15. The 48-inch-diameter Sweco separator shown is screening lint from wastewater at the Eastern Overall Company, in Baltimore, Maryland. The wastewater is fed onto the 60-mesh market-grade screen at a rate of 300 gpm. The screened wastewater is discharged to the sewer (photograph courtesy Sweco, Inc., Los Angeles, California). Used with permission from John Wiley and Sons.

rear end of the machine and out through the discharge spout. In the process, it compresses the wastes and squeezes out more liquid, which drains through the perforated hopper back into the cylinder. The water in the cylinder drains through the wire mesh and collects in a steel or wooden tank, which is part of the machine. The fine wire mesh is at all times kept free from clogging by a continuous spray pipe with jet nozzles, located above the rotating cylinder. Such scenes have been used successfully in treating wastes from meatpacking, canning, grain-washing, tanning, malting, woolen, and seafood plants.

The circular vibratory screens have been quite effective in screening wastes from food-packing processes such as meat and poultry packing or fruit and vegetable canning. Vibration is designed usually to remove solids at the periphery of the screen, although Swallow (1965) reported the use of a new center-discharge separator.

Micro-straining, a particular screening device, was first introduced by Dr. P. L. Boucher (1965) in England in 1945 for water clarification, and there were about 70 water-treatment plants in the United States using this process. It involves the use of high-speed, continuously backwashed, rotating drum filters working in open gravity-flow conditions (see cutaway picture in Figure 5.16). The principal filtering fabrics employed have apertures of 35 or 25 microns and are fitted on the drum periphery. Head loss is between 4 and 6 inches. Results in London, England, showed that micro-straining removes most of the suspended solids remaining after biological treatment (Table 5.4).

FIGURE 5.16. Cutaway view of a 7½-foot-diameter microstrainer (photograph courtesy Crane Company, King of Prussia, Pennsylvania). Used with permission from John Wiley and Sons.

The Bauer Company manufactures a perforated plate screen (referred to as a Hydrasieve), which is installed at a slight angle to the vertical. Wastewater is passed down the screen from the top, with water going through the screen and solids collecting at the bottom. The efficiency of removal depends primarily on the size of the screen opening and the wastewater application rate.

Example of Twentieth-Century Practice of Suspended Solids Reduction

Johnson and Lindley (1982) showed that a hydroclone—a cone type of settling clarifier—could be used to remove efficiently and economically suspended fish particles. The overflow is discharged as effluent, and underflow is collected and stored for reuse. Effluents—the overflow—were able to meet Environmental Protection Agency (EPA) standards for seafood processing. Although the overflow is clean enough for discharge, the underflow may need further processing for by-product recovery operation. Thus, they used a rather common type of treatment to attain the objective of suspended-solids reduction.

Suspended solids may also be an asset for the industrial waste engineer. Your author used the high suspended solids content of the Moench Tannery wastewater to form a nucleus for the aerating mixed liquor during biological treatment (Nemerow et al. 1978). In this way, separate suspended solids removal was unnecessary and biological treatment to remove colloidal and dissolved contaminants was enhanced simultaneously.

TABLE 5.4
Results Obtained on Humus Tank Effluent at Eastern Sewage Works, London, England, December 30, 1966–January 13, 1967

Characteristic	Effluent from			
	Humus Tank	Micro-Strainer	Ozonizer	Sand Filter
Suspended solids	51[a]	19	15	10
Total solids	931			928
BOD	21	13	11	9
COD	78	54	44	39
Permanganate value	16	10	6	5
Organic carbon		23	19	10
Surface-active matter				
Anionic (as Manoxol OT)	1.4	1.4	0.6	0.6
Nonionic (as Lissapol NX)	0.37		0.07	0.07
Ammonia (as N)	7.1	7.5	7.4	7.6
Nitrite (as N)	0.4	0.4	0.02	0.01
Oxidized nitrogen (as N)	26	26	26	27
Total phosphorus (as P)	8.2			7.4
Orthophosphate (as P)	6.6			7.0
Total hardness (as $CaCO_3$)				468
Chloride				98
Sulfate	212			213
Color (Hazen units)	36		4	7
Turbidity (ATU)	66		27	13
Total phenol	3.4			0.9
Temperature (°C)	8.1	8.0	7.9	7.7
Dissolved oxygen (% saturation)	52	52	99	94
Conductivity ($\mu mho/cm^3$)	1,173	1,175	1,170	1,150
Langlier index	−0.08			+0.12
pH	7.4	7.4	7.4	7.5
Pesticides ($\mu g/l$)				
α BHC		0.025		0.007
γ BHC		0.035		0.030
Aldrin		0.004		0.000
Dieldrin		0.193		0.032
pp DDT		0.031		0.030

[a]All results are given in milligrams per liter unless otherwise specified.
Adapted from Diaper (1968).

Review Questions

1. What are three major methods of removing suspended solids?
2. When would you use sedimentation for removal of suspended solids?
3. When should you use flotation for removal of suspended solids?

4. Would you ever use both sedimentation and flotation together?
5. Why and when would you use screening for suspended solids removal?
6. What are the most important factors affecting industrial wastewater sedimentation?
7. In dissolved-air flotation, what is the apparent anomaly that exists because of the size of the air bubble?
8. What are the three methods by which suspended matter can be removed by dissolved-air flotation with chemical coagulant addition?
9. What are the two major types of screening devices mentioned in this chapter? What is a third type of screening device type not pictured in the chapter? What are the major advantages of each type?

References

Bewtra, J. K. 1967. Diagram for the settling of discrete particles in viscous fluids. *Water Sewage Works* 114:60.

Boucher, P. L. 1965. Micro-straining, microzon, and demicellization applied to public and industrial water supply. In: *Proceedings of Water Treatment Symposium, May 1965, Adelaide, S. Australia.*

Bramer, H. C., and R. D. Hoak. 1966. Measuring sedimentation-flocculation efficiencies. *Ind. Eng. Chem. Process Design Develop.* 5:316.

Camp, T. R. 1953. Studies of sedimentation basin design. *Sewage Ind. Wastes* 25:1.

Clark, J. W., W. Viessman, Jr. 1965. *Water Supply and Pollution Control*, pp. 274–294. Scranton, PA: International Textbook Company.

Dobblins, W. E. 1961. Advances in sewage treatment design. Paper presented to the Sanitary Engineering Division of the A.S.C.E. (Metropolitan Section) Conference at Manhattan College, New York City (May 1961).

Eckenfelder, W. W. 1966. *Industrial Water Pollution Control*. New York: McGraw-Hill Book Co.

Federation of Sewage and Industrial Wastes Association. 1959. *Sewage Treatment Design, Manual of Practice*, no. 8, *American Society of Engineers Manual of Engineering Practice*, no. 36, p. 78. Washington, D.C.

Fitch, B. 1966. Current theory and thickener design. *Ind. Eng. Chem.* 10:18.

Great Lakes–Upper Mississippi River Board of State Sanitary Engineers. 1960. *Recommended Standards for Sewage Works*. Harrisburg, PA: May 10.

Hess, R. W., et al. 1953. 1952 Industrial wastes forum. *Sewage Ind. Wastes* 25:709.

Johnson, R. A., K. L. Lindley. 1982. Use of hydroclones to treat seafood-processing waste waters. *J. Water Pollution Control Fed.* 54(12):1607.

Katz, W. J. 1959. Adsorption—secret of success in separating solids by air flotation. *Ind. Wastes* 30:11.

Nemerow, N. L., D. Warne, L. Falk. 1978. A new and effective solution for treatment of tannery wastewater. In: *Proceedings of the 33rd Annual Purdue University, Industrial Waste Conference May 9–11, 1978.*

Quigley, R. E., E. L. Hoffman. 1966. Flotation of oily wastes. In: *Proceedings of 21st Industrial Wastes Conference, Purdue University, May 1966*, p. 527. Lafayette, Indiana.

Swallow, D. M. 1965. Design and operation of the center-discharge separator. In: *Proceedings of the Seminar on Water Pollution Control, during 30th Exposition of Chemical Industries, New York, Nov. 30, 1965*, p. 20.

Villemonte, J. R. 1962. Hydraulic characteristics of circular sedimentation basins. In: *Proceedings of 17th Industrial Waste Conference, at Purdue University, Lafayette, Indiana*, p. 682.

Villemonte, J. R., et al. 1966. Hydraulic and removal efficiencies in sedimentation basins. *J. Water Pollution Control Fed.* 38:371.

Vrablik, E. R. 1960. Fundamental principles of dissolved-air flotation of industrial wastes. In: *Proceedings of 14th Industrial Waste Conference, Purdue University Engineering Extension Series*, p. 743. Bulletin no. 104, May 1960.

Suggested Reading

Sedimentation

Camp, T. R. 1946. Sedimentation and the design of settling tanks. *Trans. Am. Soc. Civil Engrs.* 111:895.

Dobbins, W. E. 1944. Effect of turbulence on sedimentation. *Trans. Am. Soc. Civil Engrs.* 109:629.

Federation of Sewage and Industrial Wastes Association. 1959. *Sewage Treatment Design, Manual of Practice*, no. 8, *American Society of Civil Engineers Manual of Engineering Practice*, no. 36, pp. 90–91. Washington, D.C.

Hazen, A. 1904. On sedimentation. *Trans. Am. Soc. Civil Engrs.* 53:45.

Rich, L. G. 1961. *Unit Operations in Sanitary Engineering*, Chapter 4, pp. 81–109. New York: John Wiley & Sons.

Flotation

Beebe, A. H. 1953. Soluble oil wastes treatment by pressure flotation. *Sewage Ind. Wastes* 25:1314.

D'Arcy, N. A., Jr. 1951. Dissolved air flotation separates oil from waste water. *Oil Gas J.* 50:319.

Rich, L. G. 1961. *Unit Operations in Sanitary Engineering*, Chapter 5, pp. 110–135. New York: John Wiley & Sons.

Screening and Micro-Straining

Boucher, P. L. 1961. *J. Inst. Public Health Engrs.* 60:294; and Boucher, P. L. 1965. Micro-straining, microzon, and demicellization applied to public and industrial water supply. In: *Proceedings of Water Treatment Symposium, May 1965, Adelaide, S. Australia*.

Boucher, P. L. 1967. Micro-straining and ozonisation of water and waste water. In: *Proceedings of 22nd Industrial Waste Conference, Purdue University Engineering Extension Series*, Bulletin no. 129, May 1967. Lafayette, Indiana.

Campbell, R. M., M. B. Prescod. 1965. *J. Inst. Water Engrs.* 19:101.

Diaper, E. W. J. 1968. Micro-straining and ozonisation of water and waste water. *Water Wastes Eng.* 5:56.

Hazen, R. 1953. Application of the microstrainer to water treatment in Great Britain. *J. Am. Water Works Assoc.* 45:723.

CHAPTER 6

Removal of Colloidal Solids

The reason that colloidal constituents in industrial waste are so important is obvious. In those constituents, wastes containing only a quarter of total solids in the form of colloids also account for as much as 50% of the total biochemical oxygen demand (BOD). In this chapter, your author examines the characteristics of colloids, their reactions—especially their response to coagulants—and their removal from plant wastes.

Characteristics of Colloids

A colloid may be defined as a particle held in suspension by its extremely small size (1–200 millimicrons), state of hydration, and surface electrical charge. There are two types of colloids: lyophobic and lyophilic. Because of the difference in their characteristics, they react differently to alterations in their environment. Table 6.1 will assist the student in understanding their properties. Colloids are often responsible for a relatively high percentage of the color, turbidity, and BOD of certain industrial wastes. Because it is important to remove colloids from wastewaters before they can get into streams, one must understand their physical and chemical characteristics.

Colloids exhibit "Brownian movement," a bombardment of the particles of the disperse phase by molecules of the dispersion medium. They are essentially "nonsettleable" because of their charge, small size, and low particle weight. They are dialyzable; that is, they can be separated from their crystalloid (low-molecular-weight) counterparts by straining through a semipermeable membrane. The colloids diffuse very slowly compared to soluble ions. Colloidal particles, in general, exhibit very low (if any) osmotic pressure because of their large size relative to the size of soluble ions. They also possess the characteristic of "imbibition" (the taking in of water by gels). In fact, it is by this very process that bacteria spores (often considered colloidal) take up water and germinate. Colloidal gels are very often used as ultrafilters, having pores sufficiently small to retain the dispersed phase of a colloidal system but large enough to allow the dispersion medium and its crystalloid solutes to pass through. For example, Perona et al. (1967) found that the formed membranes may be used to remove up to 90% of the colored material and somewhat less of the chemical oxygen demand (COD) and total dissolved solids of pulp-mill

TABLE 6.1
Types and Characteristics of Colloidal Solids

Characteristic	Lyophobic (hydrophobic)	Lyophylic (hydrophylic)
Physical state	Suspensoid	Emulsoid
Surface tension	The colloid is very similar to the medium	The colloid is of considerably less surface tension than the medium
Viscosity	The colloid suspension is very similar to the dispersing phase alone	Viscosity of colloid suspension alone is greatly increased
Tyndall effect	Very pronounced (ferric hydroxide is an exception)	Small or entirely absent
Ease of reconstitution	Not easily reconstituted after freezing or drying	Easily reconstituted
Reaction to electrolytes	Coagulated easily by electrolytes	Much less sensitive to the action of electrolytes, so more is required for coagulation
Examples	Metal oxides, sulfides, silver halides, metals, silicon dioxide	Proteins, starches, gums, mucilages, and soaps

sulfite wastes. Colloidal systems show a wide range in viscosity or plasticity. Usually, the lyophobic colloidal suspensions exhibit a viscosity only slightly higher than that of the pure dispersing medium (Figure 6.1), and this concentration increases only very slightly when the concentration of the dispersed material is increased. On the other hand, lyophilic systems may reach very high values of viscosity. With these types of colloids, a parabolic, rather than a linear, relationship exists between viscosity and the concentration of dispersed phase, as shown in Figure 6.1. Woodard and Etzel (1965) have shown that under certain conditions, one may change a lyophilic colloid in an industrial waste to a lyophobic

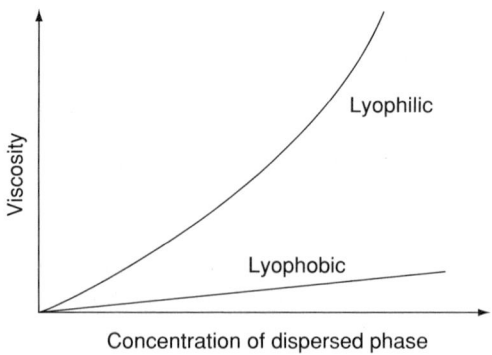

FIGURE 6.1. Effect of colloidal type on viscosity.

one. In this case, lignin was altered by the addition of acetone and sodium hydroxide to render the colloid less stable and to enhance color removal.

Many colloidal systems, especially lyophilic (gel) systems, possess the property of elasticity ("springiness" or resistance). This property enables the gels to resist deformation and thereby recover their original shape and size once they have been deformed. If a concentrated beam of light is passed through a colloidal solution in which the dispersed phase has a different refractive index from that of the dispersion medium, its path is plainly visible as a milky turbidity when viewed perpendicularly. This is known as the *Tyndall effect* (see Table 6.1).

An important property of colloidal particles is that they are generally electrically charged with respect to their surroundings. An electric current passing through a colloidal system causes the positive particles to migrate to the cathode and the negative ones to the anode.

Chemical Coagulation

The removal of oxygen-demanding and turbidity-producing colloidal solids from wastewaters is often called *intermediate treatment,* because colloids are intermediate in size between suspended and dissolved solids. The most common and practical method of removing these solids is by chemical coagulation. This is a process of destabilizing colloids, aggregating them, and binding them for ease of sedimentation. It involves the formation of chemical flocs that absorb, entrap, or otherwise bring together suspended matter, more particularly suspended matter that is so finely divided as to be colloidal.

The chemicals most commonly used are alum, $Al_2(SO_4)_3 \cdot 18H_2O$; copperas, $FeSO_4 \cdot 7H_2O$; ferric sulfate, $Fe_2(SO_4)_3$; ferric chloride, $FeCl_3$; and chlorinated copperas, a mixture of ferric sulfate and chloride. Aluminum sulfate appears to be more effective in coagulating carbonaceous wastes, while iron sulfates are more effective when a considerable quantity of proteins is present in the waste. The use of organic polymers, which can act as either negatively or positively charged ions, has made a significant impact on the efficiency of removal of colloids by chemical coagulation. These polymers, acting as a coagulant aid and applied in conjunction with the coagulant, enhance the formation of flocs and result in improved settling characteristics. Smaller dosages and the elimination of many storage problems are among the major advantages of these polymers. Dey (1965) presents results obtained in various industries where water-soluble polymeric coagulation chemicals are used to achieve improved waste solids settling. Schaffer (1964) found that these polymers were useful in maintaining higher solids concentrations in an anaerobic contact treatment process for meatpacking wastes.

The process of chemical coagulation involves complex equilibria among a number of variables including colloids of dispersed matter, water or another dispersing medium, and coagulating chemicals. Driving forces—such as the electrical phenomenon, surface effects, and viscous shear—cause the interaction of these three variables.

Coagulation by Neutralization of Electrical Charges

This can be accomplished by the following:

1. Lowering the zeta potential of the colloids (Figure 6.2). "Zeta potential" is the difference in electrical charge existing between the stable colloid and the dispersing medium.
2. Neutralizing the colloidal charge by flooding the medium with an excess of oppositely charged ions, usually hydrous oxide colloids formed by reaction of the coagulant with ions in the water. The coagulant colloids also become destabilized by the reaction with foreign, oppositely charged colloids and produce hydrous oxide, which is a floc-forming material.

From the standpoint of electrical charges, there are two predominant types of colloid in wastewaters:

1. Colloids naturally present, including several proteins, starches, hemicelluloses, polypeptides, and other substances, all possess negative charges (mostly lyophilic in nature).
2. Colloids artificially produced by coagulants, usually the hydroxides of iron and aluminum (mostly lyophobic in nature), are mainly positively charged ions.

In most scientific circles, it is believed that the charge on colloidal particles is due mainly to the preferential adsorption of ions (H^+ or OH^-) from the dispersing medium. The charge may also be due, in part, to the direct ionization of a portion of its structural groups, such as NH_2^+ and COO^-.

Hydrous aluminum and iron oxides, as well as other metal solids, can acquire both positive and negative charges. Excess Fe^{+++} makes colloids positively charged. The following expression depicts a resultant positively charged colloid:

$$FeO$$
$$FeO \bullet \times HOH\ Fe^{+++} \leftarrow \begin{Bmatrix} OH^- \\ OH^- \\ OH^- \end{Bmatrix}$$

Excess OH^- makes colloids negatively charged. The following expression depicts a resultant negatively charged colloid:

$$FeO$$
$$FeO \bullet \times H_2O\ OH^- \leftarrow H^+$$
$$O$$

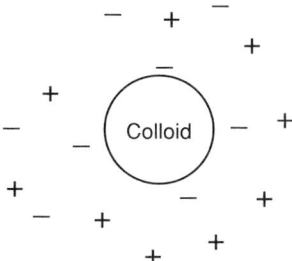

FIGURE 6.2. Stable colloid.

However, a colloid can acquire a charge by means other than adsorption. A protein dissolved in solution can be schematically illustrated as follows:

$$COO^- \text{------[protein base molecule]------} NH_2^+$$

It may become necessary to add up all the positive NH_2^+ groups and the negative COO^- groups to ascertain the final ionic charge of the solution, because of the inherent charge brought about by direct ionization of the particle. The sol is, thereby, stabilized by inherent ionization of groups within the molecule itself.

Any alteration of the type and number of double-layer ions should reduce the zeta potential to such a point that the colloid will lose its stability. Stability is defined as the ability to resist precipitation and/or coagulation into a relatively large particle. A colloid is most stable when it possesses the greatest electrical charge and smallest size. The coagulating power of ions rises rapidly as the electrical charge increases, as is stated by the Schulze–Hardy Rule. Table 6.2 illustrates the minimum concentration of various chemical coagulants required for anions and cations to complete the reaction. Ratios of concentrations of electrolytes required for valences of 1, 2, or 3 are on the order of 729:11.4:1.

Electrolytes and colloids react readily to changes in the pH of the wastewater. Most negatively charged particles, including the majority of contaminating colloids present in wastewaters, coagulate at an optimum pH value of less than 7.0. Flocculent hydroxide colloids, on the other hand, are insoluble only at pH values above 7.0 and usually more than 9.0. Lime is normally added to raise the pH and to aid in precipitation of colloids.

Alum has a pH range of maximum insolubility between 5 and 7; the ferric ion coagulates only at pH values above 4; and the ferrous ion only above 9.5. Copperas ($FeSO_4 \cdot 7H_2O$) is a useful coagulant only in highly alkaline wastes. Lime, a coagulant in itself, is often added with iron salts to raise the pH to the isoelectric point of the coagulant. At this point, the colloid has its minimum electrical charge and is least stable. Because lime is quite insoluble at pH values of 9 and higher,

TABLE 6.2
Valence and Coagulant Dosage

Electrolyte	Anion or Cation Valence	Minimum Concentration Required, mmol/liter
	Anion	
KCl	1	103
KBr	1	138
KNO_3	1	138
K_2CrO_3	2	0.325
K_2SO_4	2	0.219
$K_3Fe(Cn)_6$	3	0.096
	Cation	
NaCl	1	51
KNO_3	1	50
K_2SO_4	1	63
$MgSO_4$	2	0.81
$ZnCl_2$	2	0.68
$BaCl_2$	2	0.69
$AlCl_3$	3	0.09

coagulation with lime and copperas together increases the pH range. Aeration of wastewaters before addition of lime enhances coagulation by evolving lime (thus consuming carbon dioxide and supplying oxygen for converting iron to the oxide and hydroxide states).

Because the ferrous ion when oxidized to the ferric ion can also be used as a coagulant at low pH values, oxidation may be carried out by chlorination, as follows:

$$6Fe^{++}SO_4 \bullet 7H_2O + 3Cl_2 \leftarrow \rightarrow 6Fe^{+++} + 6SO_4^= + 6Cl^- + 42H_2O$$

Negative ions already present in wastewaters extend the useful range of pH in the acid category and positive ions extend the useful pH range in the basic category. Thus, in soft waters, the negatively charged color colloids coagulate best in the acid pH range, and positively charged iron and aluminum ions are good precipitating chemicals in alkaline waters. Prechlorination of alum-treated wastes sometimes increases color removal. Finely divided clay, activated silica, bentonite, or other coagulant aids are often used for relatively clear waters. The addition of any of these produces an effect similar to that of seeding clouds with silver iodide crystals; they provide nuclei about which the precipitate can gather, agglomerate, and flocculate, with a resultant increase in density and settling rate.

Sometimes the presence of iron and manganese in wastewaters will add to the effect of the cationic coagulants. An increase in the concentration of the coagulant shortens the time of the coagulation reaction considerably. Gentle agitation of the wastewater also enhances coagulation, by increasing the number of collisions and thus causing more rapid floc formation.

Removal of Colloids by Adsorption

A large number of compounds that are not amenable to other types of treatment may be removed from wastes by adsorption. For example, pesticides such as 2, 4-D herbicides and carbamate insecticides may be removed by adsorption onto powdered activated carbon but not onto clay materials such as illite, Kaolinite, and montmorillonite (Schwartz 1967). In addition, colloidal suspensions of DTT, chlorobenzene, and p-chlorobenzenesulfonic acid resulting in DDT production may be removed by using activated carbon (Kul'skii and Shabolina 1967). Cooper and Hager (1966) also suggested activated carbon for advanced waste treatment where reclamation is important. They present three typical activated-carbon treatment systems and a granular-carbon reactivation system (Figure 6.3). The granular carbon used in most reactivation systems in the world is made from bituminous coal. Cooper and Hager (1966) also present a summary of properties for two types of this coal (Table 6.3), and claim that this treatment is especially effective in removing biologically resistant (refractory) compounds.

Examples of industrial wastes that I encountered that contained high amounts of colloids include paint wastes, tomato wastes, textile de-sizing wastes, rag mill pulping wastes, and textile kiering wash waters. Classification of these (and other wastes) is given in the major industrial wastes section following these theory chapters (Chapter 13). More complete discussion of these wastes can be found in Nemerow and Agardy (1998).

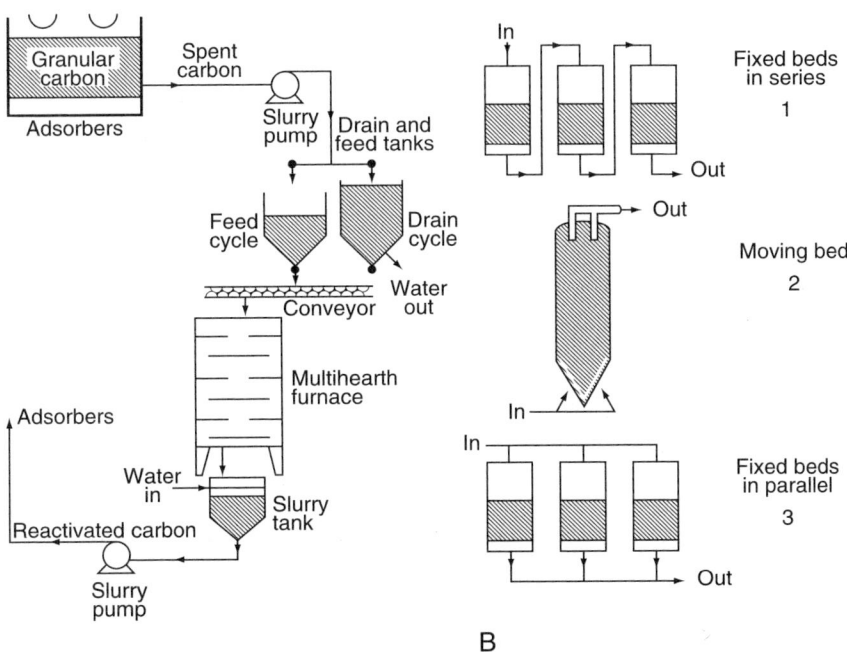

FIGURE 6.3. (**A**) Granula-carbon reactivation cycle. (**B**) Adsorber configuration for granular carbon waste treatment.

TABLE 6.3
Properties of Coal-Derived Granular Carbon for Waste Treatment

Characteristic	Type SGL	Type CAL
Mesh size, U.S. Sieve Series	8 × 30	12 × 40
Effective size, mm	0.8–0.9	0.50–0.60
Uniformity coefficient	1.9 or less	1.7 or less
Mean particle diameter, mm	1.5	0.9
Real density, g/cm^3	2.1	2.1
Apparent density		
g/cm^3	0.48	0.44
lb/ft^3	30.0	27.5
Particle density wetted with water, g/cm^3	1.4–1.5	1.3–1.4
Total surface area (N$_2$BET method), m^2/g	950–1,050	1,000–1,100
Pore volume, cm^3/g	0.85	0.94

Example of Twentieth-Century Practice of Colloidal Solid Removal

Slapik et al. (1982) reported on the use of foam flotation to remove dissolved material, primarily lead, by adsorption onto colloidal particles followed by removal of the colloid particles and its absorbed material by flotation with a surfactant. They maintain that the values of this method include its simplicity, effectiveness, and moderate costs. That a concentrated and easily handled sludge is obtained is a major advantage over the chemical precipitation method. Hence, the authors have used a novel method—previously described in this chapter—to remove very small and toxic metal particles from wastewaters.

Review Questions

1. What are colloids?
2. Why is it important to remove colloids from wastes?
3. What is the approximate size of colloids?
4. Are colloids all of one type? If of more than one type, what are the differences?
5. What importance should the industrial-waste engineer attribute to changes in density of colloids?
6. In chemical coagulation of colloids, what are the two major economic considerations?
7. What is meant by the "zeta potential" and why is this important to industrial-waste engineers in the chemical coagulation of colloids?
8. For what reasons is lime added to industrial wastes for colloidal solids removal?
9. When would you recommend the use of activated carbon for removing colloidal solids? What is the mechanism involved?

References

Cooper, J. C., D. G. Hager. 1966. Water reclamation with activated carbon. *Chem. Eng. Progr.* 62:85.
Dey, R. F. 1965. Use of organic polymers in treatment of industrial wastes. In: *Proceedings of 12th Ontario Industrial Waste Conference, June 1965*, pp. 89–104.
Fair, G. M., J. Geyer. 1958. *Elements of Water and Waste Water*, p. 616. New York: John Wiley & Sons.
Hogg, R., T. W. Healy, D. W. Fuerstenau. 1966. Mutual coagulation of colloidal dispersions. *Trans. Faraday Soc.* 62:1638.
Johnson, R. L., F. J. Lowes, Jr., R. M. Smith, et al. 1964. *Evaluation of the Use of Activated Carbon and Chemical Regenerants in the Treatment of Waste Water*. Washington, DC: U.S. Public Health Service. Publication no. 999-13.
Joyce, R. S., V. A. Sukenik. 1965. *Feasibility of Granular Activated Carbon Adsorption for Wastewater Renovation*. Washington, DC: U.S. Public Health Service. Environmental Health Service Supply and Pollution Control Publication no. 999-WP-28.
Kawamura, S., T. Yoshitaro. 1966. Applying colloid titration techniques to coagulant dosage control. *Water Sewage Works* 113:398.
Kul'skii, L. A., A. G. Shabolina. 1967. Adsorption of DDT from colloidal solutions on the iodine KAD activated carbon. *Chem. Abstr.* 67(14686y).
Middleton, A. E. 1963. Activated silica solution applications. *Water Sewage Works* 100:251.
Nemerow, N. L., F. J. Agardy. 1998. *Strategies of Hazardous Wastes Management*. Hoboken, NJ: John Wiley Publishing Co.
Perona, J. J., et al. 1967. Hyperfiltration—processing of pulp mill sulfite wastes with a membrane dynamically formed from feed constituents. *Environ. Sci. Tecnol.* 1:991.
Rudolfs, W., J. L. Belmat. 1952. A separation of sewage colloids with the aid of the electron microscope. *Sewage Ind. Wastes* 24:247.
Rudolfs, W., H. W. Gehm. 1936. Chemical coagulation of sewage. *Sewage Works J.* 8:195, 422, 537, 547.
Rudolfs, W., H. W. Gehm. 1939. Colloids in sewage treatment, 1. Occurrence and role. A critical review. *Sewage Works J.* 11:727.
Sawyer, C. N., P. E. McCarty. 1960. *Chemistry for Sanitary Engineers*. New York: McGraw-Hill Book Co.
Schaffer, R. B. 1964. Polyelectrolytes in industrial waste treatment. *Water Sewage Works* 111:300R.
Schwartz, H. G., Jr. 1967. Adsorption of selected pesticides on activated carbon and mineral surfaces. *Environ. Sci. Technol.* 1:332.
Sennet, P., J. P. Oliver. 1965. Colloidal dispersions, electrokinetic effects and concept of zeta potential. *Ind. Eng. Chem.* 57:32.
Slapik, M. A., E. L. Thackston, D. J. Wilson. 1982. Improvements in foam flotation for lead removal. *J. Water Pollution Fed.* 54(March):3, 238.
Weber, W. J. 1967. Adsorption. In: *Proceedings of Summer Institute for Water Pollution Control, Manhattan College, New York*.
Weber, W. J., J. C. Morris. 1966. *Adsorption of Biochemically Resistant Materials from Solution*. Washington, DC: U.S. Public Health Service. Publication no. 999-WP33 W62-24.

Williamson, J. N., A. M. Heit, C. Calmon. 1964. *Evaluation of Various Adsorbents and Coagulants for Waste Water*. Washington, DC: U.S. Public Health Service. Publication no. 999-WP.

Woodard, F., J. Etzel. 1965. Coacervation and chemical coagulation of lignin from pulpmill black liquor. *J. Water Pollution Control Fed.* 37:990.

CHAPTER 7

Removal of Inorganic Dissolved Solids

Minerals are as significant in industrial wastewaters as they are in water and foods for human consumption. Trace concentrations in waste enhance the growth of flora in any receiving waters just as they enhance the growth and health of humans digesting the foods and water. At the same time, an excess of minerals stresses all receiving waters and the people who drink and eat them. Therefore, it is important for environmentalists to be able to differentiate between excessive and acceptable mineral content in all of our environments, and then we must be able to recommend removal techniques to keep these minerals at the proper concentration.

The removal of dissolved minerals from wastewaters has been given relatively little attention by waste treatment engineers, because minerals have been considered less pollutional than other constituents, such as organic matter and suspended solids. However, as we learn more about the causes and effects of pollution, the importance of reducing the quantity of certain types of inorganic matter, which wastewater treatment plants and many other industries permit to enter streams, is apparent. Chlorides, phosphates, nitrates, and certain metals are examples of the more common and significant inorganic dissolved solids. Among the methods employed mainly for removing inorganic matter from wastes are: (1) evaporation; (2) dialysis; (3) ion exchange; (4) algae; (5) reverse osmosis; and (6) miscellaneous methods. Other treatment methods that remove minerals incidentally but are aimed primarily at other contaminants are discussed in Chapters 6 and 8. One should not overlook the minerals contributed by natural runoff from overland flow. The amount of dissolved solids that these natural flows contain often exceeds that contributed by wastewaters from industry.

Evaporation

"Evaporation" is a process of bringing wastewater to its boiling point and vaporizing pure water. The vapor is either used for power production, condensed and used for heating, or simply wasted to the surrounding atmosphere. The mineral solids concentrate in the residue may be sufficiently concentrated for the solids either to be reusable in the production cycle or to be disposed easily. This method of disposal is used for radioactive wastes, and paper mills have for a long time been evaporating their sulfate cooking liquors to a degree where they may be returned to the cookers for reuse.

Major factors in selection of the evaporation method include the following: (1) Economics: Does the value of the reusable residue outweigh the cost of fuel for evaporation? (2) Initial dissolved solids: Are there enough solids in the waste of a variable nature to warrant evaporation? Generally, 10,000 ppm are required. (3) Foreign matter: Is there foreign matter present that could cause scale formation or corrosion or interfere with heat transfer in evaporation? (4) Pollution situation: What effect will the minerals have on the receiving stream? For example, caustic soda kills fish, ammonium salts initiate troublesome algal growths and in some cases stimulate bacterial growth upon organic matter already present (Amberg 1955), salt interferes with water use by industries and municipalities, and so forth.

Today, many evaporators are heated by steam condensing on metallic tubes, through which flows the waste to be concentrated or evaporated. The steam is at a low pressure, usually less than 50 pounds per square inch (psi) (absolute). Most evaporators operate with a light vacuum on the vapor side, to lower the boiling point and to increase the rate of vapor removal from the evaporator. Vacuum systems are especially preferable to atmospheric evaporators when the decomposition of organic matter is involved. Care must be exercised, however, that the vacuum is not great enough to permit priming of the wastewater into the vapor.

Evaporating a waste presents many problems, which include concentration changes during evaporation, foaming, temperature sensitivity, scale formation, and the materials used in evaporator construction. In industrial-waste concentration, scale formation usually presents the major overall heat-transfer coefficient decreases, causing the efficiency to drop until it is necessary to shut down and clean the tubes—a complicated process when the scale is hard and tenacious.

Chrome, nickel, and copper acid–type plating wastes may be reclaimed from the rinse tank by evaporation in glass-lined equipment, or other suitable evaporators, and the concentrated solution returned to the plating system (Merril et al. 1949). Initial cost of equipment is high, so the quantity and value of chemicals to be recovered, plus the estimated cost of operating the treatment system if evaporative recovery were not practiced, are criteria one must use to justify purchasing such equipment.

Efficiency of evaporation is directly related to heat-transfer rate—expressed in British thermal units per hour (Btu/hr)—through the heating surface (tube wall). This rate is equal to the product of three factors: the overall heat-transfer coefficient, the heating surface area, and the overall change in temperature between the waste and the steam. It is expressed mathematically as

$$q = UA (t_s - t_w) = UA \Delta t,$$

where q is the rate of heat transfer (Btu/hr), U is the overall coefficient (Btu/ft²/hr/°F), A is the heating-surface area (ft²), t_s is the temperature of steam condensate (°F), t_w is

TABLE 7.1
Typical Overall Coefficients in Evaporators

Type of Evaporator	Overall Coefficient, Btu/ft^2/hr/°F
Long-tube vertical	
Natural recirculation	200–600
Forced circulation	400–2,000
Short-tube	
Horizontal tube	200–400
Calandria type	150–500
Coil	200–400
Agitated film	
Newtonian liquid viscosity	
1 centipoise	400
100 centipoises	300
10,000 centipoises	120

Adapted from Brown et al. (1950).

the boiling temperature of waste (°F), and $\Delta t = t_s - t_w$ is the overall temperature change between steam and waste. Typical values of U for various types of evaporators are given in Table 7.1. These figures are estimated within broad ranges, by considering the viscosity of the waste, scale formation, and operating temperatures (greater temperature differentials yield higher coefficients). Tube wall thickness also influences U: the greater the thickness, the lower the value of U.

Dialysis

"Dialysis" is the separation of solutes by means of their unequal diffusion through membranes (Eynon 1933; Lee 1935; Bassett 1938; Lovett 1938; Roetman 1944a,b; Kirk and Othmer 1950; Powell 1954; U.S. Department of Health, Education and Welfare 1963; Smith and Eismann 1964; "Reverse Osmosis" 1967). It is most useful in recovering pure solutions for reuse in manufacturing processes, for example, caustic soda in the textile industry (Michalson and Burhans 1962). Recovery involves separation of a crystalloid (NaOH) from a solution in which about 96% of the impurities is in the form of hemicellulose and the rest includes pectins, waxes, and dyes.

There are some 8–10 commercial dialyzers on the market. In our example, they all operate on the simple principle of passing a concentrated impure caustic solution upward, counter-current to a downstream water supply, from which it is separated by a semipermeable membrane (Figure 7.1). The caustic soda permeates the membrane and goes into the water more rapidly than the other impurities contained in the waste. The concentration of caustic soda is always greater in the impure solution than in the water,

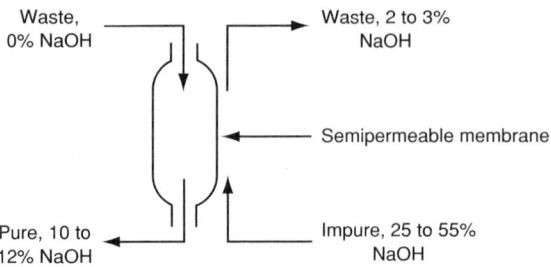

FIGURE 7.1. Typical dialysis flow diagram.

and the water that flows through the membrane into the impure caustic solution tends to dilute it. The quantity of sodium hydroxide diffusing through the diaphragm depends on the time, the area of the dialyzing surface, the mean concentration difference, and the temperature. These factors are expressed in the equation

$$Q = Kat\,(\Delta c),$$

where K is the overall diffusion coefficient, t is the time in minutes, A is area of dialyzing surface, and

$$\Delta c = \Delta c_{av} = (\Delta c_1 - \Delta c_2) / (2.3 \log_{10} \Delta c_1/\Delta c_2),$$

where Δc_1 and Δc_2 are the differences in concentration between the two solutions at the top and bottom of the diagram, respectively.

When one actually computes the weight of NaOH recovered, it becomes apparent that the quality and type of diagram are important, which is evident from the following equation:

$$W = UA\,\Delta c_{\text{log mean}}$$

where W is the weight of material passing through the membrane in a unit of time (g/min), U is the overall dialysis coefficient, and $\Delta c_{\text{log mean}}$ is the logarithmic mean concentration gradient across the membrane (= Δc_{av}). Also,

$$1/U = 1/U_1 + 1/U_2$$

where U_1 is the combined film resistance (cm/min) and U_2 is the membrane resistance (cm/min). Each diaphragm shows a different membrane resistance (U_2). The restrictive characteristics of porous membranes are due to both a mechanical sieve action and a physicochemical interaction between solute, solvent, and membrane. Cellulose nitrate, parchment, and cellophane are the principal membranes used.

Smith and Eisemann (1964) present an excellent evaluation of electrodialysis. Dialysis is an operation requiring very little operator attention, and although its main role is to conserve raw materials and to reduce plant waste, it also aids in waste treatment. With the introduction of acid-resistant membranes, dialysis has been used successfully in the recovery of sulfuric acid in the copper, stainless-steel, and other industries. Some operations can recover as much as 70–75% of the acid, but a recovery of as little as 20% may justify the process. In dialysis, the driving force of separation is natural diffusion because of concentration gradient; in electrodialysis this natural force is enhanced by the application of electrical energy. McRae (1965) found that, for a secondary effluent containing 900 ppm of dissolved solids, electrodialysis could achieve a 44% reduction, with costs ranging from 10 to 15 cents per 1,000 gallons. He found this process useful for treating the wastes from dairies.

Because of the resulting improvements in mechanical and thermal properties, ceramic-membrane filters now generally outperform polymer membranes, their established competitor (High Technology, August 1987). Ceramic membranes are made by blending dry mineral powders, such as alumina, spinel, cordierite, and zirconia, in various proportions with a solvent to form either a slurry, which is poured into a mold, or a dough, which is extruded. The resulting configurations, either hollow fibers, flat plates, or honeycombs, are then dried and fused together. Layers of supporting material are added to complete the filter. Because ceramic membranes tolerate acids and bases, they can be more thoroughly cleaned than other designs. Also, these membranes can be cleaned with steam up to 140 °C, whereas polymeric membranes cannot be steam-cleaned. Ceramic membranes can also be heated up to 500 °C to burn out impurities trapped during filtration. In addition to their relatively high cost, a major drawback is their brittleness.

Ion Exchange

Ion exchange is basically a process of exchanging certain undesirable cations and anions of the wastewater for sodium, hydrogen, or other ions in a resinous material. The resins, both natural and artificial, are commonly referred to as *zeolites*. The ion-exchange process was originally developed to reduce hardness in domestic-water supplies but has been used to treat industrial wastewaters, such as metal-plating wastes. The softening reactions may be illustrated as follows (Nordell 1951):

$$\begin{matrix} Ca\}(HCO_3)_2 \\ Mg \;\; SO_4 \\ CL_2 \end{matrix} + Na_2Z \longrightarrow \begin{matrix} & 2\,NaHCO_3 \\ & \text{or} \\ Ca\}Z + & Na_2SO_4 \\ Mg & \text{or} \\ & 2NaCL \end{matrix}$$

where Z is the symbol for the zeolite radical. When the ability of the zeolite bed to produce soft water is exhausted, the softener is temporarily cut out of service. It is then backwashed to cleanse and hydraulically regrade the bed, regenerated with a solution of common salt that removes the calcium and magnesium in the form of their soluble chlorides while restoring the zeolite to its original condition, rinsed free of these and the excess salt, and finally returned to service. The reaction may be indicated as follows:

$$\text{Ca} \} \text{ Z} + 2 \text{ NaCL} \rightarrow \text{Ca}\}\text{Cl}_2 + \text{Na}_2\text{Z}.$$
$$\text{Mg} \qquad\qquad\qquad \text{Mg}$$

Ion exchange as a means of waste treatment is only a new application of a traditional method of water softening. If the proper approach is used, it offers great potential for material and water conservation. For instance, in the treatment of metal-plating wastes (Merrill et al. 1949), rinse water is passed through beds of cationic and anionic resins selected for the particular application and the deionized water is then recycled through the rinse tank. This method may be applied on a continuous basis to the removal of contaminating metals (Merrill et al. 1949) from chromic acid solutions, permitting the return of pure chromic acid solution to the process tank. In the case of nickel- and copper-plating solutions, both the contaminating metals and the metal to be plated are cationic, so all will be extracted. Cation-exchange resins are suggested ("New Process Developed" 1965) for use in the steel industry to remove the iron from spent liquor and to recover sulfuric acid and iron oxide for further use. Unless the aim of the procedure is recovery of metals, ion exchange becomes simply a concentration method, and some treatment for the regenerated solution must be devised.

Walther (1965) reports the use of a continuous ion-exchange unit, consisting of a stainless-steel loop divided into sections by butterfly valves that successfully removed more than 700 mg/liter of dissolved inorganic solids. The unit contains about 15 ft^3 of ion-exchange resin, which moves around the loop in about 3 minutes. When the resin becomes saturated with hardness, it is removed from the loop and regenerated resin is exchanged. The spent resin is then regenerated and returned to the loop on the next cycle.

Organic matter and pH have a pronounced effect on the operation and efficiency of resin beds; the leaching of organic matter from certain resins may have a detrimental effect on the metals plated. Chemicals used for regenerating resin beds may also require special treatment before disposal.

Demineralization (ion exchange) is most useful when water of the highest quality is required, but it involves complex chemical reactions and, therefore, requires careful operation and supervision at all times. Furthermore, ion-exchange processes sometimes use chemicals that are hazardous to personnel and equipment. These are matters to think about before selecting an ion-exchanger system over an evaporator, although evaporators, too, are uneconomical in certain instances such as when the flow is light. Dialysis is normally economical and can compete in efficiency with both evaporation and ion exchange when the recovery of a pure compound is considered essential. The decision whether to use evaporation or demineralization can be intelligently made only after a thorough evaluation of the heat balance of the plant and expected operating conditions (Paulson 1952). These factors, as well as operating costs, must be considered in relation to the capital investment needed for either system.

Algae

Algae require nine minor essential elements (Fe, Mn, Si, Zn, Cu, Co, Mo, B, and Va) and seven major essential elements (C, N, P, S, K, Mg, and Ca) for their optimum growth. The use of algae for removing minerals from wastewaters has been investigated; most investigations have been carried out on sewage effluents. One such study (Golueke and Oswald 1965) involved a suburban housing-development treatment plant and used primary sedimentation, trickling filtration, and stabilization ponds. Although the sedimentation and filtration did not remove any phosphorus, the algae actively growing in the ponds caused a reduction of about 42% of the phosphate content. Other mineral concentrations were not measured.

If this method is used to remove minerals such as phosphate over a period, algae must also be removed from the effluent before this is released into a stream used for water supplies and recreation. Golueke and Oswald (1965) observed three steps in harvesting oxidation-pond algae: (1) collection and initial concentration, (2) de-watering or secondary concentration, and (3) final drying. They found chemical precipitation and centrifugation to be most economical. The harvested algae can be sold as animal feed supplements. Oswald (1960) describes *Chlorella* and *Scenedesmus* as the most active algae in stabilization ponds, because they are extremely hardy. Krauss (1956) presented the elemental composition of these two algal types to validate their fixation of minerals (Table 7.2).

As we progressed towards the end of the twentieth century, the demineralization of seawater as well as tertiary treated domestic wastewater became more general

TABLE 7.2
Elemental Composition of Green Algae

Element	*Range of Dry Weight, %*
Chlorella	
Carbon	51.4–72.6
Hydrogen	7.0–10.9
Oxygen	11.6–28.5
Scenedesmus	
Nitrogen	2.2–7.7
Phosphorus	1.1–2.0
Sulfur	0.28–0.39
Magnesium	0.36–0.80
Potassium	0.85–1.62
Calcium	0.005–0.08
Iron	0.04–0.55
Zinc	0.0006–0.005
Copper	0.001–0.004
Cobalt	0.000003–0.0003
Manganese	0.002–0.01

Adapted from Krauss (1956).

practice. In these cases, it is necessary to remove most, if not all, of the residual minerals. The most promising methods of mineral removal–especially for salts–are membrane filters and reverse osmosis. Costs for these treatments are gradually becoming comparable to those of conventional complete water treatment systems.

Table 7.2 shows the extent to which algae take up minerals from any solution in which they grow. In fact, the continued photosynthesis of algae depends directly on the ability of the culture medium (wastewater) to supply these inorganic compounds over a long period, at a rate sufficient to support the growth potential of the algae. There is some evidence that the uptake (and the algal growth) depends on the availability and the presence of inorganic nutrients. Thus, insolubility and colloidal characteristics of the nutrients may hamper algal growth, but hardness in wastewaters can contribute to it. A statistical study of Massachusetts lakes and reservoirs carried out in 1900 showed that the hard-water supplies yielded more algae than the soft (Table 7.3). Bogan (1959)

TABLE 7.3
Occurrence of *Cyanophyceae* and *Chlorophyceae* in Massachusetts Lakes and Reservoirs

Characteristic	Chemical Analysis, ppm	*Often Above 1,000/cm³*		*Below 100/cm³*	
		Cyano-phyceae	Chloro-phyceae	Cyano-phyceae	Chloro-phyceae
Color	0–30	2	2	11	0
	30–60	2	2	3	1
	60–100	3	1	7	2
	>100	0	0	1	1
Chlorides (excess above normal)	0	2	1	3	1
	0.1–0.3	1	1	10	5
	0.4–2.5	1	0	9	6
	>2.5	3	3	0	0
Hardness	0–5	0	0	6	4
	5–10	2	1	10	5
	10–20	2	1	5	2
	>20	3	3	1	1
Albuminoid ammonia (dissolved)	0–0.10	0	0	4	3
	0.1–0.15	0	0	6	4
	0.15–0.20	2	2	7	3
	>0.20	5	3	5	2
Free ammonia	0–0.01	0	0	10	4
	0.01–0.03	0	0	8	5
	0.03–0.10	3	2	4	3
	>0.1	4	3	0	0
Nitrates	0–0.05	1	0	12	6
	0.05–0.10	3	2	10	6
	0.10–0.20	1	0	0	0
	>0.20	2	3	0	0

Adapted from Walther (1965).

capitalized on the ability of algae to use phosphorus in providing tertiary treatment of the sewage from Seattle, Washington, which utilized both algal activity and lime and removed more than 90% of the phosphorus in the secondary sewage-plant effluent. Oxidation-pond usage has been increasing since the advent of lower-cost mechanical aeration.

Reverse Osmosis

"Reverse osmosis" is a membrane permeation process for separating relatively pure water or some other solvent from a less pure solution. The solution is passed over the surface of a specific semipermeable membrane at a pressure in excess of the effective osmotic pressure of the feed solution. The permeating liquid is collected as the product and the concentrated feed solution is generally discarded. The membrane must be: (1) highly permeable to water, (2) highly impermeable to solutes, (3) capable of withstanding the applied pressure without failure, (4) as thin as possible consistent with the strength requirement, (5) chemically inert, mechanically strong, and creep resistant; and (6) capable of being fabricated into configurations of high surface-to-volume ratios. A number of commercial units in practice treat brackish waters of less than 2,000–3,000 ppm of total dissolved solids.

Although several types of membranes have been developed, two types of membranes are generally used in commercial equipment. The first is a symmetric of "skinned" cellulose acetate membranes made in flat or tubular forms. Generally, the membranes are approximately 100 μm thick, with a surface skin of about 0.2 μm that acts as the rejecting surface. Typically, they operate at 40–50 atmospheres pressure and produce a water flux of 10–20 gallons of water per square foot per day with a salt rejection of about 95%. These membranes have generally exhibited a decrease in flux rate with time because of both compaction (creep) and fouling of the membrane. Therefore, operating pressures are kept low to avoid creep, and the amount of suspended solids in the feed solution is kept as low as possible to prevent fouling.

The second type of membrane is an aromatic polyamide or polyamide hydrazide. The membrane in commercial units is in the form of hollow fine fibers. The patented membrane is claimed to operate at 27 atmospheres pressure and in a water flux of 1–2 gallons/ft^2/day with a salt (NaCl) rejection of about 95%.

Some limitations of existing membranes other than the total dissolved solids concentrated are described by Lonsdale and Podall (1972) in the following manner: (1) The relatively high cost of operation could be reduced if the water flux could be substantially increased without loss in salt rejection or other properties. (2) Flux decline is serious with high-flux membranes. (3) Certain species are inadequately rejected—for example, boric acid, phenol, and nitrates. (4) For certain applications, existing membranes are not sufficiently resistant to chemical or microbiological attack, or their mechanical or thermal stability is inadequate. (5) Feed-water pH should generally be on the acidic side (pH 5–7) for best operation and to minimize membrane hydrolysis. For industrial waste treatment, some preliminary promising results have been reported with sulfite, kraft pulping, and textile dyeing wastes.

Substantial energy needed to constantly pressurize the incoming salty water is one of the drawbacks of the reverse-osmosis system. The Reliable Water Company (Figure 7.2) system uses an energy recovery mechanism that reclaims most of the fluid pressure from the brine waste as it leaves the system. Using hydraulic oil pumps, transfer barriers, and special valves, the system extracts the pressure in the brine and transfers it to the incoming waste saltwater, substantially reducing the energy requirements.

Osantowski and Ceinopoloc (1979) obtained excellent rejections of dissolved solids for desalting processes after pretreating de-inking paper-mill and slaughterhouse and meatpacking wastes. They used reverse osmosis, ion exchange, and electrodialysis. They found reverse osmosis was the most economical process for providing reusable quality water in both paper-mill and food products plants. Reuse quality requirements could be met in most cases by blending the reverse-osmosis product water with un-desalted wastewater. On the other hand, electrodialysis provided the optimal performance of the three desalting technologies investigated at an organic chemicals plant.

Ultrafiltration and Microfiltration

During the 1990s, ultrafiltration (UF) and microfiltration (MF) treatment systems began to be used in conjunction with and usually prior to reverse-osmotic systems. Ultrafilters and microfilters are membrane filters that remove larger-sized contaminants (0.001 microns and 0.01 microns, respectively) primarily by a sieving mechanism. Reverse-osmotic (RO) systems, on the other hand, remove contaminants as small as 0.00001 microns and, therefore are more useful in water-producing rather than waste-treating plants. When used in

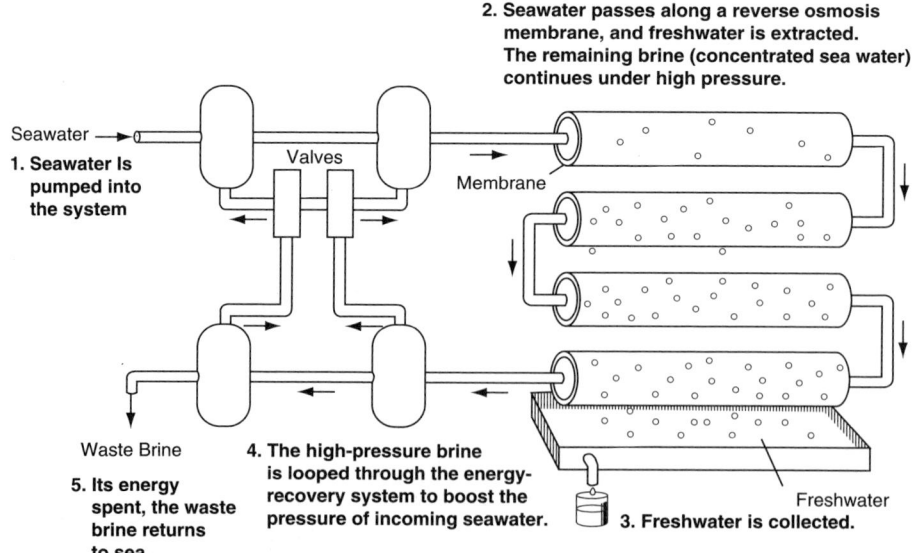

FIGURE 7.2. The desalination process using energy recovery and artificial intelligence control.

conjunction with and prior to reverse osmosis, less clogging and "plugging up" of the RO treatment units occur. This ensures longer life and a higher flux rate for the RO units.

When using any of the types of membrane filtration, one must not expect industrial operation to be comparable to municipal-water plants. Industries often treat smaller flow rates and higher solids concentrations, and they usually have less-stringent effluent requirements. The appropriateness of membranes for industry may depend on space needs, economics of plant operation, effluent discharge requirements, and any existing treatment system.

Miscellaneous Methods

Chemical precipitation or coagulation has been used to remove some inorganic matter from wastewaters. For example, elevated pH values aid in the removal of heavy metals by precipitation of the hydroxide or carbonate, and under some conditions, treatment of wastewaters with calcium hydroxide is reasonably effective in the removal of nitrogen and phosphorus.

In laboratory studies during the 1960s, we found that lime precipitation of raw sewage repressed algal growth, especially of the blue-green type. Soil percolation of sewage effluents through certain soils has also given some indication of being an effective method reducing blooms. Presumptions were that removal of either or both organic matter and limiting minerals such as phosphorus could be the critical elements in preventing excessive algal blooms (Jackson 1967).

Oxidation-reduction chemical reactions are used in certain cases to alter inorganic matter and, thus, enhance its removal. For example, chromate must be reduced, usually with ferrous sulfate or sulfur dioxide under acid conditions, to the trivalent form as a preliminary to precipitation with lime and subsequent removal as a chromic-hydroxide sludge (see reactions in Chapter 6). Likewise, cyanides must be completely oxidized, usually with chlorine under alkaline conditions, to split them up into harmless and volatile nitrogen gas and carbonate ions (see Chapter 6).

The ultimate disposal of salts has always been and still remains a major problem to the environmental engineer. One novel suggestion for the use of concentrated salt waste was proposed in 1981 (*New York Times*, p. A9, August 3, 1981). Salt solutions with at least 10% salt content can be added to the warm surface waters of lakes. The increased density of these warm surface waters tends to cause the salt solutions to sink to the bottom, where they stay and serve as a reservoir for hot-water energy. The hot-water energy is pumped out periodically to drive turbines to produce electrical energy. The hot water can also be used for space heating or for agricultural or industrial processes.

Refractories

"Refractories" can be defined for our purpose as solids in wastewater—generally in the dissolved stage—that are not susceptible to removal by the usual "primary" or "secondary" treatment techniques, including those of chemical coagulation. They may have to be removed because the increased water reuse results in a gradual buildup in water supplies downstream and deterioration of the water for its best usage, and we are learning more about the potential hazardous effects of refractories in water, including the following:

1. Fluorides causing mottled teeth
2. Nitrates causing methemoglobinemia
3. ABS interfering with surface re-aeration and adding tastes to water
4. Metals causing blood poisoning
5. Certain insecticides and benzene-structured organics causing potential nerve damage and/or carcinogenic reactions

Presently, it is feasible to remove these refractory materials by (1) membrane filtration, (2) evaporation, or (3) adsorption. The theories of these methods are described in Chapters 5 and 6. Some of the major limitations and potentials for use are given in Table 7.4.

TABLE 7.4
Refractory Containment Removal Techniques

Membrane filtration	*Evaporation*	*Absorption*
Limitations		
(a) Life of Membrane	Requires source of relatively inexpensive energy input	Cost of new carbon is relatively high
(b) Loss in flux rate	Wastewater should be very high in solids content	Can clog quite easily with finely-divided suspended solids
(c) Relatively small amount of effluent which can be removed	Wastewater could cause scaling of pipes in evaporator thus interfering with heat transfer	Cost of regenerating equipment is high, but carbon costs are reduced to $\frac{1}{10}$ of above costs by regeneration of carbon
(d) Relatively limited type of materials which can be removed		
When to use		
With a relatively small volume of primarily two component is quite valuable; when recovered in relatively pure form. Each component should be as different as posssible in molecular size and noncorrosive to membranes	With a high solids, noncorrsive, nonscale-forming waste; when an inexpensive source of heating is available	With a highly soluble, single-state waste, which however, has a high enough molecular weight to be easily removed by adsorption. Preferably, the contaminant should be organic and can either be recovered by solvent extraction if valuable, or destroyed by burning if not valuable

Example of Twentieth-Century Practice of Removal of Inorganic Dissolved Solids

Knoche et al. (1986) attempted to find acceptable treatment processes to remove inorganics so that the level of quality would be acceptable for recycle and reuse. They found that the most acceptable treatment involved the addition of powdered activated carbon in conjunction with metal ion coagulants to lower residual color levels to below 10 Pt-Co standards. However, required dosages were high. They recommended that further research be undertaken to increase the efficiency of inorganic dissolved solids removal beyond that which they obtained.

Review Questions

1. What percentage of industrial plants practice removal of minerals?
2. What are the significant minerals present in industrial wastes?
3. Name six major methods whereby inorganic ions can be removed?
4. Describe the principles and problems in using evaporation and its major use.
5. Describe the principles and problems in using dialysis and its major use.
6. Describe the principles and problems in using ion exchange and its major use.
7. What are algae used for and when are they most useful in waste treatment?
8. What miscellaneous methods are also available for removing minerals?

References

Amberg, H. R. 1955. The effect of nutrients upon the rate of stabilization of spent sulfite liquor in receiving waters. *Proc. Am. Soc. Civil Engrs.* 81:821.

Bassett, H. P. 1938. Super filtration by dialysis. *Chem. Met. Eng.* 42:254.

Bogan, R. H. 1959. Pilot Evaluation of a Tertiary Stage Treatment Process for Removing Phosphorus from Sewage. A report prepared for the city of Seattle, December 1959.

Brown, G. G., D. Katz, A. S. Foust, R. Schneidewind. 1950. Unit Operations, p. 484. New York: John Wiley & Sons.

Bryson, J. C. 1961. Control of Algae through Phosphate Control [unpublished report], Syracuse University, Syracuse, NY, September 1961.

Eynon, D. J. 1933. Operation of Cerini dialysers for recovery of caustic soda solutions containing hemicellulose. *J. Soc. Chem. Ind.* 52:173T.

Golueke, C. G., W. J. Oswald. 1965. Harvesting and processing sewage grown planktonic algae. *J. Water Pollution Control Fed.* 37:471.

Jackson, D., ed. 1967. Algae, man, and the environment. In: *Proceedings of the International Symposium, Syracuse University, Syracuse, New York, June 18–30, 1967.*

Keating, R. J., R. Dvorin. 1960. Dialysis for acid recovery. In: *Proceedings of Industrial Waste Conference, Purdue University, 1960,* pp. 567–576.

Kirk, R. E., D. F. Othmer. 1950. *Encyclopedia of Chemical Technology,* p. 5. New York: Interscience.

Knoche, W. R., D. Bhinge, E. Sullivan, G. Boardman. 1986. Treatment of pulp and papermill wastewater for potential water reuse. Paper presented at the 41st Purdue University Industrial Waste Conference Proceedings, May 13–15, 1986. Lafayette Indiana.

Krauss, R. W. 1956. Photosynthesis in the algae. *Ind. Eng. Chem.* 48:1449.

Kunin, R., F. Y. McGarvey. 1963. Status of ion exchange technology. *Ind. Eng. Chem.* 55:51.

Lee, J. A. 1935. Caustic soda recovery in rayon industry. *Chem. Met. Eng.* 42:483.

Lonsdale, H. K., H. E. Podall. 1972. Reverse osmosis membrane research in a symposium, June 1971. New York: Plenum Press.

Lovett, L. E. 1938. Application of osmosis to recovery of caustic soda solutions containing hemicelluloses in rayon industry. *Trans. Electrochem. Soc.* 73:163.

McCabe, W. L., J. C. Smith. 1956. *Unit Operations of Chemical Engineering*, p. 530. New York: McGraw-Hill Book Co.

McRae, W. A. 1965. Electrodialysis in wastewater reclamation. In: *Proceedings 2nd Water Quality Research Symposium, New York State Department of Health, April 14, 1965*, pp. 97–119.

Merrill, G. R., A. R. Macommer, H. R. 1949. *Manusberger, American Cotton Handbook*, 2nd ed. New York: Textile Book Publishers.

Michalson, A. W., C. W. Burhans, Jr. 1962. Chemical waste disposal by ion exchange. *Ind. Water Wastes* 1:11.

Nemerow, N. L., J. C. Bryson. 1963. How efficient are oxidation ponds? *Wastes Eng.* 34:133.

Nemerow, N. L, W R. Steele. 1955. Dialysis of caustic textile wastes. In: *Proceedings of 10th Industrial Waste Conference, Purdue University, May 1955*, pp. 74–81.

New England Interstate Water Pollution Control Commission. 1959. *Textile Wastes—A Review*.

New process developed to recover acid and iron from spent pickle liquor. 1965. *Iron Steel Engr.* 42:167.

Nordell, E. 1951. *Water Treatment*, p. 341. New York: Reinhold.

Ohio River Valley Water Sanitation Commission. 1953. *Methods for Treating Metal Finishing Wastes*, January 1953, p. 58.

Okey, R. W., P. L. Stavenger. 1967. Membrane technology. A process report. *Ind. Water Eng.* 4:36.

Osantowski, R., Ceinopoloc, A. 1979. An evaluation for water reuse in selected industries using advanced waste treatment processes. In: *Proceedings of the Industrial Wastes Symposium, 52nd Annual WPCF Conference, Houston Texas, 1–12 (Oct. 1979)*.

Oswald, W. J. 1960. Fundamental factors in oxidation pond design. Paper presented at the Conference on Biological Waste Treatment, at Manhattan College, New York, April 20–22, 1960. Paper no. 44.

Paulson, C. F. 1952. Chromate recovery by ion-exchange. In: *Proceedings of 7th industrial Waste Conference, Purdue University, 1952*, p. 209.

Powell, S. T. 1954. *Water Conditioning for Industry*, p. 214. New York: McGraw-Hill Book Co.

Reverse osmosis. An old concept in new hardware. 1967. *Ind. Water Eng.* 4:20.

Roetman, E. T. 1944a. Viscose rayon manufacturing wastes and their treatment. *Water Works Sewage* 91:295.

Roetman, E. T. 1944b. Stream pollution control at Front Royal, Virginia, rayon plant. *Southern Power Ind.* 62:86.

Rudolfs, W. 1937. A survey of recent developments in the treatment of industrial wastes. *Sewage Works J.* 9:998.

Smith, J. D., J. L. Eismann. 1964. Electrodialysis in waste water recycle. In: *Proceedings of 19th Industrial Waste Conference, Purdue University, 1964,* pp. 738–760.

U.S. Department of Health, Education and Welfare. 1963. *Cost of Purifying Municipal Waste Waters by Distillation,* Publication no. AWTR-6, Washington, DC.

U.S. Department of Health, Education and Welfare. 1965. *Advanced Waste Treatment Research.* AWTR-14 S Summary Report (PHS Publication NQ 999-WP-24). Washington, DC.

Volbrath, H. B. 1936. Applying dialysis to colloid-crystalloid separations. *Chem. Met. Eng.* 43:303.

Walther, A. T. 1965. LaGrange tests ion exchange unit. *Water Sewage Works* 112:212.

Whipple, G. C.. 1948. *Microscopy of Drinking Water,* pp. 214–215. New York: John Wiley & Sons.

CHAPTER 8

Removal of Organic Dissolved Solids

In earlier years, the removal of dissolved organic matter was primarily based on its oxygen-demanding characteristics. During the latter part of the twentieth century, the emphasis for its removal began to be based on its hazardous component. Unfortunately, the treatments for the two components—oxygen-demand and hazardous—are not compatible. The first is removed largely by biological means (as you will read in this chapter), and the second must usually be segregated and treated by sophisticated and often innovative techniques. The treatment of specific organic hazardous (or toxic) matter is presented in Chapter 11.

The removal of dissolved organic matter from wastewaters is one of the most important tasks of the waste engineer, and, unfortunately, one of the most difficult. These solids are usually oxidized rapidly by microorganisms in the receiving stream, resulting in loss of dissolved oxygen and the accompanying ill effects of deoxygenated water. They are difficult to remove because of the extensive detention time required in biological processes and the elaborate and often expensive equipment required for other methods. In general, biological methods have proved most effective for this phase of waste treatment because bacteria are adept at devouring organic matter in wastes, and the greater the bacterial efficiency the greater the reduction of dissolved organic matter. Microorganisms, however, are quite "temperamental" and sensitive to changes in environmental conditions, such as temperature, pH, oxygen tension (level of oxygen concentration), mixing, toxic elements or compounds, and character and quantity of food (organic matter) in the surrounding medium. It is the responsibility of the engineer to provide optimal environmental conditions for the proliferation of the particular biological species desired.

There are many varieties of biological treatment, each adapted to certain types of wastewaters and local environmental conditions such as temperature and soil type. Some specific processes for treating organic matter are: (1) lagooning in oxidation ponds; (2) activated-sludge treatment; (3) modified aeration; (4) dispersed-growth aeration; (5) contact stabilization; (6) high-rate aerobic treatment (total oxidation); (7) trickling filtration; (8) spray irrigation; (9) wet combustion; (10) anaerobic digestion;

(11) mechanical aeration system; (12) deep-well injection; (13) foam phase separation; (14) brush aeration; (15) subsurface disposal; and (16) the Bio-Disc system.

Lagooning

Lagooning in oxidation ponds is a common means of both removing and oxidizing organic matter and wastewaters. More research is needed on this method of treatment, which originally was developed as an inexpensive procedure for ridding industry of its waste problem. An area adjacent to a plant was excavated, and wastewaters either flowed or were pumped into the excavation at one end and out into a receiving stream at the other end. The depth of the lagoon depended on how much land was available, the storage period desired or required, and the condition of the receiving stream. Little attention was paid originally to the effect of depth on bacterial efficiency. In fact, reduction of dissolved organic matter was usually not anticipated or even desired, because it was presumed, and with good reason, that biological degradation of organic matter would lead to oxygen depletion and accompanying nuisances from odors. Thus, the lagoons served solely to settle sludge and equalize the flow. Modern techniques, however, have led to new theories about the stabilization of organic matter in lagoons.

We now know that stabilization or oxidation of waste in ponds is the result of several natural self-purification phenomena. The first phase is sedimentation. Settleable solids are deposited in an area around inlets to the ponds, the size of the area depending on the manner of feeding in the waste and location of the inlet. Some suspended and colloidal matter is precipitated by the action of soluble salts; decomposition of the resulting sediment by microorganisms changes the sludge into inert residues and soluble organic substances, which in turn are required by other microorganisms and algae for their metabolic processes.

Decomposition of organic material is the work of microorganisms, either aerobic (living in the presence of free oxygen) or anaerobic (living in absence of free oxygen). In a pond in which the pollution load is exceedingly high or which is deep enough to be void of oxygen near the bottom, both types of microorganism may be actively decomposing organic material at the same time. A third type of microorganism, the facultative anaerobic, is capable of growth under either aerobic or anaerobic conditions and aids in decomposing waste in the transition zone between aerobic and anaerobic conditions. It is desirable to maintain aerobic conditions, because aerobic microorganisms cause the most complete oxidation of organic matter. Anaerobic bacteria have been most effective in oxidizing dairy, textile, and other highly soluble organic wastes.

Table 8.1 gives a general scheme of the microbial degradation of the organic constituents in sewage. It also points out the difference between aerobic and anaerobic decomposition.

Algae are significant in stabilization ponds in that they complete nature's balanced plant–animal cycle. Whether seasonal or perennial, algae utilize CO_2 sulfates, nitrates, phosphates, water, and sunlight to synthesize their own organic cellular material and give off free oxygen as a waste product. This oxygen, dissolved in pond water, is available to bacteria and other microbes for their metabolic processes, which include

TABLE 8.1
Biological Degradation of Organic Constituents in Sewage

Substance Decomposing	Class of Microbial Enzymes	End-Products	
		Anaerobic Decomposition	*Aerobic Decomposition*
Proteins	Proteinase[a]	Amino acids Ammonia Hydrogen sulfide Methane Carbon dioxide Hydrogen Alcohols Organic acids Phenols Indols	Ammonia, nitrites, nitrates Hydrogen sulfide, sulfuric acid Alcohols Organic acids Carbon dioxide Water
Carbohydrates	Carbohydrase[a]	Carbon dioxide Hydrogen Alcohols Fatty acids	Alcohols Fatty acids Carbon dioxide Water
Lipids (fats)	Lipase[a]	Fatty acids Carbon dioxide Hydrogen Alcohols	Fatty acids and glycerol Alcohols Carbon dioxide Water

[a]Class of enzymes only. Dozens of enzymes may be utilized in this degradation.

respiration and degradation of organic material in the pond. Thus, we have a completed cycle in which (a) microorganisms use oxygen dissolved in the water and (b) break down organic waste materials to produce (c) waste products such as CO_2, H_2O, nitrates, sulfates, and phosphates, which (d) algae use as raw materials in photosynthesis, thereby (e) replenishing the depleted oxygen supply and keeping conditions aerobic so that the microorganisms can function at top efficiency (Figure 8.1). However, one drawback of algae should be mentioned, namely, that when they die, they impose a secondary organic loading on the pond. Another disadvantage is a seasonal one: Algae are less effective in winter.

Ice and snow cover during winter months interferes with the stabilization process in the following manner:

1. It prevents sunlight from penetrating the pond, causing a reduction in the size and number of algae present. Algae are not necessarily killed by the absence of sunlight (those known as *facultative chemo-organotrophs* can carry on metabolic processes despite darkness), but they release little or no oxygen without sunlight.
2. It prevents mixing and re-aeration by wind action.
3. It prevents re-aeration by atmosphere–water dynamic equilibrium phenomena.
4. It usually results in anaerobic conditions if it continues over an extended period.

These factors tend to result in a lowered pond or lagoon efficiency during the winter.

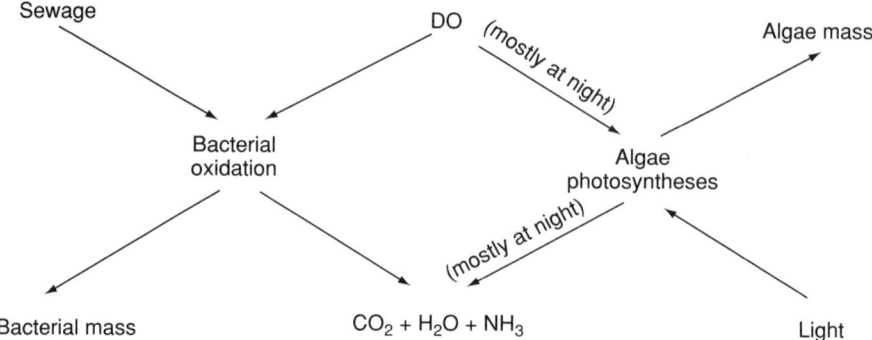

FIGURE 8.1. The role of algae in stabilization ponds (adapted from North and South Dakota Departments of Health and the U.S. Department of Health, Education and Welfare 1957).

Hermann and Gloyna (1958) (disregarding the part played by minerals) describe the reaction in high-rate ponds in which sewage is oxidizing as

$$C_{11}H_{29}O_7N + 14O_2 + H^+ \rightarrow 11CO_2 + 13H_2O + NH_{4+}.$$

The canning industry, one of the first to attempt lagooning, soon found it difficult to maintain aerobic conditions in basins; other industries experienced similar situations. As industries became aware that biological degradation occurs in lagoons, they made attempts to encourage and control the oxidation and began to refer to such lagoons as *waste-oxidation basins.*

Most modern oxidation basins have a maximum water depth of 4 feet and operate on a continuous-flow basis. Engineers try to maintain in the basin near-neutral pH, adequate oxygen concentration, and sufficient nutrient minerals for biological oxidation. Chemical neutralizers are used to alter pH values, oxygen concentrations are maintained by reducing detention times and using shallow basins, and mineral-salts nutrients may be added as needed, to accelerate biological activity. Biochemical oxygen demand (BOD) removals range from as low as 10% to as high as 60–90%.

In an interesting full-scale study, Nemerow and Bryson (1960) treated an airbase oxidation pond, at 43° north latitude with ice cover during the winter, with an elevated loading of 130 pounds of BOD in the wastewater per acre of pond area. The BOD reductions at these relatively high loadings ranged from 87.7% in August to 53% in January, with a yearly average of 69.3%. In another pilot-plant study, Nemerow (1960) achieved BOD removals in excess of 80%, using close-baffled 4-ft-deep or unbaffled 8-ft-deep basins during the critical summer period in central New York State, at elevated loadings of 312–467 lb/acre/day. A photograph of the five parallel pilot-plant basins is shown in Figure 8.2.

Oswald (1960) believes that in such heavily loaded ponds, particularly during periods when methane fermentation is either nonexistent or limited by temperature and when algal photosynthesis is not taking place in the surface layers, buildup of organic

FIGURE 8.2. Accelerated-oxidation pilot-plant basins (adapted from Nemerow 1960). Used with permission from John Wiley and Sons.

acid occurs, with a subsequent lowering of the pH level and emission of hydrogen sulfide from the pond. The author, however, did not experience these odors, even at the high loadings described earlier. Oswald (1960) offered the explanation that if methane fermentation becomes established in the bottom deposits, high rates of BOD removal may be attained without appreciable odors. He also believed that ponds in which both photosynthesis oxygenation and methane fermentation occur (facultative ponds) must be restricted to about 50 lb of BOD per acre per day, because conditions are at times unfavorable for either process. The author does not necessarily agree with these findings. Furthermore, Oswald's later high-rate oxidation ponds for treating sewage in warmer climates have been loaded to more than 600 lb of BOD per acre per day or more, being aerated for an hour each midnight. When lagooning organic wastes—which must be prevented from reaching the groundwater as well as surface bodies of water—preventative measures must be taken. These methods include making the bottom and sides of lagoons impermeable to the lagoon contents, whether or not they are partially decomposed. Banerjee (1997) gives a list of liner materials commonly used (Table 8.2).

The reader is referred to the discussion in Chapter 7 of the necessity of preventing algae growth in bodies of water that are used for water supplies and recreational activities.

Activated-Sludge Treatment

The activated-sludge process has proved quite effective in the treatment of domestic sewage as well as a few industrial wastes from large plants. In this process, biologically active growths are created, which are able to adsorb organic matter from the wastes and convert it by oxidation-enzyme systems to simple end-products like CO_2, H_2O, NO_3,

TABLE 8.2
Materials Potentially Useful for Liners of Ponds Containing Hazardous Wastes

---

Compacted native fine-grain soils
Bentonite and other clay sealments
Asphaltic compositions
    Asphalt concrete
    Hydraulic asphalt concrete
    Preformed asphalt panels
    Catalytically blown asphalt sprayed on soil
    Emulsified asphalt sprayed on soil or on fabric matting
    Soil asphalt
    Asphaltic seals
Portland cement compositions
    Concrete, with seal coats
    Soil cement, with and without seal coats
Soil sealants
    Chemical
    Lime
    Penetrating polymeric emulsions and latexes
Sprayable liquid rubbers
    Polyurethanes
    Polymeric latexes
Synthetic polymeric membranes—reinforced and unreinforced
    Butyl rubber
    Ethylene propylene rubber (EPDM)
    Chlorosulfonated polyethylene (Hypalon)
Chlorinated polyethylene (CPE)—elasticized polyolefin (3110)
Polybutylene (PB)
    Polychloroprene (Neoprene)
    Polyester elastomers
    Polyethylene (PE)
    Polyvinyl chloride (PVC)

Adapted from Banerjee (1997).

and SO_4. Biological slimes develop naturally in aerated organic wastes, which contain a considerable portion of matter in the colloidal and suspended state, but for the efficient removal of organic dissolved solids, there must be high floc concentrations to provide ample contact surface for accelerated biological activities. The flocs (zoogleal masses) are living masses of organisms, food, and slime material, and are highly active centers of biological life; hence, the term "activated sludge." They require food, oxygen, and living organisms in a delicately controlled environment.

Various degrees of efficiency are obtained by controlling the contact period and/or the concentration of active floc. The contact period can be regulated by careful design of the hydraulic systems of aeration basins, the average time of aeration being 6 hours for domestic sewage and 6–24 hours for various industrial wastes. The desired concentration of active floc is maintained by recirculating a specific volume of secondary settled sludge, normally about 20%. Higher sludge quantities lead to greater BOD removal and create a need for more air and food (organic matter) for proper balance. Also, "old" heavy sludge tends to become mineralized and devoid of oxygen, which results in a less active floc. The reverse is true of a "young" light sludge floc. The "age" of the growths, therefore, becomes an important consideration.

Busch and Kalinske (1956) summarize the situation by saying that for optimal activity, the kinetics of activated sludge require the following: a young flocculent sludge in the logarithmic stage of growth; maintenance of the logarithmic growth state by controlled sludge wastage; continuous loading of the organisms; and elimination of anaerobic conditions at any point in the oxidative treatment.

Hazeltine (1956) has said of the present status of domestic-sewage activated-sludge treatment that BOD removals are usually more than 90% when the loadings are less than 0.3 lb of BOD per pound of suspended solids in the water under aeration. Efficiencies are difficult to predict when these loadings are increased to 0.5 lb/lb. Normally, the BOD loading is related to the aeration-tank capacity; about 30–35 pounds of BOD per 1,000 ft^3 can be treated in plants with about 2,000 ppm of suspended solids under aeration.

Sawyer (1960) lists the limitations of the domestic-sewage activated-sludge process as follows: BOD loadings are limited to about 35 lb/1,000 ft^3 of tank capacity, thus requiring relatively long detention time and resulting high capital investment; there is a high initial oxygen demand by the mixed liquors; there is a tendency to produce bulking sludge; the process cannot produce an intermediate quality of effluent; high sludge/recirculation ratios are required for high BOD wastes; there are high solids loadings on final clarifiers; and large air requirements accompany the process.

The Kraus process (1945) attempts to overcome some of the sludge-bulking problems of conventional activated-sludge plants by controlling the "sludge volume index" (a measure of the volume occupied by one gram of suspended solid). The process is similar to that of conventional activated-sludge treatment, employing separate re-aeration for sludge, except that some digester sludge, digester supernatant, and activated sludge are aerated together for as much as 24 hours, in what he terms a *nitrifying aeration tank*. BOD loadings as high as 170 lb/1,000 ft^3/day have been used, with removals near 90% (Kraus 1955).

Von der Emde (1960) notes that ciliated and flagellated protozoa, as well as bacteria, are normally prevalent in activated sludge. When the BOD loading is high or very low, flagellates replace the ciliates, regardless of the level of oxygen present. Where there are short aeration periods, or when only traces of oxygen are maintained, bacteria are observed only in the sludge.

Many characteristics of industrial wastes (e.g., toxic metals, lack of nutrients required for biological oxidation, organic nondegradable matter, high temperature, and high or low pH values) give rise to problems requiring careful analysis. When the suitability of this process for a particular industrial waste is in question, laboratory and/or filed pilot plant will yield the results necessary for decision making.

Heukelekian (1949a) believed that bulking of activated sludge should result when conditions are not so unfavorable that they destroy the purification mechanism and yet sufficiently unfavorable that they bring a shift in the delicate biological balance. One of these unfavorable conditions should be an inadequate oxygen supply. If the oxygen supply in relation to the demand becomes inadequate, *Sphaeroltilus* and other filamentous organisms attain the ascendancy and the sludge becomes bulking. Biochemical activities of these organisms would bring about the purification in a way similar to the desirable sludge organisms except with lower and more efficient oxygen utilization at the lower tensions. In other words, when the sludge is diffuse and filamentous, it exposes more surface, which might enable the sludge to obtain the limited amount of oxygen present in the medium immediately surrounding it. This type of sludge, however, usually produces a sparkling effluent. Dairy wastes exhibit this problem, which has been overcome by the addition of ammonium chloride to provide a more favorable carbon-to-nitrogen ratio.

Modified Aeration

Modified, tapered, and step aeration are variations of the activated-sludge treatment. The objective is to supply the maximum of air to the sludge when it is in the optimal condition (sludge age) to oxidize adsorbed organic matter. The location of the aerator and the quantity of air supplied vary, depending on sludge solids and organic matter to be oxidized. Lower volumes of air and shorter detention times are claimed for these processes, while the mechanisms and theories of operation are similar to those of activated sludge.

Step aeration attempts to eliminate the problems encountered with plain aeration by providing 2 to 3 hours of aeration only. Highly activated and concentrated sludge floc is returned to the aeration tank at the proper location (usually the inlet); this reduces bacterial lag, accelerates logarithmic bacterial growth, and provides abundant surfaces for adsorption of new cells. The chief advantage to this process is the flexibility it offers the operator. Figure 8.3 shows that one can obtain almost any desired ratio of primary effluent to sludge seed returned.

Dispersed-Growth Aeration

Dispersed-growth aeration is a process for oxidizing dissolved organic matter in the absence of flocculent growths (Heukelekian 1949a). The bacteria (seed) for oxidizing are present in the supernatant liquor after wastes have been aerated and settled. A portion of this supernatant liquor is retained for seeding incoming wastes, whereas the settled

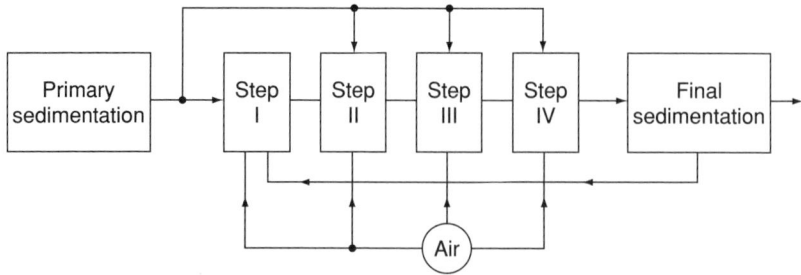

FIGURE 8.3. Schematic diagram of step-aeration treatment. Step I, high sludge seed (4,000 ppm); step II, 2,000 ppm; step III, 1,000 ppm; step IV, 800 ppm.

sludge from the secondary settling tank is digested or treated by other sludge-treatment methods. This process has been successfully used to treat many types of dissolved organic wastes (Heukelekian 1949c; Nemerow 1954, 1955, 1956, 1957). Its advantage is that it eliminates certain problems associated with sludge seeding. With many industrial wastes, it is difficult to build up any significant sludge concentration; in such cases, dispersed growth aeration (which is not dependent on sludge) finds ready acceptance. Dispersed-growth aeration does require more air to achieve the same BOD reduction as the activated-sludge process. However, when one considers that the initial BOD in dispersed-growth aeration is usually quite high, the amount of air required per pound of BOD removed is about the same as that used in the activated-sludge process, even though aeration periods to reach the same BOD reduction are normally quite lengthy (24 hours as compared to 6). Treatment by dispersed-growth aeration involves complete removal by oxidation, as in activated-sludge treatment.

Heukelekian (1949a) originally conceived this idea of seeding concentrated soluble organic wastes with dispersed, instead of flocculent, growths, when he discovered that bacteria in culture mediums normally grow in the dispersed state or in small groups and that seeding is essential for high-rate biological activity. If a waste contains only soluble material, no flocculent growth should form. In his early work on penicillin and streptomycin wastes (Heukelekian 1949b,c), he made the following claims for the dispersed-growth aeration process:

1. It is better adapted than activated-sludge methods for the treatment of concentrated soluble organic wastes because (a) activated sludge has a tendency to bulk with concentrated organic wastes, and (b) it is difficult to develop an activated sludge from a soluble waste.
2. Little sludge is formed with dispersed growths when soluble substrates are decomposed.
3. The percentage of BOD reduction decreases as the strengths of penicillin and streptomycin wastes are increased, but 80% reduction may be expected with wastes up to 3,000 ppm of BOD. Greater BOD reductions are possible when the BOD is less than 1,000 ppm and the waste is aerated for 24 hours.
4. The effluent has a higher turbidity than the raw water, and color is not removed.
5. The process may be used as a pretreatment unit for conventional biological treatment processes.
6. The seed material can readily be developed and adapted from soil or sewage within a few days.
7. Optimum results are obtained with air rates of 2–3 ft^3/gallon/hr; stronger wastes require higher air rates.
8. The initial pH of the waste does not seem a critical factor, because the pH increases during aeration. The BOD of raw waste with a pH of 6.4 is reduced as much as the BOD of the same waste adjusted to 7.2.

Nemerow and Rudolfs (1952) also found this method of treatment suitable for rag and jute paper-mill wastes. In a basic study of the oxidation of glucose by dispersed-growth aeration, Nemerow and Ray (1956) found 5 million bacteria per milliliter when

114 Twentieth Century

using nutrient broth as a medium. The two major types of bacteria found during the 24-hour aeration period were

1. Dispersed, short, thick, round-ended rods; approximate size, 2–2.5 microns × 1 micron. Some of these organisms appeared as fingerlike capsules. As the aeration period progressed, there was an apparent increase in the number of slime-enmeshed bacteria (Figure 8.4).
2. *Sphaeroltilus*-like organisms, often as unsheathed forms (Figure 8.5); these were more abundant after 6 hours of aeration and reached an apparent maximum after 24 hours.

In studying the suitability of dispersed-growth aeration for industrial wastes containing both proteins and carbohydrates, Struzeski and Nemerow (1957) found such wastes amenable to oxidation by this process. Biological oxidation was enhanced by an increase in temperature, as shown in Figure 8.6, and initial pH values up to 9.5 did not hamper it. It was also found that when soluble protein-carbohydrate wastes are to be treated by dispersed-growth aeration, units must be designed to allow ample detention time, because air rates above the critical level (1,050 ft^3 of air per pound of BOD per day) do not increase the reduction of BOD.

Contact Stabilization

Biosorption is the commercial name of one equipment manufacturer's high-rate biological oxidation process, used mainly for domestic sewage. It was originally

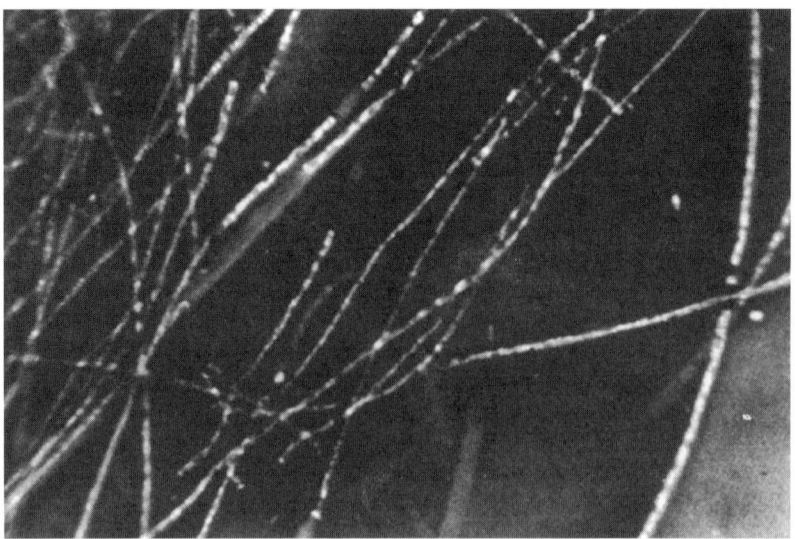

FIGURE 8.4. *Sphaerotilus*-like organism, sheathed and unsheathed (×620). Used with permission from John Wiley and Sons.

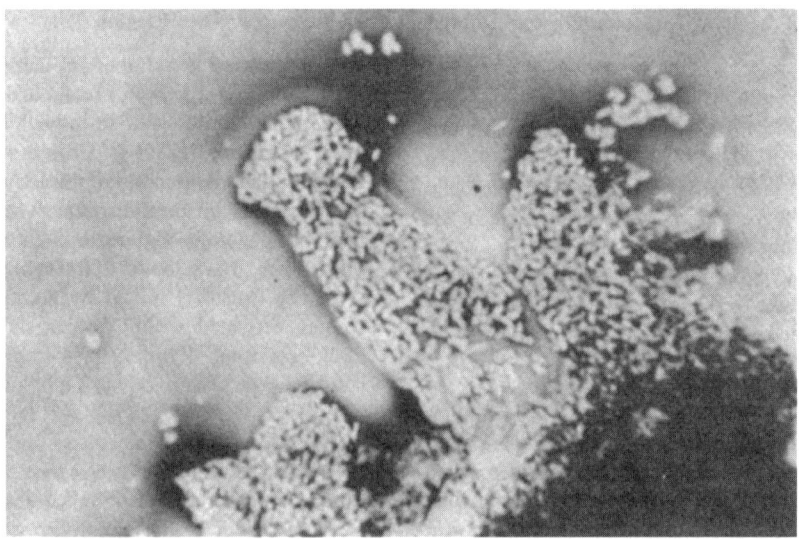

FIGURE 8.5. Round-ended rods in a capsule of slime (×620). Used with permission from John Wiley and Sons.

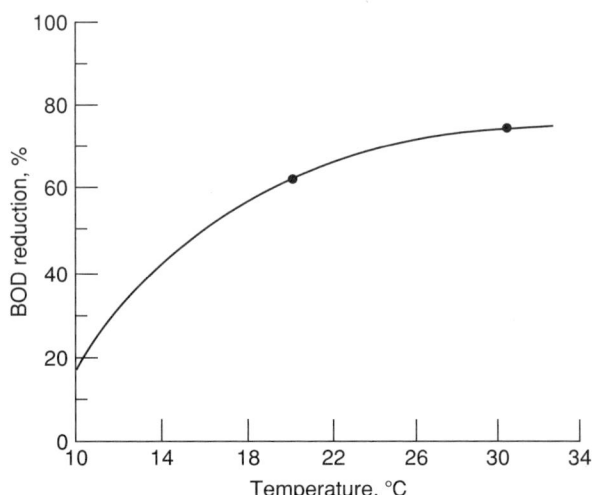

FIGURE 8.6. Effect of temperature on average BOD reduction of a synthetic protein–glucose waste, using a dispersed-growth aeration system, after 24 hours of aeration and no settling.

developed at Austin, Texas by Ullrich and Smith (1951). It is essentially a modification of the activated-sludge process and is similar in some respects to the step-aeration process but generally requires less air and plant space. In the contact-stabilization process, raw waste is mixed by aeration with previously formed activated sludge from a stabilization-oxidation tank, or aerobic digester, for a short

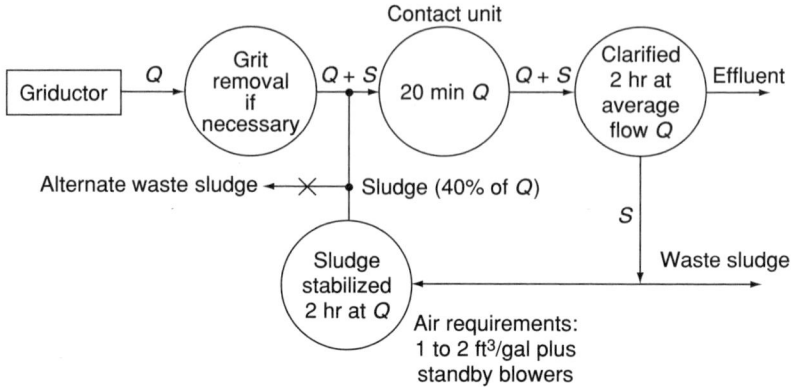

FIGURE 8.7. Schematic arrangement of contact-stabilization process.

period (15–20 minutes). This activated-sludge–raw-waste mixture is then clarified by settling for about 2 hours, after which the settled sludge (consisting of activated-sludge floc with adsorbed impurities from the raw waste) goes through intense biological oxidation in the stabilization-oxidation basin for an aeration period of 1–2 hours. It then returns to the mixing tank and is again mixed with raw waste so that it can absorb and adsorb added organic matter, and so on, in a continuous process. Excess, or waste sludge, can be taken from the system after either the clarifying or the stabilizing step, for anaerobic digestion or for dewatering on vacuum filters (Figure 8.7).

Ullrich and Smith (1951, 1957) claimed that this process requires less aeration-tank capacity than other processes, because the real aeration or reactivation takes place in the settled and concentrated sludge, not in the mixed liquor. Because the sewage and returned sludge is given only a brief mix, a small mixing compartment is needed. Pertinent pilot-plant and full-scale operating results are given in Table 8.3.

High-Rate Aerobic Treatment

High-rate aerobic treatment (total oxidation) has used since the 1950s to oxidize organic wastes (Tyre et al. 1991). This process consists of comminution of the waste, long-period aeration (1–3 days), final settling of the sludge, and return of the settled sludge to the aeration tank. There is no need for primary settling or sludge digestion, but the aeration system must be large enough to provide the required aeration period. The total-oxidation process is particularly useful in small installations, because it does not require a great deal of supervision. Little difficulty occurs with bulking on the sludge, even though the settling period is relatively short at times. In fact, because the solids resulting from this process are mostly of low volatility and therefore high in ash, the settling rate is quite fast. Return of the sludge is continous and very rapid in comparison with normal activated-sludge practice. By returning sludge at a high rate

(100–300% of flow), the system is kept completely aerobic at all times. The concentration of solids in the mixed liquor after a long period reaches a high level, and a portion of the sludge can then be wasted to reduce the concentration to 3,000–5,000 ppm. Lesperance (1965) suggests that a waste sludge can be expected equal to 0.15 lb per pound of BOD removed. The small volume of wasted sludge is then stored and further concentrated until removed by a tank car or other means to an area away from the plant.

Although it produces little waste sludge, the high-rate aerobic treatment has the disadvantages of requiring about three times as much air as conventional activated-sludge plants and of releasing some floc in the effluent. On the other hand, it needs very little operational maintenance and is well suited for shock loadings from industrial operations.

Another version of this treatment is referred to as a "completely mixed system" (McKinney et al. 1958). It operates on the assumption that if microorganisms are kept in a constant state of growth, they operate at maximum efficiency and are adapted to the particular character and concentration of the waste. This constant-growth state can be maintained only if: (1) the microorganisms and raw wastes are thoroughly and continuously mixed; (2) the organic concentration is held constant; and (3) the effluent is separated form the microorganisms at a constant rate that is equal to the waste-feed rate. Figure 8.8 depicts a typical complete-mixing activated-sludge system (McKinney et al. 1958). A loading of 60 lb of oxidizable organics per 1,000 ft^3 of aeration tank is possible with this type of treatment.

Trickling Filtration

"Trickling filtration" is a process by which biological units are coated with slime growths (zoogleal forms) from the bacteria in the wastes. These growths adsorb and oxidize dissolved and colloidal organic matter from the wastes applied to them. When the rate of application is excessive (10–30 million gallons/acre/day [mgad]) and

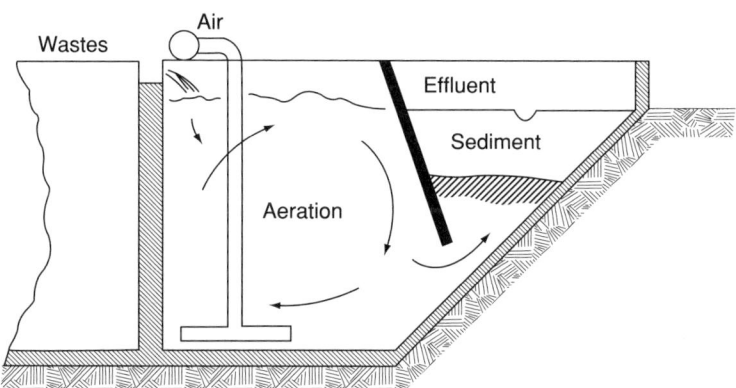

FIGURE 8.8. Complete-mixing activated-sludge system (adapted from McKinney et al. 1958).

TABLE 8.3
Summary of Disposal Systems

Company	Type of waste	Injection rate, gpm	Injection pressure, psi	Subsurface depth of wells, feet	Formation age, type, and name	Total cost of system ($)	Date started	Problems	Solutions and remarks
A	Brine; chlorinated hydrocarbons	200	500	12,045	Precambrian fractured gneiss (unnamed)	1,419,000	March 1962	Microorganisms in waste	
B	Clear 4% solution Na_2SO_4	300	45	295	Sandstone	—[a]	June 1951	None	
C	Masic waste, pH±10	70	1,000	6,160	Cambrian sandstone	250,000	Nov. 1960	Inadequate filtration	Larger filter planned
D	Magnesium; calcium hydroxides	200	Vacuum	400	Permian salt bed (Hutchinson)	—[a]	—[a]	None	
	Manufacturing waste, pH may change from 1 to 9 in hr	400	Vacuum	4,150					
E	Lachrymator waste from acrolein and glycerine units	700	150–170	1,960	Ordovician vugular limestone (Arbuckle)	500,000	Dec. 1957	Corrosion and water hammer	Heavier tubing planned
F	Aqueous solution—phenols, mercaptans, and sulfides	215	30–90	1,795	Pleistocene	135,000	1956	Sand incursion increased injection pressure	Back-washing every 4 months
					Pleistocene	30,000	Sept. 1959	Sand incursion	Periodic back-washing
					Pleistocene	—[a]	March 1960	Sand incursion	Periodic back-washing
G	Phenols; mercaptans; sulfides; brine	100	40–100	1,980	Devonian vugular limestone (Dundee)	—[a]	1950	None	

ID	Waste				Formation		Date	Problems	Treatment
					Devonian vugular limestone (Dundee)	—[a]	None		
H	Phenols; chlorinated hydrocarbons	200	450	4,000	Silurian sandstone (Sylvania)	25,000	Aug. 1956	High wellhead pressure	Acidizing and fracturing
	Brine	200	150	4,000					
	Phenols; mercaptans; sulfides	50	—[a]	4,000					
I	Coke oven phenols; quench water	50	300	563	Devonian sandy limestone (Dundee); Traverse, Dundee and Monroe	400,000	1954	None	
J	Organic wastes	60	500	1,472	Permian sandstone	562,000	Jan. 1960	Microorganisms decreased injectivity	Formaldehyde
K	Sulfuric acid waste	400	Vacuum	1,830	Ordovician vugular limestone (Arbuckle)	300,000	Feb. 1960	Mechanical failure of surface equipment	
L	Detergents; solvents; salts	254	280	1,807	Unconsolidated sand (Glorieta)	—[a]	April 1962	None	
M	38% HCl solution	14	10–20	1,110	Unconsolidated sand (Glorieta)	—[a]	1959	None	
N	Stripping steam condensate; cooling tower blowdown	50	50–70	1,110	Unconsolidated sand (Glorieta)	—[a]	1958	None	
	Aqueous petroleum refinery effluent	400		1,110					
O	Phenols; brine	75	400	7,650	Eocene sand and clay (Frio)	—[a]	1958	High injection pressure	Periodic acidizing

[a] Information not available.

continuous, the humus collected on the filter-bed surfaces is sloughed off continuously. Crushed stone, such as traprock, granite, and limestone, usually forms the surface material in the filter, although other materials such as plastic rings have proved very effective. The main advantages of plastic media include their light weight, chemical resistance, and high specific surface (i.e., square feet per cubic feet of bed volume). Because smaller stones provide more surface per unit of volume, the contact material must be small to support a large surface of active film, but not so small that its pores become filled by the growths or clogged by accumulated suspended matter or sloughed film. Crushed stone, 3–5 inches in diameter, is used, with the smallest stone at the top. The integral parts of a trickling-filter system are the distribution nozzles, contact surface, and underdrain units. This process may be summarized as follows:

1. An active surface film grows on the stone or contact surface.
2. Concentration of colloidal material and gelatinous matter occurs.
3. These adsorbed substances are attacked by bacteria and enzymes and reduced to simpler compounds, so that NH_3 is liberated and oxidized by chemical and bacterial means, giving a gradual reduction of NH_3 and an increase of NO_2 and NO_3 (Figure 8.9).
4. A flocculent, humus-like residue or sludge, containing many protozoa and fungi accumulates on the surface. When it gets too heavy, it will slough off and resettle (a continuous process with biofilters). Part of the oxygen is supplied by spraying waste, blowing air into the filter, or allowing waste to drip into the filter. Another portion is supplied by convection due to the temperature difference between the incoming waste and the bed. The larger the surface, the greater the number of bacterial organisms that come into contact with the liquid to be purified; the greater the number of organisms, the higher the purification of the liquid. The smaller the pieces of rock in the surface media, the greater the purification; too-small particles, however, promote clogging. To summarize, trickling filters act as both strainers and oxidizers.

Zobell (1937) first pointed out the importance of providing solid contact surfaces to further the physiological activities of bacteria growing in dilute nutrient solutions, as

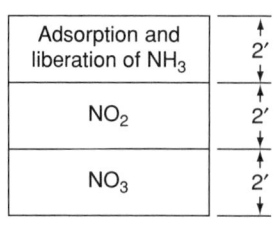
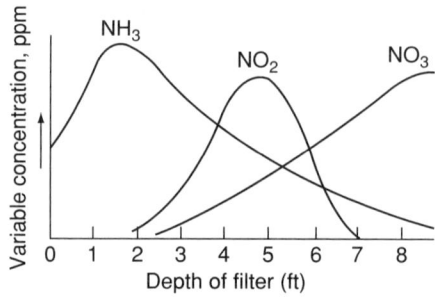

FIGURE 8.9. Changes in nitrogen occurring in filter.

most industrial wastes are. The following phenomena, according to Zobell, cause this increase in biological oxidation:

1. Solid surfaces make possible the concentration of nutrients and enzymes by adsorption to the surface.
2. The interstices between bacterial cells and surfaces act as concentration points; they retard the diffusion of exoenzymes and metabolites away from the cell, thereby favoring both digestion and adsorption of foodstuffs.
3. The interstices between surfaces and cells produce optimum conditions for oxidation-reduction and other physicochemical reactions.
4. Surfaces function as attachment points for microorganisms, which are obligatory periphytes.

A typical standard-rate, stone-bed trickling filter provides about 100 ft² of surface material per square foot of ground on which the filter is constructed. Velz (1948) proposed the performance equation for trickling filters as

$$L_d/L = 10^{-kd},$$

where L_d is the removable fraction of BOD remaining at depth D, L is total removal, k is the logarithmic extraction rate, and D is the depth of the bed. The reader will note the similarity between this equation and the monomolecular rate of decomposition of organic matter in streams:

$$L1/L = 10^{-kt}.$$

The student should realize that the contact time in a filter is relatively short, compared with an activated-sludge process. However, the organic matter (bacterial food) resides in the bed longer than computed from the detention time. Howland (1957) has contributed to our knowledge of contact time in filters. Assuming that a sheet of water is flowing steadily down an inclined plane under laminar flow conditions, he expresses the contact time as

$$T = (3v/gs)^{1/3} (1/q^{2/3}),$$

where T is the time of flow down the inclined plane, l is the length of the plane, s is the sine of the angle that the plane makes with the horizontal, g is the acceleration due to gravity, v is the kinematic viscosity of water (μ/p), and q is the rate of flow per unit width of the plane.

Howland (1957) indicates that the amount of oxidizable organic matter removed in a filter depends directly on the length of time of the flow. He recommends a deep filter containing the smallest practical media to achieve an optimum contact time and maximum efficiency. Although some researchers recommend shallower filters and larger stone to reduce both initial and operating costs, the tendency today appears to be to follow Howland's recommendations because engineers want an increasing degree of removal of BOD. Still, because of clogging and head-loss difficulties, there is a limit to how deep a bed and how small a stone size one can utilize.

Ingram (1956) suggests the following drawbacks of trickling filters: They occupy too much space; they exhibit seasonal variation in efficiency; clogging and pooling present problems; and there are limitations on the strength of sewage applied. He proposes a trickling-filtration process called "controlled filtration," which utilizes deep filters (18–24 ft). He was able to achieve greater than 70% BOD removal (the removal expected in high-rate filters loaded at a normal rate of 20 mgad and 1,300 lb of BOD/acre-ft/day) with a minimum hydraulic loading of twice and an organic loading of 1-1/2 to 10-1/2 times these normal standards. His experimental filter is shown in Figure 8.10.

Behn (1960) points out that deviations from the usual reaction rates sometimes occur because of the waste temperature and degree of filter saturation. Rankin (1953)

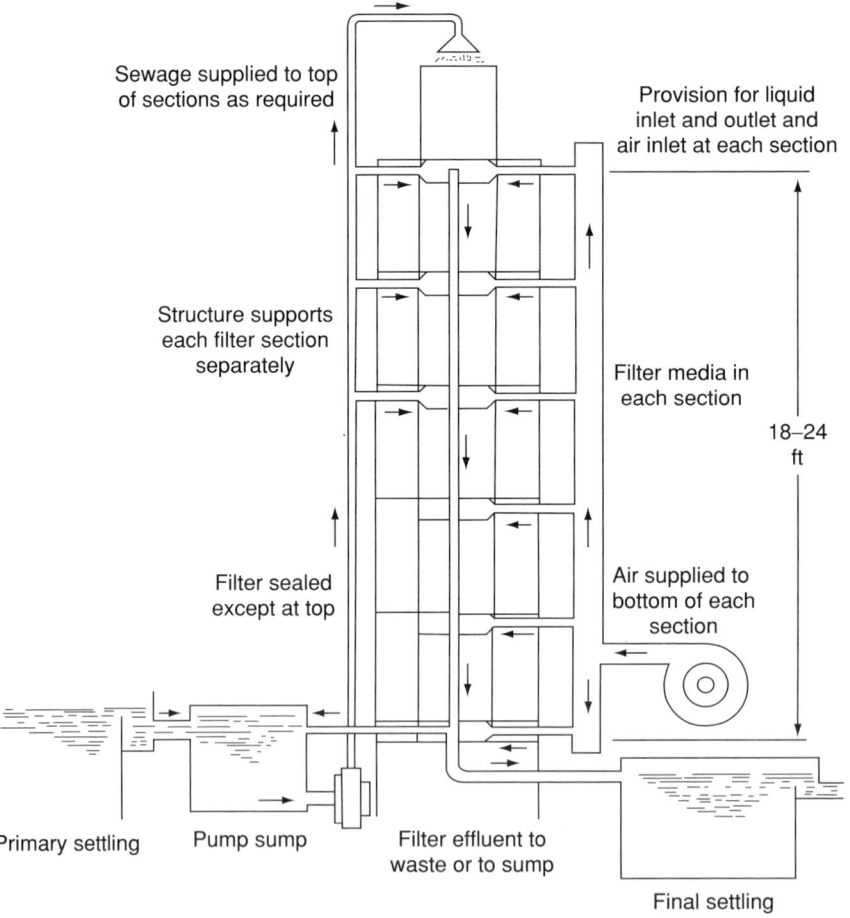

FIGURE 8.10. Diagram of experimental controlled-filtration system (adapted from Ingram 1956).

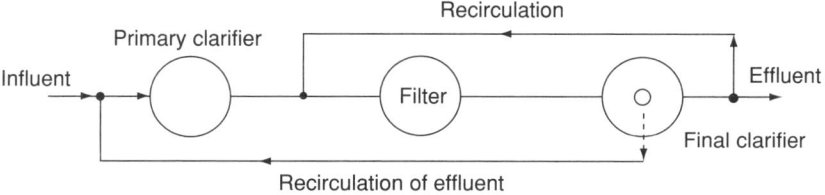

FIGURE 8.11. Single-stage trickling filter (with recirculation).

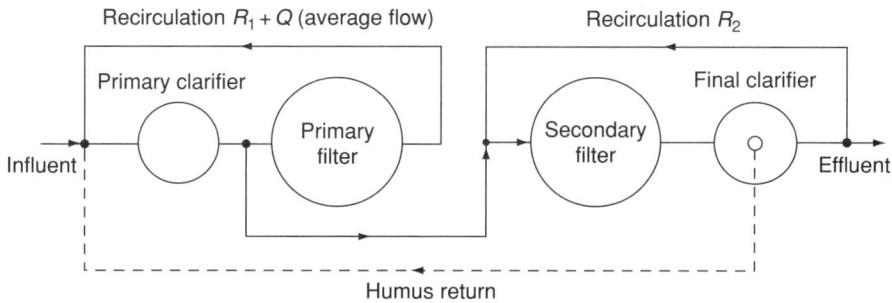

FIGURE 8.12. Two-stage series-parallel biofiltration process.

was concerned with recirculation of filter effluent and concludes, from a study of a number of treatment plants, that performance appears to depend primarily on the ratio of recirculation to raw wastewater flow, rather than on dosing rate, loading of the filter, or depth of filter (within the ranges studied). In smaller plants, with only one filter, single-stage filters appear to be most feasible, while for larger plants, where multiple filters are necessary or where stronger wastes are being treated, a two-stage series-parallel arrangement of filters yields a better effluent than single-stage filters with the same tank and filter capacity and the same volume of recirculated liquor. Diagrams of each of these systems are shown in Figures 8.11 and 8.12.

Spray Irrigation

Spray irrigation is an adaptation of the familiar method of watering agricultural crops by portable sprinkling-irrigation systems; wastes are pumped through portable pipes to self-actuated sprinkler heads. Light weight aluminum or galvanized piping, equipped with quick-assembly pipe joints, can be easily moved to areas to be irrigated and quickly assembled. Wastes are applied as a rain to the surface of the soil, with the objective of applying the maximum amount that can be absorbed without surface runoff or damage to the cover crops. A spray-irrigation system is composed of the following units: (1) the land on which to spray; (2) a vegetative cover crop to aid absorption and prevent erosion; (3) a mechanically operated screening unit; (4) a surge tank or pit;

(5) auxiliary stationary screens; (6) a pump that develops the required sprinkler-nozzle pressure; (7) a main line; (8) lateral lines; and (9) self-actuated revolving sprinklers operating under 35–100 psi nozzle pressure.

With good cover crops (dense low-growing grasses) and fairly level areas, waste to a depth of 3–4 inches can be applied at a rate of 0.4–0.6 in./hr. The process is generally limited to spring, summer, and autumn. In a study of citrus wastes, Anderson et al. (1966) found that aerobic conditions are maintained without odors to a depth of at least 3 ft.

Your author found this method of treatment quite effective for removing dissolved organic matter from vegetable cannery waste in western New York State.

Wet Combustion

"Wet combustion" is the process ("Wet Combustion of Wastes" 1955) of pumping organics-laden wastewater and air into a reactor vessel at elevated pressure (1,200 psi) (Figure 8.13). The organic fractions undergo rapid oxidation, even though they are dissolved or suspended in the waste. This rapid oxidation gives off heat to the water by direct convection, and the water flashes into a stream. Inorganic chemicals, which are present in many industrial wastes, can be recovered from the steam in a separate chamber. Heat from an external source is applied just to start the process; thereafter, it requires only 12–20% of its own heat to maintain itself. The remaining 80–88% can be utilized as process steam or to drive turbines for electrical or mechanical power. This process has a good potential where steam is essential and inexpensive enough to justify the cost of the equipment and where the inorganic chemicals in the waste are worth recovering and

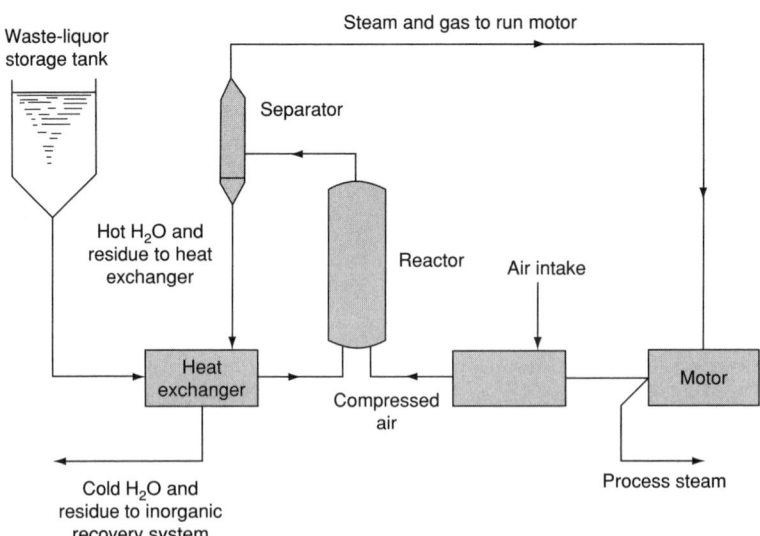

FIGURE 8.13. Schematic arrangment of wet-combustion–process units.

reusing. The wet combustion process can maintain itself only when the waste has a high percentage of organic material (usually about 5% solids and 70% organic).

Anaerobic Digestion

"Anaerobic digestion" is a process for oxidizing organic matter in closed vessels in the absence of air. The process has been highly successful in conditioning sewage sludge for final disposal. (Because digestion is primarily used for the treatment of sludge, rather than liquid wastes, the theory of its operation is described in more detail in Chapter 9.) It is also effective in reducing the BOD of soluble organic liquid wastes, such as yeast, cotton-kiering, slaughterhouse, dairy, and white-water (paper-mill) wastes. Generally, anaerobic processes are less effective than aerobic processes, mainly because of the small amount of energy that results when anaerobic bacteria oxidize organic matter. Anaerobic processes are, therefore, slow and require low daily loadings and/or long detention periods. However, because little or no power need be added, operating costs are very low. Where liquid waste volumes are small and contain no toxic matter and there are high percentages of readily oxidized dissolved organic matter, this process has definite advantages over aerobic systems. The pH in digesters must be controlled to near the neutral point.

Buswell and Hatfield (1939) proposed the following general equation for conversion of organic matter in industrial wastes to carbon dioxide and methane:

$$C_n H_a O_b + (n - a/4 - b/2)\ H_2O$$
$$= (n/2 - a/8 + b/4)\ CO_2 + (n/2 + a/8 - b/4)\ CH_4$$

In the United States, anaerobic treatment plants have been built to treat yeast, butanol-acetone, brewery, chewing gum, and meatpacking wastes. In a review of British practices, Pettet et al. (1959) found that slaughterhouse waste appeared to respond extremely well to anaerobic digestion, although up to 1959, there were no full-scale anaerobic-digestion plants in Great Britain. In the United States, BOD reductions of 60–92% have been attained with all these wastes, at loadings of 0.003–0.191 lb of BOD/ft^3 of digester per day. Concentrations of organic matter ranged from 1,565 to 17,000 ppm BOD.

Mechanical Aeration System

"Cavitation" is a typical process for mechanical aeration of wastes. The complete cavitator assembly consists of a vertical-draft tube with opening for connection to the influent pipe and a rotor assembly of the multiblade type, supported by an adjustable ball thrust bearing mounted at the motor level. The rotor is mounted on a stainless-steel shaft and the entire unit, including the draft tube, is supported by a structural steel bridge. A cross section is shown in Figure 8.14 (Schultze and Foth 1955). As soon as the rotor exceeds a certain critical speed, air is drawn in from the atmosphere through

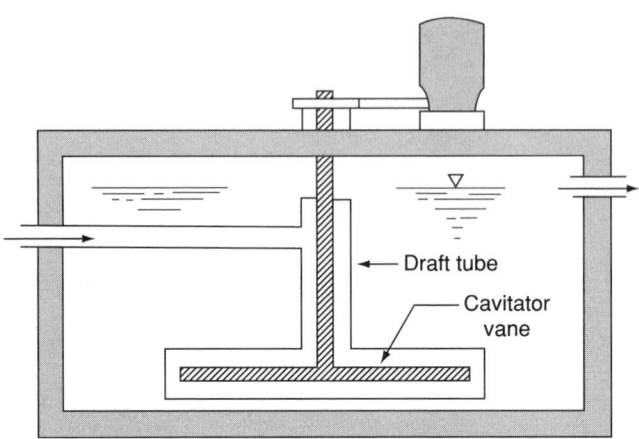

FIGURE 8.14. Typical cavitator system.

the vertical hollow tube and dispersed into the waste. The rotor creates a zone of cavitation in its turbulent trail and air moves in to fill the areas of rarefield under pressure. The amount of air that is being entrained depends on the size and shape of the rotor, the rotations per minute (rpm), and the water depth. The manufacturers claim that their system uses at least 25% of the available oxygen in the air, in contrast with conventional aeration equipment, which uses only 5%. At least one waste-treatment plant (dealing with canning wastes and sewage) attained more than 90% BOD removal with an air supply of 110 ft^3/lb of BOD per day. Operational costs (Schultze and Foth 1955) were $12.80 per day for an equivalent population of 12,000 persons. A modification to this system employs mechanical mixing by a rotor submerged (but near the surface of) the wastewater. Power costs are thus reduced, with no apparent loss of aeration or mixing efficiency. This system promises to be the most economical one for secondary treatment of wastes with a highly dissolved organic content.

Well Injection

Disposal of wastes containing dissolved organic matter by injecting them into deep wells has been successful in areas of low or nonexistent stream flow, especially when wastes are malodorous or toxic and contain little or no suspended matter. Deep-well injection has been used successfully to dispose organic solutions from chemical, pharmaceutical, petrochemical, paper, and refinery wastes; in addition, many inorganic solutions may be disposed of in this manner. To be effective, the wastes must be placed in a geological formation, which prevents the migration of the wastes to the surface or to groundwater supplies. The rock types most frequently used are the more porous ones such as limestones, sandstones, and dolomies because the porosity may help develop a filter cake that plugs the well. Other factors, in addition to geology, to be considered are depth and diameter of well, injection pressures, and the volume and characteristics of the wastes. At the end of 1966, there were 78 industrial disposal wells in the United States, most of which were used for chemical and refinery wastes (86%); most were less than 4,000 ft deep (74%), disposed of less than 400 gpm/well (87%), and operated at less than 300 psi (57%). Costs of injection disposal installations vary from $30,000 for a shallow (1,800-ft) well not requiring pretreatment to more than $1,400,000 for a very deep (12,000-ft) well with intricate pretreatment. Actual costs vary depending on depth, surface equipment, pretreatment, diameter of well, injection pressure, variability of composition of wastewater, and availability of drilling equipment.

Donaldson (1964) reported on a wide variety of industrial wastes being injected into formations ranging in age from Precambrian to modern day. In the United States up to 1964, more than 30 wells, ranging in depth from 300 to 12,000 ft, were being used for waste disposal into subsurface formations, which include unconsolidated sand, sandstone, regular limestone, and fractured gneiss. Although subsurface injection offers an economical method of final disposal where receiving water is inadequate to carry the wastewater away safely, circumstances can limit its effectiveness (e.g., the area lacks suitable underground formations for waste injection, the initial capital expense is excessive, or the pretreatment required may be too extensive and expensive).

The industrial-waste engineer must work closely with a geologist familiar with the subsurface formations in the area to select the proper waste-disposal zone. A well is drilled and core samples are analyzed for specific characteristics such as permeability and reactivity with the waste. Tests are performed to determine the injection pressure required at various wastewater flows.

Schematic drawings of typical complete subsurface waste-disposal systems are shown in Figures 8.15 and 8.16. Although cement tanks up to 50,000-gallon capacity are commonly employed within the basement of a factory, large, shallow open ponds may be used where land is available and where some settling and oxidation are required as pretreatment. The oil separator is required for petroleum refinery wastes, because oil tends to plug disposal formation and the oil can be recovered and reused. The usual

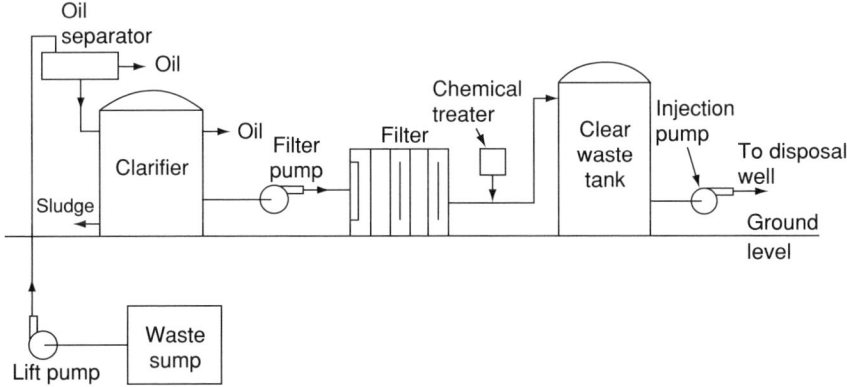

FIGURE 8.15. Typical surface equipment for deep-well waste injection from waste sump underground (from Bureau of Mines Information Circular no. 8212).

separator consists of a tank with many internal baffles to cause the oil to separate and rise. If a clarifier is then used, heavier material such as dirt, resin flocs, and suspended grease can settle out. Mechanical equipment such as sludge rakes and surface skimmers can also be used with this equipment. Because not all solids are completely removed by the treatment thus far described, filters are then used to protect sand or sandstone formations from plugging. The screens are usually metal and coated with diatomaceous earth, but in some situations, sand filters are preferred. If wastes contain slime that will form bacteria or fungi, a suitable bactericide (such as quaternary amines, formaldehyde, chlorinated hydrocarbons, chlorine, or copper sulfate) is used to control their detrimental effects. The clear-water storage tank is normally equipped with a float switch designed to operate the injection pump at certain liquid levels. The size and type of injection pump are controlled by wellhead pressure, wastewater flow, and wastewater characteristics such as pH and corrosiveness. The multiplex piston pump is most commonly used when wellhead pressures of more than 150 psi are required, whereas single-stage centrifugal pumps are used at lower pressures.

To construct the well, first a 15-in.-diameter hole is drilled to 200 ft below the deepest freshwater aquifer and a 10 1/2-in. (O.D.) casing is set and cemented to the surface. Next, a 9-in.-diameter hole is drilled to the bottom of the potential disposal formation, a 7-in. (O.D.) casing is set at the total depth of the hole, and cement is circulated in the annulus between the injection casing and the 9-in. hole to the surface (Figure 8.16). This method has been proven to seal off water aquifers from the well and

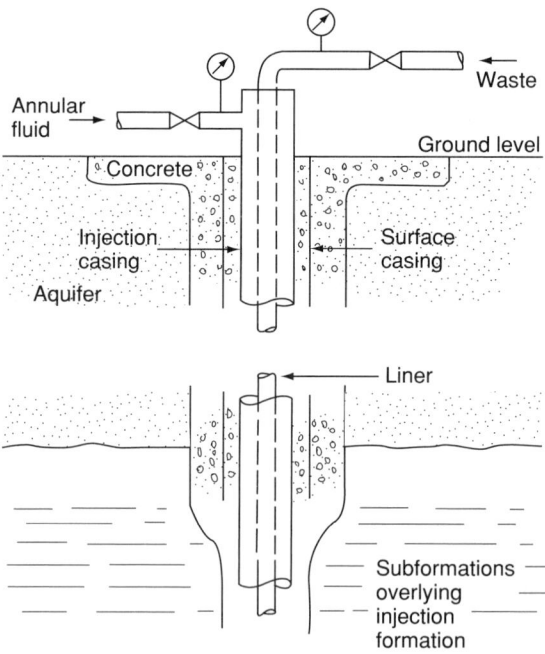

FIGURE 8.16. Typical injection well (from Bureau of Mines Information Circular no. 8212).

to protect other water resources. Table 8.4 summarizes Donaldson's (1964) findings for 20 separate installations, presenting much valuable information such as costs, associated problems, well depth, injection pressures, and formation type.

For the disposal of acid wastes, there are five requirements for deep-well disposal: (1) a satisfactory disposal horizon; (2) a horizon filled with saltwater; (3) a horizon located at a sufficient depth; (4) a suitable cap rock; (5) a waste compatible with the natural water in the disposal horizon. The possible dangers of deep-well disposal include (1) contamination of portable water supplies either by lateral migration to existing unplugged dry holes or producing wells, or by vertical migration through the subsurface, or by vertical migration due to mechanical failure; and (2) possible movements along old fault planes. The representative cost for a 4,000-ft well is $450,000, which includes

TABLE 8.4
Summary of Disposal Systems

Company	Type of Waste	Injection Rate, gpm	Injection Pressure, psi	Subsurface Depth of Wells, ft
A	Brine; chlorinated hydrocarbons	200	500	12,045
B	Clear 4% solution Na_2SO_4	300	45	295
C	Masic waste, pH±10	70	1,000	6,160
D	Magnesium; calcium hydroxides	200	Vacuum	400
	Manufacturing waste, pH may change from 1 to 9 in hr	400	Vacuum	4,150
E	Lachrymator waste from acrolein and glycerine units	700	150–170	1,960
F	Aqueous solution—phenols, mercaptans, and sulfides	215	30–90	1,795
G	Phenols; mercaptans; sulfides; brine	100	40–100	1,980
H	Phenols; chlorinated hydrocarbons	200	450	4,000
	Brine	200	150	4,000
	Phenols; mercaptans; sulfides	50	—[a]	4,000
I	Coke oven phenols; quench water	50	300	563
J	Organic wastes	60	500	1,472
K	Sulfuric acid waste	400	Vacuum	1,830
L	Detergents; solvents; salts	254	280	1,807
M	38% HCl solution	14		1,110
N	Stripping steam condensate; cooling tower blowdown	50	10–20	1,110
	Aqueous petroleum refinery effluent	400	50–70	1,110
O	Phenols; brine	75	400	7,650

[a]Information not available.

the cost of well and equipment, aboveground pumping and equipment, and holding-tank and collection equipment.

The reader may use the following checklist in the design of deep-well disposal systems:

A. State laws and legal aspects
   1. State recognition of this method
   2. Subsurface trespass
B. Geology
   1. Employment of geologist or well contractor
   2. Disposal formation
      a. Porosity
      b. Permeability
      c. Composition (sandstone or limestone)
C. Waste characteristics
   1. Volume reduction
   2. Injection flow rate
   3. Injection pressure
   4. Corrosiveness
   5. Biological effects
D. Surface equipment needs
E. Wells
   1. Number
   2. Size
   3. Monitoring
F. Economics

Following years of reporting on the use and misuse of deep-well injection of wastes, I have arrived at a conclusion that although this method of treatment may provide temporary respite from environmental concerns, it will eventually cause damage of some kind, somewhere, and to varying extents. It is truly only a temporary solution.

Foam Phase Separation

Figure 8.17 illustrates the equipment used for foam phase separation. A sparger producing small gas bubbles (usually air) causes these bubbles to rise through the liquid and adsorb surface-active solutes and suspended matter. When the bubbles reach the surface, a foam forms, which is forced out of the foamer, collapsed, and discharged as a concentrated waste.

Assume the following: (1) complete mixing in the foamer; (2) sufficient depth of liquid to reach maximum solute adsorption of the gas–liquid interface; (3) constant liquid density; (4) no bubble rupture in the foam phase; and (5) negligible volume of the liquid layer containing the surface excess of solute. At this point, the material-balance equation,

$$C_F - C_B = 1{,}000 \ G/F \ rB \ S,$$

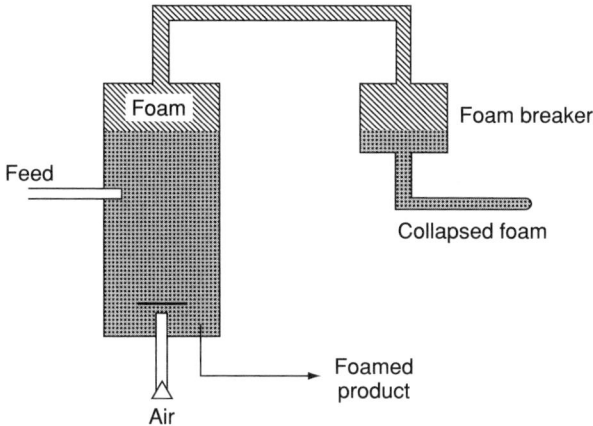

FIGURE 8.17. A column foam fractionator.

where CF and CB are feed and bottom product concentrations in mg/liter, G is the volumetric gas rate in liters/minute, rB is solute surface excess corresponding to CB in mg/cm^2, and S is the specific surface of bubbles in foam phase in cm^2/cc.

At flows of air to liquid feed of G/F = 3, it is reported that the COD is reduced by 25% and alkylbenzene sulfonate (ABS) concentrations are reduced by 50–75%. At air rates of 1.5 liters/mg of ABS in secondary effluents, removals of 0.4 ppm ABS have been reported. The success of the process, in general, depends on the foamability of the liquid waste, which is said to be of low order of magnitude. Bruner and Stephen (1965) have calculated foam separation costs (not including the foamate disposal) as follows:

mgd	cents/1,000 gal.
1	3.6
10	1.9
100	1.4

Shoen et al. (1962) used this treatment successfully to separate radium from uranium-mill wastewater. They found the pH of the wastewater very important in selecting foaming agents. An increase in foaming agent will generally produce a similar increase in foam during treatment. Grieves and Crandall (1966) also experimented with both iron and alum as coagulants using bentonite as an aid in foaming low-quality waters.

Brush Aeration

According to Pasveer (1959), the brush aeration system was evolved between 1925 and 1930 by Kessener for use in the activated-sludge process. It is essentially an extended aeration process providing more than 24 hours of aeration. Since the 1930s, it has found

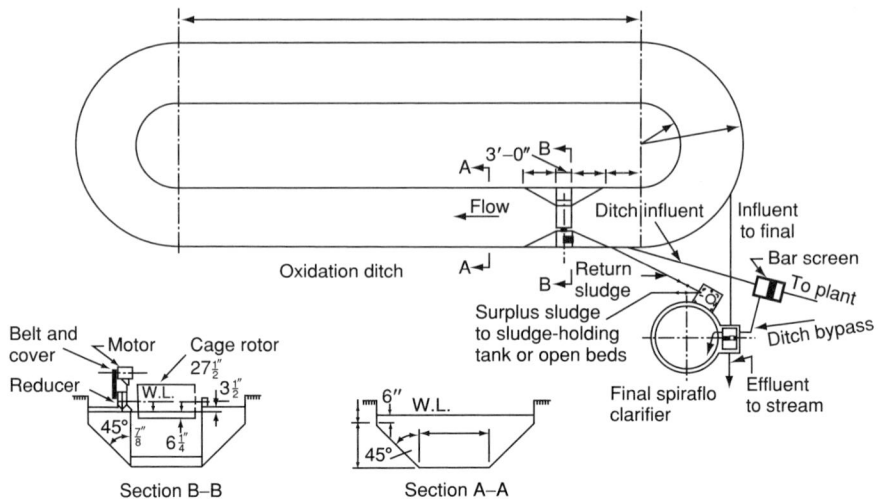

FIGURE 8.18. Typical layout of an oxidation ditch treatment plant (courtesy Lakeside Engineering Corp.).

application, particularly in the Netherlands and in a few plants in Britain, and about a dozen plants were constructed in Canada and the United States before 1964 (Ontario Water Resources Commission 1964) (Figure 8.18). Most of these aeration systems are installed in "oxidation ditches." The design of the oxidation ditch combines an aeration tank and a holding tank in a single unit; the aeration rotor circulates the mixed waste through the whole ditch by means of the rotating cage, but aeration occurs only in the vicinity of the rotor. The rotor is fixed at both ends and set transversely across the aeration ditch and rotates in the direction of waste flow. Aeration is obtained by means of long, rectangular, angle irons welded to the rotating cage. Although the results have been obtained mostly with domestic sewage, it is apparently adaptable to any organic industrial waste.

Subsurface Disposal

Three other methods of disposing dissolved organic wastes below the ground surface are injection, placement in underground cavities, and spreading. Because injection is discussed in some detail in the section "Well Injection" earlier in this chapter, and placement in underground cavities is limited to either small volumes of wastes or particular situations of subsurface formation, they will only be mentioned here as possibilities. Koenig (1964) reported, however, that in 1956, 244 cavities were used for storage, mostly for hydrocarbons and mostly in salt mines.

"Spreading" may be defined as the dispersal of liquid wastes on the ground in order to enhance their infiltration into it. Reclamation of wastewaters by spreading on land, with subsequent withdrawal of groundwater, has been extensively practiced, mostly for secondary sewage effluents. Infiltration rates govern the use of this method, while

Removal of Organic Dissolved Solids 133

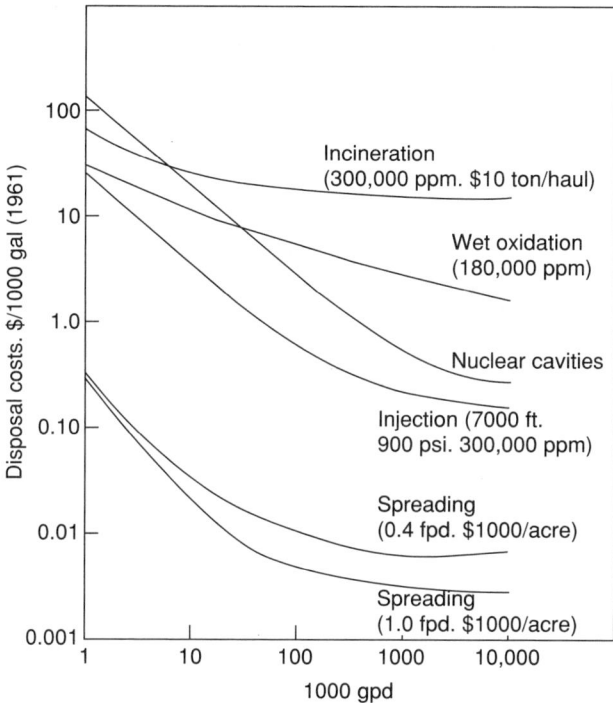

FIGURE 8.19. A comparison of unit disposal costs (1961 figures) (adapted from Koenig 1964).

ultimate effects on underground water supplies govern its acceptability. Because of numerous physical limitations, this method should be considered mainly for small volumes of concentrated organic wastes in particularly suited soils. Koenig (1964) gave comparative costs for this method of disposal and for injection, wet oxidation, and incineration (Figure 8.19).

The Bio-Disc System

The Bio-Disc system was developed independently in West Germany (by Harmann and Pöpel) and the United States (by Welch and Antonie).[1] It consists of a series of flat parallel discs that are rotated while partially immersed in the waste being treated. Biological slime covers the surface of the discs and adsorbs and absorbs colloidal and dissolved organic matter present in the wastewater. Excess slime generated by synthesis of the waste materials is sloughed off gradually into the mixed

---

[1] For references, see the section on Bio-Disc in the "Suggested Reading" section at the end of this chapter.

FIGURE 8.20. Rotating biological contractor (RBC) disc system (courtesy Allis Chalmers Company).

liquor and subsequently separated by settling (Figure 8.20). The rotating discs carry a film of the wastewater into the air where it absorbs the oxygen necessary for aerobic biological activity of the slime. Disc rotation also provides contact between the slime and the wastewater. Thus, the rotating discs provide (1) mechanical support for a captive microbial population; (2) a mechanism of aeration, the rate of which can be adjusted by changing the rotational speed; and (3) contact between the biological slime and the wastewater, the intensity of which can be varied by changing the rotational speed.

Use of closely spaced parallel discs achieves a high concentration of active biological surface area. This high concentration of active organisms and the ability to achieve the required aeration rate by adjusting the rotational speed of the discs enables this process to give effective treatment to highly concentrated wastes. At a loading of 11 lb of BOD/day/1,000 ft^2 of surface area, 90% BOD removal is obtained in 2,000 ppm BOD dairy waste. Secondary treatment of domestic sewage is accomplished with a retention time of 1 hour or less, and 90% BOD reduction is obtained at a loading of 5 lb BOD/day/1,000 ft^2. Because a buoyant plastic material is used for the discs and negligible head loss is encountered through the rotating biological contactor (RBC) itself, the power requirement for this process is very low. Its simplicity of construction and operation has demonstrated that minimal unskilled maintenance is all that is required for efficient operation.

The RBC process has gained wide acceptance in Europe. During a 10-year period, more than 400 Bio-Disc plants were constructed there, ranging in size from 24,000 to 55,000 population equivalents for treatment of domestic and industrial wastes. This process is now being introduced commercially to the United States.

Biological Treatment Design

Example

An integrated pulp and paper mill generates wastewater from its various operations with the following characteristics:

	Process A	Process B
Flow	3.5 mgd	6.0 mgd
BOD	2,500 mg/liter	1,100 mg/liter
TSS	1,000"	450 mg/liter
pH	10	6.9
Temperature: summer	25°C	25°C
winter	15°C	15°C

The proposed treatment scheme consists of flow equalization, primary clarification, activated sludge, secondary clarification, and final treatment in aerated stabilization basins. Design the treatment units given the following design criteria:

Activated-sludge effluent:
BOD = 100 mg/liter
TSS = 100 mg/liter
pH = 6.5–9.0
K = 0.0015 liters/mgd at 20°C
Y = 0.6
b = 0.01/day
O = 1.05
MLSS = 3,000 mg/liter
MLVSS = 2,500 mg/liter
F/M = 0.15–0.4
Minimum solids retention time = 10 days
Sludge recycle rate = 40%

Solutions

Equalization

Combined flow Q = 9.5 mgd
Combined BOD = (3.5 × 2,500 + 6.0 × 1,100) / (9.5)
= 1,616 mg/liter
Combined TSS = (2.5 × 1,000 + 6.0 × 450) / (9.5)
= 547 mg/liter
Detention time = 15 minutes

Volume of equalization tank = Q_t = (9.5 × 15 × 10⁶) / (60 × 24)
= 99,000 gallons
Mixing requirements: @ 0.05 hp/10³ gal
= 0.05 × 99 = 4.95 hp
Aeration requirements: @ 1.5 ft³ air/10³ gal₃
= 1.5 × 99 = 148.5 ft³

Primary Clarifications
Surface loading or overflow rate (OFR) ranges generally from 500 to 1,000 gal/day/ft². Select the loading based on settling tests:

Assume loading of 750 gal/day/ft²

Surface loading (u) = Q/A
Area A = Q/u   = (9.5 × 10⁶) / 7,540 = 12,667 ft²
Diameter of clarifier = 12,667 × 4
= 127 ft

Detention time ranges from 2 to 4 hours.
Assume detention time of 3 hours.

Volume of clarifier = Q_t

(V) = 9.5 × 10⁶ × (3 hr) / (24 hr/day)

= 1,187,500 gal

@ 7.48 gal/ft³

V = (1,187,500) / (7.48) = 158,757 ft³

Depth of clarifier = V/A

= (158,757) / (12,667)

= 12.5 ft

Depth generally ranges from 10 to 15 ft, so calculated depth is within range.

Circumference of clarifier = D

= × 127 = 399 ft.

Check effluent weir loading (ranges from 10,000 to 25,000 gpd/ft)

Weir loading = (9.5 × 10⁶) / (399) = 23,810 gpd/ft is within range

Activated Sludge
Assume 50% BOD removal in primary clarifier. Therefore, BOD in influent to activated-sludge unit is 808 mg/liter.

Calculate reaction range (k) for summer and winter temperatures:

$$K_T = K_{20}\, \varnothing^{T-20} \quad \varnothing = 1.04$$

Summer: $k_{25} = (0.0015)1.04^{25-20} = 0.00183$ liters/mg-day
Winter: $k_{15} = (0.0015)1.04^{15-20} = 0.00123$ liters/mg-day

Next, calculate volume of aeration tank for summer and winter conditions.

Summer Condition

$$F/M = k_{Sc}$$
$$= 0.00183(100) = 0.183 \text{ days}^{-1}$$

$1/SRT = Y\,(F)/M - b$ (Use this value since greater than minimum design criteria.)

$$= 0.64(0.183)^{-1} - 0.01$$
$$= 0.017 \text{ days}$$

Therefore, SRT = 9.4 days.
Use SRT = 10-day minimum per design criteria.

$$XV = [Y(S_o - S_c)\, SRT\, Q] / [1 + b\,(SRT)]$$
$$= [0.64(808-100)\,10(9.5)] / [1 + 0.01(10)]$$
$$= 39{,}133$$

@ MLVSS = X = 2,500 mg/liter

Therefore, volume of aeration tank = (39,133/2,500) = 15.6 million gallons.

Winter Conditions

$$F/M = k_{Sc}$$
$$= 0.00123(100) = 0.123 \text{ days}^{-1}$$

Use F/M = 0.15 per minimum design criteria.

$$1/SRT = Y\,(F)/M - b$$
$$= 0.64(0.15) - 0.01$$
$$= 0.086 \text{ days}^{-1}$$

Therefore, SRT = 11.6 days. (Use this value for SRT, since greater than minimum design criteria.)

$$XV = [Y(So - Sc)\ SRT] / [1 + b\ (SRT)]\ Q$$
$$= [0.64(808 - 100)\ 11.6(9.5) / [1 + 0.01(11.6)]$$
$$= 44{,}584$$

Therefore, volume of aeration tank = 44,584/2,500
= 17.8 million gallons

Thus, the larger volume required for winter conditions determines the design volume of the aeration tank.

$$\text{Detention time} = V/Q = 17.8/9.5 = 1.87\ \text{days}$$

Check F/M ratio.

$$F/M = (9.5\ \text{mgd} \times 808\ \text{mg/liter} \times 8.34) / (17.8\ \text{mg} \times 2{,}500\ \text{mg/liter} \times 8.34)$$
$$= 0.17\ \text{days}^{-1}$$

Check at maximum flow condition.

$$F/M = (9.5 \times 808 \times 8.34) / (13.3 \times 2{,}500 \times 8.34)$$
$$= 0.23$$

Both F/M are within design criteria range.

Waste Sludge Production

$$SRT = \text{lbs MLSS/lbs solids wasted/day} + \text{lbs solids in effluent}$$

Calculate waste sludge production for minimum SRT = 10 days.

$$SRT = [17.8\ \text{mgal} \times 8.34 \times 3{,}000\ \text{mg/liter}] / [Sw + (9.5\ \text{mgd} \times 8.34 \times 100\ \text{mg/liter}]$$
$$10 = 445{,}356/Sw + 7{,}923$$

Therefore, sludge wasted S_w = 36,613 lbs/day.

Calculate oxygen requirements:

lbs oxygen requires/day = y' (lbs BOD removed/day) + b' (lbs MLVSS)

lbs BOD removed/day = 9.5 mgd × (808 − 100) × 8.34
= 56,095 lbs/day

lbs MLVSS = 2,500 mg/liter × 17.8 × 8.34 × 371,130 lbs/day

lbs oxygen = 0.6(56,095) + 0.1(371,130)

$$= 70{,}770 \text{ lbs/day}$$
$$= 2{,}950 \text{ lbs/hr}$$

Aeration horsepower requirements:

Oxygen transfer rate for design conditions is calculated as follows:

$$N = No\ [(B \times cS \text{ at } -cL) / (cSc)]\ 1.024\ T{-}20$$

Calculate transfer rate based on summer conditions.

$$N = 3.0\ [0.9 \times (8.5 - 2.0) / (9.2)] 1.024^{25-20}\ (0.8)$$
$$= 1.7 \text{ lbs oxygen/hp-hr}$$
$$\text{horsepower} = 2{,}950 \text{ lbs/hp-hr}$$
$$= 1{,}735 \text{ hp}$$

Check power requirements to ensure adequate mixing in aeration tank. Assume approximately 100 hp/million gallons required for mixing.

$$1{,}735 \text{ hp}/17.8 \text{ mg} = 97.5 \text{ hp/million gallons}$$

Therefore, horsepower provided for aeration would be adequate for mixing in the aeration tank.

Final Clarifier

Solids loading to final clarifier is calculated as follows:

$$\text{Sludge recirculation} = 40\%$$
$$\text{Recirculation flow (R)} = 0.4 \times 9.5 = 3.8 \text{ mgd}$$
$$\text{Solids loading to clarifier} = (Q + R)\ MLSS \times 8.34$$
$$= (9.5 + 3.8) 3{,}000 \times 8.34$$
$$= 332{,}766 \text{ lbs/day}$$

Final clarifier area is calculated based on requirement for thickening area and clarification area.

Thickening area is calculated assuming solids loading rate of 15 lbs/day/ft² (range 10–20 lbs/day/ft²).

$$\text{Thickening area} = 332{,}766/15$$
$$= 22{,}184 \text{ ft}^2$$

Clarification area is calculated assuming an OFR of 600 gpd/ft² (range 400–1,000 gpd/ft²).

$$\text{Clarification area} = Q/OFR$$
$$= 9.5 \times 10^6/600$$
$$= 15{,}833 \text{ ft}^2$$

Therefore, the thickening area governs.

$$\text{Area} = 22{,}184 \text{ ft}^2$$

Diameter of final clarifier = 168 ft.
Assume a detention time of 3 hours.

$$\text{Volume of clarifier} = Q_t$$
$$= 9.5 \times 10^6 \times 3$$
$$= 1{,}187{,}500 \text{ gal}$$
$$= 1{,}187{,}500/7.48 \text{ ft}^3$$
$$= 158{,}757 \text{ ft}^3$$
$$\therefore \text{Depth of clarifier} = 158{,}757/22{,}184$$
$$= 7 \text{ ft}$$
$$\text{Depth of clarifier} = 158{,}757/22{,}184$$
$$= 7 \text{ ft}$$

O.K. (range 5–15 ft)

$$\text{Check effluent weir loading} = Q/d$$
$$= 9.5 \times 10^6/168$$
$$= 18{,}008 \text{ gpd/ft}$$

O.K. (range 10,000–20,000 gpd/ft)

Collection and Reclamation (Scavenging)

The scavenger hauls, treats, reclaims, and disposes of a variety of industrial wastes acquired through contract of purchase from firms that do not choose to treat their own wastes.

Scavenging firms may be as varied as the customers they serve. Some small firms may only provide hauling and land-disposal services; others specialize in a single line of waste, such as solvents, which may be profitably reclaimed. Large firms may provide a full range of treatment and consulting services.

Recycling Laboratories (Canastota, New York), for example, specializes in the reclamation of solvents, particularly chlorinated hydrocarbons such as trichloroethylene. Typically, it purchases used solvents, redistills them, and then resells them at a profit. The company also handles some alcohols and thinners and, for a fee, will accept petroleum products, which it uses in its steam plant. Recycling Laboratories is a small and fairly new firm and most of its business comes from the immediate central New York area. It envisions expanding one line at a time as it acquires experience in dealing with different sorts of wastes.

Chem-Trol Pollution Services, Incorporated, of Model City, New York (near Buffalo), represents the other end of the spectrum. It is a large firm that operates on a 240-acre site with seven separate lagoons. Lagoon storage totals 6 million gallons, while closed-tank storage totals an additional 2 million gallons. It processes wastes using the following techniques: (1) filtration; (2) thermal oxidation (incineration); (3) neutralization; (4) distillation; (5) chemical fixation; and (6) physical separation (such as centrifuging). Residues are land filled. Many of its operating practices are subject to proprietary considerations but may be considered to generally follow the usual treatment techniques.

Advantages of scavenging treatment:

1. Economy of scale exists in larger plants.
2. Operating efficiencies other than scale—specialization, neutralization, and equalization—increase as the size of the operation increases.
3. Treatment expense is involved only for the small industrial plant, rather than capital investment.
4. Resource-recovery potential exists in the larger reclamation plant.

Disadvantages of scavenging treatment:

1. Newness of the field: There is neither a trade association nor a real trade publication. Relatively little information exchange occurs, particularly with regard to proprietary practices. Regulation and standards vary with locality.
2. Transportation expense, typically by truck, may be high.
3. Batch processes are usually necessary with attendant high labor and overhead costs.
4. Dependence on secondary markets is risky for the scavenger.
5. Poor public image may exist unless a proper public relations program is used.

Nevertheless, the scavenger fills a very real need in this time of increasingly strict environmental controls and is especially useful for the treatment of low volumes of highly concentrated organic wastes. The reader is directed to Nemerow (1995, Chapter 4) for a detailed description of reuse of wastes by scavenging.

Chemical Oxidation of Organic Matter

It has been long known and sometimes implemented in waste treatment that organic matter can be completely destroyed by chemical oxidation. Oxidants such as chlorine, hydrochloric acid, chlorine dioxide, permanganate, hydrogen peroxide, pure oxygen, ozone, hydroxyl radicals, and bromine and fluorine are all effective, in varying degrees, for this purpose. Generally, chemical oxidation was considered too costly or impractical for industrial-waste treatment. Expensive and often difficult-to-handle oxidizing chemicals were consumed in the process, and further, such complete destruction of organic wastes was not usually required. The cost effectiveness of chemical oxidation was usually questioned by waste-treatment engineers.

However, today's industry faces much more stringent discharge requirements. Contamination of our groundwater has made remediation a first-order objective. And because so many groundwater supplies serve as sources of drinking water, higher degrees of purification have become a necessity. This has led to renewed consideration of chemical oxidation as a viable and cost-effective treatment alternative.

Improved analytical instrumentation, coupled with a much wider spectrum of chemical identification, allows us to identify and measure contaminants in water in lower concentrations with greater precision. This has led to broader and more stringent water-quality standards—a driving force calling for the increased use of chemical oxidants for the removal of broad-spectrum organics. Chemical oxidation is effective especially in situations in which the waste concentration of the organic matter is high and a high degree of treatment is required, making treatment cost secondary to receiving-water quality. The process of chemical oxidation, when carried to its ultimate goal, yields carbon dioxide and water and usually releases heat because the reactions are exothermic. One example of this type of application by a chemical oxidant is that of hydrogen peroxide in combination with ferrous sulfate (at a pH of 3.5–4.0). According to Bigda [20A], when H_2O_2 splits into OH– and OH+, the ultra-active OH+ is more powerful than ozone, chlorine, or pure oxygen and combines with carbon found in organic compounds, breaking the double bonds and aromatic rings and removing hydrogen molecules. The combination of about 1 part of iron to 10 parts of peroxide is referred to as *Fenton's reagent*. It has been used in the treatment of wastes containing phenol, formic acid, and other organics from paint stripper rinse waters (Tyre et al. 1991; Watts 1992). Costs for treating 1,000 gallons of waste containing 1,000 ppm of phenol was reported to be about $0.14/gal (Tyre et al. 1991), including labor and power. Other researchers and engineers recommend using ozone (Murdock 1951) and chlorine (Chamberlain and Griffin 1952) as most effective in treating phenolic-type organics.

Miscellaneous

Ozonization

Organic matter in industrial wastewaters can be completely oxidized through contact and reaction with ozone (O_3). Ozone is a gas that is produced by passing oxygen through an electrical field. Approximately 11 kwh of electrical power is required to produce 1 lb of ozone from air. From 1.5 to 2.5 lb of ozone is used for each pound of dissolved organic matter oxidized. If pure oxygen is used instead of air as a source of ozone, the power requirements are reduced by about 50%. Small-size ozone-generating units, producing up to 75 lb of ozone per day, are available commercially for use by industries with small volumes of organic wastes. The ozone produced is usually dissolved in wastewater by injection nozzles or cavitation. This form of treatment has been found especially suited for oxidizing phenolic wastes. Ozone contact time can usually be kept to less than 30 minutes, especially for final or tertiary-type treatment, where organic matter concentration is relatively low and allowable effluent limits are also low.

Photolysis

The interaction of photocatalysts with radiation below about 4,200°A produces active oxygen species that destroy organic matter by complete oxidation to CO_2 and H_2O. Certain oxides, notably zinc oxide and titanium dioxide, are known to be photosensitizers or photocatalysts. Kinney and Ivanuski (1969) reported that photocatalytic oxidation of dissolved organic matter by irradiation of slurries of zinc titanate (Zn_2TiO_2), zinc oxide (ZnO), titanium dioxide (TiO_2), and beach sand by sunlamps was effective. Dissolved organic matter in a sample of domestic sewage was reduced 50% in 24 hours and 75% in 70 hours. The reaction appears to follow first-order kinetics in most cases. Zinc oxide appears to be superior for this purpose. At concentrations of 100–200 mg/liter of organic carbon, 80% of phenol, 67% of benzoic acid, 44% of acetic acid, 40% of sodium stearate, and 16% of sucrose were oxidized in 24 hours with 10 g/liter zinc oxide catalyst. Continued illumination reduced organic carbon to a few milligrams per liter in most cases. The photocatalytic properties of illuminated beach sand, which oxidized 87% of phenol in 72 hours, strongly suggest that photocatalysts are widely distributed in nature. Further, it suggests that photocatalytic oxidation is a mechanism whereby dissolved organic matter is oxidized in the natural environment of streams and lakes. The researchers (Kinny and Ivanuski 1969) suggested that three conditions must be satisfied to achieve oxidation of an organic molecule: (1) the molecule must be adsorbed at, or be in the vicinity of, the active site on the catalyst; (2) light energy of suitable wavelength (below about 4,200°A) must impinge on the active site; (3) dissolved oxygen must be present to replace the active oxygen species displaced by the radiation. From these considerations and with everything else being equal, the higher the concentration of contaminant, the faster the rate of oxidation. Thus, it would appear that photocatalytic oxidation will find its greatest utility in problems of industrial-waste treatment where massive contamination is involved. The researchers also believed that vigorous agitation at elevated temperature would favor faster kinetics, despite lower dissolved-oxygen levels.

Pure Oxygen Treatment

The Union Carbide Corporation has refined and improved upon a biological aeration system using pure oxygen in place of air. In this system, the aeration tanks are completely covered to provide a gas-tight enclosure above the mixed liquor. Both the liquid and the gas phase are staged with a concurrent flow of the gas and liquid through the multistage system. The feed wastewater, along with the recycled sludge, is introduced into the first stage together with the oxygen feed gas. The oxygen gas is fed into the enclosure above the mixed liquor. Mechanical agitation provides the required bulk fluid motion to maintain the sludge in suspension and to ensure a uniform liquid composition. Blowers in each stage recirculate the gas in the enclosure through the mixed liquor. The gas is piped through a hollow agitator shaft and dispersed into the mixed liquor through a rotating-sparger device. The pressure of the gas enclosures above the mixed liquor is automatically controlled at a few inches of water above atmospheric pressure. The oxygen gas is fed in direct proportion to the demand of the mixed liquor. The aeration gas flows freely from stage to stage with only a slight pressure differential to prevent back-mixing of the gas between adjacent stages. The mixed liquor exiting the final stage is passed into a conventional settler for clarification. The vent gas will usually comprise only about 10% of the volumetric flow rate of the oxygen feed gas and will be about 50% oxygen. Thus, it is claimed to be 95% efficient in oxygen transfer. The manufacturer claims high treatment-rate-performance capability of the oxygenation system. It reports more than 90% reductions in both BOD and suspended solids after shorter detention times and higher loadings than the conventional aeration activated-sludge treatment systems.

Example of Twentieth-Century Practice of Organic Dissolved Solid Removal

Banerjee (1997) used a laboratory four-step cross-flow RBC to remove organic phenol. He observed that removal of phenol improved at higher input phenol concentration. Temperature increase in the range of 13–36°C improved the removal. The major group of microbes in the RBC was *Pseudomonas*. He reported that phenolic wastewater up to a concentration of 420 g/m^3 could be treated effectively by this process.

Review Questions

1. Why has removal of organic dissolved solids long been the most important and most difficult phase of industrial-waste treatment?
2. What is the basis for most treatment processes for dissolved organic removal?
3. What is necessary for optimal biological treatment?
4. What are the advantages and disadvantages of lagooning? Of activated sludge?
5. Name and describe some modifications of the activated-sludge process.
6. What is the principle of dispersed-growth aeration? What are its advantages and disadvantages?
7. What is meant by "trickling filtration" and when would you use it in preference to biological aeration?
8. Is spray irrigation a suitable method of disposing of dissolved organic matter?
9. Can you use subsurface disposal for dissolved organic solids? When?
10. What is the theory of the Bio-Disc treatment system? Does it resemble trickling filtration or activated-sludge treatment?
11. What methods are used for nonbiological oxidation of dissolved organic wastes? Explain the principles of ozonation, chlorination, pure oxygen, and photolysis treatments.
12. What is wastewater scavenging and what are its limitations?
13. Describe how chemical oxidation can be used to destroy organic matter. What chemicals are employed in this method? What special properties make these chemicals applicable to destruction of organic matter?

References

Anderson, D. R., W. D. Bishop, H. L. Ludwig, 1966. Percolation of citrus wastes through soil. In: *Proceedings of 21st Industrial Waste Conference, Purdue University Engineering Extension Series*. Bulletin no. 121, p. 892.

Banerjee, G. 1997. Treatment of phenolic wastewater in RBC reactor. *Water Res.* 31(4):705–714.

Behn, V. C. 1960. Trickling filter formulations. In: *Conference on Biological Treatment, at Manhattan College, New York, April 20–22, 1960*. Paper no. 26.

Bruner, C. A., B. G. Stephen. 1965. Foam fractionation. *Ind. Eng. Chem.* 57:40.

Busch, A. W., A. A. Kalinske. 1956. The utilization of the kinetics of activated sludge in process and equipment of design. In: *Biological Treatment of Sewage and Industrial Wastes* (J. McCabe and W. W. Eckenfelder, eds.), p. 277. New York: Reinhold.

Buswell, A. M., W. D. Hatfield. 1939. *Anaerobic Fermentations*, Bulletin no. 32, State of Illinois, Division of State Water Survey, Urbana, Ill.

Chamberlain, N., A. Griffin. 1952. Chemical oxidation of phenolic wastes with chlorine. *Sewage Indus. Wastes* 24:750.

Donaldson, E. C. 1964. Subsurface Disposal of Industrial Wastes in the United States, Information Circular 8212. Washington, DC: Bureau of Mines, U.S. Department of the Interior.

Erode, W. yon der. 1960. Aspects of high rate activated sludge process. In: *Conference on Biological Waste Treatment at Manhattan College, New York, April 20–22, 1960*, Paper no. 35.

Grieves, R., C. Crandall. 1966. Water clarification by foam separation: bentonite as a flotation aid. *Water Sewage Works* 113:432.

Haseltine, T. R. 1956. A rational approach to the design of activated sludge plants. In: *Biological Treatment of Sewage and Industrial Wastes* (J. McCabe and W. W. Eckenfelder, eds.), p. 257. New York: Reinhold.

Hermann, E. R., E. F. Gloyna. 1958. Waste stabilization ponds. *Sewage Ind. Wastes* 30:511, 646.

Heukelekian, H. 1949a. Aeration of soluble organic wastes with non-flocculent growths. *Ind. Eng. Chem.* 41:1412.

Heukelekian, H. 1949b. Treatment of streptomycin wastes. *Ind. Eng. Chem.* 41:1412.

Heukelekian, H. 1949c. Characteristics and treatment of penicillin wastes. *Ind. Eng. Chem.* 41:1535.

Howland, W. E. 1957. Flow over porous media as in a trickling filter. In: *Proceedings of 12th Industrial Waste Conference, Purdue University, Lafayette, Indiana, 1957*, p. 435.

Ingram, W. T. 1956. A new approach to trickling filter design. *Proc. Am. Soc. Civil Engrs.* 82(Paper no. 999).

Kinny, L. C., Ivanuski, V. R. 1969. *Photolysis Mechanisms for Pollution Abatement*, Taft Water Research Center Report No. TWRC-13. Cincinnati, OH: Taft Water Research Center.

Koenig, L. 1964. *Ultimate Disposal of Advanced-Treatment Waste*. Washington, DC: US Department of Health, Education and Welfare. Environmental Health Series AWTR-8.

Kraus, L. S. 1945. The use of digested sludge and digester overflow to control bulking of activated sludge. *Sewage Works J.* 17:1177.

Kraus, L. S. 1955. Dual aeration as a rugged activated sludge process. *Sewage Ind. Wastes* 27:1347.
Lesperance, T. W. 1965. Extended aeration and high rate treatment. *Water Works Wastes Eng.* 2:40.
McKinney, R. E., J. M. Symons, W. G. Shifrin, and M. Vezina. 1958. Design and operation of complete mixing activated sludge system. *Sewage Ind. Wastes* 30:287.
Murdock, H. 1951. Ozone provides an economical means for oxidizing phenolic compounds in coke even wastes. *Ind. Eng. Chem.* 41:125A.
Nemerow, N. L. 1954. Oxidation of enzyme desize and starch rinse textile wastes. *Sewage Ind. Wastes* 26:1231.
Nemerow, N. L. 1955. Oxidation of cotton kier wastes. *Sewage Ind. Wastes* 25:1060.
Nemerow, N. L. 1956. Holding and aeration of cotton mill finishing wastes. In: *Proceedings of 5th Southern Municipal and Industrial Waste Conference, Chapel Hill, NC*, April 1956, p. 149.
Nemerow, N. L. 1957. Dispersed growth aeration of cotton finishing wastes, II. Effect of high pH and lowered air rate. *Am. Dyestuff Reptr.* 46:575.
Nemerow, N. L. 1960. Accelerated waste oxidation pond studies. In: *Proceedings of Third Conference on Biological Waste Treatment, Manhattan College, New York*, April 20–22, 1960.
Nemerow, N. L. *Zero Pollution for Industry,* Chapter 4. New York: John Wiley.
Nemerow, N. L., J. C. Bryson. 1960. "Hancock Air Force Base Waste Stabilization Research Report." Syracuse University Reports to U.S. Air Force, March.
Nemerow, N. L., J. Ray. 1956. Biochemical oxidation of glucose by dispersed growth aeration. In: *Biological Treatment of Sewage and Industrial Wastes*. New York: Reinhold.
Nemerow, N. L., Rudolfs. 1952. Rag, rope, and jute wastes from specialty paper mills, V. Treatment by aeration. *Sewage Ind. Wastes* 24:1005.
North and South Dakota Departments of Health and the U.S. Department of Health, Education and Welfare. 1957. *Sewage Stabilization Ponds in the Dakotas.*
Ontario Water Resources Commission. 1964. *Evaluation of the Oxidation Ditch as a Means of Wastewater Treatment in Ontario.* Ottawa (July 1964). Research Publication no. 6.
Oswald, W. J. 1960. Fundamental factors in oxidation pond design. In: *Proceedings of Third Conference on Biological Waste Treatment, at Manhattan College, New York, April 20–22, 1960,* Paper no. 44.
Pasveer, A. 1959. New developments in the application of Kessener brushes in the activated-sludge treatment of trade-waste waters. In: *Waste Treatment* (P. Isaac, ed.), pp. 126–155. New York: Pergamon Press.
Pettet, A. E. J., T. G. Tomlinson, J. Hemens. 1959. The treatment of strong organic wastes by anaerobic digestion. *J. Inst. Public Health Engrs.* 170.
Rankin, R. S. 1953. Performance of biofiltration plants by three methods. *Proc. Am. Soc. Civil Engrs.* 79(Separate No. 336).
Sawyer, C. N. 1960. Activated sludge modifications. *J. Water Pollution Control Fed.* 32:233.
Shoen, H. M., E. Rubin, D. Ghosh. 1962. Radium removal from uranium mill wastewater. *J. Water Pollution Control Fed.* 34:1026.
Schultze, K. L., H. S. Foth. 1955. New low cost secondary treatment by new cavitation system. *Water Sewage Works* 102:74.

Struzeski, E. J., N. L. Nemerow. 1957. Dispersed growth aeration of protein-glucose mixtures. In: *Proceedings of 12th Industrial Waste Conference, Purdue University, May 1957*, p. 145.

Tapleshay, J. A. 1958. Total oxidation treatment of organic wastes. *Sewage Ind. Wastes* 30:652, Washington D.C.

Tyre, B. W., et al. 1991. Treatment of four biorefractory contaminants in soils using catalyzed hydrogen peroxide. *J. Environ. Quality* 20(October 1991).

Ullrich, A. H., M. W. Smith. 1951. The Biosorption process of sewage and waste treatment. *Sewage Ind. Wastes* 23:1248.

Ullrich, R. A., M. W. Smith. 1957. Operation experience with activated sludge-biosorption at Austin, Texas. *Sewage Ind. Wastes* 29:400.

Velz, C. J. 1948. A basic law for the performance of biological filters. *Sewage Works J.* 20:607.

Watts, R. J. 1992. Hydrogen peroxide for physiochemically degrading petroleum-contaminated soil. *Remediation* Autumn.

Wet combustion of wastes. *Power Eng.* 59:63.

Zobell, C. E. 1937. The influence of solid surface upon the physiological activities of bacteria in sea water. *J. Bacteriol.* 33:86.

Suggested Reading

Flower, W. A. 1965. Spray irrigation—a positive approach to a perplexing problem. In: *Proceedings of 20th Industrial Waste Conference, Purdue University, Lafayette, Indiana, 1965*, p. 679.

Ling, J. T. 1963. Pilot study of treating chemical wastes with an aerated lagoon. *J. Water Pollution Control Fed.* 35:963.

Luley, H. G. 1963. Spray irrigation of vegetable and fruit processing wastes. *J. Water Pollution Control Fed.* 35:1252.

Oswald, W. J., H. B. Gotaas. 1957. Photosynthesis in sewage treatment. *Trans. Am. Soc. Civil Engrs.* 122:73.

Parker, C. D. 1966. Food treatment waste treatment by lagoons and ditches at Shepparton, Victoria, Australia. In: *Proceedings of 21st Industrial Waste Conference, Purdue University, Lafayette, Indiana, 1966*, p. 284.

Deep-Well Injection

Barraclough, J. T. 1966. Waste injection into deep limestone in northwestern Florida. *Groundwater* 4:22.

Hundley, C. L., J. T. Matulis. 1962. Deep well disposal. In: *Proceedings of 17th Industrial Waste Conference, Purdue University, 1962*, p. 175.

Koenig, L. 1963. Advanced waste treatment. *Chem. Eng.* 70:210.

Powers, T. J., G. W. Querio. 1961. Check on deep-well disposal for specially troublesome waste. *Power* 105:94.

Production waste goes underground at Holland-Suco. 1966. *Mitch Water Sewage Works* 113:329.

Querio, C. W., T. J. Powers. 1962. Deep well disposal of industrial waste water. *J. Water Pollution Control Fed.* 34:136.

Selm, R. P. 1959. Deep well disposal of industrial wastes. In: *Proceedings of 14th Industrial Waste Conference, Purdue University, Lafayette, Indiana, 1959.*

Talbot, J. S. 1964. Deep well method of industrial waste disposal. *Chem. Eng. Progr.* 60:1.

Warner, D. L. 1966. Deep well waste injection—reaction with aquifer water. *J. Sanit. Eng. Div. Am. Soc. Civil Engrs.* 92(SA4):95.

Waste well goes down over two miles. 1960. *Eng. News-Record* 165:32.

Winar, R. M. 1967. The disposal of wastewater underground. *Ind. Water Eng.* 4:21.

Foam Phase Separation

Advanced Waste Treatment Research. 1955. Publication no. AWTR-14. U.S. Public Health Service, Cincinnati, OH.

Brown, D. J. 1965. A photographic study of froth flotation. *Fuel Soc. J. Unic. Sheffield* 16:22.

Eldib, I. A. 1961. Foam fractionation for removal of soluble organics from wastewater. *J. Water Pollution Control Fed.* 33:914.

Gassell, R. B., O. J. Sproul, P. F. Atkin, Jr. 1965. Foam separation of ABS and other surfactants. *J. Water Pollution Control Fed.* 37:460.

Grieves, R. B., C. J. Crondall, R. K. Wood. 1964. *Air Water Pollution* 8:501.

Rubin, E., R. Everett, Jr., J. J. Weinstock, H. M. Shoen. 1963. *Contaminant Removal from Sewage Effluents by Foaming.* U.S. Public Health Service, Cincinnati, OH. Publication no. 999-WP-5.

Bio-Disc

Antonie, R. L., F. M. Welch. 1969. Preliminary results of a novel biological process for treating dairy wastes. In: *Proceedings of 24th Industrial Waste Conference, Purdue University, Lafayette, Indiana, 1969.*

Hartmann, H. 1960. Investigation of the biological purification of sewage using the Bio-Disc filter. *Sluttgarter Berichte zur Siedlungswasserwirtschaft,* no. 9, R. Oldengourg, Munich.

Pöpel, F. 1964. Estimating construction and out of Bio-Disc filter plants. *Sluttgarter Berichte zur Siedlungswasserwirtschaft,* no. 11, R. Oldenbourg, Munich.

Welch, F. M. 1968. Preliminary results of a new approach in the aerobic biological treatment of highly concentrated wastes. In: *Proceedings of 23rd Industrial Waste Conference, Purdue University, Lafayette, Indiana, 1968.*

CHAPTER 9

Treatment and Disposal of Sludge Solids

Of prime importance in the treatment of all liquid wastes is the removal of solids, both suspended and dissolved. Once these solids are removed from the liquids, however, their disposal becomes a major problem. Unfortunately, waste engineers spend more time and money removing the solids than finally treating and disposing of them, so often a poor solids-disposal program will cause trouble in an otherwise properly designed and operated waste-treatment plant. When the solids-disposal system is poor, the solids tend to build up in the flow-through treatment units, and overall removal efficiencies then begin to decrease. Therefore, proper sludge handling enhances the overall treatment of all wastes. The following list contains most of the methods commonly used to deal with sludge solids: (1) anaerobic and aerobic digestion; (2) vacuum filtration; (3) elutriation; (4) drying beds; (5) sludge lagooning; (6) wet combustion; (7) atomized suspension; (8) drying and incineration; (9) centrifuging; (10) sludge barging; (11) landfill; (12) transporting to an acceptable landfill; and (13) miscellaneous methods.

Anaerobic and Aerobic Digestion

Anaerobic digestion is a common method of readying sludge solids for final disposal. All solids settled out in primary, secondary, or other basins are pumped to an enclosed airtight digester, where they decompose in an anaerobic environment. The rate of their decomposition depends primarily on proper seeding, pH, character of the solids, temperature, and degree of mixing of raw solids with actively digesting seed material. Digestion serves the dual purpose of rendering the sludge solids readily drainable and converting a portion of the organic matter to gaseous end-products. It may reduce the volume of sludge by as much as 50% organic matter reduction. After digestion, the sludge is dried and/or burned or used for fertilizer or landfill.

Two main groups of microorganisms, hydrolytic and methane, carry out digestion. Hydrolytic bacteria exist in great numbers in sewage and waste sludges and are capable of rapid rates of reproduction; they are saprophytic microorganisms that attack complex organic substances and convert them to simple organic compounds. Among these saprophytes are many acid-forming bacteria that produce fatty acids of low molecular weight, such as acetic and butyric, during degradation processes. In some cases, such acids are produced in quantities sufficient to lower the pH to a level at which all biological activity is arrested.

Fortunately, methane bacteria, the other group of microorganisms, are capable of using the acid and other end-products formed by the hydrolytic bacteria. Methane producers, however, are sensitive to pH changes and proliferate only within a narrow pH range of 6.5–8.0, with an optimum of 7.2–7.4; furthermore, they are few in number and reproduce slowly. Consequently, organic acids may form faster than they can be assimilated by the limited population of methane bacteria. As a result, the pH may be lowered and conditions made even more unfavorable for methane bacteria. When this happens, lime is usually added and the digestion process stopped until normal conditions return.

The proper environment for both types of bacteria requires a balance among the population of organisms, food supply, temperature, pH, and food accessibility. The following factors are measures of the effectiveness of digestive action: gas production (both quantity and quality), solids balance (total, volatile, and fixed), biochemical oxygen demand (BOD), acidity and pH, volatile acids, grease, sludge characteristics, and odor.

As mentioned earlier, fermentation (digestion) of organic matter proceeds in two stages: (1) hydrolytic action that converts organic matter to low-molecular-weight organic acids and alcohols, and (2) evolution of carbon dioxide and the simultaneous reduction to methane (carbon dioxide is actually consumed). The following general equations represent the digestion of carbohydrates, fats, and proteins:

Carbohydrates:

$$(C_6H_{10}O_5)x + x\ H_2O \rightarrow x(C_6H_{12}O_6),$$

$$C_6H_{12}O_6 \rightarrow 2C_2H_5OH + 2CO_2,$$

$$2CH_3CH_2OH + CO_2 \rightarrow 2CH_3COOH + CH_4,$$

$$CH_3COOH \rightarrow CH_4 + CO_2,$$

Fats:

$$\begin{array}{c} H_2C-O-C(=O)-R_1 \\ | \\ HC-O-C(=O)-R_2 \\ | \\ H_2C-O-C(=O)-R_3 \end{array} + 3HOH \rightarrow \begin{array}{c} H \\ | \\ H-C-OH \\ | \\ H-C-OH \\ | \\ H-C-OH \\ | \\ H \\ \text{Glycerol} \end{array} + \begin{array}{c} HO-C(=O)-R_1 \\ HO-C(=O)-R_2 \\ HO-C(=O)-R_3 \\ \text{Acid} \end{array}$$

Alpha oxidation of acids:

$$4RCH_2COOH + 2HOH \rightarrow 4RCOOH + CO_2 + 3CH_4.$$

Beta oxidation of acids:

$$2RCH_2CH_2COOH + CO_2 + 2HOH \rightarrow 2RCOOH + 2CH_3COOH + CH_4,$$

$$CH_3COOH \rightarrow CH_4 + CO_2.$$

Proteins:

$$R-\underset{NH_2}{\underset{|}{C}}H-COOH \xrightarrow[\text{Deaminase}]{HOH} NH_3 + R-\underset{OH}{\underset{|}{C}}H-COOH,$$

$$R-\underset{OH}{\underset{|}{C}}H-COOH \xrightarrow{\text{Decarboxylase}} R-\underset{OH}{\underset{|}{C}}H-H + CO_2,$$

$$2RCH_2OH + CO_2 \rightarrow 2RCOOH + CH_4,$$
$$RCOOH \rightarrow CH_4 + CO_2.$$

One hypothesis is that each molecule of methane arises from a reduction of one molecule of carbon dioxide. In other words, carbon dioxide acts as the hydrogen acceptor, while the alcohol acts as the hydrogen donor, as in the following equation:

$$\underset{\text{Ethyl alcohol}}{2C_2H_5OH} + \underset{\text{carbon dioxide}}{CO_2} + H_2O \rightarrow \underset{\text{acetic acid}}{2CH_3COOH} + CH_4 + 2H_2O$$

$$\underset{\text{hydrogen donor}}{} \qquad \underset{\text{hydrogen acceptor}}{} \qquad \underset{\text{methane}}{}$$

One can readily see that carbon dioxide is an important food constituent. In mixed cultures, carbon dioxide is produced by other organisms, and therefore becomes more available than sulfates or nitrates. Buswell and Hatfield (1939) described fermentation as a chain of reactions involving the transfer of hydrogen.

The slowest reaction in the degradation process, production of methane, is therefore the rate-controlling reaction. The essential physiological characteristics of methane bacteria follow: (1) they are obligate anaerobes; (2) they require carbon dioxide as a hydrogen acceptor; (3) as hydrogen donors, they use simple organic substances, such as calcium acetate, butyrate, and ethyl and butyl alcohols; (4) their nitrogen source is ammonia; (5) they develop at a slow rate because of low energy yields; (6) they do not form spores; and (7) they are very sensitive to changes in pH.

Buswell and Hatfield (1939) concluded that the higher the percentage of carbon atoms in the fatty acid substrate, the higher the percentage of methane in the gas. Barker (1936a,b, 1937) established the following unique features of methane fermentation:

1. It takes place in mixed or enriched cultures, and hence may be maintained continuously on a large scale.
2. It is applicable to any type of substrate except lignin and mineral oil.
3. The reaction is quantitative and converts the entire substrate to carbon dioxide and methane.
4. There is no specific temperature limitation in the range of 0–55°C, but once the culture has been acclimated to a certain temperature, a drop of 2°C may completely interrupt methane fermentation and render obstructive the accumulated acids.
5. The presence of inert solid matter is important, and thus the addition of straw or sawdust to industrial wastes may be required.
6. If the substrate concentration is too great, volatile acids build up and inhibit the fermentation, especially when their buildup occurs faster than their subsequent conversion to methane. Keeping the volatile acid level below 3,000 ppm and closer to 2,000 helps the situation, but alkali addition will not alleviate it because it is not a pH effect. Mineral salts begin to inhibit the fermentation at 4,000 ppm, and 50 ppm of nitrate nitrogen inhibit it completely.

The extent of reduction of volatile solids by digestion depends in part on the amount of volatile matter in the raw sludge. Schlenz (1937) found that when volatile

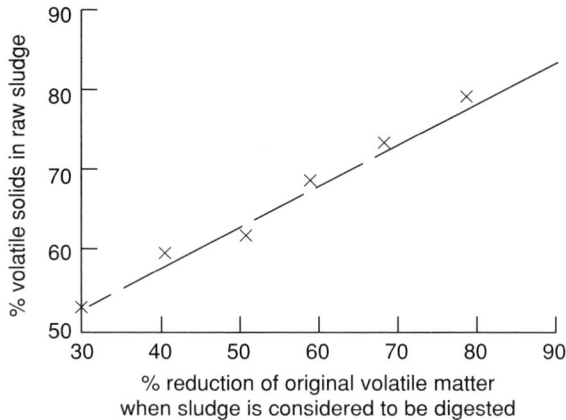

FIGURE 9.1. Reduction of volatile matter in raw sludge by digestion (from Ontario Water Resources Commission 1964).

solids in raw sludge increased from 55 to 80%, the reduction in volatile matter increased from 35 to 85%. This is shown graphically in Figure 9.1.

The usual unit-capacity requirements may be reduced, provided the operations are controlled and carried out as follows (Federation of Sewage and Industrial Waste Association 1959): (1) tank contents must be agitated to maintain an even mixture of raw and digesting solids; (2) raw sludge must be added continuously to the digestion unit; and (3) raw sludge must be concentrated or prethickened before being added to the digester. Two-stage digestion, with the first stage used primarily for active digestion and the second stage for storage and sludge consolidation, is often performed in two separate tanks. It is usually more economical in large plants with continuous operation.

With some industrial sludge wastes, it may be possible and economically feasible to either reuse the methane gas for heat/power or produce sufficient hydrogen gas (by modified digestion) to serve as an alternative fuel for automobiles. During the first stage of anaerobic digestion—where hydrolysis of the organic matter occurs primarily—hydrogen may be more prevalent than methane, while during the final stage of digestion, methane may be a primary source of additional energy.

Aerobic digestion is now playing an important role in small plants. It is claimed that less-skilled operators are required; also, air is normally available in these plants because secondary treatment of the liquid-waste fraction is becoming rather commonplace (Eckenfelder 1956).

Vacuum Filtration

"Vacuum filtration" is a means of de-watering sludge solids, and has become popular because the volume of solids for ultimate disposal is reduced and the sludge is drier than it would otherwise be, thereby improving "handleability." Large plants are increasing their use of vacuum filtration. Some plants filter chemically precipitated and/or plain settled sludge, while others filter digested sludge. In a typical vacuum-filtration unit, a porous cylinder overlying a series of cells revolves about its axis with a peripheral speed somewhat less than 1 ft/min, its lower portion passing through a trough containing the sludge to be dried. A vacuum inside the cylinder picks up a layer of sludge as the filter surface passes through the trough, and this increases the vacuum. When the cylinder has completed three-quarters of a revolution, a slight air pressure is produced on the appropriate cells, which aids the scraper, or strings, to dislodge the sludge in a thin layer. Sometimes it is necessary to add chemicals, such as lime and ferric chloride, as sludge conditioners before filtration. Filtering rates should be from 2 to 10 lb of dry solids per square foot per hour. Vacuum filters are available in diameters up to about 20 ft and in many lengths.

The quality of the filter medium (the material covering the cylinder) is important in the performance and life of the filter. In the past, woven-fabric filter media have been widely used. The physical process of solids retention on woven filters is a combination of at least three actions: (1) straining action, in which particles larger than the filter-medium openings cling to the filter; (2) adsorption, or attraction, to the filter of particles smaller than the openings in the filter medium; and (3) filtration of particles of different sizes that cling to already filtered caked material. The first two actions prevail at the onset of filtration, but as the "cake" builds up, the third is responsible for the greatest amount of solids removal. Thus, the problem arises that unless the cake is removed completely and the fiber filter medium kept clean continually, the filter will clog or "blind."

Tiller and Huang (1951) reported that there is a paucity of theory and research on filtration through porous media. Three reasons for this deficiency are (1) complexity of vacuum-filtration machinery, (2) difficulty of experimentally reproducing the precipitates found in filter beds, and (3) insufficient interest on the part of researchers. They also reported that although flow through the filter beds is almost always viscous, no reliable theory has been developed on the relation between permeability and porosity of the filter medium as affected by compressive pressure.

A major step toward lengthening the life and decreasing the operation problems of vacuum-filtration systems is the use of stainless-steel coil-spring filter media. A representative unit of this type, the Coilfilter, is shown in Figure 9.2 (Komline-Sanderson Engineering Corp.).

Elutriation

"Elutriation" is a process of improving filtration by washing the sludge. It reduces the alkalinity and, therefore, the lime coagulant demand of sludge by upgrading the biochemical quality of the sludge water before chemicals are added (Genter 1956). There are three practical methods of washing sludge solids; the equipment used in all cases is relatively simple, with upward-flow tanks frequently used.

1. Single-stage elutriation, which involves one batch at a time, is a fill-and-draw procedure; sedimentation and decantation are performed in a single step.

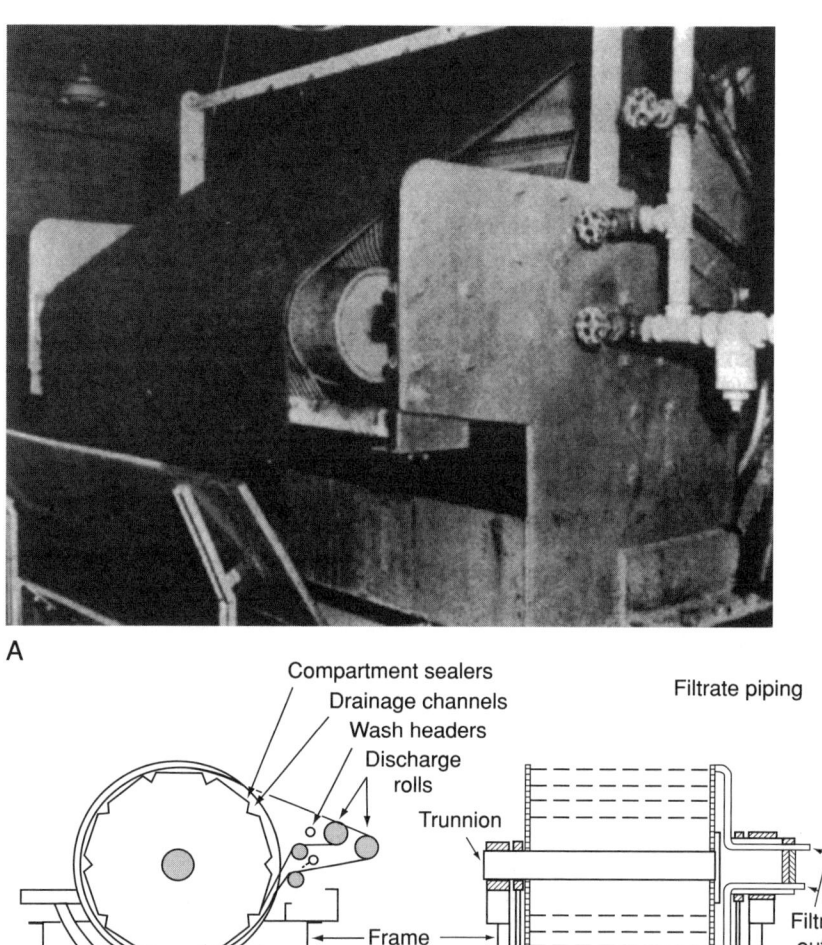

FIGURE 9.2. (**A**) The Coilfilter, a patented machine for the vacuum filtration of sludge. This particular machine, in use since 1953 at the sewage-treatment plant at St. Charles, Illinois, has filtering media made up of two layers of alloy steel coiled springs, each spring made endless by joining its two ends with a threaded plug. These springs discharge the filter cake after each revolution of the cylinder and are then washed before they reenter the vat for another cycle. The material at the left, which looks like a length of corduroy, is actually a layer of sludge (adapted from Komline-Sanderson Engineering Corp.) (**B**) Schematic drawing of the Coilfilter shown in (a) (courtesy Komline-Sanderson Engineering Corp.).

2. Two-stage elutriation involves repeating the single-stage steps on the elutriated sludge, using freshwater on the second wash. In small plants, the same settling tank may be used for both stages.
3. In larger plants (6,000–24,000 lb of solids per day), a second tank, connected in series with the first, is usually employed. Such a two-tank system can also be used for countercurrent washing. With this system, the freshwater is added only to the second-stage washing, and the decanted elutriate (or top water) from this tank flows by gravity to mix with the sludge entering the first tank.

Because the degree of chemical fouling (Genter 1946) resulting from digestion can be conveniently measured in terms of alkalinity, an elutriated sludge can be defined as one that has had the alkalinity of its water reduced by dilution with water of lower alkalinity, sedimentation, and decantation. Advantages of elutriation as a preliminary to sludge de-watering on vacuum filters include elimination of ammonia odors and of the need to use lime in sludge conditioning. Elutriation may also reduce the capacity requirements of secondary digesters (used for storage and additional digestion to ensure optimum filtration), and it is particularly helpful in that it permits small plants to use vacuum filters to advantage. Genter (1946) claimed that elutriation reduces the ratio of sludge water to the mineralized sludge solids; thus, there is a marked decrease in the chemicals required for conditioning. The savings in ferric chloride are illustrated in Figure 9.3, which is based on data from Genter (1946).

Genter (1956) also discusses a method of predicting the final alkalinity of elutriated sludge by a formula. Assuming that a equals the volumes of pure water added to one volume of fouled sludge mixture, he obtains the following relationships:

$a + 1$ = total volume of mixed sludge and clean water.
$1/(a+1)$ = fraction of original concentration of fouling agent left if solids are allowed to settle back to a washed-sludge equivalent to the original volume and the added volume of water is siphoned off.

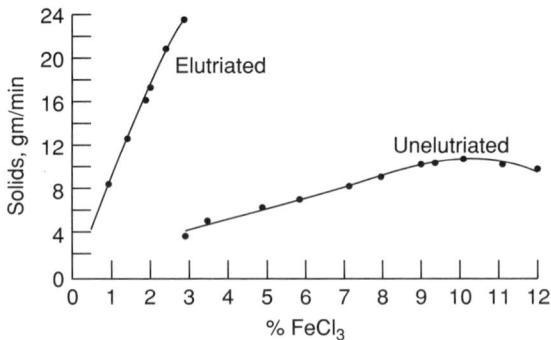

FIGURE 9.3. Effect of elutriation on $FeCl_3$ required for conditioning of sludge.

$1/(a+1)^2$ = fraction of original concentration of fouling agent if this same dilution, sedimentation, and decantation technique is repeated.

Therefore, the fraction of original fouling agent left in the final sludge is $1/(a^2+2a+1)$ if the second wash water is decanted for a new first wash and the two elutriation tanks are placed on countercurrent series. For example, if four volumes of pure water are used to wash a digested sludge of 3,000-ppm alkalinity, the alkalinity left in the elutriated sludge after countercurrent washing in two tanks is

$$(3,000) / [(4)^2 + (2 \times 4) + 1] = 120 \text{ ppm.}$$

Drying Beds

Sludge-drying beds remove moisture from sludge, thereby decreasing its volume and changing its physicochemical characteristics so that sludge containing 25% solids can be moved with a shovel or garden fork and transported in watertight containers.

Sludge filter beds are made up of 12–24 in. of coarse sand, well-seasoned cinders, or even washed grit from nearby grit chambers and about 12 in. of coarse gravel beneath the sand. The upper 3 in. of gravel particles are 1/8 to 1/4 in. in diameter. Below the gravel, the earth floor of the bed is pitched to a slight grade into open-joint tile underdrains 6 or 8 in. in diameter. These tiles may be laid from 4 to 20 ft apart on centers, depending on the porosity of the coarse gravel. Disposing of the underdrain liquor sometimes poses a problem; this should never be discharged without an analysis of its constituents and usually some form of treatment. Several smaller rectangular beds serve the purpose better than one large filter bed. These beds may be covered with glass or plexiglas when weather conditions demand, in which case ventilation must be provided to dissipate the hot wet air above the beds.

Generally speaking, raw settled sludge does not drain well on sand-drying beds. Some form of pretreatment—digestion, elutriation, and/or chemical treatment—is usually required. Well-digested sewage sludge will de-water more readily than partly digested sludge (American Society of Civil Engineers 1959). However, prolonged storage of digested sludges decreases drainability because the gases present initially permit more drainage of moisture through the filtering medium, thus reducing the evaporation cycle. A high total solids content in digested sludges naturally permits greater removal of dry solids per year from sludge beds.

Drying time is dependent on climate and dosing depth, 8 in. being generally accepted as most desirable for rapid drying. It is, naturally, short in regions of plentiful sunshine, scant rainfall, and low relative humidity, such as certain arid areas of the South where summers are long. Wind velocity also affects speed of sludge drying on the beds. In fact, all the factors enhancing evaporation will also aid in drying sludge. Cox (1940) derived the following equation for calculating the rate of evaporation of water, which may also apply to sludge-drying, although exact values of constants may vary from water to sludge water:

$$E = (e_a - e_d + 0.0016 \, \Delta T) / (0.564 + 0.051 \, \Delta T + W/300),$$

where

> E = evaporation (in./day)
> e_a = saturated vapor pressure at air temperature
> e_d = actual vapor pressure
> ΔT = difference between mean temperature of the air and that of the water
> W = velocity of the wind (miles/day).

Meyer's formulation is also widely used:

$$E = C (V-v) (1+W/10),$$

where

> E = evaporation (inches) for a given unit of time
> V = saturation vapor pressure at the water temperature (inches of mercury)
> v = actual vapor pressure of the air, 25 ft above ground
> W = wind velocity (mph), 25 ft above ground
> C = coefficient, varying with unit of time used and depth of water (varies from 0.36 to 0.50).

In addition to evaporation, the drying rate is also influenced by capillary action, which causes water to rise from the depths of the sludge to the evaporative surface.

In the case of domestic-sewage sludge, engineers estimate that approximately 20–25 lb of dry solids can be loaded onto 1 ft² of properly designed sand-base drying bed each year. Haseltine (1951) takes exception to this unit-of-loading estimate and suggests a "gross bed loading," which takes into account the number of pounds of solids applied per square foot per 30 days of actual bed use. For example, if sludge that has a density of 62.5 lb/ft³ and contains 10% solids is applied 12 in. deep and removed after 40 days, the gross bed loading is

$$(62.5 \times 0.10 \times 30) / (40) = 4.69 \text{ lb/ft}^2/30 \text{ days.}[1]$$

Haseltine (1951) also develops the following straight-line relationship between the gross bed loading (Y) and the percentage of solids in applied sludge (X) from data supplied by 14 plants for periods of operation up to 14 years:

$$Y = 0.96X - 1.75$$

The gross bed loading Y varied from 0 to 10 and X varied from 0 to 14. He concluded that, after temperature, the solids content of the sludge in drying beds is the

---

[1] The specific gravity of wet sludge assumed is 1.0.

most important factor influencing bed performance. The amount of moisture to be removed from the sludge is the third most important factor.

Sludge Lagooning

"Lagoons" may be defined as natural or artificial earth basins used to receive sludge. Lagooning is practiced when the economics of the situation (money and land) indicate its use, because it is a relatively inexpensive method of treating waste sludges. However, there are many other factors to be considered: (1) nature and topography of the disposal area; (2) proximity of the site to populated areas; (3) meteorological conditions, especially whether prevailing winds blow toward or away from populated areas; (4) soil conditions; (5) chemical composition of sludges, with special consideration given to toxicity and odor-producing constituents; (6) proximity to surface-water or groundwater supplies; (7) effect of waste materials on the porosity of the soil; (8) means of draining off the supernatant to provide more space in the lagoon; (9) fencing and other safety measures when lagoons are deeper than 5 ft; and (10) nuisances, such as weed growth, odors, and insect breeding.

Lagooning of wastes in limestone areas is particularly hazardous because of the channels and cavities found underground in these formations (Powell 1954). Ordinarily groundwater moves slowly, sometimes less than 1 ft/day, depending on the fineness of the aquiferous sand through which it percolates and the degree of saturation of the sand. In limestone country, water may travel vertically and laterally at much higher velocities, so that sludge lagooned on high ground may quickly contaminate large portions of valuable groundwater supplies. Manufacturing plants often bulldoze out a sludge lagoon every year or two, the frequency depending on sludge buildup and soil conditions.

Bloodgood (1946) stated that at least 1 lb of raw sewage solids can be digested per year per 0.17 ft^3 of lagoon capacity. However, if lagoons are to be used for both digestion and de-watering, 1 lb of raw-sludge solids requires about $0.4 \text{ ft}^3/\text{yr}$ of lagoon capacity, provided that air-dried sludge is removed as soon as it becomes ready for hauling.

Wet Combustion Process

The Zimpro process is a relatively innovative treatment for sludge. It operates on the basic principles that (1) organic matter contained in an aqueous solution can be oxidized, and whatever heat value it contains released, and (2) oxidation at this stage is more effective than if the water were first evaporated and the residue used as fuel in a conventional boiler. Because heat is liberated by a fuel only when it is subjected to combustion in the presence of air, the Zimpro process depends on air being forced into a reactor vessel. One objective of this process is the production of the maximum number of Btu's from the organic matter in a waste effluent per pound of compressed air fed into the reactor.

Because the Zimpro process eliminates conventional filters, chemicals, sludge-digestion units, incinerators, and auxiliary equipment, it reduces space and land requirements. The end-products are steam, nitrogen, CO_2, and ash. The effluent gases from the reactor, having been "scrubbed" with water, contain no fly ash and are practically odorless.

In the treatment of sewage sludge, oxidation is brought about by continuously pumping the sludge and a proportionate amount of air (both sludge and air at elevated temperatures and pressures) into a reactor vessel. Combustion occurs as the oxygen in the compressed air combines with the organic matter in the sludge to form CO_2, N_2, and steam, while the ash remains in the residual water. The reactor, and the whole process system, is automatically maintained at a constant pressure and the products of the combustion are continuously removed from the reactor. If the concentration of volatile matter is high and the sewage sludge concentration is great enough (>5%), the steam, plus the gases (CO_2 and N_2), which are products of combustion, will contain more than enough energy to run the air compressors and pumps used in the process. The residual hot water from the reactors is used in heat exchangers that raise the temperature of the incoming sludge and air sufficiently to cause oxidation to begin as soon as they come together in the reactor. In this way, once the process is started, no external heat or power is required to sustain the combustion.

Equipment required for the Zimpro process includes: compressor, air receiver, high-pressure sludge pump, sludge-storage tank with agitators, heat exchangers, reactor, separator, and cooler. A schematic drawing of the process is presented in

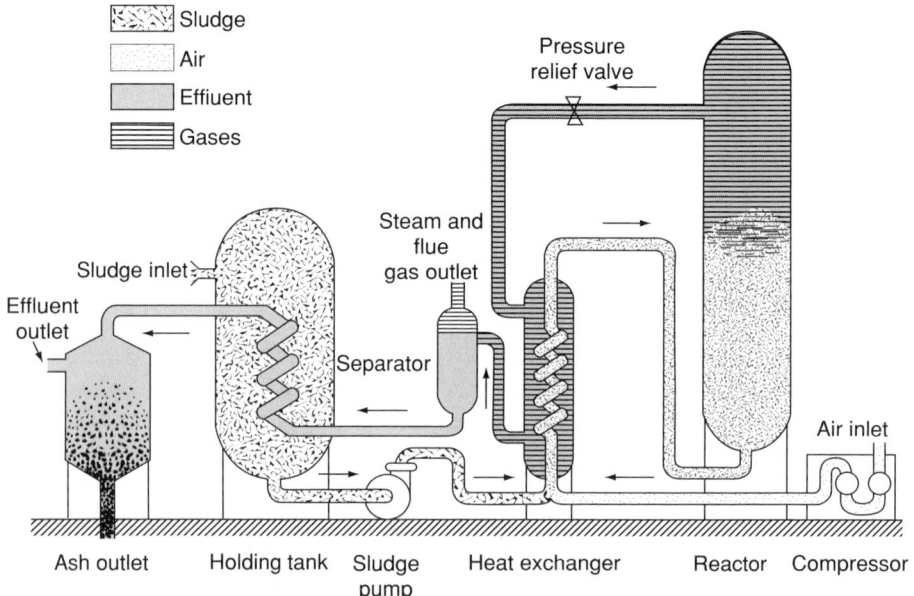

FIGURE 9.4. Schematic diagram of the Zimpro process for sewage-sludge oxidation (courtesy Sterling Drug Co.).

Figure 9.4. The manufacturer (New Zimpro Sludge Oxidation Units for Smaller Communities, Sterling Drug Co., Rothschild, Wisconsin) claimed that

> Units achieve 80 to 90% reduction of insoluble organic content of sewage sludge by oxidation without flame. Sludge is burned without de-watering or pretreating. The unit operates continuously at pressures of 500 to 600 psig and temperatures of 420°F. End products are substantially inorganic, inert, biologically stable ash; residual water; and odor-free gaseous products of combustion (carbon dioxide, nitrogen, and steam). The plant is designed for automatic operation with minimal maintenance. An air compressor and sludge pump are the only equipment components with moving parts. Power requirement is approximately 50 hp for a one-ton unit (dry weight). Building and land-space requirements are nominal.

Teletzke (1965) described the low-pressure Zimpro treatment, which operates in the range of 150–300 psi at about 300°F. He portrayed low-pressure wet-air oxidation as an economical and flexible method of producing a sterile, drainable, and completely acceptable end-product for ultimate disposal.

Atomized Suspension

The atomized-suspension technique consists of atomizing the waste liquor or slurry in the top of a tower, the walls of which are maintained at an elevated temperature by hot gases circulating through a jacket, a method described by Rabinovitch et al. (1956) and Gauvin (1957). No air, or other foreign gas, is introduced into the equipment, which sharply distinguishes this technique from spray drying. The developers claimed that in the immediate range of the nozzle, the finely divided droplets (20–25 mm in diameter) quickly decelerate from the high initial velocity and then become dispersed in the vapor produced by their own evaporation. The suspension thus created flows down the reactor in a nearly streamline motion. Evaporation, quickly completed, is followed by drying. At the end of the drying zone, dried particles can be subjected to a sequence of chemical reactions, such as oxidation, reduction, nitration, sulfonation, and so forth, through the injection of the proper internal gaseous reactants (in the presence of a powdered catalyst, if necessary). When it leaves the reactor at the bottom, the suspension consists of a solid residue (which is recovered in cyclone collectors), large amounts of steam (which is condensed and utilized), and by-product gases (which can be further processed for recovery or piped away for disposal).

Advocates of the atomized-suspension process claim that the only outside energy required is that used for pumping of the liquid, an almost negligible amount. A striking feature of the recovery flow sheet is the complete absence of blowers or compressors, although large volumes of gases and vapors are continuously flowing through the

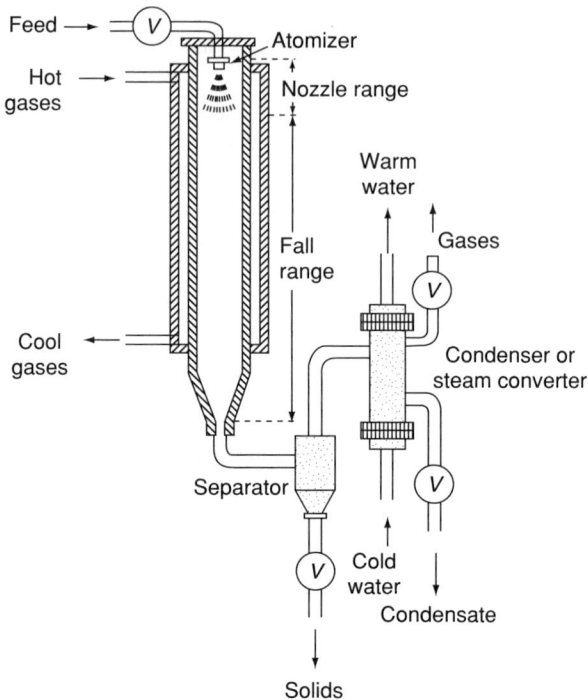

FIGURE 9.5. Apparatus for the atomized-suspension technique (adapted from Gauvin 1957 and Rabinovitch et al. 1956).

system. Need for them is eliminated by the efficient utilization of the pressure generated in the reactor during evaporation. A typical flow sheet for this process (Rabinovitch et al. 1956; Gauvin 1957) is shown in Figure 9.5.

Drying and Incineration

A large volume of sludge can be reduced to a small volume of ash, which is free from organic matter and, therefore, easily disposable, by a combination of heat drying and incineration (Dorr Co. 1941). Flash drying involves drying sludge particles in suspension in a stream of hot gases, which ensures practically instantaneous removal of moisture. When hot gases created by the drying and oxidation of the sludge itself are used directly for drying, there are no conversion losses. After the flash-drying, the gas containing sludge particles usually passes to cyclone separators, where the dried sludge is separated from the moisture-carrying cooler gases.

Flash-dried sludge is used as fertilizer, soil conditioner, or for other valuable purposes. Unused dried sludge can be incinerated by blowing it through a duct to a burner in the combustion chamber of a furnace. The sludge blower, in addition to conveying the sludge to the furnace, also supplies the major portion of the air required for combustion. To eliminate odors, preheated gases after combustion are returned to

Treatment and Disposal of Sludge Solids   163

the combusting sludge. To eliminate fly ash, the cooled gas after combustion is drawn through an ash collector by induced-draft fans, and the fly ash settles out by centrifugal action and is discharged automatically into the furnace bottom. This ash can be removed from time to time, either by shoveling or by mixing it with water and pumping it out to be used as landfill.

Whether the ultimate aim is to dry the sludge for use as a soil additive or to incinerate it to a sterile ash, it is necessary first to evaporate the free moisture from the solids, remove it in the form of a gas, and discharge it to the atmosphere. This gas is referred to as the *evaporator load*. Only high-temperature (1,200–1,400°F) deodorization is effective in controlling odors from sludge incinerators.

When sludge is to be incinerated, the heat released in the furnace is also important; the furnace volume should be ample to allow a heat release of X Btu/ft^3 of furnace (generally held at 12,000 Btu/ft^3 of furnace volume per hour to ensure long life of walls and furnace). The heat input is determined by multiplying the pounds of dry solids to be incinerated per hour by the gaseous products of the volatile-solids content and their heat value. The furnace volume required can, therefore, be computed by dividing this heat input by 12,000. Thermal efficiencies of 30–60% can be expected from incinerators. The lower the stack temperature, the higher the thermal efficiency (Leet et al. 1959). This relationship is shown in Figure 9.6. In Figure 9.7, a flow diagram is presented (Leet et al. 1959) to show heat balance for a flash-drying and incineration system.

To calculate the rate of drying during the constant-rate period (after the temperature of the material adjusts itself to the drying conditions), either the mass-transfer or the heat-transfer equation may be used (McCabe and Smith 1956):

(mass transfer) $W = k'_y(H_i - H)A;$

(heat transfer) $W = [h_y(t - t_i)A]/\lambda_i,$

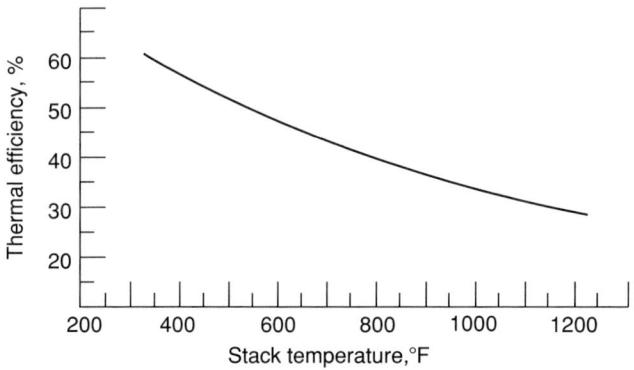

FIGURE 9.6. Effect of stack temperature on thermal efficiency (adapted from Leet et al. 1959).

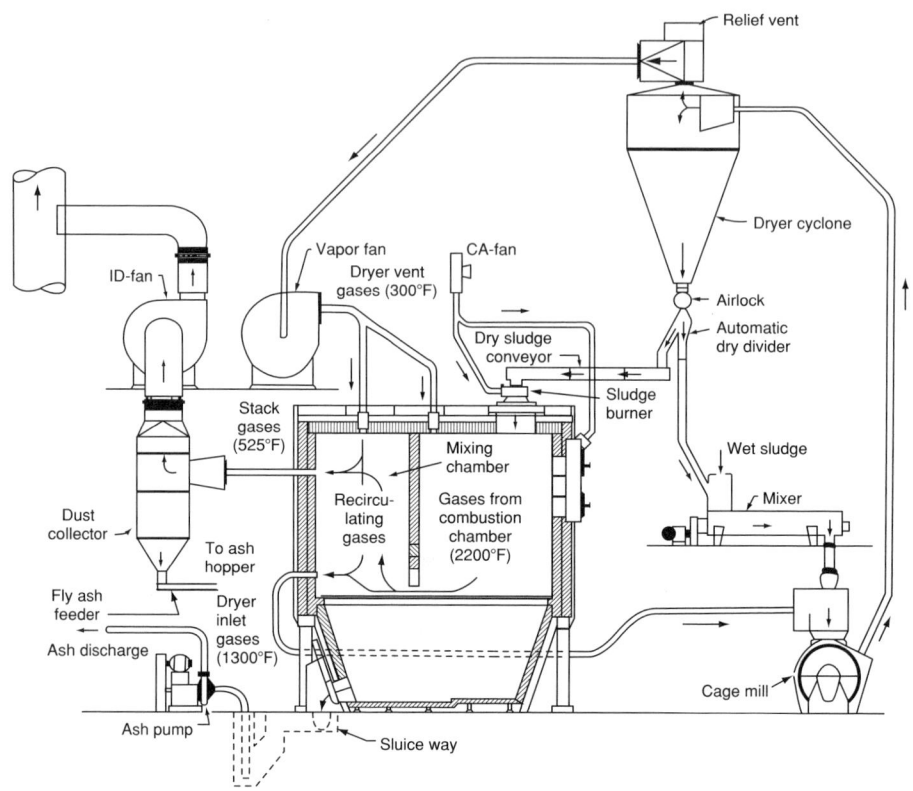

FIGURE 9.7. Flow diagram for heat balance for a flash-drying and incineration system (Leet et al. 1959).

where

W = evaporation rate (lb/hr)
A = drying area (ft^2)
h_y = heat-transfer coefficient (Btu/ft^2/hr/°F)
k_y = mass-transfer coefficient (lb/ft^2/hr for a unit of humidity difference)
H_i = humidity of air at interface (lb water/lb dry air)
H = humidity of air (lb water/lb dry air)
t = temperature of air (°F)
t_i = temperature at interface (°F)
λ_i = latent heat at temperature t_i (Btu/lb).

The heat-transfer coefficient, h_v, is estimated to be about $0.128\, G^{0.8}$ when air flows parallel to the sludge surface and about $0.37\, G^{0.37}$ when air flows perpendicular to the sludge surface (G = the mass velocity in lb/ft^2/hr).

Pit incineration has been used to dispose of certain solid and semisolid wastes. The incinerator consists of a rectangular pit lined with firebrick, to which air is supplied to retain particulates and to allow complete combustion. This disposal method is simple in concept and operation and is especially adaptable to situations in which the waste requires batch incineration. It has been used for disposal of synthetic organics and has been studied for disposal of paint sludges in the automotive industry (Balden 1967).

Dunn (1975) proclaimed that it was no longer a question of "shall we dump or shall we burn?" but "can we reclaim a product or recover waste heat from the waste?" He recommends a temperature in excess of 1,470°F (800°C) to oxidize carbon particles and 1,650–2,190°F (900–1,200°C) to remove odors from incinerator flue gas. Whether you burn solids, liquids, gases, or sludges, it is necessary to ascertain (1) the daily volume and weight, (2) whether wastes are batch or continuous, (3) what collection methods are used, (4) what methods of feeding the incinerator are required, (5) what the intended daily firing period is, and (6) whether waste heat recovery is required (or used elsewhere in the plant). Dunn (1975) acknowledged that submerged combustors were used to successfully recover HCl. When the price of caustic soda rises, this method of combustion becomes useful for the disposal of large quantities of chlorinated hydrocarbons (Santoleri 1972). Dunn (1975) also pointed out the increasing number of centralized incinerator disposal facilities in Europe because of the need for control over the handling and disposal of wastes.

Centrifuging

"Centrifuging" is a method of concentrating sludge to enhance final disposal. One of the factors that made centrifugal concentration unacceptable in the earlier installations was its low efficiency; large amounts of fine particles were returned to the system with the supposedly clarified effluent. Newer installations (Bradney and Bragstad 1955), using 20-hp built-in drive motors, can handle 3,000–4,000 gallons/hr of waste sludge, containing 0.5–0.75% solids on a dry basis. Only 11 hp are required once the centrifuge reaches operating speed (6,100 rpm). The resulting sludge is concentrated to about 5% solids and the effluent contains about 300 ppm solids. The centrifugal force throws the denser solid material to the wall of the centrifuge bowl, where it is discharged through nozzles located in the periphery. One bowl (Bradney and Bragstad 1955) is equipped with 12 nozzle openings, so that various numbers of discharge nozzles can be used depending on the amount of solids in the feed liquor and the results desired. Use of the centrifuge for higher concentrations is limited by the capability of the pumps, which discharge concentrated sludge from the centrifuges. The effluent from which the solids are separated travels toward the center of the centrifuge bowl through intermediate discs; as it discharges from the upper cover, it is claimed to average approximately 300 ppm solids. Centrifuged sludge is discharged from the lower cover of the centrifuge into a sump, from which it can be pumped to a digester or other final sludge-treatment units. Figure 9.8 is a schematic diagram of a centrifuge bowl. Blosser and Caron (1965) expect the costs of centrifuging paper-mill sludges to vary from $4 to $20 per ton of dry solids, including the hauling of the cake.

Ambler (1961) reviewed the theory of centrifugation. When a force is applied to a particle, the particle is accelerated ($F = ma$) until it reaches a velocity along the line of the force at which the resistance to its motion equals the applied force. In a settling tank, this

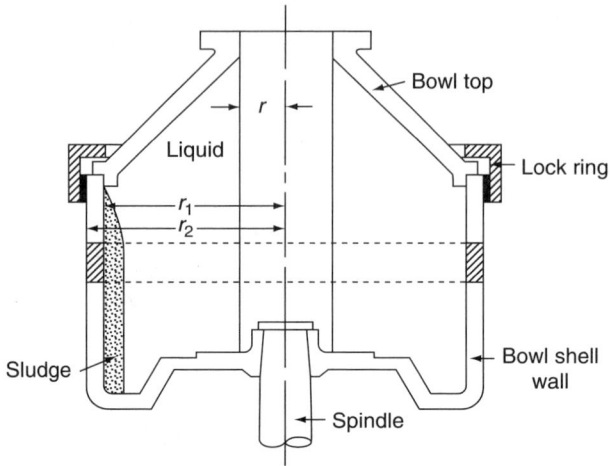

FIGURE 9.8. Centrifuge bowl schematic.

is the force of gravity. In a centrifuge, it is the centrifugal field, w^2r. The two differ only in direction and order of magnitude. The gravitational field is along a radius normal to the axis of rotation and may be upward of 60,000 times that of gravity for continuous-flow centrifuges. The velocities of particle movements are generally proportional to the square root of the diameter of the particle. The effective force acting on the particle is

$$F = (m - m1)w^2r,$$

and for a sphere, it is

$$F = \pi/6(d^3) \Delta p w^2 r.$$

The force opposing sedimentation, according to Newton's draw law in laminar flow, is

$$F = 3\pi\mu d.$$

At equilibrium (Stokes' law),

$$vs = [\Delta p \, d^2 \, (w^2)r] / 18\mu.$$

In the simplest form of a continuous centrifuge, vs is the velocity with which the particle, if it is heavier than the fluid, approaches the bowl wall. If X is the distance the particle will travel,

$$X = t - [\Delta p \, d^2(w^2)r(V)] / 18\mu \, Q$$

If X is greater than the initial distance and the given particle is from the wall of the rotor, the particle will be deposited against the wall and be removed from the system. In an ideal system (X = s/2), half the particles of diameter d will be removed. This may be considered the cutoff point at which

$$Q = (\Delta p \, d^2 / 9\mu) \cdot (V \, w^2 r / s),$$

where Q is volume of flow per unit of time.

Because the term $\Delta p d^2/9\mu$ is concerned only with the parameters of the system that follow Stokes' law and the term $V(w^2)r/s$ with the parameters of the rotor, the previous equation may be written as

$$Q = 2Vg\Sigma$$

in which

$$Vg = [\Delta p(d^2)g] / 18\mu \text{ and } \Sigma = [V(w^2)r\emptyset] / gs\emptyset,$$

where $r\emptyset$ and $s\emptyset$ are the effective radius and settling distance, respectively, of the centrifuge, and Σ is an index of centrifuge size that has the dimension of (length)^2 and is the equivalent area of a settling tank theoretically capable of doing the same amount of useful work as the centrifuge.

Ambler (1961) used the previous theory to formulate an index of centrifuge sizes for various centrifuge types as follows:

1. For the laboratory test-tube or bottle centrifuge,

$$\Sigma = (w^2)V/4.6 \log [2r^2/(r_1 - r_2)].$$

2. For the tubular-bowl centrifuge,

$$\Sigma = \frac{\pi/\omega^2}{g} \frac{(r_2^2 - r_1^2)}{\ln\left[2r_2^2 / (r_2^2 - r_1^2)\right]}.$$

3. For the disc-type centrifuge,

$$\Sigma = \frac{2\pi n \omega^2 \, (r_2^3 - r_1^3)}{3gC \tan \theta},$$

where

Σ = equivalent area of the centrifuge
w = angular velocity (rad/sec)
V = volume

r = radius from axis of rotation
r_1 = radius to inner surface
r_2 = radius to outer surface
l = light-phase discharge radius
g = gravitational constant
n = number of spaces between discs
C = concentration of solute
Ø = half-included angle of the disc.

In each of these cases, Ambler (1961) bases his calculations on the behavior of a single particle under conditions of unhindered settling and on the assumption that this particle is always in equilibrium with the force field of the centrifuge under the conditions defined by Stokes' law.

Sludge Barging

Sludge barging or ocean disposal is one of the means of final disposal of sludge that has been practiced by some cities. There is little theory involved in this method of treatment. Raw, precipitated, digested, or filtered sludge solids are pumped into a waiting barge and transported to a suitable site from the shore, where it is discharged, usually by pumping out deep under the water surface. There are some advantages of this method of disposal, such as relatively lower operating costs and reduced land demands. However, experience has shown that this method of disposal results in several environmental concerns: (1) long-term adverse effects on the ecology of the receiving water, (2) sludge floating matter rising to the surface, (3) public objection, and (4) potential for sludge residues carried to the shore during tidal cycles and causing public health impacts. Based on these concerns, this method of disposal has been discontinued and is not a recommended practice.

Sanitary Landfill

Sanitary landfill is used to bury garbage, refuse, and sludge in a planned and methodical manner (Salvato 1958). It is a relatively simple, effective, and inexpensive method for disposing of dry matter such as refuse, but sludge is usually too liquid for this procedure. However, mechanically de-watered or sand-bed–dried sludge can be disposed of in this matter.

The area proposed for the sanitary fill (Salvato 1958) should be easily accessible yet remote from sources of water supply and recreational areas while also being on land that is not too costly. The suitability of the soil and possible future use of the property are also important considerations.

For municipal refuse, the land area required is estimated at about 1 acre/yr for 10,000 persons, when using 6-ft-deep compaction. Sanitary landfills should be located above the groundwater level and no closer than 500 ft to any sources of water supply, particularly when the soil is sandy, gravelly, or of limestone derivation. The area should be staked out for trenches and benchmarks established, giving the elevation to which the finished fill is to be carried and the depth to which excavations are to be dug. Normally a trench is about 15 ft wide and about 4 ft deep. At the end of each day's dumping, the sludge should be covered and compacted by a bulldozer or tractor. Bacon (1967) suggested using sludge to reclaim land as an economic method of disposal, especially in marginal lands and coal strip-mining areas.

One of the major concerns for sludge disposal in a landfill is the presence of hazardous or toxic constituents in the sludge. Depending on the type of constituents, the soil type underlying the landfill, and the depth to groundwater, these constituents could leach out of the landfill and migrate and contaminate the groundwater. Heavy metals, precipitated during the industrial wastewater treatment, are some of the commonly found toxic constituents in sludge. Other toxic organics including petroleum-based and other chlorinated solvents could also be present in the sludge. Recently promulgated hazardous waste regulations by the Environmental Protection Agency (EPA) on land ban of hazardous chemicals have restricted the disposal of industrial wastes in landfills. In the context of these regulations, it is important to determine whether the constituents present in the sludge are banned from landfill disposal and whether the sludge is considered a hazardous waste based on special characteristic criteria specified by the EPA such as ignitability, corrosivity, reactivity, and leaching tests (EP Toxicity or TCLP) before deciding on the landfill-disposal option. Also, because of long-term potential liabilities of contaminating the groundwater, it is difficult to find landfills off-site that will accept the industrial sludge. On-site landfills may be constructed after obtaining permits from local regulatory agencies. Generally, it is required and recommended that landfills be constructed with appropriate liners to prevent the migration of the leachate containing contaminants to the subsurface and groundwater. These issues are discussed in more detail in Chapter 11.

Transporting to Approved Landfills

Transporting sludges to approved landfills was used with greater frequency during the 1990s. The reasons for this change in usage include increased liability and cost of ultimate disposal by on-site treatment. Safe and suitable landfills are designed to prevent groundwater seepage and surface overflows. In addition, these landfills are operated to prevent the inclusion of industrial sludges that are potentially hazardous.

Under current Resource Conservation and Recovery Act (RCRA) practice, both the manifesting and the analysis of sludges have become more complete and, therefore, definitive. In addition, the ultimate cost of facility closure, monitoring, and, if it proves necessary, post-closure mitigation are factored in the operation cost base of the landfill.

Examples of these industrial sludges include precipitated ferric hydroxide from steel mills and machine water fines from paper mills.

Miscellaneous Methods

Other methods for disposing of sludge solids include sludge concentration, flotation, and thickening. Biological means, aided only by temperature and time controls, are used to induce flotation of sludges (Laboon 1952). The resultant solids, in concentrations of 20%, do not require the addition of chemicals when they are subsequently de-watered on a vacuum filter. Optimum results with this method of concentration were found to exist at 35°C after a detention period of 120 hours (Laboon 1952). However, certain types of sludge (e.g., activated sludges) are not amenable to this treatment. Aside from time and temperature controls, the chief factors in the flotation method for concentration of raw sludges appear to be volatile content and pH.

In 1953 a method was developed by Torpey (1954) for thickening sludge on a continuous basis without the addition of chemicals. Generally, the flow pattern permits dilute sludge—from the primary clarifiers alone or combined with secondary sludge—to be fed to the center feedwell of a thickener. A schematic drawing of a typical thickener is shown in Figure 9.9. The solids settle, thicken in a definite "blanket" zone, and are drawn away from the bottom of the tank. The excess liquid is decanted by a peripheral weir. The thickeners also contain a mechanism with vertical pickets attached to the rake arms. The pickets are V shaped, and their channeling action allows entrapped water (water that is caught in sludge) and gases to escape to the surface. The degree to which the sludges can be thickened depends on several factors, the chief one being the source of the sludge (Brisbin 1956). The nature of the sludge is also most important. Some sludges are of a gelatinous and voluminous nature, which impedes thickening beyond a certain limit, regardless of detention time. Others are more granular and release

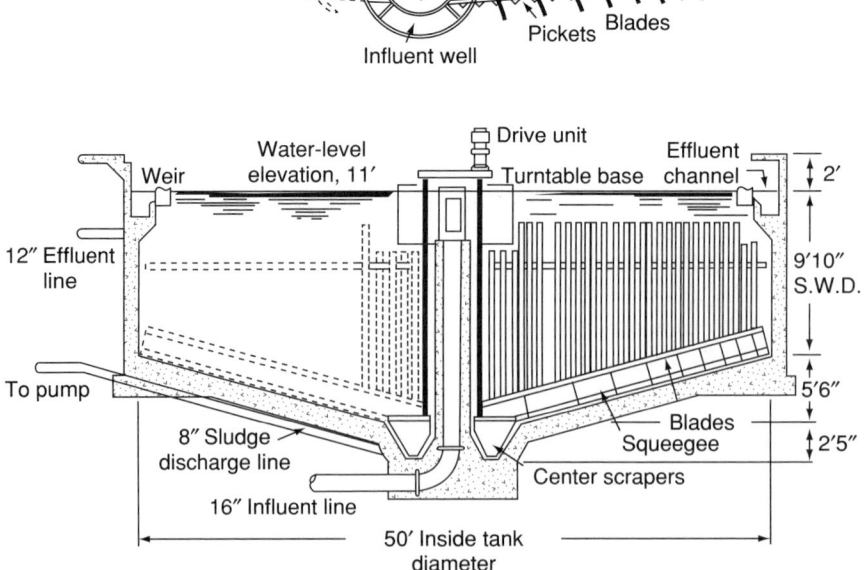

FIGURE 9.9. Schematic plan of thickener mechanism and section of tank (from Torpey 1954).

entrapped water when subjected to physical action, such as the slow mechanical mixing provided by the rotating pickets and rake arms.

Composting, a method of steeping solid wastes that contain 30–70% water in large piles and allowing microorganisms to decompose the organic fractions, has been used to some degree for solid wastes from industry. The process is accelerated when the piles are turned regularly by mechanical means. Mercer et al. (1962) found that the solid wastes of apricots and clingstone peaches were amenable to this form of treatment and that aerobic conditions were maintained by an initial daily turning for 5–6 days, followed by turning on alternate days until the process was complete.

The following includes six types of industrial sludges, along with their possible origin and more specific characteristics.

Industrial Types	Sources	Character
(1) Metal hydroxides	Plating Wastes	$Cr(OH)_3$ $Ni(OH)_2$ $Zn(OH)_2$
(2) Organic residues	Paper mill	Fines
	Tannery	Hair, skins, lime
	Cannery	Pulp, seeds, skins, fruits, and vegetables
	Textile	Fibers
	Winery	Dregs
	Sugar refinery	Lees
	Poultry and meats	Feather, innards, fat
(3) Precipitated colloids	Steel mill Pickle liquor	$Al(OH)_3$ $Fe(OH)_3$
(4) Inorganic	Cement mill Steel mill	Sand Iron
(5) Alkaline or neutral residues	Fertilizers Sugar	Gypsum + impurities Steffans Sludge
(6) Organic residues from land use	Agriculture Crop debris Animal dung	Bagasse, corn Stalks, peanut hulls Cow, pig, sheep manure, duck and chicken droppings

Example of Twentieth-Century Practice of Sludge Solids Removal

Shimp et al. (2000) reported that operating experience is inadequate in the newer variations in anaerobic digestion. Therefore, much research remains to be done. They recommended that managers of publicly owned treatment works considering any advanced digestion alternative should contemplate the following performance issues before deciding whether a particular system makes sense for their facility:

1. Class A pathogen reduction
2. Volatile solids destruction and gas production
3. Recycle nutrient loads, residual nutrients, and ammonia toxicity
4. Difficult-to-digest solids
5. Digester loadings and volume
6. Downstream de-watering
7. Residual odor
8. Implications for facility design, operations, and maintenance

Review Questions

1. Why is sludge treatment the most vital part of industrial-waste treatment?
2. How does anaerobic sludge digestion solve the problem of sludge treatment of industrial wastes?
3. How does it differ from aerobic sludge digestion? When would you use the latter rather than the former?
4. What factors influence the vacuum filterability of sludge?
5. How does elutriation aid in sludge treatment?
6. What are the principles and limitations of drying beds?
7. When would you recommend lagooning of waste sludge?
8. Under what conditions could a wet-combustion treatment be used for industrial-waste sludges?
9. What are the essential differences between Zimpro and atomized-suspension techniques for sludge treatment?
10. Discuss centrifuging as a sludge-concentrating process.
11. What are the advantages and disadvantages of sludge barging for ultimate disposal? How does this compare to sludge pumping into oceans as an ultimate disposal system?
12. Under what circumstances can you place industrial-waste sludges in a sanitary landfill?
13. What is the principle of sludge thickening and when is it used?

References

Ambler, C. H. 1961. Theory, centrifugation equipment. *Ind. Eng. Chem.* 53:430.
American Society of Civil Engineers. 1959. *Sewage Treatment Plant Design, Manual of Engineering Practice,* no. 36, p. 265. New York: American Society of Civil Engineers.
Bacon, V. W. 1967. Sludge disposal. *Ind. Water Eng.* 4:27.
Balden, A. R. 1967. The disposal of solid wastes. *Ind. Water Eng.* 4:25.

Barker, H. A. 1936a. On the biochemistry of methane formation. *Arch. Microbiol.* 7:404.
Barker, H. A. 1936b. Studies on the methane producing bacteria. *Arch. Microbiol.* 7:720.
Barker, H. A. 1937. The production of caproic and butyric acids by the methane fermentation of ethyl alcohol. *Arch. Microbiol.* 8:415.
Bloodgood, D. E. 1946. Sludge lagooning. *Water Sewage Works* 93:344.
Blosser, R. O., A. L. Caron. 1965. Centrifugal dewatering of primary paper industry sludges. In: *Proceedings of 20th Industrial Waste Conference, Purdue University, Lafayette, Indiana, May 4, 1965*, p. 450.
Bradney, L., R. E. Bragstad. 1955. Concentration of activated sludge by centrifuge. *Sewage Ind. Wastes* 27:404.
Brisbin, S. G. 1956. Sewage sludge thickening tests. *Sewage Ind. Wastes* 28:158.
Bruemmer, J. H. 1965. Use of oxygen in sludge stabilization. In: *Proceedings of 20th Industrial Waste Conference, Purdue University, Lafayette, Indiana, May 1965*, p. 544.
Buswell, A. M., W. D. Hatfield. 1939. *Anaerobic Fermentations.* Urbana, IL: Illinois State Water Survey. Bulletin no. 32.
Cox, G. N. 1940. *A Summary of Hydrologic Data: Bayou Duplantier Watershed, 1933–1939.* Baton Rouge: Louisiana State University. University Bulletin.
Dorr Co. 1941. *Sludge Drying and Incineration.* Stamford, CT: Dorr Co. Bulletin no. 6791.
Dunn, K. S. 1975. Incineration's role in ultimate disposal of process waste. *Chem. Eng.* October:141.
Eckenfelder, W. W. 1956. Studies on the oxidation kinetics. *Sewage Ind. Wastes* 28:983.
Federation of Sewage and Industrial Waste Association. 1959. *Sewage Treatment Plant Design, Manual of Practice,* no. 8, p. 214. Washington, DC: Federation of Sewage and Industrial Waste Association.
Gauvin, W. H. 1957. The atomized suspension technique. *TAPPI* 40:866.
Genter, A. L. 1946. Computing coagulant requirements in sludge conditioning. *Trans. Am. Soc. Civil Eng.* 111:635.
Genter, A. L. 1956. Conditioning and vacuum filtration of sludge. *Sewage Ind. Wastes* 28:829.
Harding, J. C., G. E. Griffin. 1965. Sludge disposal by wet air oxidation at a five mgd plant. *J. Water Pollution Control Fed.* 37:1134.
Haseltine, T. R. 1951. Measurement of sludge drying bed performance. *Sewage Ind. Wastes* 23:1065.
Komline-Sanderson Engineering Corp. Bulletin no. 102, 5-54. Peapack, NJ: Komline-Sanderson Engineering Corp.
Laboon, J. F. 1952. Experimental studies on the concentration of raw sludge. *Sewage Ind. Wastes* 24:423.
Leet, C. A., C. W. Gordon, and R. G. Tucker. 1959. *Thermal Principles of Drying and/or Incineration of Sewage Sludge.* New York: Combustion Engineering, Inc.
McCabe, W. L., J. C. Smith. 1956. *Unit Operations of Chemical Engineering,* p. 891. New York: McGraw-Hill Book Co.
Malina, F. H., Jr., H. N. Burton. 1964. Aerobic stabilization of primary wastewater sludge. In: *Proceedings of 19th Industrial Waste Conference, Purdue University, Lafayetle, Indiana, 1964,* p. 716.
Mercer, W. A., W. W. Rose, J. E. Chapmann, A. Katsuyama, F. Dwinell, Jr. 1962. Aerobic composting of vegetable and fruit wastes. *Compost Sci.* 3:3.

Miller, D. R. 1958. World's deepest submarine pipeline. *Sewage Ind. Wastes* 30:1426.
Powell, S. T. 1954. Industrial wastes. *Ind. Eng. Chem.* 46:95A.
Rabinovitch, W., P. Luner, R. James, W. H. Gauvin. 1956. The automized suspension technique. Part III. *Pulp Paper Mag. Can.* 57:123.
Rawn, A. M., F. R. Bowermann. 1954. Disposal of digested sludge by dilution. *Sewage Ind. Wastes* 26:1309.
Salvato, J. A. 1958. *Environmental Sanitation,* p. 288. New York: John Wiley & Sons.
Santoleri, J. J. 1972. Chlorinated hydrocarbons waste recovery and pollution abatement. *Nat. Incinerator Conf. Proc. A.S.M.E.,* 1972.
Schlenz, H. E. 1937. Standard practice in separate sludge digestion. *Proc. Am. Sec. Civil Engrs.* 63:1114.
Shimp, G. F., J. R. Stukenburg, J. Sandino. 2000. The future of solids treatment. *Water Environ. Technol.* November:37–39.
Sylvester, R. O. 1962. Sludge disposal by dilution in Puget Sound. *J. Water Poll. Control Fed.* 34:891.
Teletzke, G. H. 1965. Low pressure wet air oxidation of sewage sludge. In: *Proceedings of 20th Industrial Waste Conference, Purdue University, May 4–6, Lafayette, Indiana, 1965,* p. 40.
Tiller, F. H., C. J. Huang. 1951. Theory of filtration equipment. *Ind. Eng. Chem.* 53:529.
Torpey, W. N. 1954. Concentration of combined primary and activated sludges in separate thickening tanks. *Proc. Am. Soc. Civil Engrs.* 80(separate no. 443).
West, L. 1952. Sludge disposal experiences at Elizabeth, N.J. *Sewage Ind. Wastes* 24:785.

CHAPTER 10

Joint Treatment of Raw Industrial Waste with Domestic Sewage

Introduction

Industrial-waste problems should be solved using systems engineering, which when applied to industrial wastes implies combining the artful and scientific factors toward optimum solutions to treatment problems. When an engineer considers only the scientific or art factors involved in any waste-treatment problem, the solution is less than ideal. The more informed engineer uses as much of both art and science as possible in solving these problems. Even so, the result may be less than an optimum solution because of the interdependencies of many factors. Today's enlightened and progressive waste-treatment engineer takes both social and physical factors into consideration.

These factors can usually be applied to the major question of "which path to follow." As seen in Figure 10.1, industry must decide whether to treat its own waste or to contract with the municipality to accept, treat, and dispose of its wastes. Twelve potential combinations of alternatives, or side paths, are also available to an industrial plant. For example, an industrial plant may decide to partially treat its waste (alternatives 2, 4, 7, or 10) and then deliver the residual-waste volume and load to the municipal treatment plant for final treatment and disposal. It is vital then to point out and discuss the many factors of both types that will assist us in selecting the correct alternative path to follow, and hence the overall solution to these problems.

Before examining the factors that affect joint treatment decisions, the following four rules are useful in selecting an alternative.

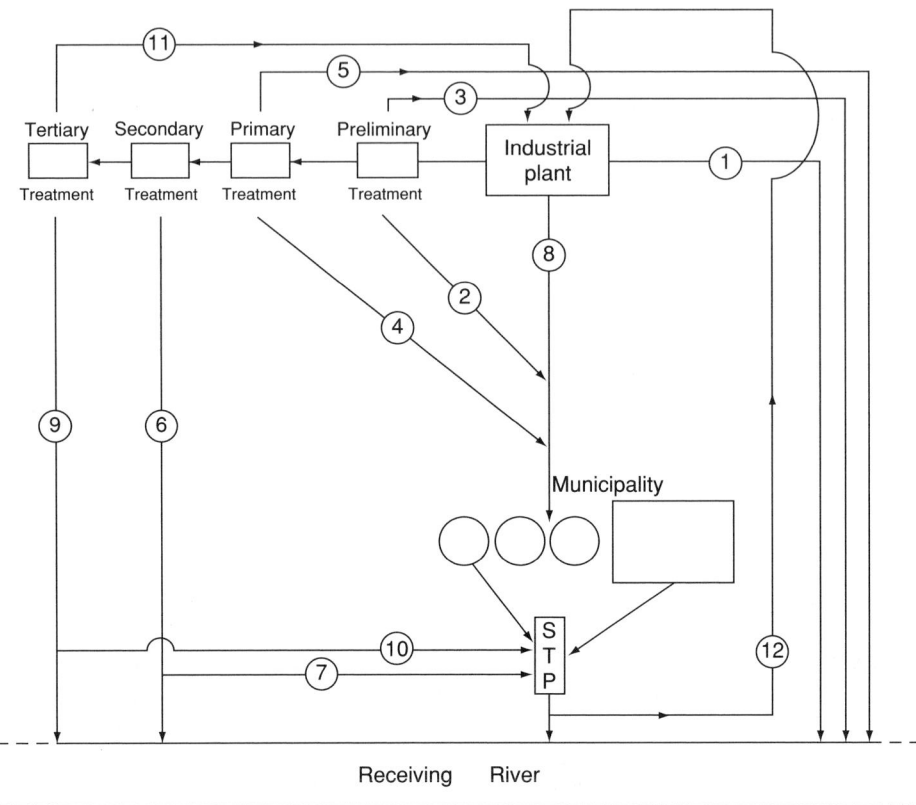

FIGURE 10.1. Twelve alternatives of industrial-waste treatment systems.

	Industry	Municipality
(1) Objective	To produce the best product at least cost	To provide best living conditions for people at lowest cost
(2) Conflict	Production causes increased municipal costs (air, water, and land environmental protection)	People depend upon industry for work and money, but it destroys the environment at the same time
(3) Solution	Produce goods in amounts and types so as to minimize adverse environmental impact	Only purchase goods that are produced and sold economically and that do not cause environmental degradation
(4) Procedure	Select lowest-cost alternative that protects the environment	Inform people of the environmental costs of production and adverse environmental impacts

Elements of the art and science involved in determining the optimal alternative are listed below:

A. Art in industrial wastes
   1. Precedent
   2. Social relationship of industry and municipal officials
   3. Political compatibility of industry and municipality
   4. Sewer service charge
   5. Location of industry, especially in relation to municipal plant and receiving stream
   6. Competence of municipal plant operator
   7. Permanence of industrial production
B. Science in industrial wastes
   1. Type of municipal sewage treatment
   2. Character of industrial waste
   3. Required water quality of receiving stream
   4. Volume of industrial waste in relation to that of municipality
   5. Economics of alternatives

In the following, I will consider the art or social factors first, because they have been less publicized and are perhaps less well known and understood by engineers.

Art in Industrial Wastes

Precedent

People are most inclined and even biased towards that which they are most familiar with or that which has been done rather uniformly and continuously in the past. There is a form of security in knowing that what is being suggested is in keeping with past practice and that its effects have been previously experienced. If, in the past, the practice has been to include all types of industrial waste regardless of volume or character in the municipal treatment system, this practice may be expected to be repeated. On the other hand, if separate treatment of industrial waste and municipal sewage has been encouraged, it will be difficult to change. Any change in policy will require considerable effort, education, and enthusiasm of all parties concerned. The engineer must recognize that, even with proper theoretical justification, it may be impossible to overcome and alter that which has become practice by precedent.

Social Relationship of Industry and Municipal Officials

The consulting engineer must be aware of and appraise the social state of the relationship between the appropriate members of the industrial and municipal community. For one reason or another, this relationship may have been enhanced by a previous event that makes future compatibility for joint treatment possible. Likewise, some past experience, sometimes minor or seemingly insignificant, may have negatively affected the relationship and made future negotiations for joint treatment rather difficult. A designing engineer must develop the art of detecting and evaluating this compatibility at the onset of project planning. An oversight at this stage or even an inappropriate consideration or evaluation can be disastrous in terms of the search for an overall optimum path of solution to the joint problems.

Political Compatibility of Industry and Municipality
When plant managers and major municipal officials share political outlooks, leaning, cooperative solutions to waste treatment are facilitated. The engineer must determine this relationship and consider the likelihood of a change in municipal officials in an election year during the progress of the project.

Sewer Service Charge
Municipalities are composed of elected officials, each possessing a philosophy on the subject of patronage or subsidy for local industry. Some prefer to accept all wastes at a minimum charge or flat fee, or on a water-use basis. Others believe that industry should solve its own environmental problems, and should it want to have the municipality treat its waste, it should be ready and willing to pay a sewer service charge that correctly and adequately covers capital and operating costs. The engineer should ascertain which type of feeling prevails in the municipality prior to establishment of such changes. Industry usually does not object to a reasonable charge but often objects to the philosophy of a comprehensive, complicated, or involved charge that may lead to many unnecessary confrontations. It might prefer to treat its own waste, regardless of "on-the-surface" economics, rather than face potential disagreements with public officials.

Location of Industry
The location of the industrial plant relative to both municipal sewers and treatment facilities and the final receiving stream plays a vital part in influencing industry for or against combined treatment. At first consideration, one might presume that economics alone is affected by plant location. True, the cost of connecting to and transmitting through existing or potential municipal sewers for a given distance directly affects costs. However, accessibility to sewers and proximity to treatment plants and receiving streams is important from a psychological viewpoint as well. Questions such as "why have two treatment plants so close to each other?" or "why pump and maintain a private pumping station and connecting sewer systems when we can treat our own wastes and dispose of them by gravity nearby?" are difficult to answer only in terms of dollars. Proximity enhances joint solutions regardless of economic considerations.

Competence of Municipal Plant Operator
The experience, congeniality, and prominence of supervisor and/or operator of a municipal treatment facility will enhance or hinder cooperative arrangements between industry and government. These qualities of the operating personnel are especially important in the initial stages of decision making. For example, industry usually considers combined treatment feasible and even desirable when the municipal operator is held in high esteem by associates on a state or national level. The basis for this feeling is deeply rooted from a psychological standpoint, but most probably stems from a feeling of security in knowing that the wastes will be treated in an optimally effective manner. An "easy-to-get-along-with" operator will enhance initial and continued discussions and negotiations. Combined treatment always manifests problems and difficulties of one type or another. Most can be solved with some effort, ingenuity, and cooperation. When

industry knows a priori that it is dealing with a person or people possessing these characteristics, combined treatment becomes practical. An experienced operator usually knows what plant operating changes to make during critical situations in order to avoid malfunctioning. This is reassuring to both industrial and municipal officials. Conversely, the lack of those operator characteristics demands greater consideration of separate industrial-waste treatment.

Permanence of Industrial Production
Industry must show a profit for its owners in order to stay in business. The future of any given industrial facility depends primarily on the whims of consumers. When whims change, a given industrial plant has three alternatives: (1) reduce production, hopefully cutting costs so that net profits do not deteriorate; (2) change production to meet consumers preferences; and (3) cease all production to avoid deficit financing. In all three alternatives, the type or quantity of waste is altered considerably. Many industrial operations have found it necessary to resort to any or all of these alternatives over a 20- to 40-year period (during which time bonds for waste-treatment facilities are usually amortized). Industry has begun to recognize this as a fact of doing business.

Because technology is changing so rapidly and with it the needs and demands of customers, industry hesitates to enter into long-term contractual commitments, especially those concerned with waste-treatment facilities. Thus, industry is now recognized as being relatively impermanent when compared to municipal operations. Certain industries are more permanent or less apt to cease production entirely (as evidenced from past records only). For example, DuPont Chemical Corporation or U.S. Steel's main production facilities could be expected to continue operation—with some modifications—with greater assurance for longer periods into the future than a small leather tannery or a peach cannery. The former types of industries would be more likely to consider joint ownership of a treatment facility than the latter. The cannery or tannery would be inclined towards private ownership of its separate facilities. There are exceptions and even variations within the latter industries, mostly depending on the size and type of ownership of the industry, as well as specific past experiences that it may have had in waste-treatment problems.

Some of the more important scientific factors associated with joint industrial-waste treatment are discussed in the following section.

Science in Industrial Wastes

Type of Municipal Sewage Treatment
A secondary biological treatment plant, if adequately sized, can best be used to treat a readily decomposable organic-laden industrial waste. Typical examples include dairies, canneries, slaughterhouses, and tanneries. However, each of these wastes, as well as other typical organic wastes, contain contaminants that can interfere with effective treatment when combined with domestic sewage. For example, dairy wastes often turn to acid extremely fast and the lowered pH can affect biological oxidation. Many

canneries have an extremely alkaline lye-peel waste, which when discharged in slug loads can hamper biological oxidation. Slaughterhouse wastes contain slug loads of grease and blood that could hinder physical and biological processes. Tannery wastes contain chromium, sulfides, and lime, which are not compatible with normal sewage treatment. Proper pretreatment and plant operations, however, can remedy these problems. In some cases, it has been shown that trickling filters can handle industrial waste with fewer upsets than activated-sludge systems.

Little would be gained by either party if an industrial waste (such as from a dairy or textile mill) was treated in a municipal treatment plant consisting solely of primary treatment.

Industrial plant managers must make an assessment of the objectives of treatment required and then examine the municipal plant available or being planned. Mutually compatible objectives and treatment facilities would be conducive to combined treatment.

Characteristics of Industrial Waste

When considering the treatment of wastes from a tissue-paper mill, industry needs a municipal plant that concentrates its equipment units on the removal of the finely divided suspended-solids area of waste treatment. It is of little benefit to the tissue-paper mill if the municipal plant possesses only a high-rate trickling filter primarily designed for biochemical oxygen demand (BOD) removal. Along similar lines, a metal-plating waste would not be a welcome addition to the high-rate trickling filter plant because of its potentially toxic metals and acids, as well as its lack of organic matter—hence, the need for BOD reduction.

The waste engineer must carry out a complete analysis of the industrial waste to ascertain its compatibility for treatment by various possible methods. Some analyses often overlooked by the traditional sanitary engineer include the k-value, the waste deoxygenation rate, ultimate oxygen demand, toxic chemicals and metals, temperature, grease content at reduced temperatures, refractory organic matter, phosphates and nitrates and other algae nutrients, and so on.

Receiving-Stream Water Quality

It is a foregone conclusion that a stream which must be maintained in a high water-quality state requires maximum offshore waste treatment. Generally, this means a minimum of the equivalent to secondary treatment. But what is secondary treatment to an industry whose waste contains varied types of contaminants? Often the conventional biological treatment system will not adequately remove sufficient amounts of the contaminants. Sometimes specific treatment such as chemical precipitation followed by adsorption on activated carbon may remove more industrial contaminants than a secondary type of trickling-filter plant. Industry has inherited the moral, if not the legal, obligation of treating its waste in a manner so as to maintain the highest possible quality of water in the receiving stream. This cannot always be accomplished by following established state or federal rules. In many instances, ingenuity must be used and sacrifices must be made to remove the proper amount of specific contaminants. In one specific waste problem, it was necessary for management to render a decision to completely eliminate one of its three raw materials. This particular raw material was relatively inexpensive, and thus a satisfactory product generated high profits. However, the waste resulting from using this particular raw material accounted for more than 80% of the total plant contamination. To remove the contamination by conventional treatment proved extremely costly, and after much deliberation, the industry reluctantly decided to discontinue use of the cheap raw material.

It is even more disconcerting for industry to be faced with a legal decision to install so-called *complete treatment* of its wastes when the receiving stream, even under critical conditions, shows little deterioration from the waste load. In this situation, industry should endeavor to prove to regulatory agencies its dilemma so that it can operate at optimum economic efficiency. The latter is essential for the long-term benefit of society, as long as valuable river resources are not degraded in the process.

Volume Ratio of Industrial to Municipal Wastes

A relatively small volume of industrial waste can usually be assimilated in a municipal sewage-treatment system regardless of its contaminants. Municipal plant operators generally react optimistically towards small volumes of any industrial wastes, agree to try to treat them, and end up accepting them with or without certain preconditions. If the ratio of waste volume to sewage had been greater, an attempt might never have been made to handle the waste, regardless of the potential of acceptability or treatability. Thus, when ratios are high, industry usually builds its own treatment plant despite the potentially favorable economics or the potential compatibility for joint treatment. There are exceptions to this generality, but they usually exist when a new facility is being contemplated by both municipal and industrial officials simultaneously.

Economics of Alternatives

Industry tends to select the least costly alternative, especially when other conditions are equal. How does industry select the least costly system? Usually industry prefers to compare alternative system costs on the basis of total capital expenditures. This method can often be misleading and even erroneous over the longterm. First, it does not take into consideration the annual cost of operation and maintenance required for effective treatment. Often, the least costly alternative can be the most expensive to maintain, especially when perfection is required in efficiency of operation.

Furthermore, the cost of obtaining loans for capital spending is not taken into account when considering only capital costs. This is significant during high-cost periods and when one or more alternatives entail public rather than private borrowing. It is well known that municipalities can usually borrow money at lower interest rates than private corporations.

And last, the financial rating of the borrower is a significant and often overlooked factor in comparing capital costs only. Certain public authorities may possess such poor financial ratings that many prominent profitable industrial firms may be able to borrow money as cheaply as or cheaper than the municipality.

Therefore, economic alternatives should be based on net annual costs, which include operation, maintenance, and amortization of capital costs. The latter should be selected on the realistic basis of the financial rating of the appropriate borrower, as well as on the basis of a reasonable bond amortization period, depending on the borrower's ability to repay and the expected life of the specific equipment required in each alternative.

The question of whether the optimum alternative selected is justified on any economic basis warrants consideration and up to now has been avoided by sanitary engineers. We have blindly presumed that what is required in the way of waste treatment by state or federal edict is required at any cost. This may be so, but the case can be strengthened by proper economic evaluation. We should compute the total net benefits to the surrounding immediate society of the specific waste-treatment alternative selected. Heretofore, we have lamented that these benefits were intangible and hence unmeasurable. (See Chapter 21 of this book for more information about benefits.)

Benefits that are affected by water quality include: (1) recreational use; (2) land use; (3) water withdrawal; (4) waste treatment; and (5) in-place water use.

Industrial Use of Municipal Sewage Plants

It is often possible and advisable for an industry to discharge its waste directly to a municipal treatment plant, where a certain portion of the pollution can be removed (Nemerow 1951). A municipal sewage-treatment plant, if designed and operated properly, can handle almost any type and quantity of industrial waste (Hubbel 1935). Hence, one possibility that should be seriously considered is the cooperation of industry and municipalities in the joint construction and operation of a municipal wastewater treatment plant. There are many advantages to be gained from such a joint venture:

1. Responsibility is placed with one owner while the cooperative spirit between industry and municipality is increased, particularly if the cost sharing is mutually satisfactory.
2. Only one chief operator is required, whose sole obligation is the management of the treatment plant. The operator is not encumbered by the miscellaneous duties often given to the industrial employee in charge of waste disposal, and the chances of mismanagement and neglect that may result if industrial production people operate waste-treatment plants, are eliminated.
3. Since the operator of such a large treatment plant usually receives higher pay than separate domestic plant operators, better-trained people are available.
4. Even if identical equipment is required, construction costs are less for a single plant than for two or more. Furthermore, municipalities can apply for state and/or federal aid for plant construction, which private industry is not eligible to receive.
5. The land required for plant construction and for disposal of waste products is obtained more easily by the municipality.
6. Operating costs are lower because more waste is treated at a lower rate per unit of volume.
7. Possible cost advantages resulting from lower municipal financing costs, federal grants, and municipal operation can be passed on to the users and may permit

higher degrees of treatment at a cost to each participant no greater than the cost for separate treatment at lower removal levels.
8. Some wastes may add valuable nutrients for biological activity to counteract other industrial wastes that are nutrient deficient. Thus, bacteria in the sewage are added to organic industrial wastes as seed material. These microorganisms are vital to biological treatment when the necessary BOD reduction exceeds approximately 70%. Similarly, acids from one industry may help neutralize alkaline waste from another industry.
9. The treatment of all wastewater generated in the community in a municipal plant or plants enables the municipality to ensure a uniform level of treatment to all users of the river, and even to increase the degree of treatment given to all wastewater to the maximum level obtainable with technological advances.
10. Acceptance of the joint treatment project and relinquishment of individual allocations would give the municipality full control of river resources and permit it to use the capacity of the river to the best advantage for the general public. The municipality has greater assurance of stream protection, because it has the opportunity for closer monitoring of effluent quality.
11 Public relations are good for the municipality.
12. Land is generally more available.
13. No permit for discharge is needed except for a contractual agreement between the two parties.

Among the many problems arising from combined treatment, the most important is the character of the industrial wastewater reaching the disposal plant. Equalization and regulation of discharge of industrial wastes are sometimes necessary to prevent rapid change in the environmental conditions of the bacteria and other organisms that act as purifying agents, to ensure ample chemical dosage in coagulating basins, and to ensure adequate chlorination to kill harmful bacteria before the effluent is discharged to a stream.

Two factors in particular have focused on the subject of combined treatment for sewage and industrial wastes: increased interest in stream-pollution abatement, and the phenomenal growth of industry in the postwar years with the subsequent increase in demand for water.

Because most sewage plants use some form of biological treatment, it is essential for satisfactory operation that extremes in industrial waste characteristics be avoided and the waste mixture be (1) as homogeneous in composition and uniform in flow rate as possible and free from sudden dumpings (shock loads) of the more deleterious industrial wastes; (2) not highly loaded with suspended matter; (3) free of excessive acidity or alkalinity and not high in content of chemicals that precipitate on neutralization or oxidation; (4) practically free of antiseptic materials and toxic trace metals; (5) low in potential sources of high BOD, such as carbohydrates, sugar, starch, and cellulose; and (6) low in oil and grease content.

Combined municipal and industrial waste treatment is the most desirable arrangement, and at the same time, the most difficult to achieve. It is the author's contention that the difficulty is usually not a scientific one but one of human compatibility and understanding. In numerous cases, the economics of the situation were overwhelmingly in favor of combined treatment and yet separate waste-treatment installations were finally used, primarily because of personality clashes and lack of sympathetic understanding.

What can municipalities do to assist industry in waste-treatment practices?

1. If the municipal treatment plant is new or being enlarged, it should be designed to serve the entire community, with all the existing and planned industry as members of a "corporation." Combined meetings, lectures, and actual visits to the treatment-plant site will enhance mutual understanding of the problems involved.
2. If the plant is already in operation, municipal officials should meet with industrial representatives and discuss the advantages and disadvantages of accepting industrial wastes into the system. Adequate safeguards such as a literature review, and pilot-plant experiments should precede administrative decisions, and any decision to accept the waste should be accompanied by a substantial and specifically detailed contract between the owner and user. Methods of sampling, analyses, charges, and waste characteristics should be clearly stated in this contract.
3. A municipality can purchase land, build a treatment plant for its industry, float bonds, and receive rent from industry for use of the plant and amortization of the bonds. In this way, the rental becomes an operating expense that industry can deduct from income *before* taxes rather than the long-term depreciation of capital assets involved in having its own plant. This feature is particularly attractive to industry.
4. Most important, perhaps, is the understanding that a municipality must have for its industry. All members of the city council should be in agreement that without industry, municipal survival is doubtful and its growth potential is nil. An industry located within the city limits is contributing taxes and intangible benefits to the city. Although it is common for new industry to locate outside city limits, where it can purchase sufficient land at a reasonable cost for future expansion, these industries, too, are an indirect but valuable addition to the community.
5. A municipality can design its treatment plant so that it will handle an industry's waste without pretreatment by the industry. An ideal arrangement is to take the industry into the business of waste disposal as a "member of the municipal corporation." Industry should pay for this service but not at the same rate as an individual householder, because a large contributor deserves concessions solely on the basis of lower unit costs for larger volumes. In addition, industry's intangible benefits to the community should be assessed and its share of the capital cost reduced proportionately.
6. Municipal controls of the influent from the industrial plant are costly and difficult to establish. Instead, it is recommended that industry control its own effluent so that the "corporation" disposal system operates efficiently. When, and if, the

system ceases to function as designed, a corporation meeting should be called to decide what measures should be taken to correct the situation. In this manner, the expenses of sampling and billing, as well as ill feelings caused by policing, are eliminated.
7. A good corporation will continually try to improve its efficiency of operation by conducting research on new methods of treatment. Research has been proven to pay off in the long haul and the lack of it has often led to plant and process obsolescence. Industry, with its research experience, could well lead the way in this connection by supporting continued research in specific combined treatment processes.
8. The designing environmental engineer should be selected by both the municipality and the industry for his or her competence and ability to work with equal ease with both groups. His or her fee should be on a lump-sum basis, approximating the sliding-scale percentage of estimated construction cost but not necessarily tied to this; rather than being penalized for reducing the capital costs (and, hence, his or her fees), he or she should be rewarded financially for economizing on construction and improving plant efficiency.

In summary, a municipality can assist its industry by encouraging mutual understanding, embarking on a program of education, and designing its plants to handle industrial waste. Other methods of assisting industry depend on the formation of a "corporation treatment plant" with joint responsibility for efficient operation. Municipal ownership and reduction in charges based on intangible and tangible industrial benefits should also be considered.

The Water Pollution Control Federation (WPCF) (1976) published a rather detailed (34 pages) pamphlet on all aspects of joint treatment. The reader is directed to this pamphlet for practical and economic aspects of this subject.

Municipal Ordinances

Although there are many types of municipal ordinances, all are designed to place an upper limit on the concentration of various constituents in waste. Sometimes this upper limit is zero, because any quantity of a certain contaminant would be detrimental to the plant or its component parts. In addition to their obligation to abide by municipal ordinances, many industries enter into separate contracts with the city. Generally, such contracts include the obligation of the municipality to construct, operate, and maintain the treatment facilities and to finance the overall project by means of some type of bond; a declaration on the part of the industry as to the maximum quantity of flow, BOD, and solids; the percentage by volume of industrial waste as compared to municipal waste; the amount the industry will pay each year to cover operation and maintenance; provision for a penalty if stated limits are exceeded; and any other pertinent matters involving joint usage of the treatment system.

The following are dangers of inadequate sewer-use control (WPCF 1976): (1) explosion and fire hazards; (2) sewer clogging; (3) overloads of surface water (storm- or cooling-water pollution); (4) physical damage to sewers and structural damage to treatment plants; and (5) interference with sewage treatment.

A comprehensive sewer ordinance (Federation of Sewage and Industrial Wastes Association 1957) usually consists of the following principal parts: introduction; definition of terms; regulation requiring use of public sewers where available; regulations concerning private sewage and waste disposal where public sewers are not available; regulations and procedures regarding the construction of sewers and connections; regulations relating to quantities and character of waters and wastes admissible to public sewers; special regulations; provision for powers of inspectors; enforcement (penalty) clause; validity clause; and signatures.

Because industrial wastes vary so greatly in character, only broad limits can be established in any model ordinance, and ordinances should always be based on recommendations of the consulting engineer (Table 10.1). Most ordinances (Federation of Sewage and Industrial Wastes Association 1957) for the control of waste substances other than sanitary sewage act in the following ways:

1. They prohibit the discharge to the public sewers of flammable substances or materials that would obstruct the flow.
2. They state that industrial wastes will be admitted to the public sewers only by special permission of a stated municipal authority.
3. They ban all wastes that would damage or interfere with the operation of the sewage works, except when such wastes have been adequately pretreated, and even then their admission is to be at the discretion of a stated municipal authority.
4. They enumerate in detail, in a separate ordinance, the procedures outlined in no. 3.
5. They give detailed regulations to supplement the procedure in no. 3, stating specific limits for objectionable characteristics of industrial wastes.

A model ordinance (Federation of Sewage and Industrial Wastes Association 1957) may spell out in detail the following regulations relating to quantities and character of water and wastes admissible to public sewers:

Section 1. No storm water, roof runoff, cooling water, groundwater, etc., will be allowed in the sanitary sewer.

Section 2. Storm water or other uncontaminated drainage will be discharged to sewers that are designated combined or storm sewers only.

Section 3. No person shall discharge any of the following wastes to sanitary sewers except as hereinafter provided: (a) any liquid or vapor having a temperature higher than 150°F; (b) any waste containing more than 100 ppm by weight of grease; (c) any gasoline, etc., or other flammable or explosive liquid, solid, or gas; (d) any garbage that has not been properly ground; (e) any ashes, metals, cinders, rags, mud, straw, glass, feathers, tar, plastics, wood, chicken manure, or other interfering or obstructing solids; (f) any

TABLE 10.1
Industrial Contaminants and General Limiting Values for Discharge into Municipal Sewerage Systems[a]

Contaminant	Concentration Generally Limiting for Municipal Sewerage Systems	Reason for Limitation	If Contaminant Is Excessive, Acceptable Pretreatment Generally Required
1. Flow	50% of municipal sewage flow	Causes sewage treatment system to react differently from its normal pattern as designed for municipal sewage. Unequalized or unproportioned industrial flow is especially troublesome	1. Equalization and proportioning 2. Recirculation and reuse within industry to reduce flow 3. Redesign sewage treatment plant to react more specifically to industrial waste
2. BOD 5, 220°C	300 ppm	Exerts a disproportionately high percentage of oxygen-demanding organic matter to municipal wastewater	1. Change in industrial manufacturing process 2. Equalization 3. Biological pretreatment plant
3. Color	Visible in dilutions of 4 parts sewage to 1 part industrial waste	Color is normally not removed by domestic sewage treatment plants, will appear in the combined, treated effluent, will be readily detected and visually undesirable from an aesthetic standpoint	1. Change in industrial manufacturing process 2. Chemical pretreatment to remove color 3. Equalization and/or proportioning
4. Suspended solids	350 ppm	Overloads disproportionately normal domestic sewage treatment plants	1. Change in industrial manufacturing process 2. Equalization 3. Sedimentation pretreatment plant
5. pH	5.5–9	Corrosion of sewers and treatment plant equipment causes a diminution or malfunction of biological treatment units	1. Equalization 2. Neutralization 3. Change in industrial manufacturing process

TABLE 10.1 (continued)

Contaminant	Concentration Generally Limiting for Municipal Sewerage Systems	Reason for Limitation	If Contaminant Is Excessive, Acceptable Pretreatment Generally Required
6. Grease	100 ppm	Interferes with plant operating equipment—including aeration, primary sedimentation, etc. Overloads sludge-handling treatment units such as scum collection, digestion, and sludge-drying beds	1. Change in industrial manufacturing process 2. Install grease traps or remove pretreatment units
7. Heavy metals Cr, Sn, Pb, Zn, Hg, Cu, Ni, etc.	1 ppm Cu, Cr, 5 ppm Zn, N	Inhibits biological action in municipal sewage units, such as activated sludge, and trickling filters, and especially sludge digesters	1. Equalization 2. Chemical pretreatment and sedimentation pretreatment
8. Nonorganics and other toxic chemicals	None so as to be toxic to bacteria serving the treatment plant or people or animals working in or near the sewage plant	Exhibit toxicity towards biological treatment units and cause health hazard to people and animals	1. Change industrial plant process 2. Use of advanced pretreatment wastewater techniques
9. Inflammable liquids, foaming agents, rags, solidifiable greases, ashes, metals, cinders, mud, straw, glass, feathers, tar, plastics, wood, chicken manure, etc.	None in such quantities that will cause either a hazard to the environment or a nuisance to plant operation	Causes a nuisance and interferes with the normal operation of the domestic sewage-treatment plant	1. Removal by process change or physical means such as screening

10.	Temperature	150°F	Hastens corrosion, drives out dissolved oxygen, volatilizes hazardous gases such as H_2S	1. Change in industrial process 2. Use of cooling water systems
11.	Storm water	None resulting from direct connections or faulty sewer construction	Occupies valuable volume capacity of domestic and industrial sewers	1. Construct separate sewer 2. Use better construction procedures
12.	Refractory organic matter	None	Contaminates municipal sewage plant effluent for possible reuse downstream for water supplies	1. Change in industrial manufacturing process 2. Carbon adsorption pretreatment
13.	Refractory mineral matter	Boron 0.7 ppm NaCl 1000 ppm	Contaminates municipal sewage plant effluent for possible reuse for irrigation waters	1. Change in industrial manufacturing process 2. Pretreat industrial plant wastewater by membrane separation or distillation

[a] All industrial wastes can be treated in some manner at some cost with some relative effectiveness so as to render the contaminants suitably low in concentration or changed in character with the result that they may be discharged safely into the environment either directly or indirectly (through some municipal system). It is suggested that a literature survey be made of the effect of each contaminant at its expected level of concentration in the resulting domestic wastewater before acceptance or rejection of the industrial waste into any municipal system.

In cases where doubt exists about the effect of the industrial waste based upon personal experience or literature survey, laboratory and/or field prototype studies should be made to ascertain the precise effect on municipal sewage-plant systems. Pretreated effluents from industry can be examined in a similar manner.

The exact values of each contaminant are based on the best evidence available to the author, but exceptions can be found in each case. The user of this table must be prepared to render to it some flexibility in both allowable concentration and type of contaminants as more knowledge becomes available.

Source: Nemerow (1987).

wastes having a pH less than 5.5 or higher than 9.0 or having other corrosive effects; (g) any toxic wastes that may be hazard to sewage plant, persons, or receiving stream; (h) any suspended solids that the treatment of which at the sewage plant may involve unusual expenditures; and (i) any noxious gases.

Section 4. There shall be installations of interceptors for grease, oil, and sand when necessary.

Section 5. These installations shall be maintained by owner.

Section 6. This section establishes the conditions pertaining to the admission of any wastes having (a) a 5-day BOD greater than 300 ppm, (b) more than 350 ppm suspended solids, (c) any of the quantitative characteristics described in Section 3, and (d) an average daily flow greater than 2% of the average daily flow of the city.

Section 7. Where preliminary treatment facilities are provided for any wastes, they shall be maintained by the owner at his or her own expense.

Section 8. When required, the owner of any property served by a sewer carrying industrial wastes shall install a suitable manhole for observation, sampling, and measuring.

Section 9. All measurements and analyses of the characteristics of waters and wastes referred to in Section 3 or 6 shall be determined in accordance with standard methods (American Public Health Association 1955).

Section 10. No statement contained in this article shall preclude any special agreement or arrangement between the city and any industry.

Municipal ordinances change with time as a result of changing federal and state regulations and the cost of service. Table 10.2 contains the 1994 sewer-use ordinance from the city of Palo Alto, California. The reader is advised to compare the ordinance provisions shown in Table 10.1 with those published by Palo Alto. Selected discharge requirements from Palo Alto are found in Table 10.3.

Sewer-Rental Charges

Sewer-rental charges are necessary to help meet the city's budget and to ensure that industry pays a fair share of the cost of disposing its wastes. Several methods can be used to charge for sewer service: (1) an *ad valorem* tax on property, which is the traditional method in more than 80% of U.S. communities and is successful in small towns and villages; (2) special assessments, with charges set according to front footage; (3) sewer-rental charges (approximately one-sixth of municipalities with treatment plants use this method); (4) special contracts negotiated with industry; and (5) combination of two or more of the previous methods. In many cases, a municipality charges the industry or industries solely on the basis of water consumption. Although this may not always prove equitable, it has several advantages to the municipality and to the industries. First, the billing system is simplified, omitting the need for detailed and time-consuming cost procedures. Second, the system eliminates the need for measuring flows from the industries and their strength characteristics. Thus, the municipality treats industries just as it does householders, rather than as "culprits."

TABLE 10.2
Sewer Use Ordinance, City of Palo Alto Regional Water Quality Control Plant (September 1995)

The City of Palo Alto Sewer Use Ordinance reflects the current thinking and concerns expressed by the Environmental Protection Agency as well as the State of California, and therefore contains many pages and great detail. Elements include:

Industrial waste discharge permit and procedure
Compliance schedules
New and existing sources (of discharges)
Reporting requirements
Personnel orientation
Modification, suspension, or revocation of industrial wastes discharge permit
Permit issuance, denial, modification, revocation, or suspension hearing
Discharger monitoring
Trucker's discharge permit
Limitations on point of discharge
Confidentiality
Accidental discharge prevention
Storage of hazardous materials
Discharger self-monitoring
Prohibitions
Copper-based root control chemicals
Grease disposal and pretreatment
Strom drains: threatened discharges
Standards

Toxicant	Instantaneous Maximum Concentration Allowable
Arsenic	0.1 mg/L
Barium	5.0
Beryllium	0.75
Boron	1.0
Cadmium	0.1
Chromium, hexavalent	1.0
Chromium, total	2.0
Cobalt	1.0
Copper	2.0
Cyanide	1.0
Formaldehyde	5.0
Lead	0.5
Manganese	1.0
Mercury	0.05
Nickel	0.5
Phenols	1.0
Selenium	1.0
Silver	0.25
Zinc	2.0

TABLE 10.3
Maximum Allowable Discharge Limits for Wastewater

Pollutant	Maximum Concentration Allowable (mg/L)
Arsenic	0.1
Barium	5.0
Beryllium	0.75
Boron	1.0
Cadmium	0.1
Chromium (Hex)	1.0
Chromium (Total)	2.0
Cobalt	1.0
Copper[a]	2.0
Cyanide	1.0
Fluoride[b]	65
Formaldehyde	5.0
Lead	0.5
Manganese	1.0
Mercury	0.05
Nickel[b]	0.5
Phenols	1.0
Selenium	1.0
Silver[b]	0.25
Single toxic organic[b]	0.75
Total toxic organics[b]	1.0
Zinc	2.0
Oil/grease[b]	200
pH[b]	5.5–11.0
Suspended solids[b]	3,000
Total dissolved solids[b]	5,000

[a] For discharge >50,000 gpd, the maximum concentration will be one-half of value listed.
[b] Copper limit –0.25 mg/L, effective 7/1/98.

In considering charges, fixed costs such as operation, maintenance, and debt retirement should all be taken into account. A portion of each of these three costs can be charged to all users of the sewer system, and the remaining portion to property owners having access to the system. This is done by itemizing the cost of each component unit of the sewer system and then allocating percentages of the annual cost of each unit to users and the rest to property owners. Total annual charges to users and property owners are determined from the summation of unit costs. The total share allocated to property owners may now be prorated according to individual property valuations (or sometimes front footages). The user's share necessitates additional prorating based on the following waste factors: volume, suspended solids, BOD, and (sometimes) chlorine demand. This is carried out in the computations of the user's share for each unit. If the unit is designed solely on a volume basis, such as the main sewage pumps, the entire user's share is charged to volume contributors of suspended solids and 10% to contributors of BOD. If the volume of sewage is based on water consumption and the supply is private (wells or river water), a meter is normally supplied by the industry for flow measurement.

When all user charges attributed to volume, solids, and BOD are added, one obtains the total user cost for each category. The total of the three in turn represents the user's share of the annual sewer costs, and the total of the user's and property owners' shares represents the complete annual sewer costs.

Schroepfer (1951) uses the following example to illustrate a fair allocation of costs. The total annual cost of operating a sewage-disposal system in a certain town consists of the following:

1. Fixed charges:
   - Intercepting sewers[1] $35,000
   - Treatment plant[1] $75,000
2. Operating and maintenance costs: $70,500
   - Total $180,500

Table 10.4 shows the allocation of the fixed charges for the sewers and treatment plant, Table 10.5 the allocation of the operation and maintenance costs, and Table 10.6 the allocation of fixed and operation charges. Figures 10.2 and 10.3 illustrate the prorating of the fixed charges of the sewers and treatment plant.

Property owners' charges should be distributed according to assessed evaluation, which in the example under discussion is taken to be $20,000,000. Therefore, $2.88 per $1,000 of property valuation will be charged to property owners. User charges depend on flow, solids, and BOD, as stated previously. Hence, the annual flow and the quantities of each type of these waste loads must either be determined after the first year's operation or be estimated prior to establishing the users' charges for the year. The third column of Table 10.7 lists the unit rates obtained from the data given in columns 1 and 2.

As the twentieth century came to an end, municipal services and treatment facilities became more complete and sophisticated. As a result, joint treatment with industrial wastes also became more complicated, demanding of industry, and protective of municipalities. It still occurred, but with many more safeguards to protect the overall joint system. Also, both parties became more acutely aware of the net result of the joint system on the receiving environment and the sewage system.

---

[1]Capital investment for intercepting sewers, $700,000, and for treatment plant, $1,500,000. Debt retirement is 5% per year (total interest and principal).

TABLE 10.4
Allocation of Fixed Costs

Units	Total Fixed Costs, $	Chargeable to Property Owners		Chargeable to Users, $	Users' Share Chargeable to					
		%	$		Volume		Suspended Solids		BOD	
					%	$	%	$	%	$
Intercepting sewers	35,000	64	22,300	12,700	100	12,700				
Treatment plant										
Main pumping station										
Equipment	1,500	40.5	600	900	100	900				
Structures	1,250	64	800	450	100	450				
Screen and grit chambers	1,500	64	950	550	60	330	40	220		
Preliminary sedimentation tanks	4,500	40.5	1,800	2,700	85	2,300	15	400		
Trickling filters	30,000	25	7,500	22,500	10	2,250			90	20,250
Final sedimentation tanks	9,000	30	2,700	6,300	50	3,150			50	3,150
Receiving pumps	750	25	200	550					100	550
Chlorination tanks and equipment	2,000	35	700	1,300	40	520			60	780
Digester tanks and receiving filters	8,000	30	2,400	5,600			100	5,600		
Subtotal	58,500	30.3	17,650	40,850	24.2	9,900	15.2	6,220	60.6	24,730
Main control building	7,500	30.3	2,300	5,200	24.2	1,310	15.2	790	60.6	3,100
Plant water supply	2,500	30.3	800	1,700	24.2	410	15.2	260	60.6	1,030
Roads and grounds	2,500	30.3	800	1,700	24.2	410	15.2	260	60.6	1,030
Plumbing and heating	4,000	30.3	1,200	2,800	24.2	680	15.2	430	60.6	1,690
Total plant costs	75,000	30.3	22,750	52,250	24.2	12,710	15.2	7,960	60.6	31,580
Total fixed costs	110,000	41	45,050	64,950	39.2	25,410	12.2	7,960	48.6	31,580

Source: From Nemerow (1987).

TABLE 10.5
Allocation of Operation and Maintenance Costs

Unit	Total	Chargeable to Property Owners		Chargeable to Users, $		Volume	Users' Share Chargeable to			
							Suspended Solids		BOD	
	Operation and Maintenance Costs, $	%	$	$	%	$	%	$	%	$
Intercepting sewers	2,200	60	1,300	900	60	500	40	400		
Main pumping station	9,200	17	1,600	7,600	100	7,600				
Preliminary treatment	6,700	50	3,400	3,300	50	1,700	50	1,600		
Secondary treatment	13,500	15	2,000	11,500	10	1,200			90	10,300
Effluent chlorination	5,200	15	800	4,400	10	400			90	4,000
Sludge disposal	17,500	5	900	16,600			100	16,600		
General	5,000	15	800	4,200	25	1,000	43	1,800	32	1,400
Supervisory	6,200	15	900	5,300	25	1,300	43	2,300	32	1,700
Collection and billing	5,000	15	800	4,200	25	1,000	43	1,800	32	1,400
Total	70,500	17.8	12,500	58,000	25.4	14,700	42.1	24,500	32.5	18,800

Source: Nemerow (1987).

TABLE 10.6
Summary of Allocation of Fixed and Operating Cost

	Chargeable to			
	Users		Property Owners	
Fixed Costs	%	$	%	$
Sewers	36	12,700	64.0	22,300
Treatment plant	69.7	52,250	30.3	22,750
Operation and maintenance costs	82.2	58,000	17.8	12,500
Totals		122,950		57,550
Averages	68.1		31.9	

Source: Nemerow (1987).

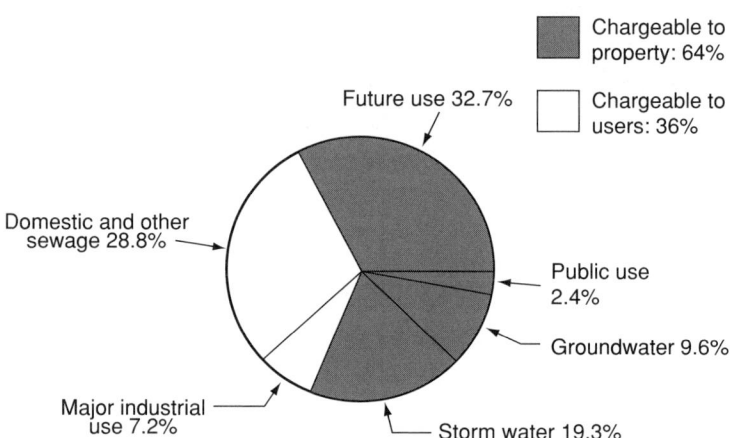

FIGURE 10.2. Allocation of fixed charges on the intercepting sewers. (Adapted from Schroepfer 1951)

Case History of Project for Joint Disposal of Untreated Industrial Wastes and Domestic Sewage

For the purposes of this discussion, I consider the case of two relatively small municipalities containing 27 small industries (mostly tanneries) that require adequate and effective treatment of their wastes. The problem presents a challenge in engineering, economics, and administration.

Existing Situation

Cayadutta Creek rises in the central part of Fulton County in New York State, flows generally south for about 14 miles through the cities of Gloversville and Johnstown, and

Joint Treatment of Raw Industrial Waste with Domestic Sewage 197

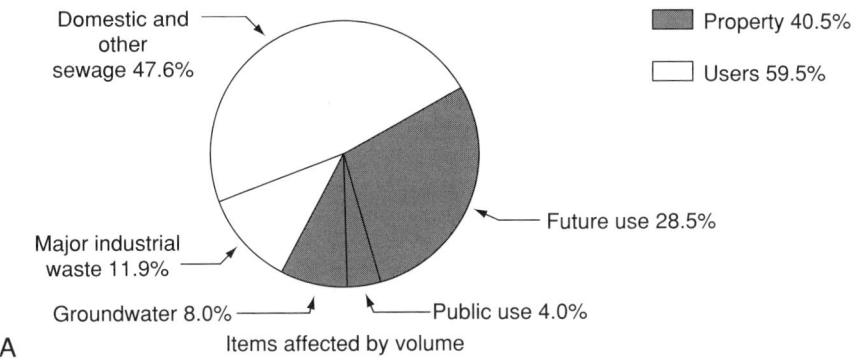

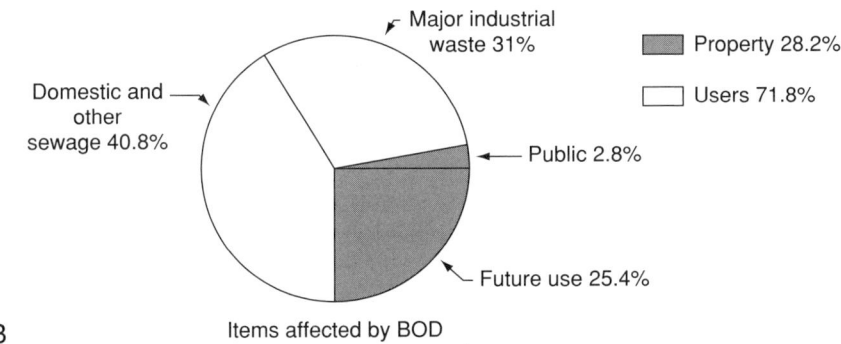

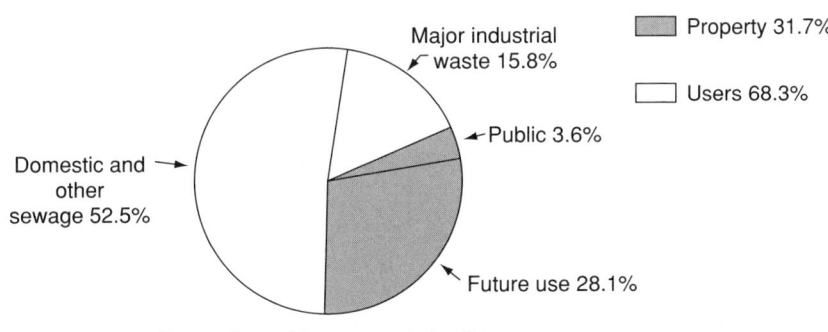

FIGURE 10.3. Allocation of fixed costs for treatment plant. (A) Items affected by volume. (B) Items affected by BOD. (C) Items affected by suspended solids. (Adapted from Schroepfer 1951.)

TABLE 10.7
Calculation of Users' Charges Based on Three Factors

	Annual Quantity	Total, $	Unit Rate, $
Volume of flow	1,370 million gal	40,000	2.93/1000 gal
Suspended solids	3,647,000 lb	32,460	0.89/100 lb
BOD	3,847,000 lb	50,380	1.40/100 lb

Source: Adapted from Schroepfer (1951).

enters the Mohawk River at Fonda (Figure 10.4). The total catchment area covers 62 square miles above Station 6. There are no official gauging stations on this stream, but approximate flow data for a comparatively short time (1898–1900) are available. This creek has been characterized by an expert state hydrologist, a member of the U.S. Geological Survey, as similar to that of Kayaderosseras Creek, which is located in Saratoga County and drains into the Hudson River basin.

Ninety-one percent of the population of the Cayadutta Creek catchment area is concentrated in the cities of Johnstown and Gloversville and the village of Fonda. In 1952, New York State cited this creek as "one of the most grossly polluted streams in the state." From within the city of Gloversville to the junction with the Mohawk River, the stream was entirely unsuited for the support of fish life, whereas formerly it was trout water throughout its entire length. It has been stated (M. Vrooman, *March 10, 1950, Report to City of Gloversville*) that the dry weather flow of Cayadutta Creek is higher than the average for streams in the state, because of the nature of the watershed, the sandy soil, and the larger wooded area. Vrooman also stated that "the average daily flow of the creek at the Gloversville sewage-treatment plant is 17 million gallons and the low measured dry weather flow is 4.2 million gallons." These figures were evidently obtained from separate, independent, and unofficial flow measurements. The tanning industry is an old one in the United States and has a record of contributing to the damaging pollution of Cayadutta Creek. The National Tanners Association (private communication) estimates that the Fulton County area has been losing about one plant every 4 years. However, there has been more glove- and garment-leather demand as the population of the United States rises. They predict, therefore, that the overall demand for glove and garment leather (produced in Fulton County) will continue slowly upward, but it will be met by fewer plants with increased production.

The sewage and wastes from the cities of Gloversville and Johnstown are discharged into Cayadutta Creek. In 1960, Gloversville had a population of 21,741 while that of Johnstown was 10,390 (Figures 10.4 and 10.5). Almost the entire urban population is served by public sewer systems, but only half of the system is tributary to a sewage-treatment plant, which serves the people and industry of Gloversville. It consists of a bar screen, grit chamber, two antiquated Dortmund-type primary settling basins, a fixed-nozzle trickling filter, one final Dortmund-type settling basin, and some sludge-drying beds. The plant was built in the early 1900s and is incapable of handling more than half the wastewater at the present flow rates.

The tanning industry retained a New York City consulting firm to represent its interests in this problem. The two cities retained a local consulting engineering firm,

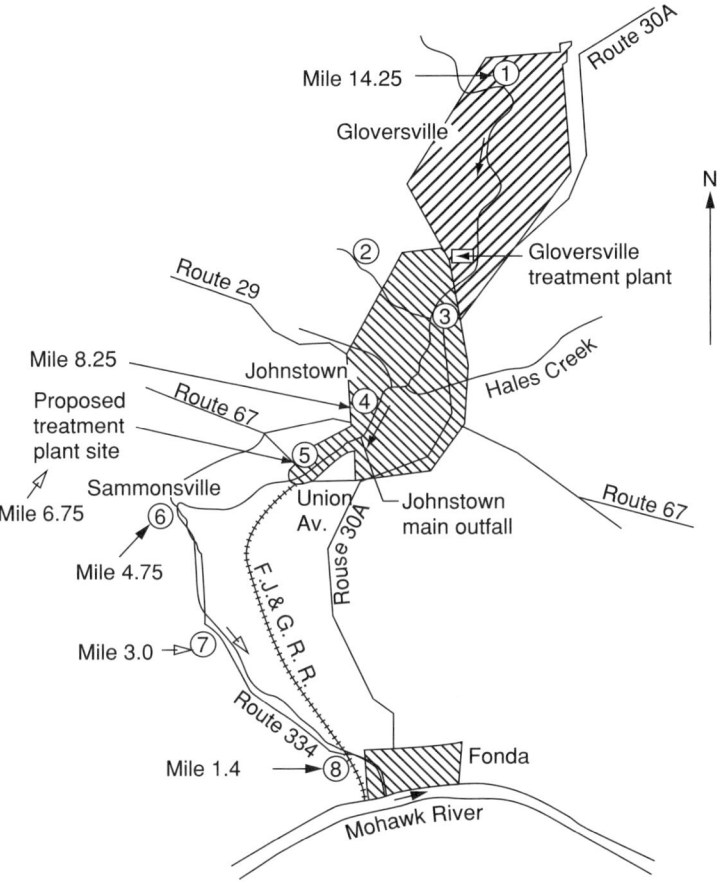

FIGURE 10.4. Cayadutta Creek.

which in turn retained the author to advise them on study procedures and solutions to the pollution problem in Cayadutta Creek.

Stream Survey

A stream survey is an essential part of any well-conceived waste-treatment study. Ideally, a survey designed to study the oxygen-sag curve should be carried out during extremely hot weather, extremely low stream flow, and typical high organic matter loading. It is seldom possible to conduct a stream survey under all of these "ideal" conditions. In this study, I was particularly fortunate to collect stream samples during extremely low-flow conditions—comparable to those that may be expected to occur for a 7-day period only once in 10 years—while the municipal and industrial pollution loads were considered to be above average. Although stream temperatures were not high (11–14°C), these values are never very high because of the relatively cold mountain water diluting

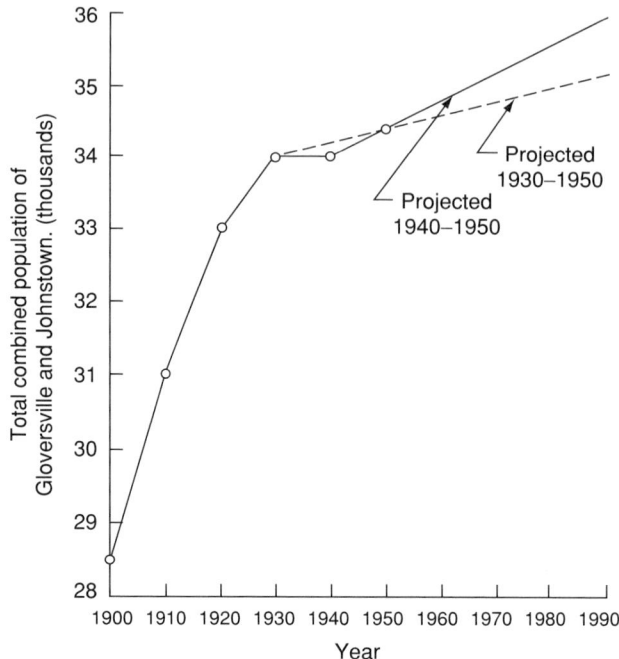

FIGURE 10.5. Population of Gloversville and Johnstown, 1900–1990.

the wastes. For example, during the state survey in 1951, samples collected on August 22 and 23 at Stations 5 and 6 showed temperatures of only 15–19°C.

During October 1964, the creek was visited and examined at various locations and dissolved-oxygen values were determined in order to locate the sag curve points. After an initial appraisal and a trial survey, the creek was sampled at the eight stations shown in Figure 10.4. After the first day samples were collected only from Stations 1, 5, and, 6 on six days at different times during each day. Composite samples were analyzed for dissolved oxygen, BOD, and temperature. In addition, the creek flow was measured on each sampling date at Station 6. These results are shown in Table 10.8. Flow times are shown in Table 10.9, a summary of the BOD and flow data in Table 10.10, the BOD curves for Stations 5 and 6 in Figure 10.6, flow data from a similar gauged creek in Table 10.11, the probability of the minimum flow data occurring in Table 10.12 and Figure 10.7, and a multiple-regression technique analysis of the stream data in Tables 10.13 and 10.14.

Cayadutta Creek Analysis
Using the multiple-regression method, the following three equations lead to dissolved oxygen solutions. When solved simultaneously, the three equations will yield the "best" equation that relates the dissolved-oxygen sag to the BOD, flow, and temperature at the bottom of the sag (Station 6).

TABLE 10.8
Cayadutta Creek Analyses in October 1964

Station and Milage	Date	Time	DO, ppm	Flow, cfs	Day 1	Day 2	Day 3	Day 4	Day 5	Water Temp., °C
1. Bleeker St. Bridge, Gloversville (clean H$_2$O site) (14.25)	10/8		10.12						8.0	8
	10/12		9						8.5	13
	10/13		10.0						9.5	12
	10/14		10.5						10.0	10
	10/15		9						8.5	11
	10/17		9.5						9.3	12
	10/18		10.0						8.9	11
4. Main St., Johnstown (8.25)	10/8		1.0						310	12.5
	10/12	1:30	1.9						240	15
	10/13	10:15	0.5						290	13
	10/14	9:00	5.1						340	10
	10/15	4:00	2.7						280	12
	10/17	11:05	6.2						240	12
	10/18	1:30	6.0						30	12
5. Harding property below Johnstown (6.75)	10/8		2.75		140	230	260	290	340	11.5
	10/12	2:00	2.8		110	180	220	260	290	15
	10/13	10:25	5.0		100	130	130	140	150	14
	10/14	9:30	3.1		120	190	170	180	200	14
	10/15	4:30	3.0		100	140	160	210	220	13
	10/17	11:25	5.4		100	70	130	150	160	14
	10/18	2:05	5.5		70	70	100	140	140	15

TABLE 10.8 (continued)

Station and Milage	Date	Time	DO, ppm	Flow, cfs	BOD, ppm Day 1	Day 2	Day 3	Day 4	Day 5	Water Temp., °C
6. Sammonsville Bridge (4.75)	10/8		2.2	59.4	130	200	270	300	330	13
	10/12	2:25	1.8	51	140	170	250	230	240	14
	10/13	10:45	1.1	44	40	70	60	80	140	12
	10/14	9:45	4.6	39	110	150	140	150	180	11
	10/15	4:50		39	40	110	150	160	160	12
	10/17	11:55	4.8	12.2	40	30	100	100	70	12
	10/18	2:40	4.8	12.2	40	30	50	50	40	13
	10/31			34						
	11/1			42						
7. Rt. 334, adjacent to Peresse Rd., Berryville Cross (3.0)	10/8		1.25					190	11	
	10/12	2:45	0					200	14	
	10/13	11:10	4.1					140	12	
	10/14	10:05	5.0					170	11	
	10/15	5:15	3.9					80	10	
	10/17	12:15	1.8					70	11	
	10/18	3:10	4.0					50	11	
8. Rt. 334, 1 mile north of Fonda next to Cannarella house (1.4)	10/8		5.1					180	10.5	
	10/12	3:05	3.5					90	13	
	10/13	11:30	5.1					140	11	
	10/14	10:20	5.1					160	11	
	10/15	4:30	5.0					80	12	
	10/17	12:45	3.6					100	11	
	10/18	3:30	4.5					40		

TABLE 10.9
Time of Flows from Station 5 Downstream Obtained 1 Week Before Stream-Sampling Program in October 1964

Site	Distance Between Points, miles	Time	Fall, ft
Start at Harding farm (Station 5)			
To old power dam	1	45 min	70
To bridge at Sammonsville (Station 6)	1	2 hr	20
To bridge at Fonda Ave.	1.75	2 hr	80
To railroad bridge	1.50	2 hr 40 min	55
Begin slack water of Mohawk River	1	2 hr 30 min	50
Total	6.25	9 hr 55 min	275

TABLE 10.10
Summary of 7-Day Sampling of Cayadutta Creek in Dry Period, from 10/8/64 to 10/18/64

Station	Reading	Average of Seven Samples	Range
1	DO, ppm	9.73	9–10.5
	Temp., °C	11	8–13
	5-day BOD, at 20°C, ppm	8.96	8–10
5	DO, ppm	3.93	2.75–5.5
	Temp., °C	13.8	11.5–15
	5-day BOD, at 20°C, ppm	214	140–340
	L	270	
		(projected factor 1.26)	
	1-day	106	70–140
	2-day	144	70–230
	3-day	167	100–260
	4-day	196	140–290
6	DO, ppm	3.17	1.1–4.8
	Temp., °C	12.4	11–14
	5-day BOD, at 20°C, ppm	166	40–330
	L	230	
		(projected factor 1.39)	
	1-day	77	40–140
	2-day	109	30–200
	3-day	146	50–270
	4-day	153	50–300
	Flow,[a] cfs[b]	36.7	12.2–59.4

[a]Time of travel between Stations 5 and 6 was 2 hr 45 min (0.115 days).
[b]Flow for 10/12/64 computed as arithmetic average of flow on 10/8/64. This flow represents approximate value of 7-day consecutive low flow likely to occur in Cayadutta Creek below Johnstown once in 10 years: $^{62}/_{90} \times 20 = 13.8$ cfs for entire Cayadutta Creek (slightly less below Johnstown). The 20 cgs in the equation is the minimum 7-day flow for Kayaderosseras Creek (*New York State Upper Hudson River Drainage Basin Survey Series Report no. 2*, p. 244).
Source: Nemerow (1987).

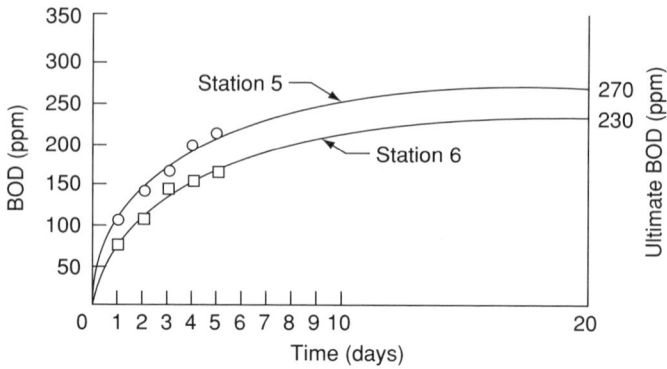

FIGURE 10.6. Laboratory BODs for Stations 5 and 6 at 20°C. Each point represents an average of seven samples collected from the creek on 7 different days at different times of day, all during a drought flow period (October 8–18, 1964).

$$b_1 \Sigma X_1^2 + b_2 \Sigma X_1 X_2 + b_3 \Sigma X_1 X_3 = \Sigma X_1 Y, \quad (1)$$

$$b_1 \Sigma X_1 X_2 + b_2 \Sigma X_2^2 + b_3 \Sigma X_2 X_3 = \Sigma X_2 Y, \quad (2)$$

$$b_1 \Sigma X_1 X_3 + b_2 \Sigma X_2 X_3 + b_3 \Sigma X_3^2 = \Sigma X_3 Y. \quad (3)$$

From Table 10.15, we can substitute numerical data in order to find b_1, b_2, and b_3. From Equation (1), we obtain

$$364.78 b_1 + 4.58 b_2 + 125.62 b_3 = -43.78. \quad (4)$$

Dividing Equation (4) by 364.78 yields

$$b_1 + 0.01256 b_2 + 0.3435 b_3 = -0.1200. \quad (5)$$

Multiplying Equation (4) by 0.01255 gives

$$4.58 b_1 + 0.05748 b_2 + 1.5728 b_2 = -0.5494. \quad (6)$$

We next apply Equation (2) and obtain

$$4.58 b_1 + 10.68 b_2 + 0.14 b_3 = 3.02. \quad (7)$$

Subtracting Equation (6) from (7) yields

$$10.62252 b_2 - 1.4328 b_3 = 3.5694. \quad (8)$$

TABLE 10.11
Minimum Flow Data of Measured Creek Compared with Cayadutta Creek

Year	Minimum Day Flow of Kayaderosseras Creek,[a] cfs	Equivalent Minimum Daily Flow[b] of Cayadutta Creek (Station 6), cfs
1927	13	8.9
1928	18	12.4
1929	24	16.5
1930	19	13.1
1931	19	13.1
1932	20	13.8
1933	19	13.1
1934	20	13.8
1935	34	23.4
1936	20	13.8
1937	21	14.5
1938	20	13.8
1939	20	13.8
1940	23	15.9
1941	15	10.3
1942	23	15.9
1943	29	19.3
1944	21	14.5
1945	26	17.9
1946	20	13.8
1947	22	15.2
1948	18	12.4
1949	14	9.6
1950	23	15.9
1951	32	22
1952	31	21.4
1953	20	13.8
1954	23	15.9
1955	20	13.8
1956	32	22
1957	19	13.1
1958	20	13.8
1959	18	12.4
1960	25	17.2

[a] Hydrologically similar to Cayadutta Creek.
[b] Calculated by dividing the drainage area of Cayadutta Creek (62 sq.mi.) by that of Kayaderosseras Creek (90 sq.mi.) and multiplying by the known rate of flow for the latter, that is, $\frac{62 \times 13}{90} = 8.9$.

Source: Nemerow (1987).

TABLE 10.12
Normal Probability Distribution Analysis of Data (1927–1960) from Table 16.9

Flow, cfs	Magnitude (M)	Plotting Position[a]
8.9	1	0.0286
9.6	2	0.0572
10.3	3	0.0858
12.4	4	0.1143
12.4	5	0.1430
12.4	6	0.1715
13.1	7	0.2000
13.1	8	0.2290
13.1	9	0.2570
13.1	10	0.2860
13.8	11	0.3140
13.8	12	0.333
13.8	13	0.371
13.8	14	0.400
13.8	15	0.428
13.8	16	0.458
13.8	17	0.486
13.8	18	0.515
13.8	19	0.544
14.5	20	0.571
14.5	21	0.600
15.2	22	0.629
15.9	23	0.658
15.9	24	0.685
15.9	25	0.715
15.9	26	0.744
16.5	27	0.770
17.2	28	0.800
17.9	29	0.829
19.3	30	0.858
21.4	31	0.887
22.0	32	0.916
22.0	33	0.945
23.4	34	0.974

[a]Calculated from the formula $M/(N+1)$, where M = magnitude in decreasing order of drought severity and N = number of values.

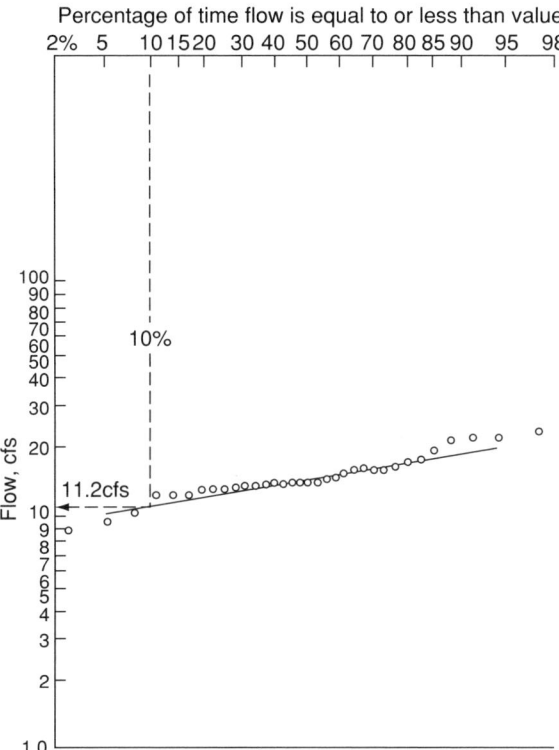

FIGURE 10.7. Minimum flow of Cayadutta Creek and expected recurrence of this level.

TABLE 10.13
Summary of Data Required from Cayadutta Creek Analyses in October 1964 for Churchill Method of Analysis

Date	DO, ppm		Drop in DO, ppm (Y)	BOD at Sag, ppm	Temp. at Sag, °C	Flow at Sag, cfs
	Station 1	Station 6				
10/8	10.1	2.2	7.9	330	13	59.4
10/12	9.0	1.8	7.2	240	14	51
10/13	10.0	1.1	8.9	140	12	44
10/14	10.5	4.6	5.9	180	11	39
10/15	9.0	2.9	6.1	160	12	39
10/17	9.5	4.8	4.7	70	12	12.2
10/18	10.0	4.8	5.2	40	13	12.2

DO, dissolved oxygen.

TABLE 10.14
Churchill Analysis Applied to Cayadutta Creek Data

Date	Dissolved Oxygen		Drop in DO, ppm (Y)	BOD at Sag, 1000/ppm (X_1)	Temp. at Sag, °C (X_2)	Flow at Sag, 100/cfs (X_3)	Y^2
	At Station 1, ppm	At Station 6, ppm					
10/8/64	10.1	2.2	7.9	3.03	13	1.68	62.41
10/12/64	9.0	1.8	7.2	4.17	14	1.96	51.84
10/13/64	10.0	1.1	8.9	7.14	12	2.27	79.21
10/14/64	10.5	4.6	5.9	5.56	11	2.56	34.81
10/15/64	9.0	2.9	6.1	6.25	12	2.56	37.21
10/17/64	9.5	4.8	4.7	14.29	12	8.20	22.09
10/18/64	10.0	4.8	5.2	25.00	13	8.20	27.04
Totals			45.9	65.44	87	27.43	314.61
Means			$\bar{Y}$	$\bar{X}_1$	$\bar{X}_2$	$\bar{X}_3$	
			6.56	9.35	12.4	3.92	$n\bar{Y}^2$
Corrected items[a]							301.21
Corrected totals							13.40

[a]n = number of samples (7).

TABLE 10.15
Summary of 24-Hour Sampling Results

Date	YX_1	YX_2	YX_3	X_2^1	X_1X_2	X_1X_3	X_2^2	X_2X_3	X_2^3
10/8/64	23.94	102.7	13.27	9.18	39.39	5.09	169	21.84	2.82
10/12/64	30.02	100.8	14.12	17.39	58.38	8.17	196	27.44	3.84
10/13/64	63.55	106.8	20.20	50.98	85.68	16.21	144	27.24	5.15
10/14/64	32.80	64.9	15.10	30.91	61.16	14.23	121	28.16	6.55
10/15/64	38.13	73.2	15.62	39.06	75.00	16.00	144	30.72	6.55
10/17/64	67.16	56.4	38.54	204.20	171.48	117.48	144	98.40	67.24
10/18/64	130.00	67.6	42.64	625.00	325.00	205.00	169	106.60	67.24
Totals	385.60	572.4	159.49	976.72	816.00	381.18	1087	340.40	159.39
Means	nYX_1	nYX_2	nYX_3	$n\bar{X}_1^2$	$n\,X_1\bar{X}_2$	$n\,X_1\bar{X}_3$	$n\bar{X}_2^2$	nX_2X_3	$n\bar{X}_3^2$
Corrected items[a]	429.38	569.38	180.04	611.94	811.51	256.56	1076.32	340.26	107.56
Corrected totals	−43.78	3.02	−20.55	364.78	4.58	125.62	10.68	0.14	51.83

Dividing Equation (8) by 10.62252, we obtain

$$b_2 - 0.1349b_3 = 0.3360. \qquad (9)$$

Multiplying Equation (4) by -0.3435, we get

$$-125.32b_1 - 1.573b_2 - 43.047b_2 = +15.038 \qquad (10)$$

Next, we multiply Equation (8) by $+0.1443$:

$$1.533b_2 - 0.20675b_3 = 0.51506 \qquad (11)$$

Finally, we apply Equation (3) and obtain

$$125.32\ b_1 + 0.04b_2 + 51.83b_3 = -20.55. \qquad (12)$$

Adding Equations (10), (11), and (12), we have

$$8.5763b_1 = -4.99694, \qquad (13)$$
$$b_3 = -0.5826. \qquad (14)$$

From Equation (9),
$$b_2 - (-0.07859) = 0.3360,$$
$$b_2 = +0.2574. \qquad (15)$$

From Equation (5),

$$b_1 + (0.00323) + (-0.20012) = -(0.1200),$$
$$b_1 + 0.00323 - 0.0012 = -0.1200,$$
$$b_1 = 0.0769. \qquad (16)$$

As a check, we substitute the following in Equation (12):

$$125.32\ (0.0769) + 0.04\ (0.2574) + 51.83\ (-0.5826)$$
$$= -20.55$$
$$9.6471 + 0.010296 - 30.1962 = -20.55$$
$$-20.5488 = -20.55$$

$$Y = a + b_1\bar{x}_1 + b_2\bar{x}_2 + b_3\bar{x}_3$$
$$a = \bar{Y} - (b_1\bar{X}_1 + b_2\bar{X}_2 + b_3\bar{X}_3)$$
$$a = 6.56 - [(0.07769 \times 9.35) + (0.2574 \times 12.4) + (-0.5826 \times 3.92)]$$

$$a = 6.56 - [0.7190 + 3.1918 - 2.2838]$$

$$a = 6.56 - 1.6270$$

$$a = 4.9330.$$

Therefore, the Cayadutta Creek equation is

$$y = a + bx_1 + b_2x_2 + b_3x_3$$

$$Y = 4.9330 + 0.0769X_1 + 0.2574X_2 - 0.5826X_3,$$

where

X_1 = 5-day BOD (1,000/ppm) 20°C at Station 6
X_2 = temperature (°C)
X_3 = flow (100/cfs)
Y = DO drop from Station 1 to Station 6 (ppm).

To verify this stream equation, we can substitute the actual observed stream values for X_1, X_2, and X_3 and obtain a calculated Y value, which can then be compared with the observed value for accuracy:

Date (1964)	Observed Y (ppm)	Calculated Y (ppm)
10/8	7.9	7.53
10/12	7.2	7.72
10/13	8.9	7.25
10/14	5.9	6.70
10/15	6.1	7.01
10/17	4.7	3.37
10/18	5.2	5.42

During the October 1964 study, the lowest flow was 12.2 cfs, the highest temperature 14°C, and the DO sag allowed was between 9.0 and 2.0 (i.e., 7.0) ppm.

To find the BOD load at sag, we calculate as follows:

$$Y = 4.9330 + 0.0769X_1 + 0.2574X_2 - 0.5826X_3$$

$$7.0 = 4.9330 + 0.0769X_1 + 0.2574$$

$$-0.5836 \,(100/12.2)$$

$$7.0 = 4.9330 + 0.0769X_1 + 3.6036 - 4.7773$$

$$3.2407 = 0.0769X_1$$

$$X_1 = 42.1417 = 1{,}000/\text{ppm}.$$

$$\text{ppm allowed} = 1{,}000/42.1417 = 23.73 \text{ ppm}. \tag{14}$$

The average BOD at sag was 166 ppm during the entire 7-day survey. Therefore, the BOD reduction required in the stream at Station 6 was

$$[(166-23.73)/166] \times 100 = 85.70\%.$$

Although the Streeter–Phelps method yielded values of k_1 and k_2, which gave a Fair's f of about 35,[2] the results are not reliable because of the variability of wastes from one moment to the next as well as the multiple entrances of wastes into the stream. The only reliable procedure for evaluating the oxygen-sag characteristics is to collect many stream samples under these critical conditions and statistically correlate the data in order to obtain a stream equation. This method is commonly referred to as the *Churchill multiple-regression technique*. The stream equation represents the line that best fits the data for the conditions under which the samples were collected. Projection of the line beyond this range of conditions is not recommended, but extrapolations to different conditions within the range of existing data can be made with a reasonable degree of certainty. The stream equation for Cayadutta Creek, developed by extensive analysis, can be used to compute the BOD reductions necessary to maintain a certain minimum dissolved-oxygen level at a given temperature. These calculations are given below and in Figures 10.8 and 10.9.

To find the waste-treatment requirements, we use the stream equation developed during the low-flow critical period for Cayadutta Creek during the October 1964 study:

$$Y = 4.9330 + 0.0769\, X_1 + 0.2574\, X_2 - 0.5826\, X_3,$$

$$X_1 = 1{,}000/\text{ppm BOD}$$

$$\text{when } X_2 = \text{temperature} = 12.4\ (^\circ\text{C})$$

$$X_3 = 100/\text{cfs} = 100/12 \text{ (see Table 10.12)} = 8.33$$

$$Y = \text{DO sag from Station 1 to Station 6} = 9.73 - 2.00 = 7.73$$

$$7.73 = 4.9330 + 0.0769\, X_1 + 0.2574\, (12.4)$$

$$- 0.5826\, (8.33)$$

$$7.73 + 4.85 - 4.9330 - 3.200 = 0.0769\, X_1$$

$$12.58 - 8.133 = 4.447/0.0769 = X_1 = 57.75$$

$$\text{ppm} = 1{,}000/57.75 = 17.3 \text{ ppm at sag.}$$

---

[2] $f = k_2/k_1$ = re-aeration rate/deoxygenation rate.

The BOD values for the 7 days of the survey at Station 5 were 340, 290, 150, 200, 220, 160, and 140 ppm, giving an average of 214 ppm. At Station 6, the values were 330, 240, 140, 180, 160, 70, and 40, with an average of 166 ppm. The percentage decrease in BOD between Stations 5 and 6 was

$$(214 - 116)/214 = 48/214 = 22.5\%.$$

The BOD (in pounds) being discharged at Station 5 on the 7 days of the survey was a total of 142,500 (see Table 10.11) and an average of 23,750 lb/day.

Taking $23,750 \times 77.5\% = 18,400$ lb/day as the BOD left at Station 6, with no treatment at 12 cfs, 12.4°C, and an allowable DO deficit at the sag point of 8.65 ppm $(10.65 - 2.0)$, we obtain an allowable 17.3 ppm BOD at the sag or

$$[(17.3 \times 12) / (1.54)] \times 8.34 = 1,122 \text{ lb/day}.$$

Since $23,750 - 0.225 = 18,400$ lb/day, the BOD reduction required is

$$[(18,400 - 1,122) / (18,400)] \times 100 = (17,278) / (18,400) = 93.8\%.$$

At 12.4°C and the same DO sag (7.73 ppm), but at the increased stream flow of 20 cfs, we obtain an allowable BOD at the sag of

$$Y = 4.9330 + 0.0769\, X_1 + 0.2574\, X_2 - 0.5826\, X_3$$

$$7.73 - 4.9330 - 3.2000 = 0.0769 X_1 - 0.5826\, (100/20)$$

$$[(7.73 + 2.4130 - 8.1330)] / (0.0769) = X_1 = (2.0100)/(0.0769) = 26.05 \text{ ppm}$$

$$= 1,000/26.05 = 38.4 \text{ ppm}$$

at 38.4 ppm $\times$ 8.34 $\times$ (20/1.54)

$= 4,155$ lb BOD at Station 6

BOD reduction required $= 14,245/18,400 \times 100 = 77.5\%.$

At the same temperature (12.4°C) and the same DO sag (7.73 ppm), but at the average stream flow at the sag of 36.7 cfs (October 1964 survey), we obtain an allowable BOD at the sag of

$$Y = 4.9330 + 0.0769\, X_1 + 0.2574\, X_2 - 0.5826\, X_3$$

$$7.73 = 4.9330 \times 0.0769\, (X_1) + 0.2574\, (12.4°C)$$

$$-0.582\, (100/36.7)$$

$$7.730 + 1.590 - 4.9330 - 3.2000 = 0.0769\, X_1$$

$$(9.3200 - 8.1330) / 0.0769 = 1.1890/0.0769 = 15.45 = X_1$$

$$\text{ppm} = 1,000/15.45 = 64.8$$

at 64.8 ppm × 8.34 × 36.7/1.54 = 12,850 lb BOD at Station 6

BOD reduction required = [(18,400 − 12,850) / (18,400)] × 100

= 5,550/18,400 = 30.2%.

Figures 10.8 and 10.9 indicate that to maintain 2 ppm of dissolved oxygen (a preselected safe value for this class of stream) of loadings of 23,750 lb BOD/day at the bottom of the sag at a temperature of 12.4°C, BOD reductions of 65–94% at 23,750 lb/day loading and 77–97% at 35,019 lb/day would be required for critical stream flows of 12–25 cfs.

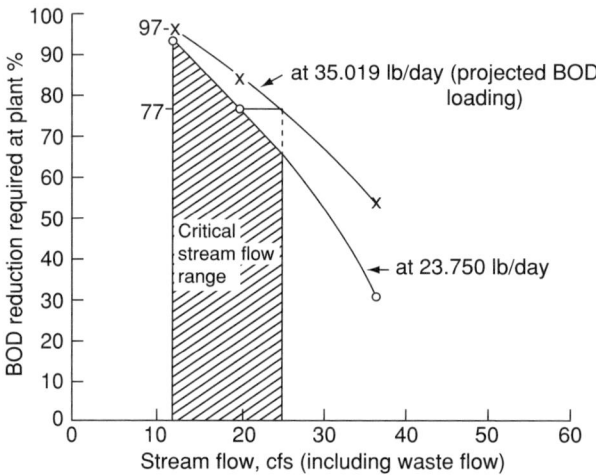

FIGURE 10.8. BOD reduction required at Station 5, at 12.4°C and 2 ppm DO remaining and based on a waste discharge of 23,750 lb/day.

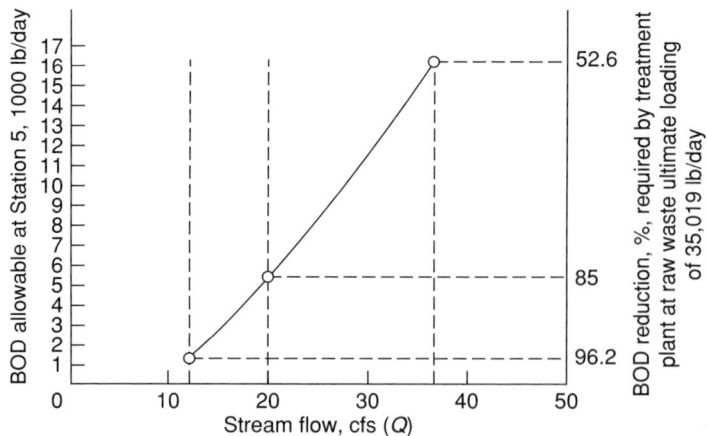

FIGURE 10.9. Special design curve for computing treatment plant requirements at 12.4°C and 2 ppm DO at Station 5.

Composite Waste Sampling

Waste samples were collected hourly for a 24-hour period from three sources: the Johnstown 30-in. sewer (main), the Johnstown 8-in. sewer (Tynville), and the Gloversville sewage plant, on November 17 and December 3, 1964. Similar samples were collected for a 24-hour period on January 21, 1965, except that the 8-in. Johnstown sewer was eliminated as being relatively insignificant. Weirs were installed in the Johnstown lines to record the total flows from Johnstown and from Gloversville. Samples were collected and composited according to the rate of flow at the hour of sampling. A summary of the proportionate pollutional loads and volumes for these 3 days is shown in Table 10.16. Additional 24-hour composites from each line were collected according to flow and analyzed on February 18, March 30, April 21, and May 6, 1965 (Table 10.17).

Composite Waste Analyses

The hourly data revealed several significant findings:

1. *Slugs*, which can be defined for this situation as instantaneous discharges of high volumes of waste, concentrated acid or alkali, or BOD, are apparently not a major problem. The flow increases by about 100% of the daily average, for

TABLE 10.16
Summary of 24-Hour Sampling Results

			BOD Load			Flow		Suspended Solids		
Source	Date	Pounds/ Day	% in Peak, Period[a]	% of Daily Total	mgd	Ratio of Peak Period to Daily Average	% of Daily Total	Pounds/ Day	% in Peak Period	% of Daily Total
Gloversville	11/17/64	13,350	45.2	67.5	3.80	1.38	66	27,550	80.8	87.5
	12/3/64	13,350	50.2	56.3	3.67	1.34	50	11,100	61.6	53
	1/21/65	23,400	41.1	63.1	3.96	1.35	60.5	21,600	25.8	63
Johnstown Main	11/17/64	6,400	44.8	32.2	1.92	1.55	33.4	4,100	57.7	12.5
	12/3/64	15,000	27.5	42	3.60	1.46	49	18,400	42.0	45
	12/1/65	13,000	43.2	36.9	2.58	1.33	39.5	12,800	37.0	37
Tynville	11/17/64	102.4	25.4	0.53	0.0449	1.24	0.6	32.9	56	
	12/3/64	223	25.2	1.4	0.0809	0.925	1	213	17.9	2.0
Totals	11/17/64	19,852			5.7649			31,683		
	12/3/64	28,573			7.3509			29,713		
	1/21/65	37,000			6.54			34,400		

[a]The peak period was assumed 6.00 a.m. to 12 noon.

TABLE 10.17
Composite Analyses (24 Hours) of Gloversville and Johnstown Wastewater

Characteristic	February 18–19, 1965		March 30, 1965		April 21, 1965		May 6, 1965	
	Gloversville	Johnstown[a]	Gloversville	Johnstown	Gloversville	Johnstown	Gloversville	Johnstown
Flow, mgd								
24-hr average	3.41	3.05	3.05	4.88	3.63	2.84	4.3	2.07
6 a.m. to 12 noon average			4.73	6.33	4.29	3.00		
6 a.m. to 2 p.m. average								
pH	9.4	8.3					5.9	2.92
Total solids, ppm	3130	2450	2430	1970	2542	1840	3120	2962
Suspended solids, ppm	258	81	265	145	418	213	475	322
Volatile suspended solids, ppm			196	96	265	135	305	250
BOD (5-day, 20°C), ppm	405	385	300–435	285–330	371–435	371–386	485–520	540–585
BOD, lb/day	11,500	9800	(367)[b]	(307)	(403)	(378)	(502)	(562)
Settleable solids, ml/L			20.0	5.5	13.0	4.5	14.0	16.0
On supernatant								
Suspended solids, ppm			97	70	128	86	172	135
			78	52	92	60	120	110
Volatile solids, ppm			95–180 (285–330)	225–355	266–281	326–386	210–300	375–405
BOD, ppm								

TABLE 10.17 (continued)
Composite Analyses (24 Hours) of Gloversville and Johnstown Wastewater

	February 18–19, 1965		March 30, 1965		April 21, 1965		May 6, 1965	
Characteristic	Gloversville	Johnstown[a]	Gloversville	Johnstown	Gloversville	Johnstown	Gloversville	Johnstown
8-hr readings							690	423
Suspended solids, ppm							420	310
Volatile solids, ppm							405–495 (450)	405–435 (420)
BOD, ppm								
Total ash, ppm	2250	1570						
Total volatile, ppm	880	880						
Suspended ash, ppm	140	48						
Suspended volatile solids, ppm	118	33						
Suspended volatile solids, %	46	41						
Analysis of settled 2-hr sludge								
Total solids, %	1.54	0.88						
Total ash, %	34.4	37.5						
Total organics, %	63.6	62.5						

[a] 30-inch sewer.
[b] Average values are given in parentheses.

about 12 hours during the daytime. The pH becomes alkaline (8–10) during the same period but returns to normal (7–8) during the 12 night hours. The BOD varies considerably during both day and night but is generally high and is confined to a range of 300–700 ppm from 6 a.m. until midnight. There is little pattern of discharge of BOD and no apparent practical gain as far as BOD is concerned from separate equalization basins. Because the flow and pH are largely higher during the entire daytime period, equalization to level out these factors would require very large basins. The cost of such units and the potential danger of septicity seem, to the author, to far outweigh the benefits derived from leveling the flow and pH. In this instance, it seems that the great numbers of varied tanneries themselves contribute to equalization of waste simply by their diversity.

2. *The total BOD loads and flows* given in Tables 10.16 and 10.17 can be examined more easily by referring to Table 10.18. The total flow measured averages 6.724 million gallons per day (mgd) and contains an average of 23,442 lb of 5-day 20°C BOD and approximately 20,650 lb of suspended solids. These values do not include any flows or loads not connected to the Gloversville sewage treatment plant or the 30-in. sewer outfall in Johnstown. The 8-in. sewer outfall in Johnstown, though measured and sampled at the beginning, contains less than 1% of the total volume or BOD load and can, therefore, be considered insignificant in these surveys.

3. *The maximum variations in flow and load* from day to day were found to be 18% from the average flow and 22% from the average BOD. These variations are considered well within normal values and tend to substantiate the use of average daily values given under the previous section in designing waste-treatment facilities.

TABLE 10.18
Summary of Total Loads for Treatment

Date	Total flow, mgd	Total BOD, lb/day	Total Suspended Solids, lb/day
11/17/64	5.765	19,852	31,683
12/3/64	7.351	28,573	29,713
1/21/65[a]	6.540	37,000	34,400
2/18/65	6.460	21,300	9,405
3/30/65	7.930	21,925	12,700
4/21/65	6.470	21,150	17,700
5/6/65	6.370	27,850	22,700
Average	6.724	23,442	20,650

[a]Since an unusually large percentage of deerskin was tanned, this day was not considered typical of even maximum normal operation, and therefore, it was excluded from the average.

Industrial production records during the sampling days (Tables 10.19 and 10.20) demonstrated that all major industries connected to the two major sewer systems were in operation at almost full capacity during these days. This provides some measure of assurance in using the average flow and BOD values obtained during this period. Table 10.21 shows industry's percentage of the total measured flow on these days. This reveals an industrial waste problem of about 50% by volume when industry is operating near its rated capacity.

Laboratory Pilot-Plant Studies

To form more definite conclusions on the proper units to be included in the waste-treatment plant, certain small-scale laboratory studies were necessary, including sludge digestion and activated-sludge treatment.

Sludge Digestion

A mixture of primary and secondary settled sludge was collected from the settling basins at the Gloversville treatment plant. A pilot digester, consisting of a glass container and a gas-collecting system maintained at 37°C, was set up in a private laboratory in Johnstown. The raw sludge sampled selected was analyzed for organic matter at the start of the "batch" digestion period and again after 50 days of digestion, and gas volume measured almost daily (Table 10.22).

Although this was a batch-type experiment, over the 50-day period, 9.07 ft^3 of gas was produced per pound of volatile matter destroyed. Greater amounts of gas may be expected from a continuous-digestion operation maintained at optimum environmental conditions. In this experiment, more gas would have evolved after an increased digestion period, but the rate of gas production did slow down considerably after 50 days. Normal gas production for sewage sludge is about 15 ft^3/lb of organic matter destroyed. Digestion experiments on a continuous basis and over a longer period would be needed to assess whether an accumulated toxic effect exists. However, Vrooman and Ehle (1950) reported successful digestion of this waste sludge.

Activated-Sludge Treatment

The apparatus used in the study consisted of an aeration tank with two mixers and three separate air-diffuser tubes fitted with porous stones. Air flow (cubic feet of air per hour) was measured by a previously calibrated rotameter. The tank was 23.5 in. long and 8.5 in. wide and was filled to a depth that gave an aeration volume of 6 gallons.

Because settling was expected to be an integral part of a biological treatment plant of this type, various mixtures of settled tannery waste (1:1 mixture of beam-house and tan-yard wastes) and settled domestic sewage were added to the aeration tank in a semi-batch procedure to simulate continuous operation as closely as possible. The standard average aeration period of 6 hours was used and the waste mixture was added in three increments of 2 gallons each at 2-hour intervals. The tank contents (6 gallons of a mixture of tannery wastes and domestic sewage) were first settled for 1 hour. Then 2 gallons of the supernatant were siphoned off and 2 gallons of the waste mixture

TABLE 10.19
Industrial Production during Sampling Days

Company	November 17, 1964		December 3, 1964		January 21, 1965	
	Water Used, gpd	% Production[a]	Water Used, gpd	% Production	Water Used, gpd	% Production
Wood and Hyde Leather	200,000	100	225,000	100	225,000	100
Filmer Leather	41,310	50	41,310	50	41,300	50
Twin City Leather	120,000	50	120,000	100	120,000	100
Wilson Tanning	49,920	80	54,337	80	38,000	50
Leavitt-Berner Tanning	151,700	100	151,700	100	151,700	100
F. Rulison & Sons	76,300	100	76,300	100	76,300	100
Peerless Tanning	24,000	33 ⅓	48,000	66 ⅔	48,000	66 ⅔
Karg Bros.	266,000	68	254,000	65	386,000	90–95
Decca Records[b]	10,807	100	10,807	100	10,807	100
U.S. Rabbit Tanning Co.[b]	10,000	100	7,000	75	7,000	75
Gloversville Continental Mill[b]	200,000	66	190,000	100	190,000	100
Independent Leather	134,000	60	107,000	40	161,000	60
Liberty Dressing	78,950	70–75			120,803	60
G. Levor		80		80		
Framglo Tanners (1)	500,000	100	500,000	100	500,000	100
Framglo Tanners (2)		80		80		80
Rebel Dye[b]	26,250	10	30,500	10	24,500	9
Lee Dyeing[b] (Johnstown)	175,000	40	1,000	0	1,000	0
Adirondack Finishing[b]	450,000	80	450,000	80	500,000	90
Crown Finishing (Maranco Leather)	42,352	100	42,000	100	42,352	100

TABLE 10.19 (continued)
Industrial Production during Sampling Days

Company	November 17, 1964		December 3, 1964		January 21, 1965	
	Water Used, gpd	% Production[a]	Water Used, gpd	% Production	Water Used, gpd	% Production
Simco Leather	61,300	100	61,000	100	61,000	100
Johnstown Tanning	60,000	70	35,000	60	10,000	40
Napatan	21,072	100	21,000	80	21,000	100
Ellithorp Tanning	105,000	100	110,000	100	110,000	100
Johnstown Knitting[b]			100,000	45	100,000	75
Gloversville Leather						
Riss Tanning						
Industrial total, gpd	2,803,961		2,735,954		2,845,662	
Total flow, mgd	5.7649		7.3509		6.54	
Total BOD, lb/day	19,852		28,573		37,000	
Industrial portion of total water flow, %	48.7		37.2		43.4	

[a]Percentage of plant's total productive capacity.
[b]Figures are based on yearly consumption (average figure).

TABLE 10.20
Water Consumption Related to Production Percentage[a]

Industry	Water Consumption, gal		Beamhouse		Production Percentage		Tanning	
	From Meter Reading	From Other Sources	Type of Skin	Rated Potential	Compared 1/21/65	Operation Potential	Type	
Leather tanneries								
Wood and Hyde Leather	150,000	75,000	Burn–sheep	100	100	100	Combination	
Filmer Leather	90,000	Pond (in future)	Horse, cow, jacks, deer	100	75	80	Combination	
Twin City Leather	10,028		Sheep	100	100	100	Combination	
Wilson Tanning	25,215		Sheep, goat, deer	0	0	33½	Combination	
Leavitt-Berner Tanning (not contributing to sewage treatment plant)	140,000		Sheep, goat, deer			100	Combination	
F. Rulison & Sons	60,150		Horse, cow	80	80	80	Combination	
Peerless Tanning	21,766					75	Chrome	
Karg Bros.	400,000		Pig, deer	100	100	100	Chrome	
Independent Leather	230,000					100	Combination	
Liberty Dressing	85,582		Goat, calf			45	Combination	
G. Levor	Drinking water only	600,000	Pig, calf, goat	80		80	Combination	
Framglo Tanners (1)	Drinking water only	400,000	Sheep, deer			100		
Framglo Tanners (2)	Drinking water only	300,000		80		None		
Crown Finishing (Maranco Leather)	26,465					100	Chrome, some combination and vegetable	
Simco Leather	45,030							
Johnstown Tanning (1)	11,000					10		
(2)	0							

TABLE 10.20 (continued)
Water Consumption Related to Production Percentage[a]

Industry	Water Consumption, gal		Production Percentage					
			Beamhouse			Tanning		
	From Meter Reading	From Other Sources	Type of Skin	Rated Potential	Compared 1/21/65	Operation Potential	Type	
Ellithorp Tanning	248,000	From Levor				90	Chrome and combination	
Gloversville Leather	0	152,000				25	Chrome and combination	
Riss Tanning	27,970					33½	Chrome and combination	
Nonleather industries								
Rebel Dye	61,000 (8,145 ft³)		20					
Adirondack Finishing	474,310		80					
Lee Dyeing	8,800 (1,175 ft³)		0					
(Johnstown) Gloversville Continental Mill	208,000 gal		66 ⅔	Anticipate 330,000 gal due to dyeing technique but not necessarily increased production				
Johnstown Knitting	150,200 (20, 178 ft³)		75					
Diane Knitting	10,590 ft³		75					
Decca Records	17,745 (2,366 ft³)		75					
Mohawk Cabinet			80					
U.S. Rabbitt . Tanning Co	1,000		0	Water running in tubs to keep them soaked. Should be disregarded				
Total industrial flow, gal	4,019,262							
Total flow, gal	6,460,000							
Total BOD	21,300 lb/day							
Industrial flow (% of total)	62.5%							

[a] Data are for the 24 hour period from 6 a.m. on February 18 to 6 a.m. on February 19, 1965.

TABLE 10.21
Industrial Waste Flow

Date	Total Flow, mgd	Industrial Flow (Estimated by Survey), mgd	Industrial Portion of Total Flow, %
11/17/64	5.765	2.804	48.7
13/3/64	7.351	2.736	37.2
1/21/65	6.540	2.846	43.4
2/18/65	6.460	4.019	62.5
Average			ca.48

added. The tank contents were aerated for 2 hours and settled for 15 minutes, 2 gallons of supernatant were withdrawn, and 2 more gallons of waste mixture added. This procedure resulted in the addition of 6 gallons of waste mixture in a period of 4 hours, for a total aeration time of 8 hours and an average aeration period of 6 hours, as shown below:

Time, hr	Volume Added, gal	Aeration Time, hr
0	2	8
2	2	6
4	2	4
		Average: 6

The results obtained are summarized in Table 10.23 and shown graphically in Figure 10.10. Each loading represents about 1 week of aeration data with samples being taken for analysis several times during this week of adaptation and acclimation.

Laboratory experiments verified that 65–75% of this waste would degrade biologically even when loaded at the high rate of 95–115 lb BOD/1,000 ft^3 of aerator (see Figure 10.10). Higher BOD reductions (75–85%) were obtained with lower BOD loadings (60–82 lb/1,000 ft^3) and increased dilution of the tannery waste with domestic sewage.

These studies showed that the activated-sludge process or a modification of it could be used successfully in the overall treatment of the combined tannery and sewage wastes. Larger prototype field experiments would disclose whether these results can be projected directly to full-scale operation.

The present Gloversville treatment plant has experienced much difficulty because of its overloaded condition. However, from the best records available, when all the flow units of the plant were operating, about 58–60% of the BOD was removed. This reduction in BOD also shows that biological degradation by trickling filtration is possible with

TABLE 10.22
Sludge Digestion (Laboratory Study)[a]

Accumulated Gas Produced, cc	Days of Digestion at 37°C
250	1
290	2
460	3
640	4
730	5
850	6
940	7
990	8
1,040	9
1,140	10
1,140	11
1,150	12
1,190	13
1,240	14
1,290	15
1,340	16
1,380	17
1,420	18
1,480	19
1,540	20
1,600	21
1,650	22
1,700	23
1,820	24
1,900	25
1,940	26
1,980	27
1,980	28
2,010	29
2,020	30
2,030	33
2,040	36
2,050	40
2,170	49
2,200	50

[a]The results were analyzed as follows:
- organic matter (raw sludge) = 5.4533 gm
- organic matter (after 50 days) = 1.5542 gm
- loss of organic matter = 3.8991 gm (71.2%)
- gas produced: total = 2,200 cc
- per gram of organic matter destroyed = 567 cc
- per gram of volatile matter added = 403 cc

To convert results from the metric system:
$2,200 \times 0.000353 = 0.0777$ ft^3 of gas produced (total)

$\dfrac{3.8991}{454} = 0.00858$ lb of volatile (organic) matter destroyed

$\dfrac{0.00777}{0.00858} = 9.07$ ft^3 of gas per pound of organic matter destroyed

TABLE 10.23
Activated-Sludge Pilot Laboratory Studies

Waste Treated					BOD		
Origin	Quantity, gal	BOD Loading, lb/1,000 ft³	Suspended Solids Under Aeration, ppm	Air rate, ft³/lb BOD Removed	Influent, ppm	Effluent, ppm	Reduction, %
Sewage	5						
+ tannery waste mixture	1	60	2,330	2,450	239	44	81.6
Sewage	4						
+ tannery waste mixture	2	82.8	2,221	1,900	331	78	76.5
Sewage	3						
+ tannery waste mixture	3	114.7	2,768	1,735	459	165	64.1
Sewage wastes							
Gloversville	5						
Johnstown	1	70	3,386	2,000	280	39	86
Sewage wastes							
Gloversville	4.015						
Johnstown no. 1	2.112						
Johnstown no. 2	0.044	73.8	2,508	2,116	295	68	76.9
Tannery waste mixture	3						
+ tap water	3	93.0	2,646	1,070	374	91	75.6

this waste under full-scale field conditions. The exact degree of this oxidation could be determined more easily in a properly designed and operated field pilot plant.

Literature Survey

A study was also made of previous research work or reported practice dealing with combined treatment of domestic sewage and tannery waste by the activated-sludge process. The reports of Chase and Kahn (1955), Braunschweig (1965), Jansky (1961), Thebaraj et al. (1962), Snock (1928), Hubbel (1935), Pauschardt and Furkert (1936), Kubelka (1952, 1956), Fales (1928), Kalibina (1931), Furkert (1937), and Mausner (1938) provide some evidence that the activated-sludge treatment process is feasible for tannery–sewage waste mixtures. This review of previous work tends to substantiate the biological pilot studies described earlier. Most of this reported work, however, has been on a research or pilot-plant basis or of a more sewage-diluted waste. There is a definite

Joint Treatment of Raw Industrial Waste with Domestic Sewage

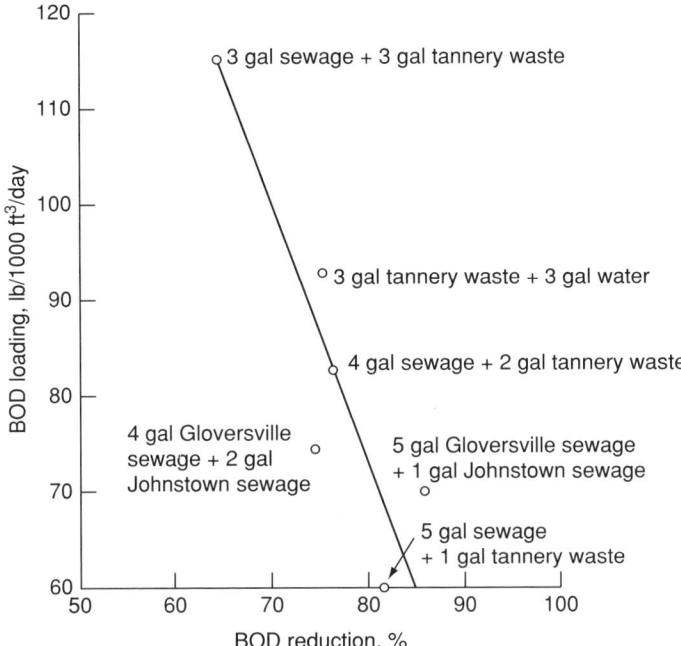

FIGURE 10.10. Activated-sludge treatment: BOD reduction related to BOD loading. (From Nemerow 1987.)

lack of reviews of full-scale biological treatment used on tannery and sewage waste mixed 50:50 (by volume).

Braunschweig (1965) disproves the notion that chromium in the newer tannery processes interferes with aerobic biological treatment. The work by Jansky (1961) is typical of that of the supporters of activated sludge as a treatment method. Thebaraj et al. (1962) found that all aerobic biological systems were effective and that the choice depended on several economic and practical considerations. Fales (1928), much earlier, was of the same opinion as Thebaraj but pointed out the higher operating costs of activated sludge compared to trickling filtration. Furkert (1937) also verifies our digestion studies and noted that except for the high H_2S content of the digester gas, the composition of the gas is normal.

Conclusions from Study

The following specific conclusions and recommendations were made as a result of this study:

1. The stream survey was instrumental in providing evidence that secondary treatment of the combined industrial and sanitary wastes of the area were required and that 65–94% BOD reduction would be needed depending on the dilution

available in the stream. Use of the curves plotted in Figures 10.8 and 10.9 would allow for a more precise selection of BOD reduction required for specific critical stream flows.
2. The existing dry-weather flow to be treated averages 6.724 mgd with peaks of two to three times this rate; about 50% of this flow originates from the industries in the area.
3. The combined area waste contains a daily average of 23,442 lb of 5-day 20°C BOD and 20,650 lb of suspended solids. These loadings are affected considerably by the type of skin tanned, deer skin being an especially significant contributor of high BOD and solids loads.
4. Laboratory pilot studies demonstrated that the conventional activated-sludge treatment process is capable of reducing the BOD of the combined waste from 65 to 85% (depending primarily on the organic loading) at loadings ranging from 60 to 115 lb BOD/1,000 ft^3 of aerator capacity.
5. A digestion batch experiment yielded about 9 ft^3 of gas per pound of volatile matter destroyed and effected a 71% reduction in organic matter.
6. A literature survey confirmed the findings of the laboratory results that is the combined wastes of this type are amenable to biological oxidation.
7. Because of the unique nature of the volume and characteristics of the tannery–sewage waste mixture, as well as the size and cost of the project, field prototype studies preceded full-scale plant construction.
8. There should be additional laboratory research on development of improved methods of aerobic biological treatment of tannery wastes to allow for greater BOD reduction at higher BOD loadings.

The decision reached as a result of this study was that the cost was too high, the risk too great, and previous reported experience inadequate for full-scale biological treatment to be recommended. A prototype in the field—preferably at the site of the Gloversville treatment plant—was to be built and operated for about 6 months to obtain greater certainty that the earlier findings were valid. This prototype should contain both trickling-filtration and activated-sludge units (as well as provision for its modification). It should also allow for experimentation with series and parallel operation of the units and both diffused and mechanical aeration. Some sludge-digestion studies should be carried out over the entire period.

A schematic drawing of this field prototype is shown in Figure 10.11. It consists of two sets of screens in series (½-in. openings followed by ¼-in. openings), pump, primary settling, trickling filter, aeration, and final settling. The plant began operation in early August 1965 and sampling was begun on August 16. Table 10.24 gives data for the first 7 weeks of operation. Table 10.25 shows how the prototype operating results influenced the final design parameters.

Overall Planning Study Conclusions

The conclusions reached through data collection, pilot and prototype plant studies, engineering evaluations, and reviews of design and operational experiences in major

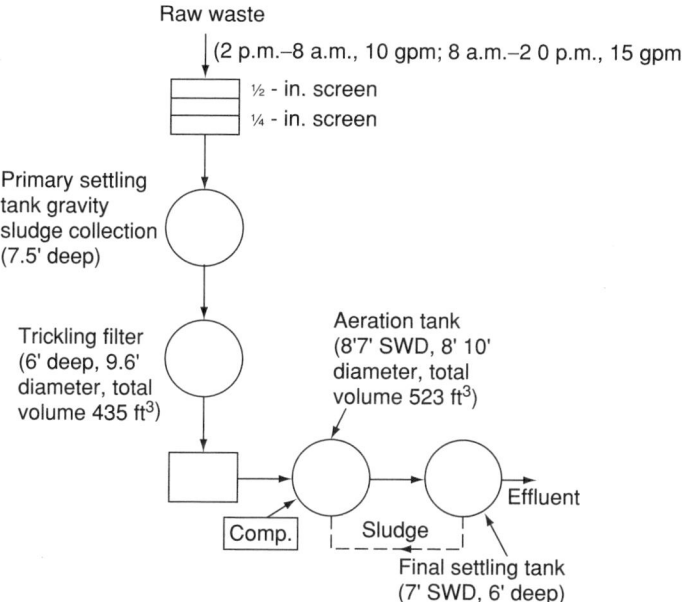

FIGURE 10.11. Field prototype of the Gloversville–Johnstown joint treatment plant. (From Nemerow 1987.)

municipal sewage-treatment plants treating large amounts of tannery waste may be summarized as follows:

1. *Degree of treatment.* Primary treatment by settling, followed by secondary treatment through biological processes, is required to meet New York State standards for plant effluent quality that may be accepted by the Cayadutta Creek under conditions of minimum dissolved oxygen content (at times of low flow and high temperature). The efficiency of treatment units and processes must be high, with an overall plant removal of approximately 85% of the incoming BOD.
2. *Pretreatment at mills.*
   a. Tanneries should remove fleshings, hair, hide pieces, and trimmings to make discharges transportable in gravity sewers. This can be accomplished by means found most efficient and economical, including primary settling tanks and/or mechanical screening. Animal greases plus petroleum solvents should be removed at the tanneries.
   b. The glue factory should remove settleable solids to make discharges transportable in gravity sewers. This can be accomplished by means found most efficient and economical, including primary settling tanks and/or mechanical screening.
3. *Sewage-treatment plant: processes (general).* The liquid wastes treatment will include pretreatment by mechanical screening, grit removal, pre-aeration of

TABLE 10.24
Prototype Operating Data

Date	Raw Waste BOD, ppm	Raw Waste pH	Raw Waste SS, ppm	Primary Effluent BOD, ppm	Primary Effluent pH	Primary Effluent SS, ppm	Trickling-Filter Effluent BOD, ppm	Trickling-Filter Effluent pH	Trickling-Filter Effluent SS, ppm	Final Effluent BOD, ppm	Final Effluent pH	Final Effluent SS, ppm	Under Aeration SS, ppm	Under Aeration pH
8/16/65	655	7.6	603	448	7.3	248	353	7.3	251	93	7.2	139	1,430	7.2
8/17/65	468	8.4	528	373	8.3	256	241	7.6	202	81	7.2	44	1,603	7.3
8/18/65	468	9.1	578	380	9.0	170	230	8.5	262	73	7.4	75		
8/19/65	563	8.0	369	390	8.0	211	268	7.6	187	85	7.2	78	1,850	7.3
8/20/65	493	8.5		408	8.7	281	256	8.3	166	73	7.4	84	2,631	7.6
8/23/65	443	7.0	382	433	7.4	145	316	7.5	243	72	7.5	43	2,313	7.4
8/24/65	555	8.1	393	425	8.1	184	330	7.8	250	90	7.4	89	2,066	7.5
8/25/65	370	7.9	392	313	7.8	169	398	7.8	394	79	7.4	197	2,487	7.5
8/26/65	408	8.3	454	360	7.8	243	231	7.8	168	102	7.3	127	2,788	7.2
8/27/65	273	8.8	337	308	8.5	135	215	7.5	155	19	7.6	53		
8/30/65	298	7.9	347	210	8.1	111	124	7.5	116	25	7.4	47	1,848	7.6
8/31/65	250	7.9	379	230	7.9	230	163	7.8	214	24	7.5	67	1,934	7.4
9/1/65	423	7.8	460	260	7.9	292	256	7.8	260	54	7.5	160	854	7.6
9/2/65	483	8.6	550	408	8.4	350	435	8.4	360	96	7.5	166	2,243	7.8
9/3/65	455	7.9	431	315	7.9	311	235	8.0	205	141	7.5	216	2,775	7.6
9/9/65	563	8.1	486	418	8.2	82	275	7.9	87	76	7.4	264	2,430	7.9
9/10/65	563	7.9	262	413	7.4	110	290	7.4	137	51	7.1	218	2,500	7.4
9/13/65	563	8.4	329	453	8.1	154	290	8.2	142	112	7.4	180	2,219	7.6
9/14/65	628	8.7	392	538	8.6	240	351	7.6	210	162	7.4	366	1,899	7.4
9/15/65	658	7.6	356	568	7.6	257	349	7.5	178	69	7.0	22	2,653	7.2
9/16/65	623	9.0	235	593	8.5	108	368	8.5	105	62	7.5	19	3,025	7.6
9/17/65	635	7.6	284	460	7.6	60	348	7.5	96	75	7.1	44	3,226	7.5
9/21/65	720	7.9	425	425	7.8	199	215	7.6	178	135	7.4	130	2,300	7.5
9/22/65	530	7.2	435	450	7.0	167	204	7.0	166	55	7.0	338	2,892	6.9
9/23/65	490	8.0	349	393	8.0	173	183	7.6	224	33	7.3	51	2,790	7.5
9/24/65	730	8.7	288	523	8.5	192	170	8.3	115	111	7.5	140	2,490	7.8
9/27/65	543	8.5	306	383	8.4	247	233	7.8	240	117	7.4	93	2,326	7.6
9/28/65	480	8.1	403	420	8.0	197	289	7.8	193	127	7.6	124	2,361	7.8
9/29/65	516		315	463		177	258		180	126		141	2,664	
9/30/65	650	8.1	307	490	7.6	145	344	7.5	195	192	7.4	157	2,096	7.4
10/1/65	435	7.8	312	363	7.6	162	280	7.6	229	158	7.2	241	1,495	7.4

SS, suspended solids

TABLE 10.25
Prototype Operating Results and Design Parameters

Unit and Effect	Prototype Results[a]	Design Parameters	Comment
Primary settling tank			Prototype tank construction and inherent limitations in small tanks resulted in lower settling efficiencies. Better results are expected in full-scale tanks with scum- and sludge-removal facilities and improved hydraulic characteristics. Additional settling-tank efficiency could be obtained by using flocculating agents if needed.
Suspended solids removal, %	49 (75)[a]	60	
Surface settling rate	390 (330)	800	
5-day BOD removal	24 (39)	30	
Roughing filter, with loading of 150 lb of BOD_5/1,000 ft^3	33% BOD_5 removal	30% BOD_5 removal	Performance of roughing filter established by test.
Aeration tank Process loading of 0.26 lb BOD/lb MLSS (8-hr peak)	81% BOD_5 removal	81% BOD_5 removal	Performance of activated sludge system established by test.
Process loading of 0.4 lb BOD/lb MLSS (24-hr average)	77% BOD_5 removal	77% BOD_5 removal	
Aeration tank (oxygen requirements, lb/day)	0.7 lb BOD_{5R} + 0.02 lb MLSS	1 lb O_2/lb BOD_5 removed/day)	Aerator capacity designed for mean peak 8-hr BOD loading with 25% present safety factor. Oxygen transfer ratio, α, to be determined in laboratory prior to final specifications on aerators.
Secondary settling tanks, surface settling rate (gal/ft^2/day)	760 gal/ft^2/day	600 gal/ft^2/day (at peak 8-hr rate of 13.1 mgd)	Selection of design overflow rates not on basis of pilot plant results. Inclusion of skimming devices on secondary settling tanks due to experience with pilot plant.
Flotation (thickening of waste-activated sludge)	Waste sludge of less than 0.5%. Solids thickened to 5% or greater at loadings greater than 2 lb/ft^2/hr	Design loading: 2 lb/ft^2/hr	Review of prototype data indicates that the design loading is suitable and that this loading should be achieved without the use of chemical conditioning.
Sludge digestion tank	Digestion studies did not develop digestion rate curves	Displacement time Primary, 25 days Secondary, 25 days	Digestion studies were not conclusive.

MLSS, mixed liquor suspended solids. [a]The data in parentheses give the results achieved by using polymer.

grease for removal in primary treatment, and pretreatment for pH control and chemical coagulation (in future); primary treatment by settling; and secondary treatment by trickling filters (high-rate roughing filters), then through activated sludge, followed by settling, with provisions for chemical precipitation for more complete solids removal in the future, with discharge to Cayadutta Creek where further treatment is by dilution and the oxygenation capacity of the stream.

Sludge treatment and disposal will include high-rate digestion with sludge-gas utilization followed by de-watering by lagoons and disposal by approved landfill methods; de-watering by vacuum filters and disposal by approved landfill methods; or de-watering by vacuum filers, incineration (multiple hearth), and disposal of ash by landfill.

4. *Sewage treatment plant: processes (recommendations).* The following treatment units and processes are specifically recommended and should be included in preliminary planning:
   a. Three mechanically cleaned barrack screens to remove large debris from the flow
   b. Two circular grit-removal units designed on surface overflow rate to remove grit and sand before primary settling (separation and washing of settled grit and organic matter by two hydrocyclone classifying devices). Before final design, consideration to be given to utilization of an aerated grit-removal unit
   c. Disposal of screenings and grit in sanitary landfill
   d. Grease removal by skimming in the primary settling tanks; grease flotation facilitated by aeration following or incorporated with grit-removal unit and immediately preceding the primary settling tanks
   e. Possible future chemical application in the aeration structure for pH control and introduction of chemicals to aid precipitation of wastes, for short periods, at times of exceptionally low flows in Cayadutta Creek
   f. Possible future addition of coagulating chemicals in the flow to the secondary clarifiers for "polishing" effluent and/or in the discharge from the secondary clarifiers for control of algal nutrients, if found necessary (structure provided for addition of coagulating chemicals in inflow)
   g. Six rectangular primary settling tanks with mechanical sludge collectors and scum skimmers
   h. Biological secondary treatment in two stages by two high-rate (roughing) filters, with stone or plastic media and rotating arm distributors, and activated-sludge treatment (in multiple units) in two sections, with mechanical aeration units directly powered by electric motors
   i. Two circular secondary settling tanks with sludge- and scum-collection mechanisms, sludge collectors to be of the "vacuum cleaner" type
   j. High-rate digestion provided through a primary digester followed by a secondary digester; floating covers on both digesters with gas-collection and holder facilities; gas utilization for heating of sludge and buildings; gas recirculation mixing provided in both digesters, with possible operation of either digester as primary

k. De-watering of digested sludge and disposal of sludge cake by approved landfill methods

Final recommendations and determinations on sludge de-watering and disposal must consider net annual costs, reflecting capital costs and operation, the physical problems involved in handling sludge as amounts increase over the years, and the future utilization of the land considered for landfill. Certain inferences of these evaluations can be made from the engineering and cost points of view. However, the Cities of Gloversville and Johnstown and the Town of Johnstown, as well as the New York State Department of Health, must also consider future land use and future disposal of both sewage-treatment plant wastes and municipal refuse. As a guide in the financial comparisons, the following is an estimate of the approximate net annual costs (capital plus operational) for the three possible methods:

A. De-watering by lagoons and disposal by approved landfill methods — $20,000
B. De-watering by vacuum filters and disposal by approved landfill methods — $50,000
C. De-watering by vacuum filters, incineration (multiple hearth), and disposal of ash by landfill — $65,000

De-watering at Fond du Lac, Wisconsin, was formerly achieved by vacuum filters. To reduce high de-watering costs, vacuum filtration was abandoned and replaced by evaporation and drainage in lagoons. De-watered sludge removed from the lagoons is disposed of in landfill. This technique has been in operation since 1962. Sewage-treatment plant loadings at Fond du Lac are only slightly less than those anticipated in the Gloversville–Johnstown design (6.4 mgd including 3 mgd of tannery wastes). Fond du Lac suspended solids are 173,000 lb/wk (compared to Gloversville–Johnstown at 195,578 lb/wk).

Based on the previous operational experience and consideration of the low annual costs compared to de-watering by vacuum filtration, de-watering by lagooning is recommended. In accordance with the requirements of the New York State Department of Health, the alternative to de-watering of the digested sludge by lagooning would be de-watering by vacuum filtration. For both methods, ultimate disposal would be by landfill. Both primary and secondary treatment units are shown schematically in Figure 10.12 and in a general view in Figure 10.13.

Solids Handling

The estimated quantity of raw solids to be handled at the plant is 170,000 lb/wk, of which approximately 70% is of industrial origin, primarily tannery wastes.

A review of existing secondary wastewater treatment plants in the United States and Canada handling a large percentage of solids disclosed four plants that were treating tannery effluents in combination with municipal wastewaters. All four plants use digestion. To dispose of the digested solids in liquid form on wastelands using tank

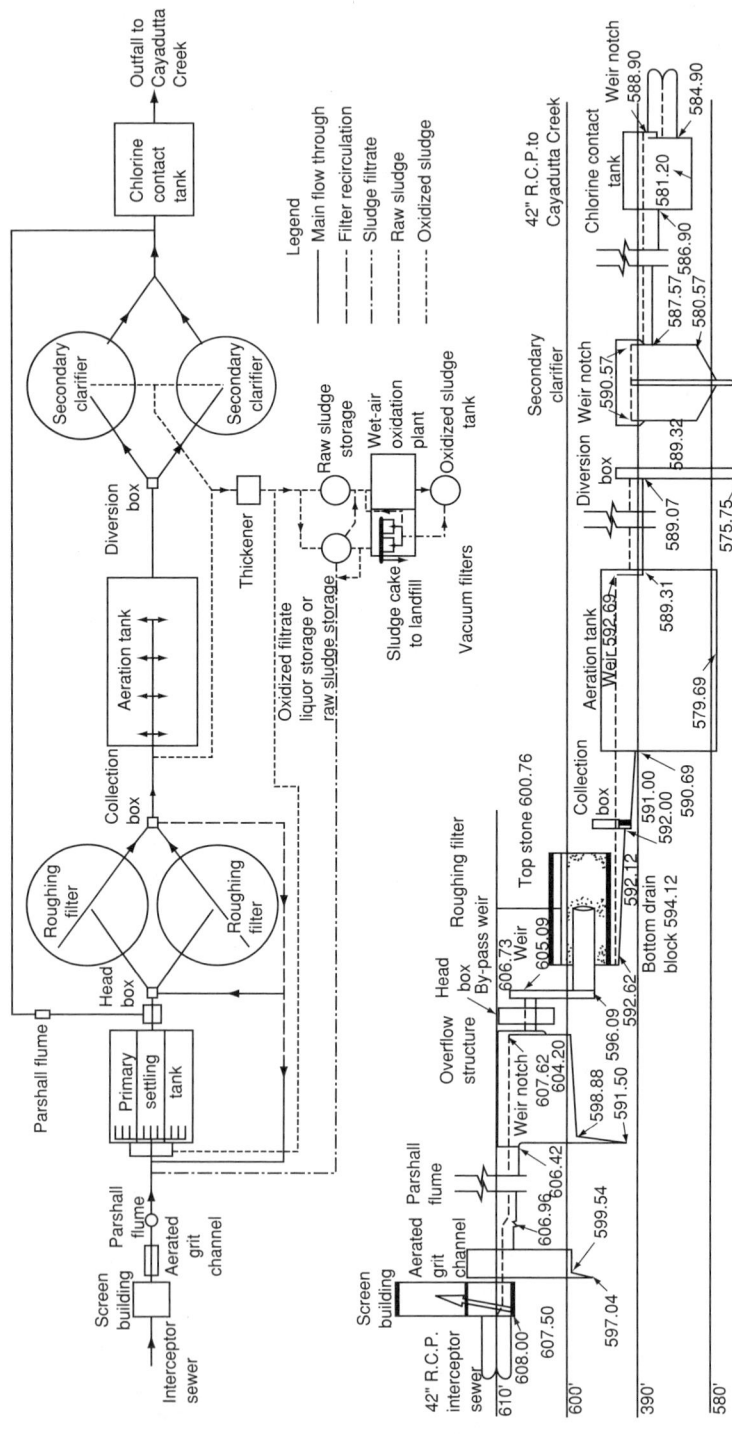

FIGURE 10.12. Line diagram and hydraulic profile of the Gloversville–Johnstown joint waste–water treatment plant. (Courtesy Morrell Vrooman Engineers.)

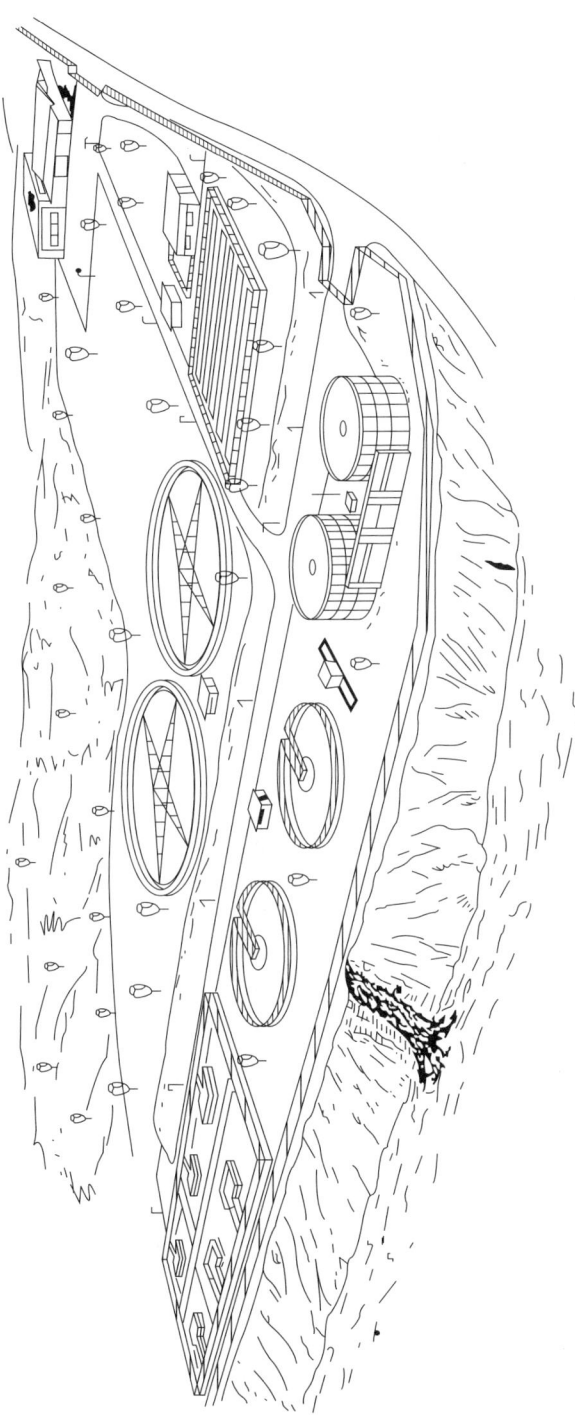

FIGURE 10.13. General view of the Gloversville–Johnstown joint treatment plant. (Courtesy Morrell Vrooman Engineers.)

trucks, one lagoons the digested solids, and one uses vacuum filtration, drying, and landfill. All of the plants are able to digest the solids effectively.

The problems with digestion of tannery-municipal solids have not been primarily chemical or biological but physical. Hair and scum have caused serious problems in the digesters themselves and in digestion-tank appurtenances. Extensive pretreatment, including fine screens, have been necessary at some plants to reduce such problems. The screens have in some cases introduced another problem, blinding.

The cities of Gloversville and Johnstown are surrounded by a rural area providing land for landfill of the final residue from the sludge-handling system. Both cities operate refuse landfills and own large areas of land designated for future landfill use. Therefore, landfill of the de-watered solids from the wastewater treatment plant could be accomplished.

A review of construction and operational costs indicated that digestion and de-watering in lagoons would provide the most economical solution to conditioning the solids. This solution was discussed extensively with the regulatory agencies and with the Gloversville–Johnstown Joint Sewer Board, who have responsibility for administering the project, but was eventually eliminated.

There was considerable interest in the wet-air oxidation system being used to condition and destroy organic solids. This system was viewed as capable of treating industrial solids without possible upset by the changing chemistry of the leather industry. The Zimpro Division of the Sterling Drug Company had considerable experience in handling similar solids at South Milwaukee and in Kempen, Germany. The wet-air process offered the added advantage of producing a solid that was readily filterable. The filterability of either raw or digested solids has been a potentially troublesome and expensive feature of the operation.

The sludge-handling process finally selected was a low-pressure (300 psi), wet-air oxidation plant, vacuum filters for de-watering, and ultimate disposal of solids in landfill. The wet-air plant as designed will reduce the nonsoluble organic solids by 40–50%. The high BOD filtrate will be pumped to a holding tank and discharged to the head of the plant during low BOD load periods. This plant will have the capacity to handle the weekly solids loading in 5 days on a 16-hour schedule or in 3½ days on a 24-hour schedule. It is basically one unit with several key items of equipment duplicated. The two vacuum filters will de-water cake from the oxidized-sludge holding tank. This tank will be equipped with overflow weirs and a sludge collector. The dense sludge pulled from the tank will be filtered during the 8-hour day shift, 5 days per week. This solution to the sludge-handling problem is not conventional, but it addresses the unique problem of these particular communities in their location and with their industries.

Final Design of the Gloversville–Johnstown Joint Sewage Treatment Plant

1. *Flow*
   24-hr mean 9.50 mgd
   8-hr mean peak 13.12 mgd

	1-hr mean peak	16.22 mgd
	1-hr maximum peak (to be used in hydraulic design of conduits)	19.3 mgd
	Maximum hydraulic capacity through primary	30 mgd
2.	*Process loading to sewage-treatment plant*	
	BOD weekday 24-hr mean	35,019 lb/day
	BOD weekday 8-hr mean	53,351 lb/day
	Suspended-solids load	195,578 lb/week
3.	*Mechanically cleaned bar screens*	
	Number of units	3
	Width of screen channel	3 ft
	Bar spacing	1-in. clear opening
	Velocity through screen at	
	5.0 mgd	1.2 ft/sec
	9.5 mgd	1.6 ft/sec
	30 mgd	2.7 ft/sec
4.	*Grit-removal units*	
	Number of units	2
	Type	Aerated
	Dimensions	7 ft wide × 30 ft long × 7 ft deep
	Particle size removed	100% of 0.2 mm at 20 mgd
	Grit-cleaning devices (cyclone with screw classifier)	2 units
	Grit disposal	Landfill
5.	*Primary settling tanks*	
	Number of tanks	6
	Flows and surface settling rates	
	8-hr peak (basic design)	13.12 mgd; 800 gal/ft^2/day
	24-hr	9.50 mgd; 580 gal/ft^2/day
	Maximum peak hour	19.3 mgd; 1,180 gal/ft^2/day
	Total surface area required	16,400 ft^2
	Surface area of each tank	2,736 ft^2
	Tank dimensions	152 ft long × 18 ft wide × 8 ft deep
	Displacement time at	
	13.12 mgd	1.8 hr
	9.50 mgd	2.5 hr
	Estimated removals	
	Suspended solids	60%
	5-day BOD	30%
6.	*High-rate (roughing) filters*	
	Number of units	2

Process loading:
    Peak 8-hr BOD   150 lb/1,000 ft^3
    24-hr average BOD   98 lb/1,000 ft^3
Diameter of filters   165 ft
Depth of filters   6 ft
Hydraulic loading
    8-hr peak   315 gal/ft^2/day
    24-hr average   227 gal/ft^2/day
Volume of filter media   250,000 ft^3
Filter media   4-in. stone or plastic
Recirculation pumps   3–3,500 gpm variable speed
Removal, 5-day BOD   30% at 150 lb/1,000 ft^3

7. *Aeration tank and equipment*
Number of tanks   1 with 2 compartments
Volume division of each tank   ¼ and ¾
Tank dimensions   260 ft long × 130 ft wide × 13 ft deep
Total tank volume   439,000 ft^3
Process loading:
    24-hr average   0.25 lb BOD/lb
    (39 lb/1,000 ft^3)   MLSS at MLSS of 2,500 ppm
    8-hr peak (60 lb/1,000 ft^3)   0.39 lb BOD/lb MLSS at
          MLSS of 2,500 ppm

Displacement time including
    33% recirculation
    24-hr flow, 9.50 mgd   6.3 hr
    8-hr flow, 13.12 mgd   4.5 hr
Aeration equipment
    Type   Mechanical
    Number of aerators   8
    Connected horsepower   800 hp
    Aerator oxygenation capacity (each)   310 lb/hr

8. *Secondary settling tanks*
Number of tanks   2
Diameter   120 ft
Side water depth   10 ft
Surface settling rate
    13.12 mgd flow   600 gal/day/ft^2
    9.50 mgd flow   434 gal/day/ft^2

9. *Chlorination equipment and contact tank*
Number of tanks   1 with 2 compartments
Dimensions   100 ft long × 60 ft wide
          (30 ft each) × 7 ft deep

	Detention time at	
	30 mgd	15 min
	9.5 mgd	47 min
	5.0 mgd	90 min
	Chlorinators	
	Number	2
	Rating	4,000 ppd
	Evaporators	
	Number	2
	Rating	4,000 ppd
	Residual analyzer number	1
10.	*Secondary return sludge pumps*	
	Number of pumps	3
	Type	Variable speed
	Maximum capacity each	2,500 gpm
11.	*Waste sludge pumps*	
	Number of pumps	3
	Type	Variable speed
	Maximum capacity each	200 gpm
12.	*Sludge thickener*	
	Number of units	2
	Type of thickener	Flotation
	Loading	2 lb/ft^2/hr
	Operation	100 hr/wk
	Total surface area required	264 ft^2
	Total surface area provided	300 ft^2
13.	*Wet-air oxidation unit*	
	Number of units	1
	Capacity	25 tons/day
	Operating pressure	300 psig
	Insoluble organic matter reduction	50%
	Operating volume and schedules	
	170,000 lb/wk	65% volatile content, 5% solids
	24 hr/day continuous operation	3½ days/wk
	16 hr/day operation	5 days/wk
	Oxidized- sludge storage	120,000 gal for 1 day
	Duplicate items	Boiler
		High-pressure pump
14.	*Raw-sludge holding tanks*	
	Number of tanks	2
	Total holding capacity	7 days

Tank proportions
    Side water depth   20 ft
    Diameter   42 ft

15. *Vacuum filtration*
    Number of units   2
    Filter area   400 ft^2 each
    Design filter rate
        Oxidized sludge   5 lb/ft^2/hr
        Raw sludge   5 lb/ft^2/hr
    Operating volume and schedules
        Oxidized sludge, 115,000 lb/wk   29 hr filter time/wk
        Raw sludge, 170,000 lb/wk   43 hr filter time/wk

16. *Ultimate disposal of oxidized filter cake*
Sludge to be landfill in city refuse areas or other selected areas. Because of the character of oxidized sludge cake, sludge need not be covered.

Estimated Costs and Financing

Estimated costs of the plant units described in the preceding list follow:

	Amount
Site development	$300,000
Screen building and grit tanks	158,000
Primary settling tank	323,000
Roughing filters (two)	437,000
Aeration tank	645,000
Secondary settling tanks	310,000
Chlorine contact tank	67,000
Recirculation building	203,000
Overflow and Parshall flume structures	10,000
Sludge building, including thickeners, vacuum filters, and wet-air plant	1,056,000
Sludge and oxidized liquor holding tanks	77,000
Yard piping and conduits	279,000
Waterline to plant site and meter pit	32,000
Fencing	11,000
Administrations building	228,000
Electrical contract	250,000
Subtotal	$4,386,000
Contingency (5%)	219,300
Total	$4,605,300

The project costs and the federal and state grants are estimated as follows:

Interceptor sewers	$1,227,000
Wastewater treatment plant	4,600,000
Subtotal	5,827,000
All other costs and contingencies	1,212,000
Total	7,039,000
Less federal and state grants	3,981,000
Net cost to community	3,058,000

It is estimated that the plant will have 20 full-time employees and that the annual operational and maintenance costs will be $262,000. The local communities will receive a reimbursement from the State of New York to the amount of one-third of this cost, leaving a local net cost of $175,000.

The estimated total annual costs for the project and their distribution between the cities are as follows:

Annual debt service (30 years at 4.5%)		$220,000
Annual net operation and maintenance costs		175,000
Total		$395,000
Distribution		
City of Gloversville (55%)	$217,250	
City of Johnstown (45%)	177,750	
Total	$395,000	

The agreement between the cities of Gloversville and Johnstown calls for a 55/45% split of capital costs and of operation and maintenance for 3 years. The division of operation and maintenance charges will be reviewed every 3 years to reflect the results of samples collected and analyzed.

The final method of allocating costs to the users has not been fully established. The two cities would like to keep the rates in the cities the same, based on the volume of water used, probably with surcharges for industrial users. It is the aim of the Sewer Board and the cities to avoid a rate structure dependent on repeated and critical sampling of industrial users. The average homeowner in the cities will pay for the service based on his or her water usage, and present estimates put the average annual cost at less than $20 per year per home.

Application of Plan in Practice

There are many lessons that this case history serves to teach us. Several of the theoretical solutions to this problem had to be abandoned because of the social, economic, and governmental situation. Table 10.26 describes five such inconsistencies, along with the reasons for deviating from theory.

TABLE 10.26
Inconsistencies between Theory of Design and Actual Practice in Design

Situation	Theoretical Solution	Actual Practice	Reason for Violating Theoretical Solution
Flow BOD	Measured as 5–7 mgd Measured as 20,000–25,000 lb/day	Designed for 9.5 mgd Designed for 35,019 lb/day	Addition of a large glue-manufacturing plant and another tannery; consideration of future loads
Sludge handling	Digestion plus lagooning was least costly and proved acceptable in the laboratory	Zimpro plus vacuum filtration plus landfill	Lack of state approval because of health hazards of lagooning digested sludge; lack of confidence in efficiency of sludge digestion
Charging for services	Incentive plan based on unit costs for a pound of BOD and suspended solids and a gallon of waste	Percentage of water bill	Ease of charging; elimination of need to sample, police, and analyze industrial wastes; strong representation on joint Sewer Board of tannery industry
Equalization	Theory would indicate that it is needed because of great fluctuations in instantaneous flow and character	In practice, not required because the great number of tanneries and length of travel in sewers provide equalization	Great number of tanneries; increase in cost of construction and operation of equalization basin
Biological treatment	Theory and laboratory results indicate that activated sludge is an excellent method of reducing high BOD	Practice shows it more suitable to use a combination of roughing filters and activated-sludge treatment in series	Activated sludge computed to be about twice the cost[a]

[a]Costs are compared as follows:

	Roughing filter	Activated sludge
Fixed charge	$0.55/lb	$0.39/lb
Operating costs	$0.03/lb	$0.72/lb
Total costs	$0.58/lb	$1.11/lb

It has been reported that the plant experienced grease problems because of ineffective pretreatment by some tanneries (1974/1975). However, more than 90% BOD reduction and about 85% suspended-solids reduction were attained.

This case study serves as an example of a twentieth-century application of joint treatment of raw industrial wastes with domestic sewage.

Review Questions

1. What specific alternatives are open to a municipal official concerning acceptance of industrial wastes? What are the alternatives open to an industrial plant manager when considering joint treatment?
2. List and discuss the 12 advantages of joint treatment.
3. Can all industrial wastes be treated in municipal sewage-treatment plants? What are the limiting concentrations of contaminants, the reason for limitation, and the acceptable pretreatment required for excessive contaminants of each type?
4. What are some of the problems of combined treatment?
5. What can municipalities do to assist industries in waste-treatment practices?
6. What methods are used by municipalities for defraying costs of sewer services and sewage-treatment costs?
7. What method of charging an industry is advocated by both the American Society of Civil Engineers and the American Bar Association? What are its main advantages?
8. In the case history of joint treatment presented in this chapter, what are the pertinent background factors related to its solution?
9. Describe how the stream survey, composite waste sampling and analysis, and laboratory pilot studies were vital to the joint treatment decision.
10. Why was a literature survey necessary?
11. What were the overall planning conclusions?
12. What are some of the inconsistencies between theory and design used in this case study?
13. Does joint treatment of raw wastes represent a reasonable or optimum solution to the problem?

References

Introduction

American Public Health Association. 1955. *Standard Methods for the Examination of Water, Sewage, and Industrial Wastes,* 10th ed. New York: American Public Health Association.

California State Water Pollution Control Board. 1955. "A Survey of Direct Utilization of Waste Waters." Sacramento, CA: California State Water Pollution Board. Publication no. 12.

Federation of Sewage and Industrial Wastes Association. 1957. *Municipal Sewer Ordinances, Manual of Practice,* no. 3. Washington, DC: Federation of Sewage and Industrial Wastes Association.

Geyer, J. C. 1937. The effect of industrial wastes on sewage plant operation. *Sewage Works J.* 9:625.

Graver Water Conditioning Company. 1954. *The Treatment of Sewage Plant Effluent for Water Reuse in Process and Boiler Feed.* New York: Graver Water Conditioning Company. Technical Reprint T-129.

Nemerow, N. L. 1951. Fiber losses at paper mills: effects on streams and sewage treatment plants. *Sewage Ind. Wastes* 23:880.

Nemerow, N. L. 1956. *Water Wastes of Industry.* Bulletin no. 5, Industrial Engineering Program. Raleigh: North Carolina State College.

Schroepfer, G. M. 1951. Sewer service charges. *Sewage Ind. Wastes* 23:1493.

Veatch, N. T. 1948. Industrial uses of reclaimed sewage effluents. *Sewage Works J.* 20:3.

Water Pollution Control Federation. 1976. Joint Treatment of Industrial and Municipal Wastewaters: A Publication of the Technical Practice Committee. Washington, DC: WPCF, 1976.

Case History

Braunschweig, T. D. 1965. Studies on tannery sewage. *J. Am. Leather Chemists' Assoc.* 60:125.

Chase, E. S., and P. Kahn. 1955. Activated sludge filters for tannery waste treatment. *Wastes Eng.* 26:167.

Civil Engineering Department, Syracuse University. January 1969. "Research Report no. 10." Syracuse, NY: Civil Engineering Department, Syracuse University.

Fales, A. L. 1928. Discussion of paper by W. Howalt, Studies of tannery waste disposal. *Trans. Am. Soc. Civil Eng.* 1394; *Hide and Leather,* 75:48.

Furkert, H. 1936 and 1937. Mechanical clarification and biological purification of Elmshorn sewage, a major portion of which is tannery wastes. *Tech. Gemeindebl.* 1936;39:285, and 1937;40:11; *Chem. Abstr.* 1937;31:5076.

Hubbel, G. E. 1935. *Water Works Sewerage* 82:331.

Jansky, K. 1961 and 1962. Tannery waste water disposal. *Kozarstvi* 1961;11:327 and 355; *J. Am. Leather Chemists' Assoc.* 1962;57:281.

Kalibina, M. M. 1931. The application of the biological method of judging the efficiency of a purifying plant. Observations on the growth of organisms in an activated sludge tank. *Chem. Abstr.* 25:3422.

Kubelka, V. 1952. *Veda Vyzkum Prumyslu Kozedelnem* 1:113.

Kubelka, V. 1956. Principles of a final biological purification of tannery effluents. *Veda Vyzkum Prumyslu Kozedelnem* 1:113.

Mausner, L. 1938. Tannery waste water. *Gerber* 41:1519.

Nemerow, N. L., H. S. Sumitomo. 1936. "Pollution Index for Benefit Analysis" (unpublished).

Pauschardt, H., H. Furkert. 1936. *Stadtereinigung* 411 and 427.

Snock, A. 1928. *Collegium* 703:612.

Thebaraj, G. J., S. M. Bose, Y. Nayudamma. 1962. Comparative studies on the treatment of tannery effluents by trickling filter, activated sludge and oxidation pond systems. *Central Leather Research Institute, Madras, India* 13:411.

Vrooman, M., V. Ehle. 1950. *Sewage Ind. Wastes* 22:94.

CHAPTER 11

Hazardous Wastes

To the general public, the subject of hazardous wastes carried different meanings from its onset in the mid-twentieth century to the 1990s. At the beginning, a "hazardous" waste was simply one that would result in danger to any person or structure with which it came into contact. As more widespread understanding grew, a "hazardous waste" was considered one that was "toxic" to microorganisms that were present in domestic wastes and generally used to treat them biologically. And, finally, most environmental workers became acutely aware that certain hazardous wastes would also be toxic to the soil and water environments that they ultimately entered when almost unaltered in form composition.

It is difficult to put exact dates to when these definitions became general knowledge. Certainly the periods of their existence overlapped greatly and certainly some scientists—especially those educated in the environmental science area—understood all of these "effects" right from the beginning. However, little was done by anyone to ameliorate the adverse effects on the environment until late in the century. The reasons for the "lack of action" on treatment of hazardous waste were largely driven by industrial secrecy and economics. Action finally began when enough catastrophes occurred and industrial secrets and cover-ups were revealed.

Definitions

Generators of waste materials can ascertain whether their waste is hazardous either by referring to a published list or by complying with the U.S. Environmental Protection Agency (EPA) definition of hazardous waste. The EPA defines the characteristics of a hazardous waste in terms of ignitability, corrosivity, reactivity, and extractive procedure toxicity. These are mentioned by Nemerow (1983, Table 1) and discussed briefly in the following four sections of this chapter.

Many hazardous wastes contain toxic materials that can be classed in more than one category. In these cases, the waste will be grouped in the category of major significance. The reader must understand, however, that certain constituents of a waste may also be important and from another category.

The legal U.S. definition of a hazardous waste is extremely complicated and even controversial. At the time this book was written, an interested person was urged to refer to the EPA Regulations for Identifying Hazardous Wastes—40-CFR261, 45-FR33119, May 19, 1980, and revised July 1, 1983, and amended on 10 occasions from then until July 25, 1985.

The reader can ascertain the status of a certain waste by referring to the following four figures published by the Bureau of National Affairs, In. Washington, D.C. 20037. The waste must first receive a legal definition of a Resource Conservation Recovery Act (RCRA) solid waste (Figure 11.1). It can then be classified as "hazardous" and handled as such by referring to Figure 11.2. Regulations covering hazardous wastes are

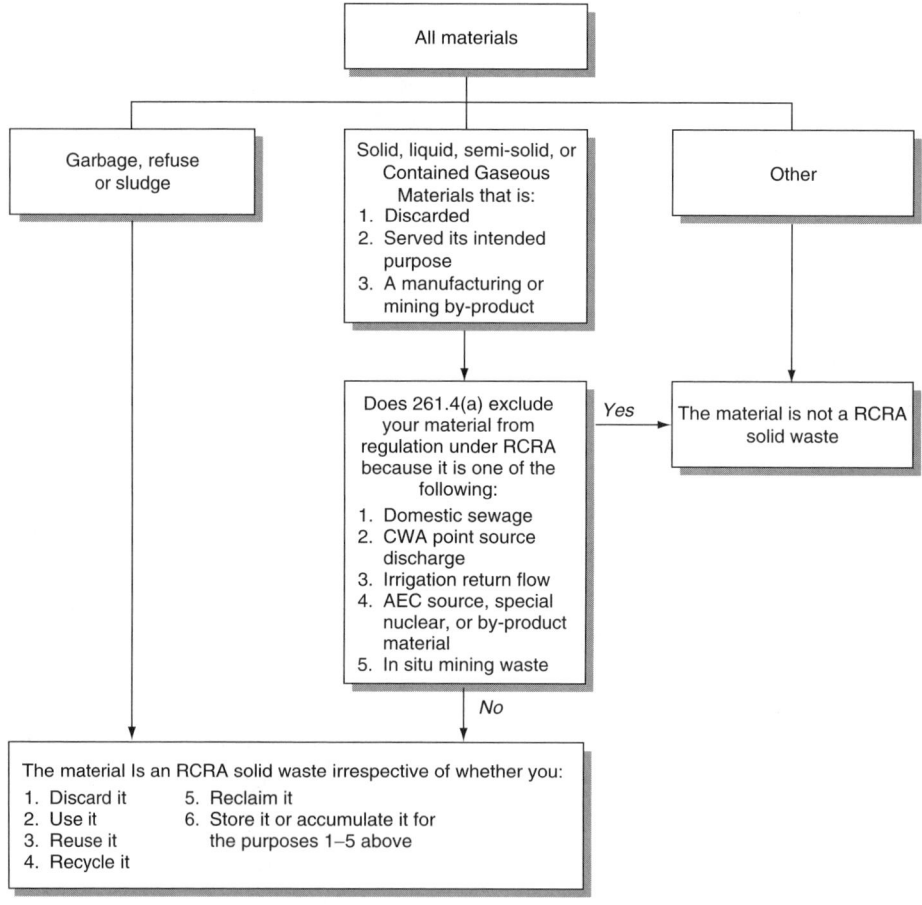

FIGURE 11.1. Definition of a solid waste.

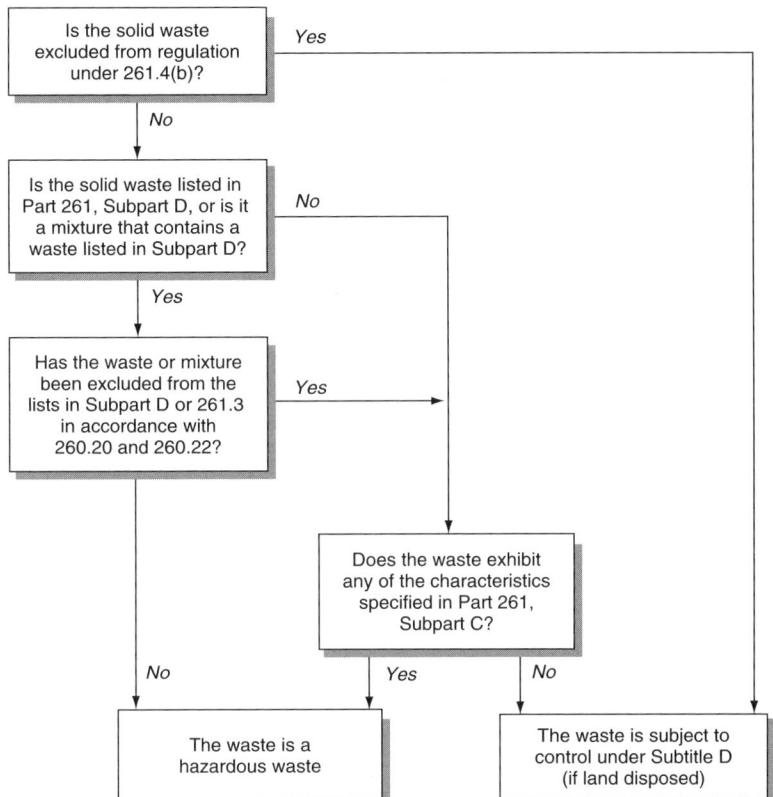

FIGURE 11.2. Definition of a hazardous waste.

shown in Figure 11.3 and for those subject to control under Subtitle C of EPA regulations in Figure 11.4.

Dangers

Once a decision has been made to manufacture a potentially toxic chemical or product, the door is open to catastrophic release of either to the environment. Careful control must be maintained by the industrial manufacturer and product users from the very beginning. The dangers of toxic wastes being released to the environment are real and originate from three sources: (1) production of the toxic chemical, (2) use of the toxic chemical in subsequent manufacturing of a product, and (3) release of a toxic product (or other chemical) from the use or wastage of the product. In addition, there is always the potential hazard of escape of any of these toxic materials during transportation from any location to another for use or disposal. In reality, all of these dangerous emissions or releases of toxic material to the environment occur accidentally. No one

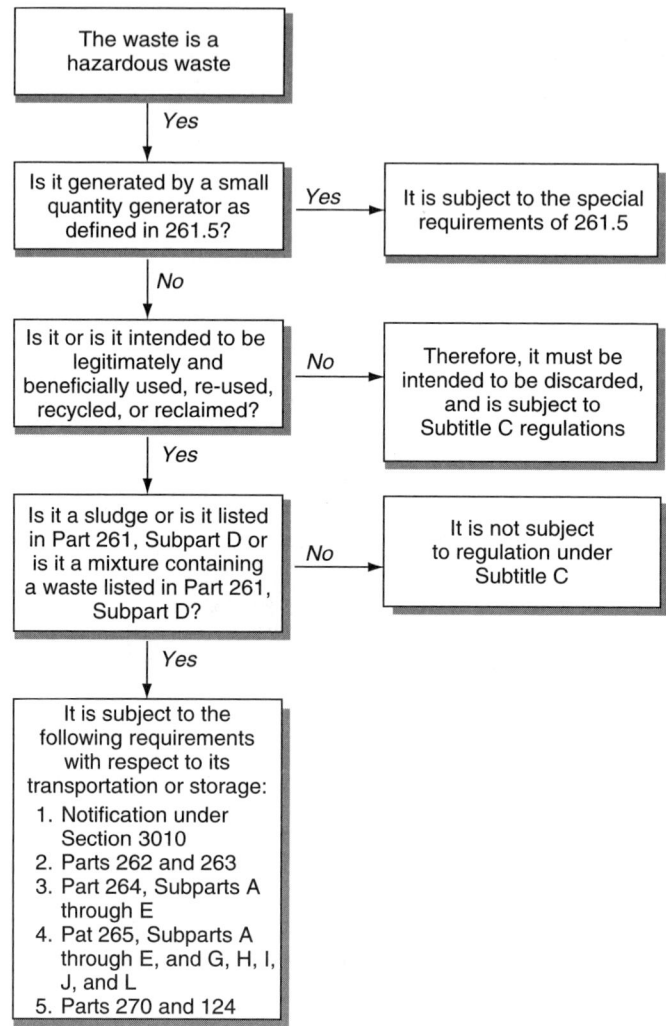

FIGURE 11.3. Special provisions for certain hazardous waste. (Amended by 48 FR 14153, April 1, 1983.)

intends to cause these catastrophes, but accidents occur frequently enough to be considered a usual occurrence no matter how infrequently they happen. The only reasonable prevention—other than nonuse of hazardous materials—is to design fail-safe devices to operate when the accidents occur.

Scope of Problems and Costs of Solutions

Pearce (1983, 57) reports that in the industrialized nations, at least by far the largest proportion of hazardous waste generated is recycled or reclaimed by industry. He gives

Hazardous Wastes 249

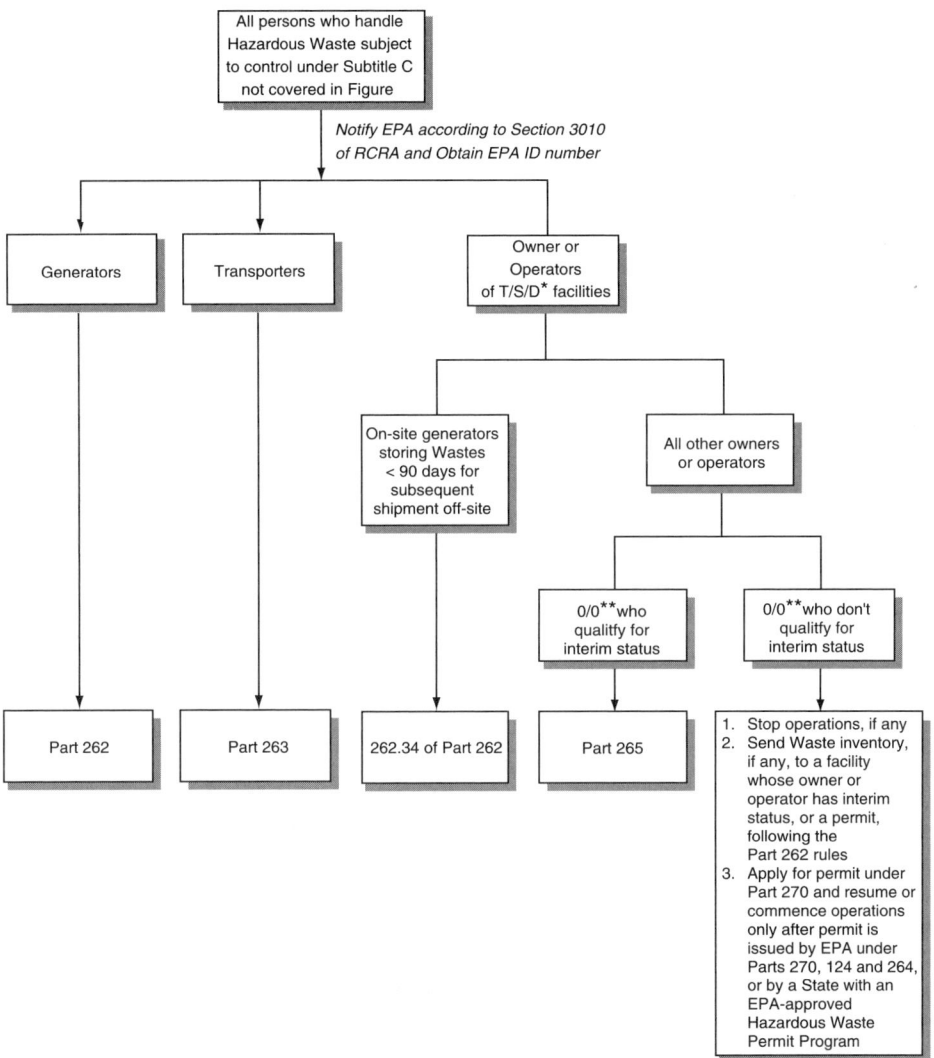

FIGURE 11.4. Regulations for hazardous waste not covered in diagram 3. (Amended by 48 FR 14153. April 1, 1983)
*T/S/D, for Treatment, Storage or Disposal
**0/0, for Owners or Operators

as an example Great Britain, where about 10 million tons per year are recovered by industry, whereas only 3.4 million tons are rejected for disposal.

The EPA (1980) determined that costs of hazardous waste disposal were major factors in attaining final solutions to the problem. Although costs vary widely, the EPA gave ranges for various treatments, as shown in Table 11.1.

Another EPA (March 1985) rough general cost comparison for hazardous waste treatment gives the following:

TABLE 11.1
Cost of Hazardous Waste Disposal Practices

Technology	$/Metric Ton
Land spreading	2–25
Surface impoundment	14–180
Chemical fixation	5–500
Secure chemical landfill	50–400
Incineration (land based)	75–2,000

Source: U.S. EPA (1980).

Type of Waste	Type or Form of Waste	Price 1981, $/Metric Ton
Management		
Landfill	Drummed	168–240
	Bulk	55–83
Deep-well injection	Oily wastewater	16–40
	Toxic rinse water	132–164
	Cyanides/heavy metals, and highly toxic waste	66–791
Land	Liquids	53–237
Incineration	Solids and highly toxic liquids	395–791

Effects on Environment

Hazardous wastes pollute the air, water, and land drastically. They often affect human beings directly and most certainly indirectly. The deterioration of the environment is extensive and often irreversible, or reversible at heavy cost in both money and people-power. Effects are caused by wastes that burn, corrode, react violently, are toxic, and are acutely hazardous to the health and welfare of society.

It is becoming increasingly evident that traces of hazardous chemicals, including suspected carcinogens, are getting into the food chain via surface-water and groundwater supplies. People are exposed through drinking water, air, and food such as fish originating in rivers and lakes. Toxic pollutants find their way into surface-water supplies from waste-treatment plants, coal burning, refuse incineration, and atmospheric fallout. An example of this dangerous effect is that occurring in the U.S. Great Lakes basin (Great Lakes Study 3, 1985) in which 1,065 hazardous or potentially hazardous substances were identified in the lakes, with the highest concentration at the southern edges. Besides the recreational and commercial fishing in these lakes, many large cities, such as Syracuse, New York, take in raw water from them (Lake Ontario) for drinking-water purposes. Drinking water must now be carefully examined and analyzed for potentially hazardous chemicals.

Philosophies of Solutions

Problems caused by hazardous wastes must be solved by (1) not producing them in the first place, (2) collecting and reusing every bit of them in some other product that does affect the environment, or (3) destroying them in ways that either there will be nothing left or their residues will be compatible with and harmless to the environment.

Any of these solutions costs money to instigate and requires ingenuity and research to implement. They cannot usually be solved in ways heretofore practiced by municipalities and industries in rather conventional manners.

One of the major objectives of this book is to explore potentially feasible means of solving what is now a major problem for society: disposal of our hazardous wastes. Developed countries such as the United States, Germany, France, and so on tend to locate chemical plants that use or produce hazardous chemicals in less developed countries in order to minimize production and distribution costs. A lesson rapidly being learned by developed countries is that these lower production costs cannot be gleaned at the expense of the local environment. Hazards from these plants in developing countries are just as dangerous and intolerable as they are in developed nations. It was originally thought possible by developed nations to export environmental damage—and acceptable to lesser developed countries in order to gain economic compensation. The Bhopal incident accentuates the fact that this kind of thinking on anyone's part no longer exists.

Varied Types of Industrial Hazardous Wastes

In 1974, the EPA listed 15 major industries that discharge wastes containing hazardous substances. It indicated the specific hazardous materials in each waste (Table 11.2).

In addition to private section industries, the U.S. Navy has voiced concern over its contribution to the hazardous waste problem (LaPue 1996). The Navy spent $3.2 million to scrape up and dispose of $7,900$ yd^3 of soil deposited between the years 1943 and 1975, contaminated with "everything from mercury in torpedoes to lead paint and since-banned

TABLE 11.2
Hazardous Waste Industries

Industry	Hazardous Substances
1. Mining and metallurgy	As, Cd, Cr, Cu, Cn, Pb, Hg, Se, Zn
2. Paint and dye	Cd, Cr, Cu, Cn, Pb, Hg, organics, Se
3. Pesticide	As, Cl-hydrocarbons, Cn, Pb, Hg, organics, Zn
4. Electrical and electronic	Cu, Cl-hydrocarbons, Cn, Pb, Hg, Se
5. Printing and duplicating	As, Cr, Cu, Pb, organics, Se
6. Electroplating-metal finishing	Cd, Cr, Cn, Cu, Zn
7. Chemical manufacturing	Cl-hydrocarbons, Cr, Cu, Pb, Hg, organics
8. Explosives	As, Cu, Pb, Hg
9. Rubber and plastics	Cl-hydrocarbons, Cn, Hg, organics, Zn
10. Battery	Cd, Pb, Ag, Zn
11. Pharmaceutical	As, Hg, organics
12. Textile	Cr, Cu, organics
13. Petroleum and coal	As, Cl-hydrocarbons, Pb
14. Pulp and paper	Hg, organics
15. Leather	Cr, organics

fluids from torpedoes." Between 15,000 and 110,000 gallons of drummed waste were drained at the site of the Navy yard in San Diego, California, before it was paved, along with about 10,000 gallons of transformer fluid laden with polychlorinated biphenyl (PCB). The Navy expects to excavate and remove 1.5 to 6 ft of soil underneath the asphalt pavement, with special precautions taken to prevent dust and airborne contamination associated with the removal procedures.

Sundaresan et al. (1983, 70) group industries producing toxic and hazardous wastes into two categories:

1. Conventional solid wastes generated in large quantities and stored near the factory such as phosphatic fertilizers and thermal power plants.
2. Other semisolid and solid residues including liquid wastes resulting from the organic and inorganic chemical industries. These wastes are not voluminous, but some are highly toxic and potentially hazardous.

Yakowitz (1988) presents a list of 14 types of hazardous wastes (Table 11.3). It appears to be the most comprehensive breakdown of specific general characteristics available at the time.

TABLE 11.3
List of Hazardous Characteristics

Code Number	Characteristics
H1[a]	Explosive
	An explosive substance is a solid or liquid substance (or mixture of substances) that is in itself capable by chemical reaction of producing gas at such a temperature and pressure and at such a speed as to cause damage to the surroundings.
H2[a]	Oxidizing
	Substances that, while in themselves are not necessarily combustible, may, generally by yielding oxygen, cause or contribute to the combustion of other materials. (Organic substances that contain the bivalent-0-0-structure are thermally unstable substances that may undergo exothermic self-accelerating decomposition.)
H3[a]	Inflammable
	The word "flammable" has the same meaning as "inflammable."
	Inflammable liquids are liquids, or mixtures of liquids, or liquids containing solids in solution or suspension (e.g., paints, varnishes, lacquers, etc., but not including substances otherwise classified on account of their dangerous characteristics) that give off an inflammable vapor at temperatures of not more than 60.5°C, closed-cup test, or not more than 65.6°C, open-cup test. (Since the results of open-cup tests and of closed-cup tests are not strictly comparable and even individual results by the same test are often variable, regulations varying from the above figures to make allowances for such differences would be within the spirit of this definition.) Inflammable solids are solids, other than those classed as explosives, that under conditions encountered are readily combustible, or may cause or contribute to fire through friction.

TABLE 11.3 *(continued)*

H4[b]	Irritating
	Noncorrosive substances and preparations that, through immediate, prolonged, or repeated contact with the skin or mucous membrane, can cause inflammation.
H5[b]	Harmful
	Substances and preparations that if they are inhaled or ingested or if they penetrate the skin, may involve limited health risks.
H6[a]	Toxic
	Substances and preparations, that, if they are inhaled or ingested or if they penetrate the skin, may involve serious, acute, or chronic health risks and even death.
H7[b]	Carcinogenic
	Substances and preparations that, if they are inhaled or ingested or if they penetrate the skin, may induce cancer in people or increase the incidence.
H8[a]	Corrosive
	Substances that, by chemical action, will cause severe damage when in contact with living tissue, or, in the case of leakage, will materially damage, or even destroy, other items or a means of transport; they may also cause other hazards.
H9[a]	Infectious
	Substances containing viable microorganisms or their toxins, which are known, or suspected, to cause disease in animals or humans.
H10[a]	Liberation of flammable gases in contact with water
	Substances that, by interaction with water, are liable to become spontaneously inflammable or to give off inflammable gases in dangerous quantities.
H11	Liberation of corrosive fumes in contact with air or water.
H12	Liberation of toxic gases in contact with air or water.
H13	Capable, by any means, after disposal, of yielding another material, e.g., leachate, which possesses any of the characteristics listed above.
H14	Ecotoxic
	Substances which, if released, present or may present immediate or delayed adverse impacts to the environment by means of bioaccumulation and/or toxic effects upon biotic systems.

[a]Definition from *Transport of Dangerous Goods*, Recommendations of the United Nations Committee of Experts on the Transport of Dangerous Goods, Third Revised Edition, United Nations, New York, 1985.
[b]Definition from Article 2 of the European Communities Council Directive of 18 September 1979 amending for the sixth time Directive 67/548/EEC on the approximation of the laws, regulations and administrative provisions relating to the classification, packaging, and labeling of dangerous substances (Directive 79/831/EEC.)
[c]Guidance with regard to this characteristic may be obtained by consulting the lists of known and strongly suspected carcinogens published periodically by the International Agency for Research on Cancer.
Source: Adapted from Yakowitz (1988).

The American Chemical Society (ACS) is especially concerned with the various types of hazardous wastes. It published a list of the types of these wastes of concern to its members (ACS 1984) (Table 11.4).

TABLE 11.4
RCRA-Regulated Hazardous Wastes

Waste Group Generated	% of Establishments Generating
Spent solvents—halogenated and nonhalogenated[a]	51.0
Ignitable wastes[a]	43.4
Corrosive wastes[a]	33.4
Spilled, discarded, or off-specification commercial chemical products or manufacturing chemical intermediates	28.8
EO toxic wastes[a]	27.8
Electroplating and coating wastewater treatment sludges and cyanide-bearing solutions and sludges	16.4
Statutory hazardous wastes (i.e., not listed or regulated as hazardous wastes by EPA or state)	12.2
Listed industry wastes from specific sources	10.2
Acutely hazardous wastes	10.2
Reactive wastes[a]	7.1

Note: The list, taken from an EPA survey, provides estimates of the percentage of generators producing specific types of RCRA-regulated hazardous wastes during 1981. The sum of the total exceeds 100% because a generator may have produced more than one type of waste, and a given waste stream may have been a mixture that consequently was reported under multiple characteristics.

[a]These waste groups include wastes that, while not specifically listed in EPA's list of hazardous wastes, exhibit one of the hazardous characteristics, such as high flammability, high or low pH levels, toxicity, and volatility or violent reactive tendencies with other substances.

Ignitable Hazardous Wastes

The EPA (1980) classifies these hazardous wastes in four categories:

1. A liquid that has a flash point less than 60°C (140°F); exemption—aqueous solution with less than 24% alcohol
2. A waste that is not a liquid but is capable under standard temperature and pressure of causing fire through friction, absorption of moisture, or spontaneous chemical changes, and, when ignited, burns so vigorously and persistently that it creates a hazard
3. An ignitable compressed gas
4. An oxidizer

From these definitions, such ignitable hazardous wastes can originate from all three physical states: gases, liquids, and solids.

Waste Oils

Waste oils not only are objectionable aesthetically in any of the three environments (air, water, or land) but also are classified as toxic or hazardous because they are ignitable. Burning by itself is a hazard to health and property. Further, the products of combustion can also be toxic to breathing. For all these reasons, oils must be treated and ultimately disposed of so as to prevent toxicity in any environment.

Automotive crankcase used waste oil constitutes a large proportion of waste oils. Collection, recovery, and re-purification of waste oils represents the ideal solution to disposal problems. However, economics must justify such methods. When oil prices rise, justification for recovery is also improved. For evaluating recovery of waste oils, the physicochemical analysis of average waste oils is necessary. A typical analysis is given in Table 11.5. The main obstacle to recovery is that of collection. The use of strategically located collection stations where individuals and garages can bring their oils will enhance recovery. Other deterrents to recovery include additives in the oils, environmental laws on reprocessing, and removal of tax incentive laws for reuse of waste oils.

Other possibilities for treatment include (1) mixing waste oils with lighter new fuel oil for burning in power plants, and (2) incorporating them with municipal solid wastes and tree and bush cuttings or beach littoral collections for composting. No data are currently available on either of these methods.

Spent Oil Emulsions

Spent oil emulsions originate mainly from metal fabricating and power plants where mixtures of oils and waters are used to cool moving parts. Because these oils are contaminated with water and grit or cuttings, they can be centrifuged or skimmed for concentration before transportation away from the site of origin. Lubricating oils can also be heated, coagulant added, and centrifuged before reuse in the same plant.

Other Oily Wastes

Other waste oils originate from ship discharges, either accidental or planned. Oil water mixtures from bilges are usually discharged to the sea when the bilge water level gets too high. Sometimes bilge water is coagulated and filtered before discarding the water. In that case, a residual sludge remains to be disposed of. In proper proportions, and with good

TABLE 11.5
Typical Automotive Oil Waste Composition

Variable	Value
Gravity, °API	24.6
Viscosity at 100°F	53.3 Centistokes
Viscosity at 210°C	9.18 Centistokes
Flash point	215°F (C.O.C. Flash)
Water (by distillation)	4.4% volume
BS&W	0.6 by volume
Sulfur	0.34 weight %
Ash, sulfated	1.81 weight %
Lead	1.11 weight %
Calcium	0.17 weight %
Zinc	0.08 weight %
Phosphorus	0.09 weight %
Barium	568 ppm
Iron	356 ppm
Vanadium	5 ppm

premixing and feeding devices, it may be burned in special furnaces, even aboard ship. Usually, it is held in a collection tank and pumped out at an onshore station. Such stations can handle larger quantities of oily sludges more effectively and economically. Furnaces or incinerators can be designed specifically to burn these wastes. One company, Haz-Mat Response Technologies, has developed a product to clean up hazardous waste spills (Findle 1996). The product is a mixture of hydrocarbon polymers called "Rubberizer." The product absorbs and transforms petroleum products and solvents into a rubber-like and essentially inert material. This makes the spill cleanup easier because the hardened material can be lifted easily out of the water or other surfaces such as roads and airport runways. The hardened matter can be buried or placed safely in a sanitary landfill.

Onshore oil sludge storage tanks can also be used to receive tank-cleaning wastes. These wastes are usually more dilute, containing mixtures of condensed steam and oil.

Power stations, both stationary and moving, generally use centrifugation to purify their fuel oil before burning, to remove the last traces of water. The latter contaminant may reduce boiler efficiency. It is possible to use these or other standby centrifuges to recover oil from tank cleanings or even bilge waters. The recovered oil could be burned in the power plant's boiler or re-purified, if necessary and and sold or reused for lubricating purposes.

Oily Shipboard Waste or Refinery Sludges

Waste oils can be mixed with soils and decomposed by naturally adapted soil bacteria. Dotson et al. (1974) report that bacteria of the genus *Pseudomonas* grew most rapidly in a soil–oil mixture. These bacteria continued to grow until this food source was exhausted. Despite this finding, Dotson et al. (1974) blame lack of information on full-scale studies as the reason for not using this disposal method. A mixed population such as that of soils is more efficient than a single species or genus. Petroleum hydrocarbons vary in susceptibility to decomposition. High molecular weight, viscosity, and crystallinity are properties that inhibit biological oxidation and decomposition. Straight-chain medium molecular weight hydrocarbons such as kerosene and light motor oils oxidize readily, whereas aromatic types are more resistant. Some evidence of oil decomposition in soil is provided by the oxidation of the asphalt coatings of underground pipes and the disappearances of oil leakages in lakes and shore areas attributed to microbial assimilation.

Corrosive Hazardous Wastes

The EPA (1980) considers corrosive any wastes falling into either of the following two classes:

1. An aqueous waste that has a pH of either 2 or less or 12.5 or more
2. A liquid that corrodes steel at a rate greater than 6.35 mm (0.25 in.) per year at 55°C

These corrosive wastes could be either very acid or very alkaline and thereby hazardous to any environment into which they are emitted.

Acid Wastes

Acid wastes may be discharged by any plant manufacturing materials, apparel, and chemicals. Most important and hazardous are those coming from chemical plants such as those producing dyes, explosives, pharmaceuticals, and silicone resins. The major ones are dilute wastes from hydrochloric, sulfuric, and sometimes nitric acid. Regardless of the degree of acidity or the origin of acid wastes, the main method of treatment is neutralization.

Alkaline Wastes
Alkaline wastes are considered corrosive at pH values greater than 12.5 and, hence, are classified as hazardous. As such, they must be treated properly before release to the environment. Like acidic wastes, they too must be neutralized, as described in Chapter 3.

Reactive Industrial Wastes

The EPA (1980) rather broadly defines reactive wastes into eight distinct categories:

1. Normally unstable, readily undergoes violent change without detonating.
2. Reacts violently with water.
3. Forms potentially explosive mixtures with water.
4. When mixed with water, generates toxic gases, vapors, or fumes in dangerous quantities.
5. Cyanide- or sulfide-bearing waste when exposed to pH conditions between 2 and 12.5 can generate toxic gases, vapors, or fumes in dangerous quantities.
6. Capable of detonation or explosive decomposition or reaction at standard temperature or pressure.
7. Forbidden explosives, class A or class B explosives.

Wastes Containing Cyanides and Isocyanides
Wastes containing cyanides and isocyanides are generally recognized by the public as being very "reactive," and thereby are classified as hazardous. They are classified as generating toxic gases, vapors, or fumes in dangerous quantities. They are used and released from many chemical industries and metal-plating plants. These compounds aid in solubility and plating of protective metals or machined parts varying from typewriters to silverware. They also can serve as a basic starting chemical for nitrogenous fertilizers. However, when they are allowed to reach the environment without treatment and in sufficient concentration, they are deadly.

Other chemicals containing cyanides are also used and occasionally released to the environment. Such a situation occurred accidentally at Bhopal, India, in December 1984, when a tank of methylisocyanate (MIC) leaked the gas and killed more than 2,000 persons. MIC has a formula of

$$CH_3NCO \text{ or } \underset{\underset{H}{|}}{\overset{\overset{H}{|}}{H\!C}}\!-\!N=\overset{\|}{\underset{O}{C}}$$

and is reported by Sax (1979) to be highly dangerous when exposed to heat, flame, or oxidizers. It is highly irritating to the skin, eyes, and mucous membranes, can cause pulmonary edema, and can be absorbed via the skin. This chemical is an important starting compound for the manufacturing of nitrogenous fertilizers. Because it exists as a liquid and boils at the relatively low temperature of 39.1°C, it must be kept quiescent and cool until used or oxidized to degrade the cyanide.

Sulfide Residues
Sulfide residues or sludges may result from the natural or artificial precipitation of sulfur-bearing wastes from a variety of chemical wastes such as sulfur dyes from textile plants or acid sludges from oil refinery processing.

Oil Refinery Sludges
Spent caustic wastes from catalytic polymerization and alkylate washers in oil refineries contain considerable amounts of sulfides. Acidification of these caustic wastes (with H_2SO_4) removes some of the objectionable compounds. The acid sludges may be used as a source of fuel or to produce byproducts such as oils, tars, asphalts, resins, fatty acids, and chemicals. Some refineries recover sulfuric acid from the acid sludges for their own use. Without reuse or recovery, these acid sludges still contain high amounts of precipitated sulfides, which, in turn, could be released to the environment if the pH or other conditions caused leaching of the sludge.

Trihalomethanes
Organic compounds reaching drinking-water supplies mainly from waste-cleaning solutions contain toxic chemicals such as trichloroethylene, tetrachloroethylene, methyl chloride, vinyl chloride, and carbon tetrachloride. These chemicals are all relatively volatile—reverting from the liquid to the gaseous state readily—and can be removed from the water supply by either granular activated carbon adsorption or air stripping.

Halogenated methane or ethane compounds are highly reactive and unstable and can release toxic gases. They are, therefore, considered hazardous. Such compounds can be released in their manufacture in the petrochemical industry or in the resulting solvent used in dry-cleaning establishments.

Major methods of treatment include prevention of their release by collection tanks for accidental overflow spills and rejects. Most of these compounds are colorless liquids with ethereal odors, relatively low boiling points, and high vapor pressures, and very irritating to the conjunctiva and hence dangerous to the health of humans and animals.

EP Toxic Industrial Wastes

Toxicity to humans and animals as described in this chapter can affect different organs of the body. For example, carcinogenic toxicity causes cancer in living cells and tissues, while genotoxicity causes genetic damage to these same cells and tissues. On the other hand, neurotoxicity causes toxic effects only on the nervous system, nephrotoxicity on the kidneys, and hepatoxicity on the liver. It is not the object of this book to discuss details of toxicity, but only to point out that there are various types and each may be caused by many different chemicals at different concentrations for different periods of exposure.

The EPA (1980) considers a solid waste toxic if, when extracted by the EP method (Nemerow 1983, p. 116), the leachate contains ions of constituents 100 times or more the Primary Drinking Water Standard. If the wastes contain less than 0.5% filterable solids, then the filtrate is considered exact.

Mercury-Containing Wastes

Caustic chlorine plants use mercury cells and waste mercury. Sundaresan et al. (1983) report a plant producing 60 tons of alkali per day and discharges 1,400 m^3 of wastewater with a pH of about 11 and from 3 to 5 mg/liter of mercury. Brine sludges are also reported to contain from 18 to 20 mg Hg/L. Leaching of mercury into water supplies with subsequent biomagnification causes serious diseases of the nervous system of humans eating fish from such waters.

Mercury, a toxic chemical, is also a naturally occurring element on Earth. However, about 10,000 tons of it are extracted from HgS, cinnabar ore, in the United States each year. About half is released directly into the environment through the discharge of industrial wastes.

The accumulation of mercury in predatory fish, such as tuna and swordfish, constitutes a major danger to humans eating the fish. Many catastrophes, such as that in Minimata Bay, Japan, have resulted in death from eating fish contaminated with mercury. Other industries besides the chlor-alkaline plants discharging mercury-laden wastes are electric wire and equipment and paper processing.

Treatment of mercurous wastes by industry primarily involves recovery within the plant and better housekeeping practices to eliminate wastes.

Metal-Containing Sludge

Many types of industrial- and municipal-originated sludges contain metals that constitute a hazard to the environment receiving them.

Municipalities contribute two main sludge-type wastes that contain metals: (1) wastewater (sewage) sludge, and (2) refuse solids or sludge. The concentrations of metals in both depend on the types of industries within the collection system and the habits and practices of these plants and people in the district.

Domestic-sewage sludge often contains various amounts of heavy metals. The latter originates in wastes discharged mainly from small electroplating shops and garages located within municipal limits. Quantities are difficult to control even though many cities have enacted ordinances limiting allowable concentrations of metallic ions. Generally, no more than 1 ppm of copper, cyanide, or chromium, and 2–5 ppm of zinc or nickel should be allowed in sewage plant influents. If the quantities exceed these values, concentrations in the sludges resulting from sedimentation or biological treatment contain excessive amounts of metals, which hinder anaerobic digestion processes.

Chemical precipitation at elevated pH values aids in the removal of heavy metals by precipitation of the hydroxide or carbonate. Refer to Chapter 6 for treatment details.

Other Inorganic Wastes
In general, the presence and removal of inorganic/dissolved minerals from wastewaters have been given relatively little attention by environmental engineers. But, with the passage of the TOSCA legislation of 1976 and RCRA legislation of 1976, these minerals, which are deemed toxic, and the type and concentration are given serious consideration for treatment and removal.

Chlorides, phosphates, nitrates, sulfates, and certain metals are examples of the more common and significant inorganic dissolved solids. Among the methods employed mainly for removing inorganic matter from wastes are (1) evaporation, (2) dialysis, (3) ion exchange, (4) algae, (5) reverse osmosis, (6) chemical precipitation, and (7) certain oxidation-reduction reactions (see Chapter 7 for more detailed descriptions).

Pickling Liquors
Before applying the final finish to steel products, the manufacturer must remove dirt, grease, and especially iron oxide scale, which accumulates on the metal before and during fabrication. Normally this is carried out by plunging the steel material into dilute sulfuric acid (~20% by weight). The process, known in the trade as "pickling," produces a hazardous waste called *pickling liquor* containing primarily ferrous sulfate.

Galvanizing and Metal-Plating Wastes
A good description of the origin, characteristics, and treatment of these wastes in can be found in Chapters 5 and 6 (Nemerow and Agardy 1998). After metals have been fabricated into the appropriate sizes and shapes to meet customer specifications, they are finished to final product requirements. Finishing is accomplished by stripping, removing undesirable oxides, cleaning, and plating. In plating, the metal to be plated acts as the cathode while the plating chemical metal in solution acts as the anode. The overall liquid wastes are relatively small in volume but are extremely toxic, as confirmed by the EP toxicity test. Most significant toxic metals are chromium, zinc, copper, nickel, and tin. Acids, cyanides, alkaline cleaners, grease, and oil are also found in these wastes and must be considered when planning treatment of metals.

The management policy in the United States for eliminating this type of hazardous waste has been waste minimization. This subject is treated elsewhere (Nemerow 1995). Duke (1994) evaluates the effectiveness of minimization of waste technologies in the San Francisco Bay area. Out of 52 actions, he reports that 25 are treatments, 13 are maintenance or housekeeping, and the remaining 14 are process changes. He concludes that minimization of hazardous wastes "has not fully penetrated the industry" (Duke 1994).

Graphic and Photographic Waste Liquors
Wastewater from large-scale film-developing and printing operation contains "spent" solutions of developer and fixer, which carry thiosulfates and silver. The solutions are usually alkaline with various amounts and types of organic reducing agents along with

the silver. The silver metal is toxic in certain concentrations and, therefore, qualifies this waste as a toxic one.

Treatment to eliminate toxicity involves removal and recovery of valuable silver metal. Three methods are generally used for silver recovery: metallic replacement, electrolysis, and precipitation.

Salts

Salts are generally considered nontoxic primarily because in most water resources such as lakes, streams, and groundwater, their concentration is still relatively low. However, salt concentrations of more than 2,000–3,000 ppm are generally toxic to plants. Salts gradually build up in receiving waters, especially those being reused for irrigating croplands where rainfall is low. Here, salts may never flush out of the root zone of crops. Where drainage from these croplands is hampered by underlying clay layers, salt pockets in the groundwater will build up.

Citing figures by the United Nations Food and Agricultural Organization (FAO), Earthscan reported that "about 120 million hectares (300 million acres)—half of the world's irrigated land—suffer from reduction of crop yields due to salinization" (Raloff 1984).

One way of controlling the toxicity of salts in groundwater caused by continual reuse of irrigation waters with inadequate land drainage is planting of alfalfa or in certain cases grasses. Alfalfa is an extremely thirsty and deep-rooted perennial. It will soak up the high-salt water table and cause it to fall below the normal crop root level. After a number of seasons or years, the original crops of barley or wheat can be replanted without danger of salt toxicity. In the meantime, the alfalfa or grasses can be harvested for other uses.

When certain industries such as pickle processing and textile finishing discharge heavily salted wastes, they may increase the salt concentration in receiving waters to toxic levels. In such cases, these wastes should be segregated and concentrated further by solar evaporation or membrane filtration, and the salt then recovered and reused by industry.

Without improved salt removal from industrial wastewaters, our water resources will eventually become salt toxic. When and if that occurs, civilization as we know it will begin to deteriorate and dry up.

A possible solution is to develop more salt-tolerant crops as foods. Raeburn (1985) described plants known as *halophytes* that grow like weeds along seacoasts and estuaries. One of them, salicornia, a plant that grows to a foot or two in height, looks like a bunch of green pencils. The plant produces oil high in polyunsaturates at a greater yield than soybeans, a high-protein food product, and a principal source of vegetable oil. They also found that two varieties of another crop, atriplex, yield as much animal feed as alfalfa and can be harvested several times a year. They report much work progressing on developing hybrid strains of salt-tolerant wheat, rice, barley, and tomatoes.

Phosphatic Fertilizer Sludges and Ore Extraction Wastes

Wastes from the phosphate industry arise from (1) mining the rock, and (2) processing the rock to elemental phosphorous and other pure chemicals. In mining, the main wastes originate from the washer plant when the rock is separated from the water solution, as well as from the flotators, where phosphate particles are separated from the impurities retained on the screens. In processing the phosphate, the major source of waterborne waste is the condenser water bleed-off from the reduction furnace.

The wastes are high in volume and contain fine clays and colloidal slimes, as well as some resinous oil from the flotators. Both mining and processing wastes contain small concentrations of radon emitted from uranium rock associated with the phosphate rock. Flotation wastes contain some resinous oils. Mining slimes also contain other heavy metals. Toxic constituents are classed as radioactivity, oils, and heavy metals.

The major hazard of these sludges comes from leaching of the toxic contaminants into water supplies or exposing humans to radioactivity from reused gypsum as building material.

Krishnan et al. (1996) report on the use of an environmentally balanced industrial complex for handling this phosphogypsum sludge and, thereby, realizing an overall savings of 42% in real production costs.

Ore Extraction Wastes

Copper, lead, and zinc can be extracted from their ores and concentrated to obtain pure metals. In copper processing, about 4 tons of concentrated ore produces 1 ton of pure copper metal. Slag, which develops as a waste from metal smelting, contains silicon, iron, and magnesium, as well as 6–8% zinc, 0.5% manganese, 0.4% copper, and 1.7% sulfur. Reused spent acid from the electrolyte is occasionally wasted and contains small amounts of toxic metals considered hazardous to the environment.

Power Plants

The operation of steam power plants generally involves the generation of heat from coal, oil, or uranium fuel to produce steam with exceptionally pure water. The steam is used to drive turbines, which, in turn, are coupled to generators.

Toxicity of residuals arising from the production of electricity in the aforementioned manner originates in the following:

1. Hot concentrated water salines from boiler and evaporator blow-down. These may include various metals and hexavalent chromium used as a corrosion inhibitor. These are hazardous.
2. Acid and alkaline chemical solutions used in cleaning power-plant equipment. These are hazardous.
3. Acid water drainage from coal storage and ash ponds. These also are hazardous.
4. Leakage or discharge of nuclear-fired power-plant cooling waters. These are hazardous.
5. Fuel (nuclear) reprocessing plant wastes. These are also hazardous.

Nitrogenous Fertilizers

In all nitrogenous fertilizers, ammonia is the basis and by far the most significant compound in wastewaters. In the fertilizer effluents from the production of ammonium phosphates, a considerable amount of fluorine gas is driven off or elemental and fluorosilicious acid fluoride is found in the condensates. This element and traces of arsenic are the main hazardous wastes. They are toxic.

Hazardous Wastes 263

The principal method of treatment of the toxic components has been storage of the wastewaters in lagoons, from which they are equalized and proportioned to discharge over longer periods, coagulation of the fluoride with lime and/or alum-polyelectrolytes, or recovery of the fluorine directly as fluorosilicious acid.

Wood-Preserving Wastes

Almost all of the 400 U.S. wood-preserving plants use pressure processes to preserve wood (for long-term exposure to soil, water, and air) with either creosote or pentachlorophenol or both. Air-dried timber is first steamed at about 20 psi for up to 12 hours while the condensate is continuously removed. Following evacuation and raising the pressure with air to 30–90 psi, the retort is filled with the creosote or pentachlorophenol oil. When all the air has been expelled by the preservative, pressure is maintained at about 200 psi for 2–8 hours. Entrapped air and preservative bubble out after the pressure is released. The preservative escaping is recovered and returned to a storage tank for reuse. Excess oil is removed by a vacuum before steaming. This condensate is the source of the major portion of toxic waste (Figure 11.5).

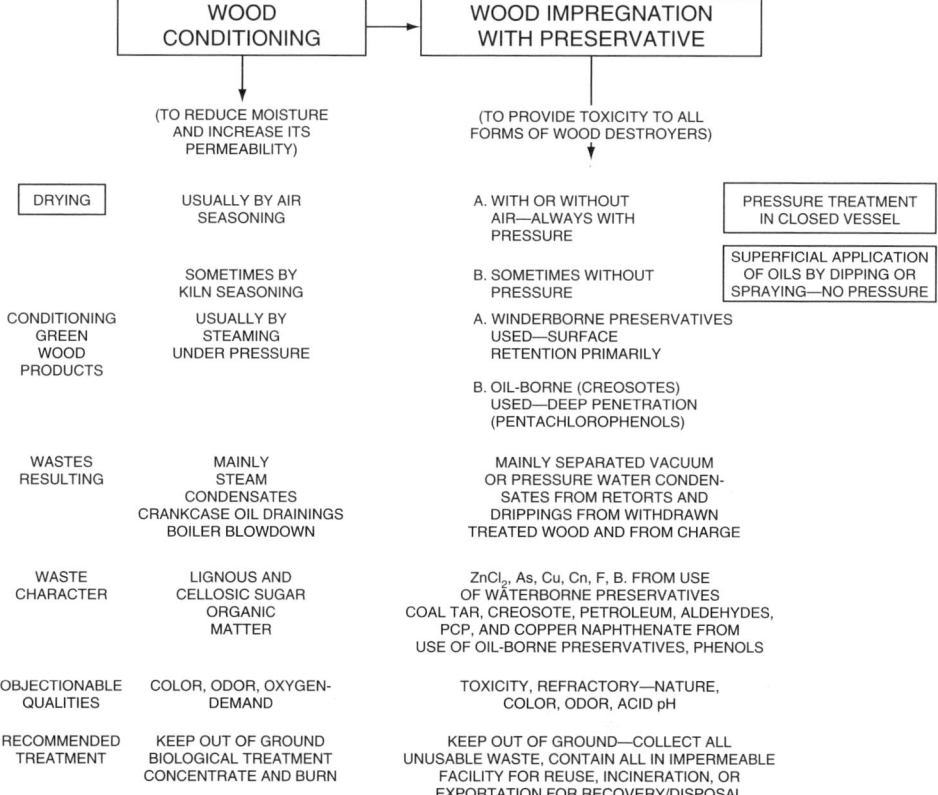

FIGURE 11.5. Wood-preserving processes.

Acute Hazardous Industrial Wastes

The category of acute hazardous industrial wastes includes industrial wastes that are a danger to humans and animals in very small concentrations when released into the environment. They affect the health of all living beings in an insidious manner, usually invisible and undetectable without sophisticated sampling and measuring equipment. Such wastes include radioactive, biological, and asbestos-laden air and water.

Radioactive Wastes

Major origins of radioactive wastes are from the use of nuclear energy by power plants and hospitals. Natural fallout from nuclear explosions is also an ever present possibility and has occurred in the past. In power plants, radioactive wastes can be released in contaminated cooling waters, leaks and subsequent wash-downs. In addition, fuel elements need reprocessing to renew the uranium rods periodically. Acid reprocessing wastes containing high levels of radioactivity can be released either at the power plant site or at designated national reprocessing sites. For example, a tank filled with radioactive gas ruptured at a uranium reprocessing plant on January 4, 1986, in Gore, Oklahoma. The container was being heated at the time ("Worker killed, dozens hurt" 1986) when it burst, releasing about 14,000 lb of slightly radioactive uranium hexafluoride gas, which breaks down into toxic hydrogen fluoride and low-level radioactive uranyl fluoride particles. One person died of acid burns and 14 were hospitalized for skin and respiratory system exposure.

American homes contaminated with radon gas could be causing as many as 30,000 lung cancer deaths per year. The gas rises to the land surface from any source of uranium and can travel horizontally for miles before penetrating cracks in house foundations. If the air within the building is not exchanged continually with outside air, the radon concentration can build up to dangerous levels.

Because these wastes cause genetic changes, usually in the form of some type of cancer in humans and animals coming in contact with them, proper disposal is mandatory. Generally, they are either concentrated and contained or diluted and dispersed. In the United States, the plans are for using the former method on all high-level radioactive wastes and burying them in governmentally selected and monitored, safe, deep underground sites. Low-level radioactive wastes are retained until safe levels of radioactivity are reached or are diluted with other wastes and discharged to sewers or streams.

Radioactive wastes may be released to the air, land, or water environments from many varied sources. Some of these include hospitals, power plants, and various industrial laboratories. Its ultimate site for disposal determines—in many cases—the hazard to the environment.

One unusual source of radioactive contamination occurred in several loads of steel construction bars shipped to Arizona from a Mexican plant. The steel was contaminated with Cobalt 60, a human-made radioactive material produced in nuclear reactors. It is also used to treat cancer tumors and in gauging devices. In this case (*Miami Herald* 1984), the steel bars were already in place in several construction sites in Arizona when the radioactivity was discovered. Some bars contained up to 350 millirems/hr; the permissible dosage for U.S. workers in the nuclear industry is 5,000 millirems/yr.

As another step in using underground burial of these wastes, the Department of Energy (1986) named 12 areas in seven states as prime candidates for the nation's second nuclear waste dump, tentatively scheduled for operation shortly after the year 2000. The sites proposed are given in Box 11.1.

Three sites in Nevada, Texas, and Washington State were chosen for the first nuclear waste depository scheduled for construction beginning in about 1996. Important criteria besides the political and social acceptability of these disposal sites are sufficient bedrock formations, ability to withstand earthquakes, and the proper composition to prevent contamination and subsequent escape of groundwater.

As of 1986, about 10,000 metric tons of spent fuel from 98 nuclear power generators is stored in water pools near the reactors.

The nuclear industry is still protected from excess environmental liability by the Price-Anderson Act of 1957. This legislation was intended to spur investment in commercial nuclear power. In recent years, there has been public pressure not to renew this act (Welch 1986). The main argument against the provisions of the act is that it also encouraged recklessness and negligence that may lead to catastrophic disasters.

Under the provisions of the Nuclear Waste Policy Act of 1982, the Public Service Electric and Gas (PSE&G) Company of New Jersey, along with other operations of nuclear plants, has signed fuel-disposal contracts for its nuclear plants with the U.S. Department of Energy. These contracts require the federal government to ultimately take title and provide necessary services to transport, package, and place the spent fuel in underground repositories (Public Service Electric and Gas Co. 1984). Utilities are required to pay a fee of 1 mill per kilowatt-hour of nuclear energy produced to fund the disposal program.

BOX 11.1. Seven Proposed Waste Storage Sites

1. Georgia: Lamar, Monroe, and Upison counties, 214 square miles.
2. Maine: Hancock, Penobscot, and Washington counties, 52 square miles; Androscoggin, Cumberland, and Oxford counties, 385 square miles.
3. Minnesota: Marshall, Pennington, Polk, and Red Lake counties, 300 square miles; Norman and Polk counties, 113 square miles; Benton, Mille Lacs, Morrison, and Sherburne counties, 397 square miles.
4. New Hampshire: Cheshire, Hillsborough, Merrimack, and Sullivan counties, 78 square miles.
5. North Carolina: Franklin, Johnson, and Wake counties, 142 square miles; Buncombe, Haywood, and Madison counties, 105 square miles.
6. Virginia: Bedford, 209 square miles; Halifax and Pittsylvania counties, 307 square miles.
7. Wisconsin: Langlade, Menominee, Marathon, Oconto, Portage, and Shawano counties, 1,094 square miles.

In the case of an "all-out" nuclear war or an accidental drop of a nuclear bomb, the environment (at the least) would become a gross hazard to all living beings. Chemist John Birks (paper presented at A.C.S. meeting, Miami Beach, May 1985) calculated that approximately 200 million tons of chemical-laden smoke would pour into the atmosphere after a widescale nuclear exchange. This smoke, Birks says, would include carcinogenic asbestos fibers released from burning ceiling tiles and asphalt shingles, large amounts of poisonous COs, NOs, and CH_4 from forest fires, and millions of tons of noxious chemicals from the combustion of rubber, plastics, petroleum products, and industrial chemicals.

Biological Wastes
Wastes from decomposing organic foods or bodies of humans or animals represent a hazard to people who make contact with them. The dangers are largely biological, that is, the diseases associated with decomposing organic matter. The infectious diseases can be carried by flies, mosquitoes, and rodents. Nemerow (1984) lists many diseases by these carriers. My main concern in this text is hazardous wastes of a biological nature contributed in hospital wastes. Also of concern since the 1980s are substances that interfere with reproduction (genotoxins). Certain hazardous chemicals can cause low sperm counts, structural sperm abnormalities, testicular cancer, and birth defects (Castleman 1985). Some examples of these chemicals include the following:

> Dibromochloropropane (DBCP): temporarily sterilized all men who handled it during manufacturing
> Dioxin: severe reproductive deformities in mice and monkeys
>> birth defects in Vietnamese children whose mothers were exposed to Agent Orange
>> disease to American soldiers serving in Vietnam exposed to Agent Orange
> 2, 4, 5T: high rates of miscarriage in women near sprayed areas
> DBCP, Kepone, and Dioxin: sterility and other genotoxic effects in mice and rodents and probably humans
> Ten antibiotics including penicillin and tetracycline: suppress sperm production temporarily
> Tagamet (stress-related drug): reduced sperm counts by 43%

In addition, radioactivity and x-rays have been known to cause both sterility and structural abnormalities in sperm.

Hospital Wastes
Although wastes from hospitals include domestic, contaminated, and special wastes, we are concerned primarily with the pathological waste component. This biological waste is hazardous because of contamination with pathogens. Tubercular lungs are particularly hazardous because of the potential for airborne release of pathogens. The organism *Mycobacterium tuberculosis* has been found especially prevalent and dangerous in these wastes, as have bacteria of *Pasteurella, Brucella,* and *Psittacosis*

groups as well as certain viruses. Surgical and autopsy wastes contain pathogens and present dangers if not handled properly. Other contaminated wastes may also be classed as hazardous because they contain blood, pus, and sputum, which can infect the air with microbiological agents. Some wastes also are contain radioactive, such as C14, Cr51, AU198, I151, Fe50, P32, and Na24. These must be segregated and handled as radioactive. Pathological and contaminated wastes usually comprise less than 10% of the total hospital wastes.

Incineration is the safest method of disposal of these hazardous hospital wastes (Nemerow 1984). However, because this is costly unless done collectively and often includes dangers and malpractices, such systems must be designed and operated with care. Separation of these wastes from other hospital wastes, transportation, and storage must precede proper incineration.

An incidence of improper practice serves to illustrate some of the problems involved with disposing of hospital hazardous wastes ("Incinerator belches foul smoke" 1985). A malodorous pall hung over an area of northwest Dade County, Florida. The source of the odor was Metro Waste Services, Inc., where leak-proof metal containers of infectious hospital waste were trucked in over a gravel path, hauled up a concrete ramp, and fed into a smoky incinerator that even its owner admits is a rust-eaten "piece of junk." The state and county officials have cited the operator, claiming it was not burning the waste completely and that contaminated runoff was seeping into the ground. The two major concerns—even when the incinerator was repaired—were (1) the smoke itself, when it hangs close to the ground and penetrates the building and surrounding area, is obnoxious, and (2) the potential that live organisms could exist in both the smoke and drainage from the stored waste. This situation is typical of many incinerator operators who attempt to burn these wastes on a small scale.

Although high-temperature incineration is conceded by most environmentalists to be the preferred method of hospital waste disposal, inefficient or incomplete combustion of halogenated plastics, bactericides, and hazardous pharmaceuticals may yield toxic dioxin, HCl, and chlorine gas. Doyle et al. (1985) claim that alkaline scrubbing systems can effectively neutralize and remove acid gas and toxic air contaminants. Hospital wastes contain about 20–30% plastics (contrasted with about 5% for municipal wastes) and generally average 7,500–10,000 Btu/lb of heating value (compared to 5,000 Btu/lb municipal refuse). There are many hazardous materials in these wastes that must be considered when incinerating them. Infectious components (~10% of the total) can usually be destroyed completely by proper incineration. However, burning does not destroy inorganic constituents like mercury, some cytotoxic agents used in chemotherapy, and antineoplastic agents. Although dioxin and hydrochloric acid are the major toxic components following incineration, furans and PCBs are also present. Hospital incinerators are usually relatively inefficient in that they operate at lower temperatures and contain shorter stacks than larger commercial or industrial ones.

In addition to hospitals, laboratories, doctor and dentist offices, and small medical treatment centers contribute potentially infectious wastes. These include needles, syringes, gauze, dressing gowns, intravenous tubings, and used blood-storage bags. As of 1988, 31 states had laws requiring hospitals, at least, to disinfect (steam-sterilize or incinerate) such waste before they bury it in landfills. However, both methods have limitations and, hence, may not be satisfactory. Further, tracing illegally dumped wastes of these types is very difficult. More illegally dumped medical wastes occur today because of the banning of small and scattered amounts in conventional landfills, and burning is costly. However, some companies are working on small disinfection units for individual medical waste suppliers and on incinerators capable of burning these wastes without creating air contamination from gases resulting from plastic burning.

Electron beam treatment has been proposed ("First plant in U.S." 1996) to treat medical waste (as well as domestic sewage and biologically and toxic-contaminated soil). The treatment comprises exposing ("zapping") wastes with electrons accelerated (by about 8 million V) under high-voltage electricity. The electrons that hit the viruses in the "zapper" disrupt atoms and molecules, rendering both bacteria and organics nontoxic immediately. Estimated costs include 7–9 cents/lb for beam treatment plus the cost of residuals disposal in landfills.

Biological Wastes and Other Contaminants

There were 4,742 reports of *Salmonella* sp. infection in Illinois, Michigan, Indiana, Wisconsin, and Iowa according to the Illinois State Inspector General's office during April 1985. At least three deaths were also linked to the outbreak. Salmonella is an infection of the gastrointestinal tract that usually causes cramps, vomiting, diarrhea, and fever. It causes death in 1 of every 1,000 cases, and the ones most at risk are the elderly and children. It is usually caused by contaminated food; in the aforementioned situation, it was traced to contaminated milk from Hillfarm Dairy in suburban Melrose Park, Illinois. Normally, most dairies use pasteurization, a high-temperature heating process, to kill Salmonella and other bacteria. Among the hypotheses put forward to explain the Illinois outbreak were (1) incomplete pasteurization, (2) contamination introduced after pasteurization, (3) heat-resistant Salmonella bacterium, and (4) sabotage.

Viruses growing in polluted waters have been proven to be hazardous to human health even though not ingested directly in drinking water. For example, eating uncooked or even poorly cooked shellfish represents a clear risk ("Eating raw shellfish" 1986) of gastrointestinal tract illness and hepatitis A infection from eating raw or steamed clams and oysters contaminated with viruses or bacteria apparently originating in polluted water. A study reported in the *New England Journal of Medicine* documents 103 outbreaks, in which 1,017 people became ill during 8 months in 1982, and concluded that the illness usually was associated with raw clams or oysters but that there was an "unexpected high attack rate" among people who ate steamed clams that had not been cooked sufficiently.

Nausea, vomiting, diarrhea, or abdominal cramps were the principal ill effects. Closing polluted waters to taking of shellfish is the most effective preventative of this health hazard, but often closing occurs too late to prevent all illnesses. In addition, many shellfish lovers resist ample cooking required to kill viruses.

Asbestos-Laden Wastes

Asbestos is a mineral fiber that has historically been used for insulating and fireproofing in a wide variety of industrial and construction materials and equipment. Asbestos dust can also originate in municipal refuse from discarded auto-brake linings and from demolition wastes. In 1989 the EPA proposed a ban on the use of asbestos in its most common applications, but this was struck down by the courts. Sax (1979) gives excellent examples of uses of asbestos, adapted in Table 11.6.

Asbestos is a health hazard when the microscopic fibers are inhaled or ingested. Exposure can result in asbestosis (a respiratory disease), lung cancer, or mesothelioma (cancer of the lining of the chest or abdominal cavities). Exposure is severely aggravated by smoking. The primary victims of asbestos exposure were World War II shipyard workers who handled large amounts of loose asbestos insulation.

Such practices have been corrected, and protective measures have been implemented. Worker protection is achieved by using inhalation masks, work-zone air monitoring, high-efficiency particulate filters, and moisture-laden air systems during product manufacturing, building demolition, or removal of asbestos-containing building materials.

Treatment of wastes primarily has been by sedimentation and/or filtration of the fibers and landfill disposal. I report using chemical coagulation treatment in Chapter 6 with iron salts or polyelectrolytes followed by filtration to remove better than 99.8% of fibers from water containing 12×10^6 fibers/liter. Disposal of asbestos waste employs landfilling using wet-type operations that eliminate dust.

Asbestos waste is not a listed RCRA waste, but there are many restrictions on its handling and disposal. Asbestos is a hazardous air pollutant under NESHAPS. The Asbestos Hazard Emergency Response Act (AHERA) of 1986 requires the EPA to address the transportation and disposal of asbestos; this has not yet happened. Asbestos is a hazardous substance under the Comprehensive Environmental Response,

TABLE 11.6
Sources of Asbestos Wastes

Hazardous Sources of Asbestos Usage	*Nonhazardous Sources of Asbestos Usage*
1. Thermal lagging and delagging (special risk from "blue" asbestos mainly from steel and power plants)	1. Fillers combined with other ingredients, mainly in linoleum, floor tiles, rubber paints, plastics, adhesives, roofing, and motor assemblies.
2. Furnace insulation, mainly from heavy industry	2. Grinding in assembly of brake and clutch parts
3. Heat and sound insulation, mainly from locomotive, railroad car, boiler, chemical plants, and gas works manufacturing	3. Asbestos washers and gaskets
4. Asbestos-cement sheets and boards of insulation, mainly from building trade plants	

Source: Adapted from Sax (1979).

Compensation, and Liability Act (CERCLA). Thus, generators of asbestos wastes to landfills are subject to CERCLA liability. Finally, a number of states regulate asbestos as hazardous waste or as an air pollutant. These regulations may go beyond federal rules.

Toxic Industrial Wastes

Although hazardous materials are those that are included in all hazard classes, toxic materials are those that cause toxic effects. Therefore, all toxic materials are hazardous, but not all hazardous materials are classed as toxic. These definitions are also verified in this book. While Chapter 11 deals with hazardous materials, it also describes toxic materials and their treatment. Special treatments of toxic materials may also be found in the following section.

Several of the newer chemical industrial plants manufacture organic chemicals for sale to producers of other products discharge some of these wastes to the environment. Many are extremely toxic to aquatic life, animal life, or humans in very small quantities. Most of these chemicals were either unheard of or seldom used 10 to 20 years ago. Few of them were measurable in small concentrations with laboratory scientific equipment until the last 10 years. Even now the long-term potential toxic effects of many of these modern chemicals are unknown. In the following section, I discuss the production effects and treatment of such chemicals as those that are contained in organic-laden wastes, as well as pesticides, pharmaceuticals, and laboratory chemicals.

Organic-Laden Wastes

Chemical plants containing organic chemicals such as PCBs, phenol, formaldehyde, spent solvents, halogens, organic sulfurs and nitrogens, paints and varnish residues, organic acids, dyes and pigments, and explosive and defoliant chemicals are considered in this section. They are so diverse as to defy classification except that they are all organic and potentially toxic.

PCB-Laden Oil/PCB-Containing Wastes

Polychlorinated byphenols originate in transformers and power capacitors, hydraulic fluids, diffusion pump oil, heat transfer applications, plasticizers for many flexible products including printing plates, and numerous other products. The major properties of heat resistance without decomposition and nonflammability are hard to duplicate in other substitute chemicals. Despite this, their use has been banned in the United States for all but transformers and high-voltage capacitors.

Reclass 50, a mobile operation conducted by a two-person crew, can bring a high-voltage transformer from 60,000 to less than 50 ppm faster than any other system (Graham 1985). In this process, the unit is put through three soaking steps over a period of 7–15 months, depending on the transformer size, PCB concentration, and operating temperature. Except for a short period, the transformer remains in service. First, the original askarel is drained and replaced with two flushings of TF-1, a proprietary material, which in turn is replaced with a final soak of silicone. The last step is filling with Carbide's silicone dielectric fluid, which will be used for the remainder of the transformer's service life.

PCBs are oily, colorless organic liquids in the same chemical family as the pesticide DDT. They have been linked to birth defects, reproductive failures, liver problems, skin lesions, cancer, and tumors.

Before 1972 PCBs were used in all nondrying microscopic immersion oils because they possess great stability, and optical properties are still said to be unmatched today.

Cargille (1983), whose company makes PCB-containing chemicals, recommends that customers send used and unused chemicals back to Cargille for reuse. However, a new labeling regulation of the EPA "forcing disposal" hindered this solution.

Fox and Merrick (1983) met PCB-discharge standards of 1.0 ppb by the following three operations and then conventional wastewater treatment:

1. Eliminating discharge into the storm and sanitary sewer system by source control.
2. Rerouting PCB-contaminated wastewater to the existing combined industrial–sanitary conventional treatment plant.
3. Flushing contaminated systems and storing PCB-contaminated oil and debris in tanks before disposal.

Treatment of PCBs has occured primarily by banning their production or use. Where found, they must be recovered and incinerated, pyrolyzed, or reused.

Sunohio developed a PCBX process that chemically destroys PCBs in transformer mineral oil. It is described in some detail in Figure 11.6.

Following the Fullers Earth Dual Filters shown in Figure 11.6, the treated fluid flows through a fourth-stage filter, a vacuum degasser, and a final fifth-stage submicron filter. As the treated fluid flows through the sequence of filters, all the spent reagent and reaction byproducts are removed, yielding a clean reusable fluid with a PCB concentration of less than 2 ppm. The reaction byproducts, spent reagents, and exhausted filter mediums are collected in approved containers, solidified, and disposed of in an appropriate landfill.

According to Sunohio, because PCBX treatment reclassifies transformers and allows for continued use of the oil, capital expenditures are avoided. Faulted transformers can be rebuilt after PCBX treatment instead of disposing of them. PCB oil leaks and spills, which often result in high cleanup costs, operation disruption, extremely adverse public relations, and potential EPA fines, can all be avoided by proper PCB treatment.

Another method of detoxifying PCB wastes—especially those of sludges such as collected in lake or river bottoms—is the fluidized bed combustion process. American Toxic Waste Disposal, Inc., of Waukegan, Illinois, uses this system for the PCB muds dredged from Lake Michigan (*The Amicus Journal* 1985). It involves grinding the sediments into fine particles and forcing hot air ($1,200°$ F) through them on a perforated plate of bed until the PCBs vaporize. A cyclone separator and scrubbers then isolate the PCB gases, which are converted to liquid form. Then the PCBs can be either incinerated or adsorbed in activated carbon filters for later disposal in a relatively smaller space than that required for the untreated waste muds. One study, however, found it required

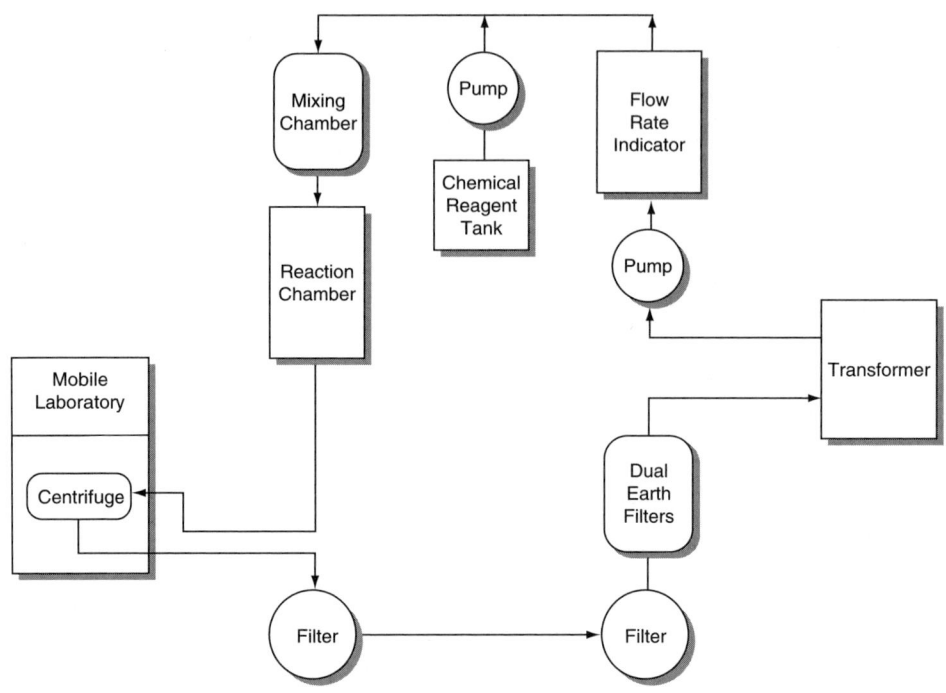

FIGURE 11.6. Mobile PCBX unit.

two-thirds of a barrel of fuel oil to burn each cubic yard of sediment. This represents a significant cost to the reclaimer/disposer.

Phenols and Formaldehydes

Phenol is an acutely toxic chemical containing the benzene ring (C_6H_5OH). In acute poisoning, the main effect is on the central nervous system. Also, absorption from skin contact of phenol may be very rapid and death results from collapse within 30 minutes to several hours. In industry phenols are mainly used or produced in integrated steel mills, synthetic textile mills, and resin (plastic) manufacturing. As early as 1943 (Ohio River Pollution Control-Supplement D), it was acknowledged by technicians that phenol was detrimental to public water supplies in very low concentrations. In higher concentrations, phenols were known to be detrimental to fish life; in fish phenol produces paralysis of the neuromuscular mechanism and hemolyses of the blood.

Phenol and formalin are used as starting chemicals for manufacturing the phenolic-derived resins. The phenolics and formalin, together with catalysts, are fed into an insulated and heated reaction tank. After reacting, draining following water evaporation, and cooling, the resin is used for product manufacturing.

Cooling waters, condensates, and blow-down wastes result in about 7 lb of wastewater per 100 lb of resin. These wastes usually contain high phenol contents (>1,000 ppm) and high biochemical oxygen demand (BOD) (>10,000 ppm). Treatment has been by phenol extraction, lagooning, thermal incineration, and sometimes discharge (and dilution) into municipal sewage systems.

Phenolic wastes, though bactericidal, have been degraded biologically under the proper environmental conditions usually involving bacterial adaptation periods. For example, Zoltek (1984) took groundwater already contaminated with phenols and terpenes allegedly from a nearby plant, which treats wool with coal tar creosote, and contracted for pumping it to the sewage treatment plant. His laboratory results showed the positive action of municipal sewage in degrading these compounds during biological treatment.

Spent Solvents

Organic solvents are used in many chemical industries for assisting in production. Dry-cleaning and paint industries are typical major users of solvents. Solvents used in paint, lacquer, and varnish manufacturing include ketones, aromatics, aliphatics, alcohols, glycol ethers, glycol esters ethers, glycols, glycol esters, terpines, and so on. The solvents aid in keeping pigments dissolved and are eventually evaporated either during the manufacturing process or during product use. Waste solvents discharged following condensation are potentially toxic to receiving waters.

Solvents of the chloromethanes are found in the widespread dry-cleaning industry establishments, as well as in dewaxing of oils, refrigerant preparation, and medicine formulation. Most are made by chlorinating methane in the presence of oxygen and high velocities under a temperature of about 700°F with product recirculation for heating. The exact product obtained can be varied somewhat to satisfy market demand. Where these chemicals are used in sufficient quantities to warrant environmental concerns, they must be collected and treated (usually by an outside contractor) and not discharged untreated to the local air, ground, or water. Treatment is usually by concentration by adsorption and/or incineration.

Dedert (1985) offered a "Supersorbon" process to recover solvents. Hazardous solvents are recovered from the contaminated air in various chemical plants by drawing the air through activated carbon and regenerating the carbon by back-flushing with steam and subsequent decanting off the condensed water, as shown in Figure 11.7. Solvents recovered are toluene, naphtha, lactol spirits, and similar blends of aromatic, aliphatic, and paraffinic hydrocarbon solvents. Acetates, alcohols, ketones, and chlorinated hydrocarbons can be recovered with added steps such as distillation and chemical drying. Claims are made of more than 90% yield of solvent. Some operational data indicate that 2–4 lb of steam, 0.07–0.20 of electricity, and 3–5 gallons of cooling water are required per pound of solvent. In addition, from 1 to 2 lb of activated carbon must be replaced for each ton of recovered solvent.

Another source of solvent turpentine originates from the manufacturing of naval stores. In this industry, shredded wood chips, usually pine, are located in a battery of extractors into which live steam and another solvent—usually naphtha—permeate and added and maintained at about 80 psig (pounds per square inch gage). Most of the naphtha is separated from the turpentine pine oil and rosin in a concentrating evaporator. Some turpentine escapes condensation and/or lost during purification of pine oil and rosin residues.

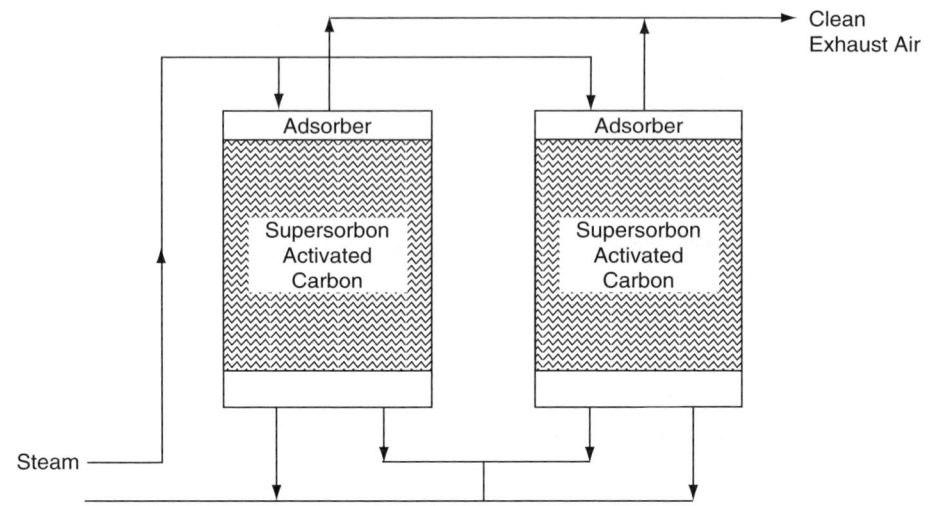

FIGURE 11.7. Supersorbon solvent recovery.

Still another source of solvents such as xylene, benzene, toluene naphthalene, acetylene, ethylene, propylene, and butene is the petrochemical industry. These solvent chemicals are actually derived from coal, oil, or natural gas and are precursors of finished products such as rubber, plastics, and fibers.

By distilling, compressing, fractionating, and reforming natural gas, petroleum, and coal under a variety of specific operating conditions (such as time, temperature, and pressure), these organic chemicals (solvents) are obtained. Inadvertently, some solvent is lost to the environment during the conversion or in the use of them to produce final useful products. Most of these solvents are known to be toxic, and usually carcinogenic, in certain concentrations.

Organic Wastes Containing Halogens, Sulfur, and Nitrogen
A variety of chemical and manufacturing industries produce wastes containing organic halogens, sulfur, and nitrogen. Sometimes they are discharged with other contaminants and difficult to separate, such as the explosives manufacturing plants that produce TNT wastes, and highly acid and colored textile-finishing plants that may use sulfur or azo dyes and produce wastes which are colored alkaline and replete with suspended and colloidal solids. Organic halogens are also produced by disinfection with halogen, usually chlorine, or organic waste residues. For example, chlorobenzene, an important intermediate for sulfur colors and a solvent used to make other products such as aniline, phenol, and chloronitrobenzene, among others, is produced by passing dry chlorine through benzene in iron vessels using ferric chloride as a catalyst and keeping the reaction at 140°F. The same compound could be produced with less efficiency as a result of disinfection of benzene-containing wastewater. In both cases, some chlorobenzene would be wasted into the environment and contribute to toxicity of the medium.

Special Treatments for Hazardous Wastes

In this section, I report, describe, and evaluate the many types of treatment systems being used or suggested for counteracting the negative impact of hazardous wastes on the environment. Although the conventional physical, chemical, and biological systems are covered, I do not go into basics very deeply, as this information can be found elsewhere (Nemerow 1987). On the other hand, I cover more extensively newer more sophisticated treatments that are directed specifically at handling hazardous wastes. For example, waste solidification, regional exchanges, plasmolysis, and recovery and reuse are explained with greater care because of their implications in hazardous wastes filed. Examples are given of treatment of specific hazardous wastes by appropriate methods whenever possible.

General Treatment

Hazardous wastes are incompatible not only with the environment, but also with other more common wastes of industry. Because of the direct and insidious dangers of hazardous wastes, they must be considered separately as infringements upon our normal industrial practices and upon every citizen's right to live and use environmental amenities. Those industrial wastes that burn, corrode, over-react, or are toxic are considered hazardous and are given special priority in treatment preference and type.

Pearce (1983, 57) stated that of the 3.4 million tons of hazardous wastes produced by industry in the United Kingdom, slightly less than 80% is landfilled, while less than 12% is disposed of in the ocean, 3% incinerated, and the remaining 5–6% is treated in other ways.

The EPA (1980) determined that only 10% of the treatment methods used for hazardous wastes are "acceptable." These are detailed in Table 11.7. Van Noordwyk et al. (1980) classified hazardous wastes by quantity generated and disposal method used. Organic chemicals (SIC 2861, 2865, and 2869) were categorized as shown in Table 11.8.

TABLE 11.7
Hazardous Waste Disposal in United States, circa 1980

Disposal Method	Percent of Total
Acceptable	
Controlled incineration	6
Secure landfills	2
Recovered	2
Unacceptable	
Unlined surface impoundments	48
Land disposal	30
Uncontrolled incineration	10
Other	2

Source: U.S. EPA (1980).

TABLE 11.8
Organic Chemical Treatment

Method of Treatment	Quantities On Site[a]	(M's/yr) Off Site
Landfill	483,000	113,000
Incineration	2,250,000	51,000
Controlled	699,000	—
Uncontrolled	1,550,000	—
Deep well	6,540,000	—
Biological treatment or lagoons	565,000	—
Recovery	267,000	
Total	10,100,000	164,000

[a]1977 data, basis.

Nemerow (1984) reported that large quantities of commercial hazardous wastes are collected and treated in the ways presented in Table 11.9. In a National Research Council staff digest (Reducing Hazardous Waste Generation 1985), reported that the RCRA law of 1976 defines as hazardous certain wastes because they may (1) cause or significantly contribute to an increase in mortality or an increase in serious irreversible or incapacitating reversible illness; or (b) pose a substantial present or potential hazard to human health or the environment when improperly treated, stored, transported, disposed of, or otherwise managed. It arrived at four general principles that should govern efforts to reduce the generation of hazardous waste.

1. No single approach to encouraging waste reduction will be most effective in all circumstances.
2. Reductions in the generation of hazardous industrial wastes can be expected to occur through a series of loosely defined and overlapping phases. It is desirable to reduce the generation of hazardous wastes.

TABLE 11.9
Commercial Hazardous Waste Disposal Methods

Waste Management	Capacity Wet, 1,000 Tons per Year, (1981)	Volume Received Wet, 1,000 Tons
Landfill	37,372	1,965
Land treatment and solar evaporation	1,400	282
Chemical treatment	1,305	734
Deep-well injection	1,095	475
Resource (recovery)	341	83
Incineration	102	80
Total handled by nine commercial waste—treatment firms	41,615	3,610

3. The costs of alternative methods of waste disposal should reflect the social costs of protecting public health and the environment.
4. Regulation will continue to play a crucial and central role in the overall waste-management effort, but future waste reduction is more likely to be fostered by nonregulatory methods, such as information dissemination programs and economic incentives.

There are many processes in various stages of development for treating and destroying all types of hazardous wastes. The processes are compiled from response to two national solicitations for new hazardous waste treatment ideas, from several literature reviews, and through contact with experts in the field (Office of Policy, Planning and Evaluation 1985). These processes are summarized in Table 11.10.

Torricelli (1985) informs us that the Office of Technology Assessment cited 26 systems or products that could effectively and permanently destroy toxic sites. Among these, he chose to bring out four: chemical detoxification, biodegradation, encapsulation, and supercritical water oxidation.

Segregation

Segregation involves the removal of hazardous industrial wastes from other process wastes for subsequent special handling. Segregated hazardous waste can either be treated by the industrial plant at its own facility or be picked up by a service company for transporting to a central specially designed treatment plant for hazardous waste.

In the late 1970s, Rollins Environmental Services located in Logan Township, New Jersey, was generally conceded to be the largest waste-processing plant in New Jersey (Civil Engineering 1979). Rollins required information about the nature of the waste so the company can segregate and label wastes for pickup at its plant site. If wastes are not segregated and instead mixed, serious problems could occur. For example, a mixture of chlorinated hydrocarbons and metal sludge waste will corrode clay or plastic liners in landfills (because of the chlorinated hydrocarbons) or release metals to the atmosphere if incinerated (because of heavy metals in sludge).

Location of Facilities

The problem of gaining acceptance for a site for any sort of treatment, including storage, has always been a difficult one for conventional waste treatment. The situation becomes even more sensitive and one faces public objections even at the mere mention of "toxic" or "hazardous" waste. An ideal site where no people live but which is situated near suppliers does not exist in reality. Any site will take a great deal of time, patience, and tact in gaining its acceptance for use as a location for storing, treating, and/or disposing of hazardous waste.

Treatment Required-Risk Evaluation

The type and extent of hazardous waste treatment required for protection of the environment is related to the amount of risk one is prepared to accept. An acceptable

TABLE 11.10
Emerging Alternative Technologies, circa 1985 (From Assessment of Incineration)

Technology and Description	Suitable Wastes	Capacity per Unit of Time	Cost per Unit of Waste	Environmental Data	Commercial Status
Wet air oxidation: —Process of oxidizing organic compounds in water at temperatures of 350–650°F	Very dilute organic and inorganic aqueous waste except highly refractory organics	10 gal/min	5–10 cents a gallon for 10-gallon system Costs for larger system not available	Limited data available EPA currently evaluating a unit on cyanide, sulfides, and nonhalogenated waste	Technology currently available Widespread use expected in next 2 years on aqueous waste
Molten glass incineration: —High temperature in furnace (2,300°F) destroys organic waste streams —Combustion gases pass through ceramic filters —Glass slag encapsulates inorganic residues	Any combustible waste Degree of halogenation not a consideration Scrubbers required for HCL	Existing glass manufacturing process 100–21,000 lb/hr	Cost figures not available	Data not available Testing needed Ceramic tiles and residue encapsulation in slag indicate minimal impact	Not reported
Supercritical water: —Inorganics, insoluble in supercritical water, are removed from waste streams —Organics rapidly oxidized	Aqueous waste streams with high levels of inorganics and toxic organics Treatment of highly halogenated material not yet demonstrated	Currently treats 1,000–2,000 gal/day	Cost figures not available	Bench-scale test on various wastes indicates DREs of 98.5–99.8%	Commercial scale units will be available in 1–3 years

Technology	Throughput	Cost	DREs/Emissions	Availability
High-temperature electric reactor: —Two companies: Thagard Research Huber Corporation —Vertical reactor heated by electrodes implanted in walls to pyrolize organic wastes	Process initially designed for solid waste Also suitable for liquid refractory waste streams	75–125 lbs of solids per hour No figures available for liquids, but it is assumed throughput would be less	Cost information not available Huber claims cost comparable to conventional incineration	DREs far in excess of the 99.99% RCRA requirement	Mid-summer 1985
Molten salt reactor: —Burning and scrubbing (900°C) oxidize carbons of organic matter to carbon dioxide and water —Byproducts (phosphorus, arsenic, sulfur, halogens) retained in melt as inorganic salt	Designed for solid and liquid waste Especially applicable to highly toxic and halogenated combustible waste with low percentage of ash	Pilot-scale facility processes 80–200 lb/hr	Cost data not available	No emissions; organic salts are the only byproduct DREs for organics, pesticides, and chemical warfare agents 99.99–99.99999%	Currently available for commercial use but none operating on that scale to date
Plasma arc: —Uses high temperature of plasma (50,000°F) to destroy hazardous waste —All hardware in 45-ft mobile trailer	Highly toxic liquid waste streams Degree of halogenation not a consideration	600 lb/hr in commercially sized unit	Cost data not available	Current demonstration in New York State will provide data on DREs and emissions	Commercial scale unit to be operating in 1–3 years

risk depends not only on the cost of protection involved, but also on who will be required to pay for the environmental protection. Such risk analysis becomes simpler if the cost is borne directly by the one who will benefit from the treatment protection. In matters pertaining to hazardous waste treatment, however, it is more likely that industry will be required to pay for protecting an environment and its users that are far removed from the industry itself. Economists refer to such situations as "external dis-economies."

Even when governments are forced to grant funds for such expenditures to protect certain people specifically involved, they are somewhat reluctant to do so. Beneficiaries of governmental actions and expenditures on hazardous waste are difficult to identify and the effects are difficult to quantify.

For years the most acceptable practice of risk analysis and evaluation has been the use of cost–benefit relationships. That is, there must be a dollar benefit for every dollar spent to protect the environment from the adverse effects of hazardous wastes. With these wastes, costs are not only huge, but also difficult to predict reliably. Benefits are even more troublesome to quantify because they usually relate to human health, the quality of life, and even a value of human life itself. Costs and benefits often can be manipulated by the person or group presenting the case for or against remedial action depending on personal goals.

One device reported by Douglas Ginsburg (Shabecoff 1985), administrator for information and regulator affairs at the Office of Management and Budget, is to "try to guide and question and kibitz the process to get as much risk reduction as possible for the expenditure." This procedure is laudatory but is still highly subjective and will vary from one person to another and one case to another. What is needed is a dependable consistent objective method. Perhaps it might be better to state publicly that we will seek the maximum risk prevention that the public will buy, or that any environmental risk is too great to accept where hazardous wastes and people are involved. We have almost done this where the operation of nuclear power plants is concerned.

However, as Shabecoff (1985) points out, "There is no consensus on what society is willing to pay to avoid risk." For example, a life saved has been valued at various amounts ranging from $400,000 to $7.5 million. Much lower life values have been set where many lives are involved. For example, each life lost in the Bhopal accident has been valued at only $30,000 by a preliminary offer of settlement.

The administrator of the EPA (1985), Lee Thomas, was quoted as suggesting that "we are a long way from the point at which decisions can or should be made by mechanistic application of any acceptable risk or cost 'criterion.'" However, your author still believes that the public would benefit more from an established, open, scientific process for risk evaluation even if it was not perfect and always open to some debate. At least its parameters would be known and its values could be seen and evaluated by both violators and recipients of hazardous waste effects.

Dr. Steven Kelman of Harvard University believes that "we should hold the government responsible not for a risk-free society, but for a higher standard of behavior than people hold themselves" (Shabecoff 1985). This is probably because each person's chance of death or disease is less than that of government, which represent more people.

Unfortunately, governments seldom make decisions and laws in a vacuum and are usually influenced largely by individuals and their lobbyists. Therefore, government's decisions may appear to benefit certain individuals or groups rather than society as a whole. This fact is yet another reason and argument for objective risk–benefit procedures rather than subjective ones.

Land Treatment

The use of vacant land as a receptacle for hazardous waste is recommended only when the land can be upgraded from its previous use and when no additional adverse environmental consequence results. Qualifying lands should be far away from underground water supplies and residential communities where odors percolate or contact with hazardous gases, liquids, or drying solids might take place. Because these ideal conditions seldom are found, we suggest underlying the entire land area with an impermeable membrane to prevent leachate from entering the water supply. Leachate should be collected and treated, especially for heavy metals and toxic organics such as those originating from pesticides.

Land reuse for parking facilities and one-storied warehouse or parks are examples of upgraded uses following proper disposal of hazardous wastes.

Whenever possible, hazardous wastes—usually after some form of pretreatment—are returned to lands that are compatible with the residues. For example, many lands are low lying and unusable in their present condition. Such lands could be used as receptacles for such wastes. Sundaresan et al. (United Nations Environmental Protection 1983) states that DDT wastes are neutralized with lime and the lime sludge dumped in low-lying areas of land.

The Japanese are mixing solid wastes, excavated earth, gravel, and sludge, as well as ash from incineration plants to form a source of new land ("Land-poor but trash rich" 1986). They haul the mixture to Tokyo Bay where it is placed and has already served as real estate for Haneda Airport and Tokyo Disneyland, an oil terminal port, a power plant, an apartment complex with 4,500 units, an industrial park, and even a sewage-treatment plant. Of course, the competition for space has caused obstacles with other users of the bay. Shellfish have been reported to have been killed by the change in habitat caused by the newly created land, but the pressures upon their society for "living room" have dictated reuse of solid wastes and a change in the environment.

Thermal Treatment

Heating hazardous wastes to some elevated temperature can cause changes in the physical, chemical, and biological nature of some wastes and render them innocuous. This treatment is not to be confused with complete burning in the presence of excess air, as in incineration (Figure 11.8).

Perhaps one of the most promising and used thermal process is pyrolysis, which involves burning residue at about 900°C in the absence of air. The process has been used in the chemical industry and is often called "destruction distillation." It has been used for treating municipal refuse with the production of several byproducts such as light oil, gas, ammonium sulfate, and tar, as well as solid waste residue. The gas energy is usually more than ample to supply the heat for pyrolysis. Whether pyrolysis of a hazardous waste will be effective in destroying the toxicity has not been fully resolved.

Should an industrial hazardous waste be organic and easily decomposable at relatively low (<900°C) temperatures and yielding nontoxic products, pyrolysis may be a reasonably effective treatment process.

One form of thermal treatment is called "plasma technology." It developed from knowledge obtained from space-flight research, which required superelevated laboratory temperature to simulate spaceship nosecone reentry conditions. Electrical energy (high voltage) is converted with high (85–90%) efficiency to heat energy. One such process of plasma technology developed by SKF Steel Company (Herlitz 1983) recovers iron, zinc, and lead from steel mill bag-house dust. Without previous sintering, Herlitz reported that the fine material is pneumatically injected with coal powder into the lower part of a coke-filled shaft furnace provided with plasma generators where the material is injected. The oxides are instantaneously reduced, and liquid and gaseous metals are formed. The gases are extracted in a normal condenser. One such plant to produce 97% liquid iron, zinc, and lead from 80,000 tons of this dust per year was commissioned in Sweden in mid-1984. The process appears feasible for other solid wastes containing otherwise hazardous waste metals.

Another form of thermal treatment described by Worthy (1982) in the pilot-plant stage is the "molten salt process." He described a reactor tank that contains a constantly moving bed of molten sodium carbonate. The movement is maintained by a constant

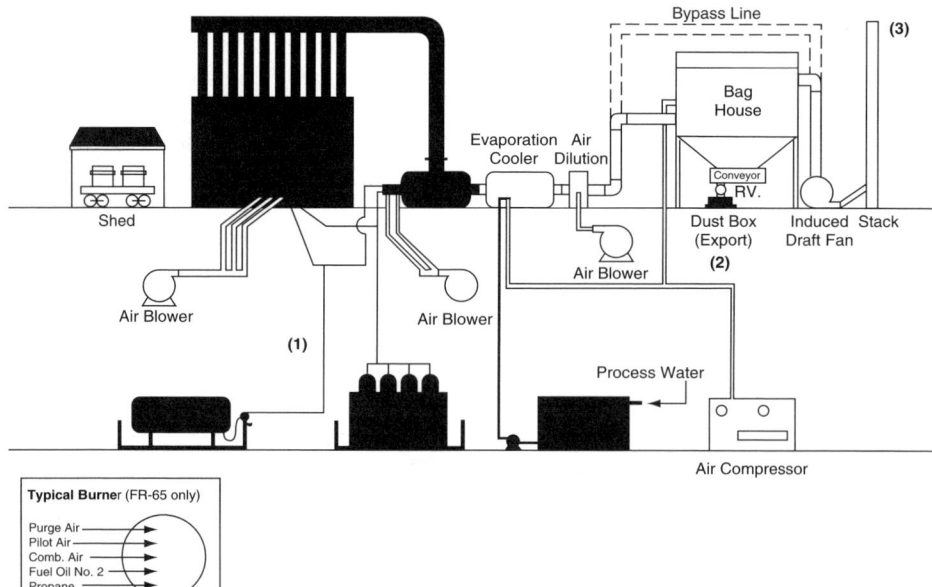

FIGURE 11.8. Thermal decontamination process flow diagram.

supply of wastes and air being fed beneath the melt surface, which is maintained at temperatures of between 750°C and 1,000°C. Destruction efficiencies are reported from bench-scale studies to be more than 99.99% for several organic hazardous wastes including toxic chemicals used in warfare, pesticides, and PCBs. This process is being developed by Rockwell International.

Worthy (1982) also reveals that the Thagard Research Corporation has reported developing a high-temperature fluid wall-type reactor. It consists of a tubular core of porous refractory carbon, which is externally heated by carbon electrodes to give off ample radiant energy.

This energy activates reactants (wastes) that descend through the tube. Radiant energy provides rapid and immediate heating of the contaminating reactants. Gas of an inert type (usually nitrogen) circulates around the porous core, which reduces contact of the reactants with the wall and prolongs the life of the core. Temperatures above 2,200°C are reported in the reactor.

Worthy (1982) reported developing a fluidized bed turbulence to promote combustion of hazardous organic contaminants. It is based on the principle that in supercritical water (374°C and 218 atmospheres), certain insoluble organic compounds become highly soluble and complex organics are converted to low-molecular-weight compounds. When these resultant products are reacted with oxygen, hydrogen, and carbon, they are rapidly oxidized, and halogen, phosphorous, sulfur, and metals form insoluble salts and settle out. A nonhazardous stream of supercritical water (500°C and 252 atmospheres) is used directly for process heat or indirectly to drive turbines for generating power. In bench-scale studies, such toxic organics as chlorobenzenes, DDT, and PCB, as well as chlorinated ethanes have been destroyed.

ENELCO/Von Roll Environmental Elements Corporation (1984) uses a process of thermal destruction for disposing of industrial wastes. It is essentially a modernization of older incineration systems under highly controlled conditions. Control of the waste-combustion process involves accounting for the waste heat content and oxidation rate of each waste and then scheduling the mixing and feed rate of the waste so that the combustion heat release is reasonably constant. Pumpable waste is injected by its system at a predetermined rate resulting in a constant release of heat. Batch-fed waste, injected at predetermined regular intervals, releases additional heat, and because of the nature of the batch-feed process, this heat release will not be constant. Of the 23 Von Roll industrial waste disposal plants in operation, four are regional hazardous waste disposal plants. One serves Nyborg, Denmark, and can handle more than 60,000 metric tons of solvent, chemical byproducts, resins, paint and varnish residues, waste oil, plastics, halogenated hydrocarbons, and inorganic sludges and wastewaters totaling 1,200 metric tons/wk. The waste can be fed directly in steel drums. Kiln discharges of molten slag and flue gases generate steam to provide most of the heating requirements of the town of Nyborg. Recoverable materials are sold.

DuPont Company has provided a hazardous waste service since 1975 by thermal decontamination. It uses a furnace temperature of 600–750°C and is claimed by DuPont to be the largest furnace in the world. Three sources of residual waste form this process (Figure 11.8) and include residual furnace noncombustibles (carted away to a smelter), bag-house solids to landfills, and stack gases discharged to upper air.

Incineration

Incineration is a high-temperature burning process whereby combustible wastes are reduced to inert residues (ashes). Incineration provides an economical nuisance-free clean method of ultimately disposing of municipal refuse. However, gases and residual ashes remain potential sources of pollution. In addition, when hazardous chemicals or materials are also incinerated, their combustion products must be evaluated to ensure that toxicity has been removed by the burning.

Incineration takes place in a furnace consisting of an enclosed refractory-lined structure equipped with grates and supplied with excess amounts of air in which the burning takes place (see Chapter 9 for more details). Temperatures in the furnace approximate 1,750°F. The combustion chamber is an enclosed refractory-lined structure, sometimes combined with the furnace, in which more complete burning of residual fly-ash material and gaseous materials occurs. Following this is a "subsidence chamber," a large insulated chamber allowing the combustion gases to expand and reduce their velocity during which particles can settle prior to gas emission to the chimney (stack).

Fly-ash should be prevented from reaching outside air environment by settling and/or filtration and sometimes scrubbers. This collected ash can be combined with the residual grate-ash for final landfill or reuse in making other materials, such as bricks or cement blocks. The amount of residual ash depends primarily on the composition of the original hazardous waste, exit gas velocity, and fly-ash removal systems. Some of the fly-ash may contain unburned organic matter and must be tested further for residual toxicity before ultimate discharge. If residual toxicity remains, furnace temperatures may have to be increased, gas velocities reduced, or detention time increased in the combustion chamber, or all three used to reduce the discharge of toxic materials.

Final capital costs and operating costs may be relatively high as compared to other techniques. Therefore, design and operating conditions must be carefully controlled. Generally, we recommend both laboratory and pilot-plant experimental burning to obtain proper design and operating procedures.

For coastal producers of toxic wastes, land incinerators may be dangerous to inhabitants of those areas. In these cases, incineration on ships on the ocean can serve to eliminate potential land air pollution.

Waste Management Company (Chicago, Illinois) has converted a cargo ship, the *Vulcanus,* to an ocean-going incinerator. The company proposes to burn 3.6 million gallons of highly toxic PCB wastes on the high seas in the Gulf of Mexico.

Ship incineration has been purported to disperse acid wastes directly into the ocean, thereby avoiding costly scrubbing. Water, carbon dioxide, and vaporous hydrochloric acid falls out into the ocean within a short distance from the ship. Safe collection and transportation of these toxic wastes to the ship could be a major problem. Any accident involving the incinerator ship during its travel to the burning site could result in major environmental damage.

LeRoy (1983) reports incinerator capacity in France for burning 260,000 tons of hazardous waste per year including phenolated water, hydrocarbons, and chlorinated waste. He acknowledges relatively high costs for incineration of these wastes because of (1) technological complexity, (2) operating difficulties, and (3) the need to purify gas and smoke.

Freeman (1981) limits hazardous waste incinerators to two types: liquid injection and rotary kiln systems. In the former, liquid wastes are fed along with support fuel and air for combustion into the incinerator, which is maintained at 820–1,600 °C. Similar temperatures are also used in the latter rotary kilns, which handle large volumes of both solid and liquid wastes. Selection of the proper type of incineration and for the proper reasons is imperative. Freeman (1981) presents tables of advantages and disadvantages of the two types of incinerators (summarized in Table 11.11).

Noland (1984) demonstrated that a "transportable" incineration system could be disassembled, transported approximately 1,000 miles, be reassembled, and fully operational within 2 weeks. He reports destruction of "pink-water" explosive plant waste by incineration of more than 99.99% (based on primary kiln ash analyses), no detectable explosive wastes in the stack gas, and stack emissions were in compliance with all federal and state regulations (including SO_2, HCl, nitrogen oxides [NOx] CO, and particulates).

TABLE 11.11
Incineration

Rotary Kiln Liquid Injection	
Advantages	*Disadvantages*
Can be used for a wide variety of both liquids and solids (mixed or separate)	High capital costs for installation especially when using low feed rates
No problem when melt occurs	Careful operation in order not to damage refractory
Drums or bulk containers with various feeds are used	Airborne gases or particles may exit prior to complete oxidation
Good mixing and air for solids	Spherical or cylindrical solids may pass through faster and avoid complete burning
Continuous ash removal does not interfere with burning	Excess air required due to leaks, which lowers fuel efficiency
No moving parts in kiln.	
Wet scrubber can be added	
Rotational speed of kiln can be varied to control residence time in burning	If drying grates are used prior to kiln they may become plugged with heat
No preheating, mixing needed	High particulate loadings
Can be operated at very high temperatures of 1,400 °C to ensure destruction of toxic chemicals	Relatively low thermal efficiency
	Liquid must be able to be atomized
Able to handle a broad range of liquids.	Liquid must be heated sufficiently or supplemental fuel added
No ash removal needed from incinerator bottom	Must be capable of complete combustion without flames hitting refractory
Can handle small amounts of liquids	Liquid waste may clog burner nozzles
Responds fast to waste temperature changes	Requires sophisticated instrumentation
Almost no moving parts	
Low maintenance	

Source: Adapted from Freeman 1981

The EPA concluded (1985) that (1) incineration, whether at sea or on land, is a valuable and environmentally sound treatment option for destroying liquid hazardous wastes, particularly when compared to land disposal options now available, and (2) there is no clear preference for ocean or land incineration in terms of risks to human health and the environment.

Incineration at sea represents a potential solution to counteract the objections of people because of environmental land pollution. On November 27, 1985, the EPA gave tentative approval to a hazardous waste burning trail aboard ship at sea (Taylor 1985; Shabecoff 1985). The test of burning about 700,000 gallons of PCBs and perhaps dioxin took place in the spring of 1986 aboard a Chemical Waste Management ship in the Atlantic Ocean, 140 miles east of the mouth of Delaware Bay. The company provided $60 million in financial guarantees to cover a possible accident during the test burn. The entire test burn cost about $1.8 million. The wastes will be transported by railroad from a landfill in Emile, Alabama, and loaded aboard the ship in Philadelphia. Therefore, the hazards of storage and transportation described in the section "Other Oily Wastes" (earlier in this chapter) prevail also in this potential method of disposal.

Cheremisinoff (1988) recommends that "high temperature processes for specific applications should be particularly considered where available land is scarce, stringent requirements for land disposal exist, destruction of toxic materials is required, or the potential exists for energy recovery. He lists seven potential advantages and considerations of high-temperature processes: (1) maximum volume reduction, which reduces ultimate disposal requirements; (2) detoxification, which destroys or reduces toxics; (3) energy recovery from combustion of waste products; (4) costs, which are generally higher than for other disposal alternates (these are hardly advantages of high-temperature treatment but seem valid); (5) operating problems create high-maintenance requirements (once again, not an advantage but an important consideration); (6) personnel required are highly skilled and experienced; and (7) environmental impacts can result if air and solid effluents are not included in the treatment scheme.

The IIT Research Institute offers a new hazardous waste treatment process that essentially uses high temperature as the treatment vehicle. It is accomplished by inserting tubular electrodes into organic-laden landfills. Radiofrequency energy charges the electrodes to heat the soil to 200–1,000°F. The vaporized hazardous pollutants are collected in the same tubular pipes and evacuated to a separate vapor treatment system.

In August 1996 the U.S. Army (Mims 1996) began the systematic incineration of hazardous chemical weapons containing nerve and blister agents. The Army claims this is a "safe and secure way" of eliminating some 30,000 tons of weapon agents. The incineration activity at the Tooele, Utah, plant was expected to take several years.

At Tooele, where the weapons are stored in 208 earth-covered concrete bunkers, 1,500 leaks have been reported since 1967 (Nemerow 1974). The 6,000 tons of chemical weapons at Tooele were left over from the World War II–era stockpile and include nerve gas, sarin, three kinds of mustard agents, five kinds of nerve agents, and the blister agent lewisite. The incinerator used to destroy these hazardous wastes contains automated equipment that punches holes in the weapons, drains the chemicals, and sends/transports them to five furnaces where they are incinerated at 2,700°F. Problems with this type of disposal arise from the potential of releasing cancer-causing dioxin (formed during the incineration process), as well as accidental breakdown in the plant, which would release toxic chemicals to the environment and surrounding population.

The release to the atmosphere of hydrocarbons, particularly chlorinated and fluoridated hydrocarbons from chemical plants, petrochemical plants, and oil refineries, continues to be a problem. At present there is no "practical" way to measure these emissions except through mass balance calculations, which are less than precise. Thermatrix, Inc., of San Jose, California, has been installing its flameless thermal oxidation technology at plants throughout the United States and in Europe and Asia. Installations include petroleum, chemical/petrochemical, pharmaceutical, and pulp and paper industries, as well as PCB cleanup. Units have also been employed in soil and groundwater remediation applications. This technology ("Make polluters pay" 1995; Agardy and Wilcox 1990), evolving from early investigations into oil shale recovery, has proven to be a "complete solution technology" in that no measurable residuals appear in the effluent stream. Typical destruction removal efficiencies exceed 99.99%, with di minimis formation of nitrogen oxides (NOx), carbon monoxide, and products of incomplete combustion (PICs), partially as a result of the thermal destruction taking place at temperatures generally below 1,800°F. In addition, NOx emissions rarely exceed 2 ppm. Not only has this technology demonstrated wide-scale applicability, but it uses considerably less energy than other typically employed incinerator technologies and, depending on the Btu content of the waste stream, can and does have a heat-recovery capability. Figures 11.9 and 11.10 show unit/assembly schematics, while Figures 11.11 and 11.12 show typical system installations.

Chemical, Physical, and Biological Treatments

Chemical, physical, and biological treatments are designed purposely to render the industrial hazardous wastes free from toxic chemicals and materials.

Chemical treatment aims at removing smaller elements of toxic nature, generally dissolved or colloidal solids. In cases where removal of toxic dissolved organic solids is the objective, chemicals used are powerful oxidizing agents such as chlorine, potassium dichromate, or ozone. The oxidation proceeds rapidly, depending on the concentration of organic matter and amount and contact time of oxidant used. End products are usually harmless gases such as carbon dioxide, nitrogen, and water. This type of treatment is costly because of the continuous consumption of oxidizing chemicals and detention time involved. The advantage of using this process and especially chlorination is that any harmful bacteria and/or viruses present are also killed. Photolysis is another form of chemical oxidation that destroys organic matter to carbon dioxide and water. Certain heavy metal oxides, notably zinc oxide and titanium dioxide, are photocatalytic and, in the presence of dissolved oxygen matter and beach sand, when irradiated, will decompose the organics by about 75% in 3 days.

Chemical treatment aimed at removing the larger colloidal solids is known as *chemical coagulation*. This is a process of destabilizing the colloids, aggregating them, and binding them for ease of sedimentation. It involves the use of chemical flocs, which absorb, entrap, or otherwise coalesce suspended matter that is too finely divided or

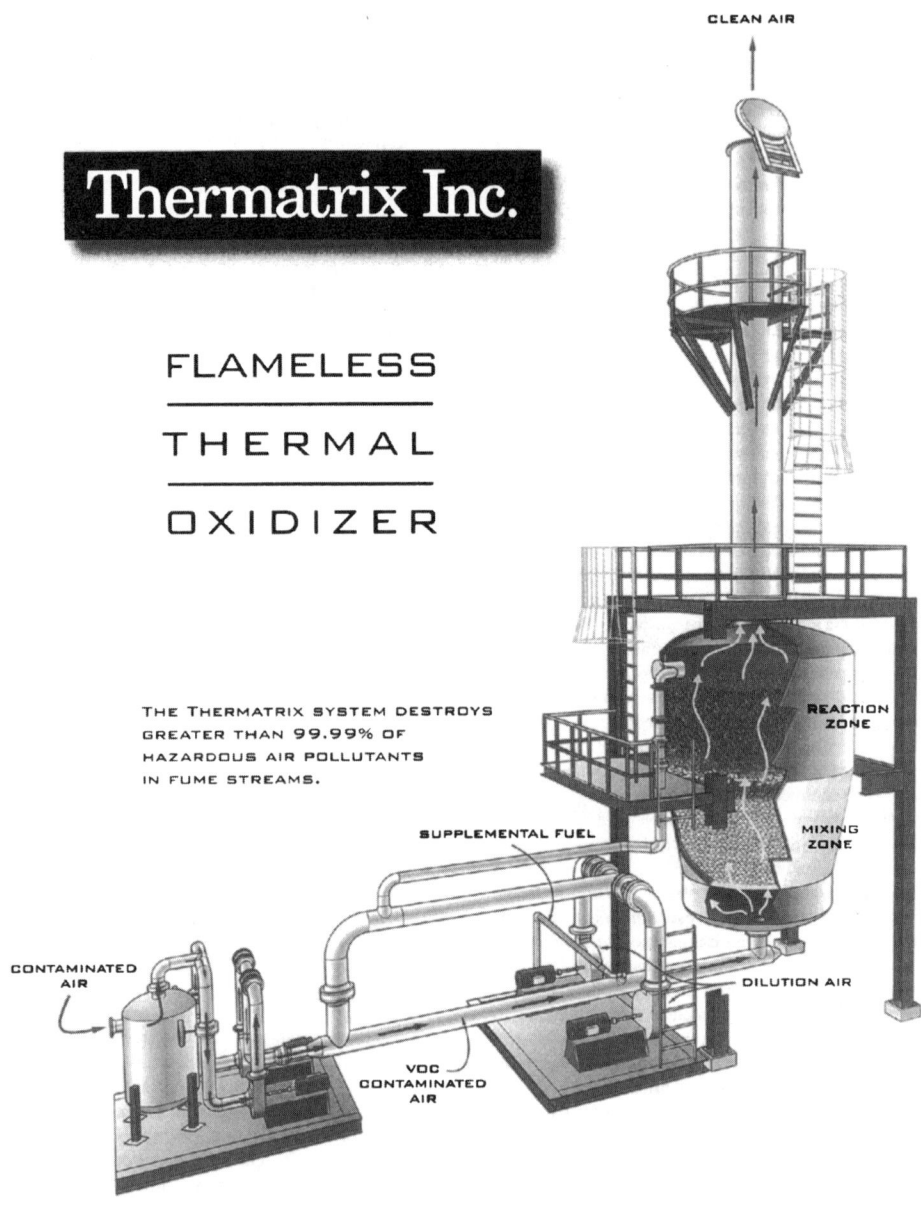

FIGURE 11.9. Flameless thermal oxidizer. Used with permission from John Wiley and Sons.

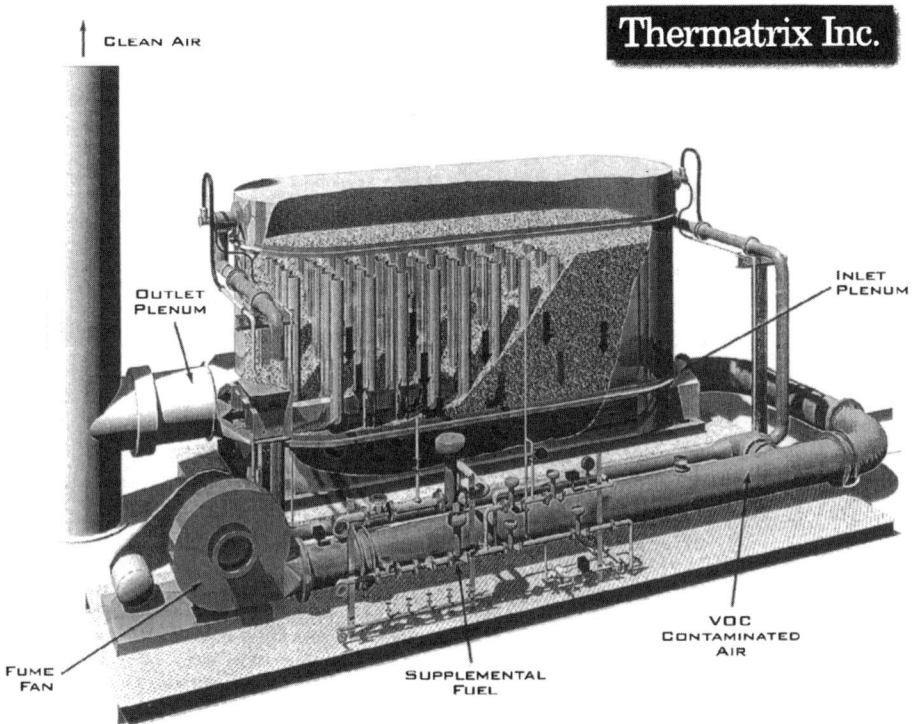

FIGURE 11.10. Typical system. Used with permission from John Wiley and Sons.

small to be settled alone. The chemicals most widely used are alum ($Al_2[SO_4]_3$) and ferric sulfate ($Fe_2(SO_4)_3$) (for theories of this treatment, see Chapter 5).

Physical treatment of hazardous materials in industrial wastes include: (1) settling; (2) filtration; (3) adsorption or absorption; and (4) flotation (refer to Chapter 5 for theories of these treatments).

The removal of dissolved organic matter by *biological* means has long proven successful in the treatment of domestic sewage and certain industrial wastes. Adapted cultures of microorganisms are placed in contact with dissolved organic matter of these wastes under the proper environmental conditions of oxygen, temperature, and pH. Organic matter is adsorbed and decomposed by the bacteria involved to yield carbon dioxide and water and some cell growth matter. The cell growth matter is later settled out and a portion returned to the process for biological enhancement, the remainder wasted and treated as sludge for ultimate disposal.

Some hazardous chemicals such as phenol from coking plants or synthetic textile mills can be degraded by certain adapted microorganisms to yield harmless end products. Efficiency of biological degradation depends on adaptation of the microorganisms,

290 *Twentieth Century*

FIGURE 11.11. Typical system. Used with permission from John Wiley and Sons.

contact time, oxygen available, and the proper environmental conditions. In actual practice, two types of biological pass the waste over a fixed surface containing the microorganisms and allowing permeation of air at the surfaces as well. There are many variations of each process in existence today (refer to Chapter 8 for these variations).

Zoltek (1984) successfully used biological treatment at an existing sewage treatment plant to digest groundwater contaminated with phenolic compounds from an old pine tar plant.

Flathman and Githens (1985) cleaned up a soil and groundwater spill of isopropanol, acetone, and tetrahydrofuran by a combination of biological and physical techniques. These resulted in 90% removal of IPA and THF within 3 weeks. Acetone, an intermediate oxidation product of IPA metabolism, was removed by the end of the sixth week. The spill originated from several buried tanks that leaked contents into a 12-ft basin of sand and pea gravel.

Hazardous Wastes 291

FIGURE 11.12. An installation. Used with permission from John Wiley and Sons.

Neutralization

Because excessive acid (low pH) or alkaline (high pH) in industrial wastes is considered toxic, neutralization before discharge into the environment is a must. In addition, neutralization is usually considered desirable prior to most forms of biological treatment should the wastes contain large quantities of biodegradable organic matter as well. Eight major methods of neutralization of either acid or alkaline wastes are described in Chapter 3. Your author suggests defining the titration curve of the potentially toxic industrial waste. This will allow the polluter to determine the exact quantity of neutralizing agent required to obtain any amount of neutralization obtained. Then the polluter is in a position to decide whether what amount of neutralization is cost effective. In order to do this, one must also determine the damage costs involved in non-neutralization of the wastes.

Underground Injection

In locations where the subterranean environment is suitable, it is possible to inject potentially hazardous wastes underground. Deep-well injection has been used to dispose of organic solutions from refinery petrochemical, chemical, paper, and pharmaceutical plants. The ultimate fate of the hazardous waste should be ascertained prior to selecting this method of disposal. The acceptable waste should be free from suspended or clogging solids and material that will react adversely with the substrata into which it is being injected. The suitable underground environment should be permeable to a point of final residence and isolated there from any potential water resource. Chapter 7 describes these details.

To determine the suitability of the underground, a well is drilled and various depth core samples are analyzed for specific characteristics relating to its permeability and reactivity with the waste. Tests will also confirm the injection pressure required at various waste flow rates. Underground soils may be treated to improve permeability to reduce the injection pressure required.

The discharger must be aware of the dangers of (1) contaminating potable water supplies by lateral migration of the hazardous waste or even by vertical migration though open subsurfaces or mechanically failed soil structures, and (2) causing movement of geological faults leading to potential earthquakes in the area. In this case, the waste acts as a lubricant between two adjacent underground slabs of faulted rock causing them to move apart more easily.

Land Farming

The filling in of farming land with stabilized hazardous waste sludge is sometimes used as a disposal/utilization technique. Only in cases where the harvestable cover crop is not harmed by the sludge or is not used as a food material is this method acceptable. These constraints limit the use of this method for most hazardous wastes. However, in situations where flammable or ignitable wastes are involved, this method may be feasible. In this case the oils and other associated organic matter may be decomposed by soil bacteria using the matter for food and enhancing the surface cover crop growth. Bacterial/viral-contaminated waste sludge may also be landfarmed in certain instances where the cover crop will not be harvested in the near future. This may give ample time for the bacteria/virus to become inactivated. Stabilized and/or solidified hazardous waste may also be landfarmed in certain instances where the waste serves primarily as a soil stabilizer and a preventer of leaching into groundwater.

One type of industrial waste being landfilled is that from the construction industry. It is common practice to bury construction demolition and other debris such as damaged tree limbs, damaged plywood, wood forms, old concrete blocks and bricks, and so on. These materials may slowly be broken down under the ground without interfering with growing surface crops. A land cultivation report (EPA 1978) states that 3% of all industrial waste can be disposed of by land cultivation. They found that the amount of industrial solid waste disposed of by this process was limited by soil texture, drainage, permeability, and the waste pH, bulk density, soluble salt and metal concentration, and nitrogen and phosphorous contents. Costs for this process vary widely from \$2–\$18/m^3 of industrial waste. Transportation costs must be added to this amount.

Bahorsky (1983) concludes that land application of textile wastes can be less expensive in capital extended aeration systems, less expensive to operate, requires less training, produce no sludge problems, be significantly less energy intensive, and give consistently high-quality effluents. However, he makes no recommendation that the land be reused subsequently for farming. It does seem possible to farm on this land after a certain period of waste treatment.

Disposal in Natural Storage Areas

A commonly recommended solution to the disposal of hazardous wastes is to return them to their natural origin or to areas of the land compatible with them. For example, coal overburden and fly ash and unburned carbon could be returned to coal mines before renovating the natural land.

Sundaresan et al. (United Nations Environmental Protection 1983) recommend returning neutralized, precipitate, and de-watered lime sludge from titanium dioxide wastes to abandoned ilemenite mines.

Permanent Landfills

In some cases, hazardous wastes are disposed of either with or without some form of pretreatment storage. Naturally many precautions must be taken in using this technique. Most important of these include the security of the waste in the landfill; it must not be reached from the surface of the land and it must not escape into surface or underground water supplies.

One such use is described by Sundaresan et al. (United Nations Environmental Protection 1983) for disposing of carbamate pesticide wastes. After neutralization as a pretreatment, the wastewater is solar evaporated in polyethylene-lined ponds. The resulting lime sludge from neutralization is disposed of as landfill. The authors state that because the sludge may contain carbamates, "adequate safeguards have to be taken." Presumably this means sealing the sludge in a permanent landfill.

LeRoy (1983) acknowledges that because all hazardous wastes cannot be incinerated or detoxicated, some with low toxic content can be disposed of in landfill provided certain conditions are observed. Wastes are permitted in landfills based on their quantity, solubility, toxicity, and concentration. In particular, sites for industrial waste where disposal is allowed in France must be impermeable, so the waste and substances leached from it are adequately confined. LeRoy (1983) reports 14 such landfills in France at special disposal charges ranging from 80 to 500 francs/ton.

In the United Kingdom, Peter Pearce (United Nations Environmental Protection 1983) reports that of the 3.4 tons of hazardous wastes generated per year and not recovered about 80% is landfilled. The remainder is disposed of either in oceans or by incineration or other treatment. About the same percentage generated is landfilled in the United States as of 1987 (Pizzuto 1981). Pearce points out some advantages for using permanent landfills for ultimate disposal of hazardous wastes. Adjacent land areas sometimes can be used by an industry to avoid costly transportation problems. Further, incineration, the major alternative treatment of hazardous wastes, costs about two to three times that of landfilling. Incineration has the disadvantage of not being suitable for many types of wastes and of generating toxic gaseous products and unburned particles. Efficient removal of heavy metals from the exhaust and damage to the incinerator structure also poses problems. Pearce (United Nations Environmental Protection 1983) lists the following six constraints in selecting a permanent landfill site for the ultimate resting place for hazardous wastes.

1. Institutional constraints: usually legal and ownership of land
2. Geographic considerations: existing land and water use, topography, climate, hydrology, location of waste sources, population distribution, and transportation routes
3. Geological considerations: proximity to known fault zones or sink holes, potential for ground shifting causing landslides, and so on
4. Waste characteristics: property and volume of wastes determine the design and storage capacity of a landfill area
5. Management priorities: financing during the beginning phases, operating and maintaining the landfill after closing the fill
6. Environmental and social considerations: ecology, water resources, archaeological resources, sociopolitical and socioeconomic factors

Hazardous wastes can usually be landfilled unless they are known to be radioactive, highly flammable, explosive, excessively toxic, odorous, or corrosive (Department of the Environment 1978). Noncompatible or highly leachable wastes must be given special consideration in designing safe fills. Sax (1979) warns that we should watch out for the compatibility of each hazardous waste with others, which may cause spontaneous reaction in the fill, especially with water and acids. In such cases, pretreatment or segregation of different wastes may be required.

The question of how permanent are landfills is of importance. For example, consider the liners preventing leaching of potentially toxic chemicals from the fill. It is too early to predict the ultimate longevity of various liner materials for landfills. With that in mind, it is preferable and safer to design these facilities for some leachate to protect water resources.

In New Jersey, storage tanks containing volatile liquids have been painted white and covered with floating covers. Lower temperatures thus yield lower evaporation rates and ensure no loss of toxic organics to the atmosphere.

Piasicki (1983) pictorially depicts what can go wrong with permanently storing hazardous wastes in a landfill. They include (1) burrowing gophers attacking the landfill cover, (2) freezing temperatures shrinking and tearing the landfill liner, (3) mineral acid and solvents mixing and triggering a chemical fire, (4) chemicals corroding waste collection pipes or weight of waste may crush these same pipes, (5) debris may clog perforated collection outlet pipes, and (6) the landfill's protective cover can be breached by erosion, by new construction on the site, or by settling after bulk solids compact and barrels disintegrate further down in the fill. If the cover splits at the surface, rainwater can reach the waste and overload the leachate collection system or cause the entire landfill to overflow.

Sachdev et al. (1983) investigated the use of fly-ash with or without additives as a liner material for landfill disposal sites. They found the following parameters significant for use of fly-ash as liner material. They aimed for a hydraulic conductivity of 10^{-5} cm/sec or less and a leachate with a minimal impact on groundwater.

- *Boiler type*: The type of boiler and boiler temperature affect the physical and chemical characteristics of the fly-ash. Fly-ash from a pulverized boiler is generally finer than that from a cyclone or stoker fired boiler. The finer fly-ash has more surface area and is, therefore, more reactive.
- *Coal source*: The coal source may affect fly-ash characteristics. There are differences between western coal fly-ashes and eastern coal fly-ashes. Western coal fly-ashes have higher lime content and higher pH than fly-ashes. The higher lime content (>10%) fly-ashes usually exhibit self-hardening properties.
- *Fly ash handling mode*: Wet and dry fly-ash handling have different effects on fly-ash reactivity ("Great Lakes Study" 1985). Wet handling reduces the reactivity of fly-ash.
- Loss-on-ignition (LOI): LOI indicates the amount of unburned carbon in the fly-ash. High LOI (and, therefore, high carbon content) in fly-ash inhibits pozzolanic activity. The concrete industry specifies a maximum LOI value of 6% in fly-ash for use in concrete manufacturing (American Chemical Society).
- *Particle size*: The particle size of fly-ash is important. finer fly-ash offers more surface area for reaction and generally provides lower permeability than coarser material. A well-graded fly-ash is preferred over a uniform particle–size fly-ash because it would generally exhibit smaller void spaces and, therefore, lower permeability. Fly-ash used as a mineral admixture in Portland cement concrete must have a minimum of 66% by weight passing through a No. 325 sieve (American Chemical Society).
- *pH*: A higher pH (≥ 10) is required to promote the precipitation of calcium silicate, which is the source of bonding in the pozzolanic process.
- Lime content: Lime is the source of calcium, which reacts in the pozzolanic reaction. It is the "free" lime in the fly-ash that is available for reaction. Lime can be added if fly-ash has low lime content. A fly-ash with lime content greater than 10% normally exhibits self-hardening characteristics.
- *Amorphous silica content*: Amorphous silica is the noncrystalline portion that reacts with calcium in the pozzolanic reaction. A fly-ash with low amorphous silica would, therefore, have low reactivity. The amorphous silica content would generally increase with increasing boiler temperature.

However, in 1985 the U.S. government officials report that one-third of toxic waste landfills might be forced to close because of results obtained by underground monitoring and by new insurance requirements ("U.S. waste deadline" 1985). Industry appears to be moving away from dumping toxic material into landfills and toward the incineration and treatment of hazardous wastes. In fact, 1,100 toxic landfills—almost two-thirds of the nation's total—were in fact closed because they could not comply with federal environmental rules ("Toxic sites close" 1985).

Ocean Dumping

The dumping of hazardous industrial wastes into ocean areas may be practiced sometimes safely because of the following attributes of oceans: (1) they tend to dissolve and disperse the wastes three-dimensionally, as contrasted to the surface use only of land resources; (2) oceans are mostly out of contact with people except for shipping and transportation—two rather "lowly" uses of water resources; (3) some pollutants may settle thousands of feet to ocean bottoms where presumably they will cause people no harm; (4) the ocean is already considered a mixture of a multitude of chemical elements and compounds; and (5) the ocean food chains are much longer than those we experience with on-land food chains, so humans are placed farther away from hazardous chemicals.

Oceans should be used only for wastes that, when diluted fully, will offer no threat to marine life, especially the kind that serves as food for humans. The oceans should provide enough time before reuse by humans for degradation of the hazardous industrial waste. Nemerow (1985) gives a detailed description of the effects of some common ocean pollutants on the viability of ocean resources. It is important to keep in mind that the aquatic ecosystem of oceans has a finite assimilative capacity for a particular contaminant without significant deleterious effects. The assimilative capacity of any particular ocean area is determined by physical processes such as mixing currents, geomorphology, types of sediments, and types of water chemistry and biology.

The National Research Council (1984) recommended in 1984 that although oceans are suitable for discharge of industrial and municipal wastes, "certain xenobiotics (human-made organics) are so persistent that they cannot be removed in treatment plants or diffused to safe limits in the ocean."

Secure Burial, Reuse, and Chemical Fixation

Prouty et al. (1983) used a containment method of incorporating toxic sludges into bricks during their manufacturing. They claim savings in energy since bricks are lighter and more porous, as well as enhanced insulation quality. Bricks are produced by mixing finely ground clay with water, forming it into the desired shape, drying, and finally burning it. They tested metal-laden sludges and found a portion of the metals volatized during brick burning, but the remaining metals were bound in the brick and will not leach from it.

Wright and Caretsky (1981) give seven techniques for chemical fixation of toxic wastes. *Cement-based fixation* is a process that mixes toxic wastes in a slurry of water and cement. When the concrete hardens, it can be used for various building purposes—sometimes in surface coatings of asphalt, vinyl, or emulsion mixtures to increase the strength and decrease the permeabilities of the concrete. Radioactive wastes and heavy metal sludges have been "fixed" in this manner. *Lime-based fixation* involves mixing the waste with lime and fine-grained silicious waste material such as fly-ash, blast furnace slag, or cement kiln dust. The *encapsulation fixation* process encloses previously bonded waste in a covering of non-reactive inert material. Usually the toxic waste is mixed with a thermosetting plastic, placed in a mold, and heated to fuse into a hard block. Many hazardous industrial wastes originated from the extraction, use, and alteration of toxic elements in native underground sites in a form similar to that in which they were found. The concept appears reasonable. However, several obstacles must be overcome. First, the volume, weight, or mass of hazardous materials must be reduced to a point at which it would be economically feasible to return them underground. Second, the exact location to where they are returned must be suitable. That is to say, it must be available, owned by the disposer, and not likely to cause environmental problems to potential underground resources such as water, air, or land. Third, methods must be found to solidify and contain the hazardous wastes so that they will not migrate away from the burial site once they are placed underground. These are not insurmountable obstacles; however, they must be faced and settled before solidification becomes a viable solution for the safe disposal of hazardous wastes. *Self-cementing fixation* is used with waste sludge containing large amounts of $CaSO_4$ or $CaSO_3$. Small portions are de-watered and calcined under controlled conditions and remixed with the entire waste sludge and additives as necessary. The product is a hard, plaster-like relatively impermeable mass. This process is especially useful for sludges from powerplant stack gas scrubbers. The *classification process* mixes the toxic waste with sand before fusing into glass. This is used mainly for radioactive wastes. *Thermoplastic fixation* dries, heats, and mixes the toxic waste with a heated plastic material such as paraffin, bitumen, and polyethylene, cooled, and resulting solid matter stored indefinitely. Although this process is mainly used for radioactive wastes, it may be used for certain organic solvents, oxidizing salts such as nitrates, chlorates or perchlorates, and dehydratable salts.

Organic polymer fixation produces a spongy, rather than a solid, mass by blending the waste with a prepolymer and catalyst. This is done at room temperature and usually in batch basis rather than a continuous process. It is also used for industrial radioactive wastes. Francis (1984) found it much more economical and practical to treat a toxic metal land-filled sludge in place than to remove it for treatment and subsequent alternative disposal. Essentially he found that after detailed analysis of the land-filled sludge, nickel as the hydroxide was the major toxic contaminant. His plan was based on the fact that nickel hydroxide solubility in water can be effectively controlled by keeping the pH at 10.2, its point of minimum solubility. The landfill sludge site was covered with a layer of finely ground calcium carbonate to neutralize acid rain before it reached the waste. Then the site was covered with mounded, compacted, and graded earth to divert rainwater and runoff from the waste. Finally, a 12-in. layer of gravel followed by a 6-in. topsoil layer was added on the earthen cover. The topsoil is needed to provide vegetation necessary to prevent erosion. His system of containment, though costly, was estimated at about 10% of that necessary to remove sludge and to treat it other ways.

ENRECO, Inc. (1985) successfully completed the solidification phase of more than 50 sites during 1984. The company makes a careful analysis of solidification agents and then uses a unique injector to add and mix the solidification agents with the waste. Some agents that have been used include fly-ash for oil-field drilling, kiln dust for oil-field skim pits, and carbon for PCB-impounded waste sludge. In the last case, the solidified sludge was removed and transported to an acceptable landfill. In other cases, the solidified mass was buried on site. Photographs of successful operations of

(1) mixing, (2) solidifying, (3) removing, and (4) burying with land renewal were shown.

Jones et al. (1985) point out that many hazardous wastes, especially metal-plating wastes, may contain constituents that could interfere with the blinding process of solidification with type I Portland cement and fly-ash, including oil and grease, light-weight oil, phenol, sulfates, strong base, pesticides, degreaser, lead, copper, and zinc.

Exportation

Since the beginning of time, humans have developed the attitude that if we can give our problem to someone else, we are relieved of any further responsibility, or "out of sight, out of mind." Therefore, it comes as no surprise to observe industry expounding the philosophy of contracting for services to carry away hazardous wastes to a site distant from the plant where it originated. The assumption is made in these cases that it is far less expensive to pay some agent to assume the responsibility for providing the service than to assume the liability and costs yourself. It is also presumed that the disposal of the hazardous waste is safe and effective at the distant site.

Industry should shoulder the responsibility (before exporting hazardous waste) of determining not only whether this method is more economical, but also whether it is effective and safe ultimately.

In June 1983, a novel plan to export toxic PCB wastes from Florida to Honduras was uncovered ("Export plan" 1983). The plan was foiled and no criminal charges were filed. However, EPA and state officials agreed that this export plan underlined growing fears that strict regulations on toxic waste disposal in the United States could lead some American firms to try to dump their toxic materials abroad.

In March 1989, a treaty among more than 100 countries was unanimously adopted restricting shipment of hazardous waste across borders (Greenhouse 1989). The pact requires waste exporters to notify and receive consent from receiving countries before shipping the waste. In addition, the treaty requires countries exporting waste and those receiving it to ensure that the waste is ultimately discarded in an environmentally sound manner. The treaty prohibits shipments of hazardous waste to nations that have banned waste imports. It also requires that all cross-border waste shipments be packaged and labeled properly and that all companies transporting or disposing of hazardous wastes be authorized to handle wastes. Among the wastes considered hazardous are clinical wastes from hospitals, wastes from pharmaceutical factories, PCBs, compounds containing mercury or lead, and wastes from the production or use of dyes, paints, and wood-preserving chemicals. The treaty is a beginning and its nature will require additions, clarification, refining, and strengthening as time goes on.

Recovery and Reuse

Recovery and reuse of hazardous waste represents the ultimate in environmental and economical efficiency. If found feasible, such a solution prevents environmental damage and provides some monetary return to the generator of hazardous waste. Not to be overlooked are the indirect benefits of recovery and reuse to society. Most of these benefits are associated with a slower decrease in the world's natural resources, a so-called "slowing down of entropy buildup." In addition, consumer prices for products made with recovered materials should be lower in the long run.

Therefore, the goal of all industry and its environmental engineers should be that of recovering and reusing all wastes including especially hazardous ones. Chapter 15 of this book presents a novel system for doing this.

Here, we present only some examples of effective recovery and reuse of certain hazardous industrial wastes. In addition, Wilcox and Agardy (1990) present many possibilities for effective recovery and reuse of industrial wastes.

An important consideration in this chapter should be the question of whether the business of recovery and reuse of many chemical elements and compounds considered toxic to the environment represents a beneficial service to society independent of its profitability. If so, one could expect that it would be reasonable for public subsidization to some extent of such ventures. If not, then the business of recovery and reuse must stand solely on its own economic viability similar to that of other profit-making industries.

Before recovery of hazardous wastes, one should attempt to reduce the amount of these wastes to a minimum. In that connection, the Jacobs Engineering Group (1988) proposes four major incentives for what it refers to as "waste minimization."

1. Economics
   A. Landfill disposal cost increases
   B. Costly alternative treatment technologies
   C. Savings in raw material and manufacturing costs
2. Regulations
   A. Certification of a waste-management program on the hazardous waste manifest
   B. Biennial waste-management program reporting
   C. Land-disposal restrictions and bans
   D. Increasing permitting requirements for waste handling and treatment
3. Liability
   A. Potential reduction in generator liability for environmental problems at both on-site and off-site treatment, storage, and disposal facilities
   B. Potential reduction in liability for worker safety
4. Public image and environmental concern
   A. Improved image in the community and from employees
   B. Concern for improving the environment

Copper Recovery from Plating and Engraving Baths

When a hazardous industrial waste is composed almost wholly of copper as a contaminant, it usually can be recovered in quite pure form as the hydroxide. Such may be the case in electronic printing-board wastes. Even when the wastewater is contaminated slightly with other metals such as iron or zinc, copper may be precipitated and then purified, if necessary, before recovery and reuse.

Copper hydroxide is insoluble at an optimum pH of between 7 and 9. For recovery and reuse, NaOH is preferable to $Ca(OH)_2$ for raising the pH, which contaminates the recovered sludge. The pH should be kept below 9 and preferably above 8 to prevent re-solubilization of the copper hydroxide precipitate. Other contaminating metals, especially zinc and chromium, will be incompletely precipitated at this pH.

Another method of recovering copper from mixed metal wastes is that of depositing copper on brass chips in the presence of chloride. This process of copper recovery is used primarily for treating brass mill wastes where chromium, copper, and zinc are common impurities found in the effluent.

Cation exchangers may also be used on copper wastes such as those from the electronic industry in making printed circuits. Copper is present in these wastes as the sole contaminant. However, only analysis of the hazardous wastewater will reveal whether either of the latter two recovery methods is preferable to copper precipitation as hydroxide.

Cartwright (1984) describes a complete metal-plating reclamation and waste-treatment system designed, installed, and operated in Bloomington, Minnesota. The latest treatment technologies are detailed including reverse osmosis, two-bed deionization, mixed-bed deionization, activated carbon adsorption, ultrafiltration, ozoonation, slant tube clarification, and sludge de-watering. More than 90% of the rinse water from the electroplating line is purified back to 18 meg ohm quality and recycled; ultrapure makeup water is supplied from city sources and toxic wastes are precipitated as an insoluble sludge. Nickel-iron, nickel, copper, and gold metals are involved in preparing 10-cm diameter wafers of these metals on a special ceramic substrate. Cartwright (1984) describes the concept of the reclamation and minimal discharge as the "wave of the future."

Spent Caustic Soda Regeneration

Certain industrial wastes such as textile kiering and pulp-mill cooking liquors contain spent sodium hydroxide in relatively high concentration and in relatively uncontaminated form. In these cases and in other special situations, caustic soda may be recovered and reused in the same industry or sold to other plants for other uses.

Dialysis or other variations of membrane separation is a major method used in recovering pure caustic soda. The caustic permeates the membrane and dissolves in water usually much easier and faster than any other contaminants contained in the waste (see Chapter 7 of this book for theories and rate reactions for typical dialysis systems). The quantity of sodium hydroxide diffusing through the membrane depends on the time, the area of the dialyzing surface, the mean concentration difference, and the temperature. Naturally, the membrane pore size and degree of clogging will control the flux rate (permeability) of the caustic soda.

Evaporation is also used to concentrate caustic soda for recovery and reuse, especially when heat energy is relatively inexpensive and when impure caustic can be reused satisfactorily.

Mercury Recovery

Toxic mercury compounds occur in several industrial wastes, mainly in the alkali-chlorine plants using mercury cells. Small amounts of mercury escape these plants, end up in watercourses, and are picked up and biomagnified in fish from where they reenter the human food chain. Because of the extreme toxicity of mercury, it apparently reverts to especially toxic and dangerous methyl mercury form. Because mercury is found in very small amounts in the wastes from several plants using mercury cells or thermometers, it is seldom practical to design and install large complicated treatment systems. Rather, it is more realistic to separate by gravitational means (mercury is almost 14 times as dense as water) small quantities of waste mercury at the origin of its escape from the process or use and return it to reuse in the plant.

Cadmium Recovery

Cadmium accumulation in humans has been linked with hypertension, emphysema, and bronchitis. Much cadmium waste has been attributed to automobile battery manufacturers. The major treatment techniques for removing trace amounts of cadmium include chemical precipitation, cementation, reverse osmosis, ion exchange, chelating resins, and foam separation.

The nickel-cadmium cell is an excellent storage life battery. It is a secondary cell type, that is, it depends on chemical reactions that are reversible by electric energy and, therefore, does not need chemical replacements. The cell uses nickel hydroxide as the positive electrode and cadmium-cadmium hydroxide as the negative electrode as the electrochemically active materials. Potassium hydroxide is the electrolyte in an aqueous solution. When these used batteries are delivered to chemical recovery plants, they are stored, washed, and valuable nickel and cadmium recovered.

Drainings, washings, and unusable metal parts of the batteries must be kept from polluting the soil and associated groundwater during the recovery process.

Silver Recovery from Plating and Photographic Wastes

Silver is both too toxic and too valuable to allow even the smallest amount to be discharged to waste. Two major industries use and waste sufficient silver to make it imperative for them to recover it: silver plating and photographic developing.

Silver can be precipitated as chloride, sulfide, hydroxide, or insoluble silver. These are affected by the addition of sodium disulfide, sodium hydroxide, and sodium borohydride ($NaBH_4$), a powerful reducing agent. In some instances, ferric chloride is added to enhance silver solids separation. Any other heavy metal contaminants will also be precipitated and must be removed from the solids before recovering the silver.

With a lithographic film waste, Cook et al. (1979) report the borohydride system gave the best overall results based on residual metal levels, cost, and suitability of the precipitated silver for further recovery. His chemical cost was $0.50 per troy ounce of silver removed. Nemerow (1963) reports that recovering silver from a silverware-plating plant waste by precipitating it as the chloride yielded silver concentrations in the effluent of less than 3.5 ppm from an original value of up to 250 ppm.

In addition to precipitation, silver can also be recovered by metallic replacement and electrolysis. In the case of photographic wastes containing silver, spent hyposolutions are brought into contact with a metal surface such as steel stampings, zinc, or copper.

After more than two-thirds of the silver is recovered, the metal is removed from the tanks, dried, and sold to the refiner.

In the electrolysis method, both an anode and a cathode are placed in the waste silver and an electric current passed through. Silver will plate out on the cathode. A total of 32 ft^2 of cathodic area is required for treating 50 gallons of waste containing 19 ounces of silver. After 24 hours of air agitation, 98% of the silver is removed. About 600 troy ounces of silver can be recovered before the cathode has to be desilvered or replaced.

Waste Oil Recovery

Waste oils are found in industrial discharges from many types of plants and originate from uses such as lubricating, fuel, hydraulic circuits, and refining. As might be expected from such widespread uses, the qualities of the oily residues are vastly different.

Residual oils from transformers, turbines, and some hydraulic systems are sometimes referred to as "bright oils." These oils can be recovered and directly reused for purposes that do not require the highest quality (purity) oils. An example of such secondary reuse is in stripping molds in foundries.

Waste oils from motor cooling, mill hardening and rolling and drawing of steel, and other metal lubricants and cooling are called "black oil." These oils can be recovered and regenerated for reuse by re-refining by specialized companies. It may be dangerous practice to incinerate these black oils rather than regenerate them because of the impurities such as lead given off during burning. In addition, wasteful uneconomical burning is eliminated by regeneration and reuse.

Fitzpatrick (1985) discusses companies that turn bore or mill metal. A product of these operations is always some form of oil-contaminated scrap metal. The scrap has little or no value, and is a nuisance in its present contaminated form. Fitzpatrick claims that a properly designed and sized system will reclaim valuable oil, upgrade the value of metal scrap, and reduce handling and hauling costs. He also claims that these systems return investment capital at a rate comparable to other capital equipment. Scrap chips must first be collected by some units placed beneath the various machines, raised to crushers to reduce the scrap size, discharging the scrap into containers for storage for resale, and the reclaimed oil to cleaning by settling and filtration for reuse as friction-reducing and cooling agents.

Another location in the steel mill where oil may be recovered and reused is the quench water. As steel is quenched, the metal oxide and oil combine to form a grimy scale. The scale accumulates into an oily dirty sludge. One company reports using the Barrett Sludge Extractor ("Sludge extractor" 1985) to recover most of the quench oil from the sludge and saving about $30,000/yr in oil purchases. The extractor operates like a centrifuge to spin the wet sludge at high speed and compact the sludge while driving the oil towards the center of the unit. From there the oil is collected, filtered, and recovered for reuse while the dry sludge remaining is disposed of as required by local regulations.

There are two major types of fuel oil wastes: diesel train refueling oil and auto and truck waste crank-case oils. When recovering these oils for reuse, they are generally heated, coagulant is added, and the resulting oily slurry centrifuged to remove the impurities. The centrifugate is then resold to oil refiners and/or distributors for reuse.

Solvents Recovery

Solvents are universally used by many industries both in the gaseous and in the liquid phase. Examples of such industries are dry-cleaning, printing, paints and varnish, rubber, and metal degreasing. Besides objectionable odors, some solvents are photochemically reactive. Fluorocarbon solvents affect the upper layer (ozone) of the atmosphere. The recovery and reuse of waste solvents can be compared in philosophical and economical terms to that of waste oils. Solvents must be regenerated to separate them from their impurities. Regeneration is carried out by either steam stripping or rectification. Because regeneration equipment is expensive mainly from a capital investment standpoint, regional or collective facilities for several industries are recommended.

The Miami Herald (Oct. 27, 1983) reports of the Gold Coast Oil Company, which distilled and recycled chemical wastes, subsequently closed due to pressure from Dade County, Florida, but responsible legally for any residual toxic wastes. The company left behind about 2,500 fifty-five gallon drums of unidentified waste and unrecyclable residues and solvents. It was gooey, sticky, and discolored. The drums leaked and the material tested for "alarmingly high concentrations" of heavy metals. The lesson here is that even recycling plants may run into environmental problems resulting from additional residual wastes. Changing the solvent to one less hazardous is a well-established procedure, but it, too, can have drawbacks. For example, the refrigerant chlorofluorocarbon (CFC), widely used for the past 40 years or more, is now being replaced by a class of compounds known as hydrofluorocarbons (HFCs). However, these HFCs, which are replacing CFCs (which reacted with and destroyed the ozone layer, thereby increasing ultraviolet rays to the earth's surface) are now found to influence (and enhance) the production of acid rain, another pollutant.

Vara International of Vero Beach, Florida, is one of a number of companies that specialize in solvent recovery. It has used a special palletized activated carbon to adsorb solvents for more than 40 years. The choice of carbon is very important; special concern should be uniform particle size, low ash content, high retentivity, and strength.

Water Recovery and Reuse from Industrial Laundries

Industrial laundry wastewater, though not generally considered hazardous, may contain priority pollutants such as heavy metals. At the least, it is a complex and variable mixture of high concentrations of organic materials and suspended solids. Van Gils et al. (1984) give a consolidation of analyses of typical industrial laundry wastes (Table 11.12). They reported that lime coagulation, settling, high-rate ultrafiltration, and fixed-bed carbon adsorption proved effective in providing a reusable water at an operating cost of $3.80 per 1,000 gallons of wastewater processed. No mention is made of the disposition of the contaminants removed. Presumably, they would be landfilled or burned during regeneration of the carbon.

TABLE 11.12
Typical Industrial Laundry Wastewater Constituent Concentration

Constituent	Concentration, mg/litre
BOD	1,300
COD	5,000
Susp.solids	1,000
Oil and grease	1,100
Lead	4.5
Zinc	3.0
Copper	1.7
Chromium	0.88
Nickel	0.29
Chloroform	3.3
Benzene	2.5
Perchlorethylene	9.1
Toluene	5.2

Source: Adapted from Van Gils et al. (1984).

Space is often a limiting factor in smaller laundries for normal physical–chemical treatment required. However, with the increasing size of industrial laundries and more automated adsorption-filtration systems now encountered, these treatments and the reuse value of the water are more feasible.

High-gradient magnetic separators consisting of a filter bed packed with a fibrous ferromagnetic material such as stainless steel or wool screens and magnetized by a strong external magnetic field that surrounds the bed have been described and recommended as useful for treating laundry wastewater (Oberteuffer 1973; DeLatour 1973). This treatment is said to be extremely fast, with filtration rates as much as two orders of magnitude higher than conventional sedimentation-filtration processes. Once again, contaminants removed must be disposed of ultimately by burning or land-filling. Other treatments for laundry wastes can be found in Nemerow (1998).

Dye Recovery and Reuse from Textile Wastes

Textile dye wastes were reported as far back as 1952 (Nemerow 1952) as being pollutional in nature and varying in treatability depending on their chemical nature. In 1957, Nemerow described and classed commercial dyes according to their structure. At this time it was proposed to alter the color of a major class of textile dyes, the azo group, to remove the color contaminant characteristic. The azo chemical structure was altered by reduction with stannous chloride and salting with NaCl.

From that time to the present, Nemerow has been advocating recovering and reusing textile dyes. Support for this thesis was based on the difficulty and high cost of color removal from these wastes, as well as the economic value of the recoverable dyes.

In 1982 Nemerow suggested to textile industries the removal of dyes on fine filter material such as clay or fine weave fabric. The dyes can be concentrated in up to 80% solids before evaporating with waste heat the remaining 10–15% water. In considering the economics of dye recovery by this or any other method, one must consider the avoidance of damage costs or other destructive treatment costs as negative costs of dye recovery.

Bergenthal et al. (1984) demonstrated the technical feasibility of batch dye bath reconstitution and reuse at a carpet mill. They overcame several technical problems that are common to a wide variety of textile mills such as the following:

1. Selecting product styles and shades that include dye bath reuse
2. Reforming dye recipes to use a single dye group to dye many shades
3. Demonstrating the feasibility of dye bath reuse on a portion of the product while the remainder continued with normal production
4. Producing high-quality product with recycled dye baths
5. Adapting dye bath reuse to conform to the mill's standard dyeing procedures

They observed a 25–50% reduction in both pollutants and water use. Economic benefits were termed attractive and especially when dye bath dilution with steam condensates and overflow cooling waters was eliminated. This latter finding lends even more credence to recommendations that dry dyes should be recovered, stored, and reused by adjustments with fresh dyes at appropriate times in the production schedule.

Nemerow (1988) found that textile dye wastes can be treated economically and effectively by ultrafiltration. Membrane filtration studies are in progress.

Detoxification

Hazardous wastes can be detoxified in place or before disposal in another site by a variety of processes depending on the type of waste and special situation. Any of these processes portends a residual waste for ultimate disposal. However, the residual waste would no longer possess toxic properties. Detoxification at the site of hazardous waste origin is highly recommended. Not only are the toxic waste generators better able to detoxify their own waste as it originates, but they also are protecting the final disposers from accidental spills or malfunctioning systems. Also, the generators are legally and morally responsible for making their own waste safe for further transportation and/or treatment and disposal.

Sunohio (1981) developed and promoted a detoxification process consisting of a mobile chemical plant on a tractor-trailer truck. It has been used to detoxify PCBs. A chemical agent strips chlorine atoms from insulating liquids and thereby removes its toxicity. The promoters claim that the detoxified insulating fluids can then be reused for the same purpose without decreasing its effectiveness.

Piasecki (1983) reports of EPA studies of soaking toxic solvents and acids in hot baths of molten salt that destroy the toxic chemicals. The wastes are injected into a pool containing sodium carbonate heated to about 1,650°F. The hazardous hydrocarbons are burned off and converted to CO_2 when oxygen is supplied or available, whereas sulfur and chlorine react with the salt and end up in the ash residue of only less than 1% of the original volume. One advantage of this process is that it allegedly takes place at temperatures lower than those of conventional incinerators. This minimizes the production and evolution of nitrous oxides, precursors of acid rain. Rockwell International offers a unit that can detoxify 1 ton of waste per hour.

Piasecki describes the process as pumping the contaminated oil (PCB) into a chemical soup that contains metallic sodium. The sodium strips the chlorine from the PCBs and combines with it, creating a form of sodium chloride. The biphenyls are removed as a harmless sticky residue, and the oil can be returned to the transformer, used as fuel oil, or dumped without special handling. The process can reduce PCB contamination from a level of 10,000 ppm to less than 2 ppm. Acufex Corporation was developing a truck-treatment process to re-purify contaminated soils of this PCB and return the soil to its original site.

LeRoy (1983) describes a collective detoxification establishment to consist mainly of the following:

1. A workshop for removing cyanide by oxidation
2. A workshop for removing chromate
3. An acid–base neutralizing workshop

He adds that machining fluids are also processed by physicochemical methods such as ultrafiltration and evaporation.

Hanson (1984) describes an air stripper that removes more than 99% of total chloroform extractables from a well-water supply. The air-stripping system is costing the town of Hartland, Wisconsin, about 4 cents/1,000 gal of water treated and strips about 500 parts per billion of TCE at 1,000 gal/min. Once again, however, such a treatment transfers the toxic material from the water to the air environment.

Wright and Caretsky (1981) include soil flushing, chemical detoxification, and microbial inoculations in treatment of waste dumps to detoxify them. Soil is flushed by applying water at the surface and collecting leachate with the use of shallow well points. Analysis of the leachates will indicate whether they must be treated further or may be discharged safely to water courses. This is especially useful when soils and wastes are acidic and likely to leach out heavy metals. Chemical detoxification is done the same way except some chemical agent is added to degrade or alter the toxic material to render it harmless. Seeding the soil with a microbial growth, which can degrade the contaminant(s), is also done when a suitable amount of time is available and the toxic material can be biodegraded.

The technique of "plasma arc" detoxification provides the most complete destruction with the highest energy level possible short of atomic reactors. It is claimed to have been invented by a Canadian engineer, Tom Barton. A high energy arc is transmitted between two electrodes, which heats a short chamber of air to temperatures of about 45,000°F, transforming any impurity in the chamber into a plasma. "Plasma" can be defined in this case as "ionized gas composed of electrons and positive ions in such relative numbers that the gaseous medium is essentially electrically neutral." Sometimes plasma is referred to as a "fourth state of matter."

Familiar forms of plasma-like matter are lightning bolts, fluorescent lights, and welding arcs. Barton feeds a toxic waste into the middle of the chamber where it is attacked by the plasma disintegrating the molecular bonds, and only carbon, hydrogen, oxygen, and chlorine atoms remain. He pumps the gases to a scrubber when the potential for formation and release of residual toxic gases exists. Barton's development of the plasma arc represents an improvement over the NASA implementation of the 1960s space reentry of vehicles. His electrodes are said to withstand heat and use for longer periods than the older NASA ones. Several companies in addition to the U.S. Army are researching the plasma arc method for destruction of not only toxic wastes generated directly by industry, but also for renovation and purification of contaminated soils.

Two more conventional methods of detoxifying waters containing small quantities of volatile organic chemicals such as trichloroethylenes from dry-cleaning wastes are (1) air stripping and (2) adsorption.

Air Stripping
Air stripping is accomplished by pumping water down through a packed column through which air is being blown up and discharged above the column to the atmosphere. The air strips the volatile organics from the water that is collected beneath the column. Air-to-water ratios and their temperatures are important factors in determining removal efficiencies.

An illustration of air stripping was reported by Wench and Josephson (1984). They drew polluted leachate from under a toxic landfill to a point at which it could be pumped 1,400 ft, then sprayed in an adjustable arc at a rate of about 200 gal/min. The jet mixed air with water at a ratio of 8,000 to 1. The treatment process not only eliminated toxic organics from downstream well water but also volatilized the same during spraying in the air.

Granular Activated Carbon Adsorption
In granular activated carbon adsorption, contaminated water is fed through enclosed tanks containing a packed bed of specific size and density of activated coal material. The latter holds the volatile organic contaminants while the purified water is usually discharged at the tank bottom devoid of contaminants. The bed area and contact time are two of the more important parameters affecting the degree and rate of removal of these volatile organics.

Nyer (1985) reports that Rockaway Borough used granular activated carbon to remove trichloroethane from its underground drinking water at an average increase in the monthly bill for each household of $3.00. The system handles 1.5 mgd at an influent TCE concentration of 335 ppb.

Nyer (1985) also reports that the city of Acton, Massachusetts, air strips the volatile organics from its groundwater with removal of 96–99% at a total cost of $0.053/1,000 gallons. All organics were removed to less than 1 ppb. This naturally eliminates the possibility of toxic trihalomethanes from the drinking water following disinfection.

Oil Recovery Systems, Inc. (1985) uses an air-stripping system to remove volatile organics from water. It reports excellent results in removing chlorinated solvents such as trichloroethane, trichloroethylene, and methylene chloride, as well as petroleum hydrocarbons such as gasoline and fuel oils. However, it is significant to note that the system of air stripping is often used in conjunction with activated carbon adsorption.

Weston engineers ("Clean-up of the Gilson" 1984) claim to have eradicated the first Superfund site cleanup by a combination of containment and treatment. They installed a 3-ft wide soil-bentonite slurry wall around a 20-acre dumpsite, thereby significantly curtailing exfiltrated water from the site. Then they constructed a full-scale 300-ppm treatment plant (based on successful pilot-plant results), which was completed in April 1985. The treatment consisted of chemical precipitation of the inorganics followed by pressure filtration to remove any remaining metallic floc. The waste stream was then preheated to about 90°C and passed through a stripping column. Air was passed counter-currently through the column and the off-gases burned in a vapor incinerator. The column effluent was divided, the larger portion discharged within the slurry wall, and the smaller portion subjected to biological treatment by extended aeration and then groundwater recharge outside the wall. NH_4Cl and H_3PO_4 were added to enhance biological treatment. Weston estimates that it will take about 1¾ years at the 300-ppm rate to provide the two full flushes of the contaminated dumpsite necessary to remove 90% of the contaminants held in the soil water. Capital costs approximated $5 million, with annual operating costs of about $1.4 million.

GDS, Inc. has developed a patented system for soil and groundwater detoxification, which it claims is cost and performance effective (Jhaven and Mazzaca 1983). The system involves pumping contaminated groundwater into activating tanks where the microorganisms are enriched with compounds of phosphorous and ammonia, and sometimes iron, magnesium, and manganese salts. The treated groundwater is then settled and pumped in trenches for recirculation by reinjection into the original underground site. The company claims decontamination of both soil and groundwater by this process. It admits that complete site cleanup may take years but overall results in a savings of both time and money. A schematic view of the system is shown in Figure 11.13.

The EPA (1985) reported the fungus *Phanerochaete chrysosporium* was used to degrade dioxins, DDT, benzopyrene, and two kinds of polychlorinated biphenyls, or PCBs. The EPA predicts that contaminated soil could be inoculated or mixed with the fungus grown on sawdust or wood chips. It admits, however, that it may take a long time, perhaps years, to degrade large amounts of the chemicals but claims it is a better alternative.

ATW/Calweld, Inc. (1985) has introduced a "detoxifier" system that incorporates a mobile *in situ* treatment system with advanced chemical and biochemical technologies. The system is capable of chemically detoxifying or biochemically degrading aqueous toxic substances or contaminated soils, in place, to ensure a much more economical and safe cleanup solution than traditional remedial procedures. Mounted on a heavy-duty crane (Figure 11.14), these systems include an intrusion device that is powered down into the impounded waste or soil. Computer control and feedback determine the rate of descent and speed of the device, which mixes, turbulizes, pulverizes, and homogenizes the impounded contents while selecting, feeding, and integrating chemical reagents or biocatalysts and their nutrients with the conditioned contents. Oils and greases are biodegraded or transformed into nonsoluble metallic soaps. The solidified mass will not

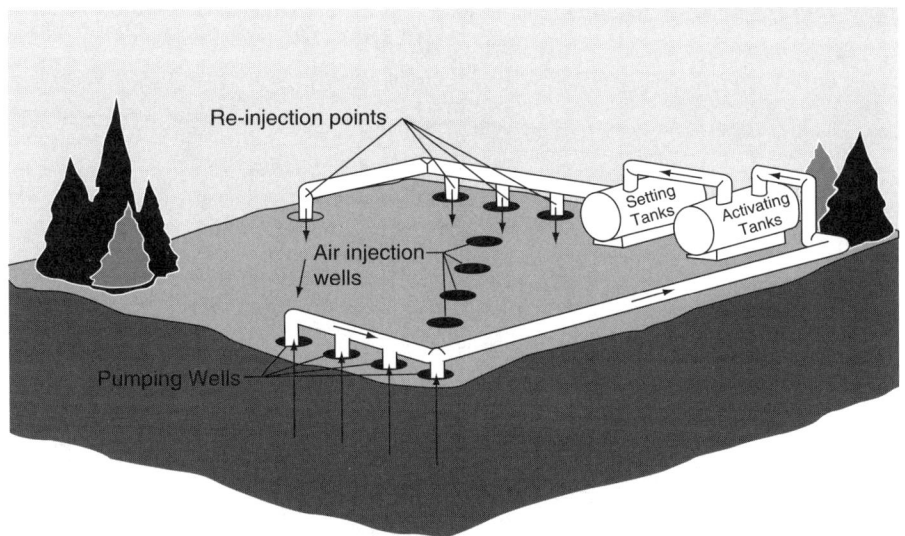

FIGURE 11.13. Soil and groundwater detoxification system.

reslurry according to the company. The company also states that any gases or vapors liberated during the process are collected and scrubbed.

NASA has shown that certain plants such as spider plants are effective in eliminating toxic gas concentrations of NO_2 and CO. Although this is perceived more as a treatment for indoor air, it could also be used under proper conditions for outdoor environments (Duke 1985).

Irvine and Busch (1979) reported the use of sequencing batch reactors (SBRs) as an excellent alternative to conventional activated-sludge biological treatment for wastewater. Irvine et al. (1984) showed that the SBR can provide substantial saving in energy and costs by removing organic compounds found in hazardous wastes biologically, rather than with activated carbon. The SBR is a periodically operated fill-and-draw reactor with five discrete periods in each cycle: fill, react, settle, draw, and idle. Herzbrun (1985) found that total organic carbon (TOC) degradation averaged 76% and phenol degradation averaged 99% during the first month of bench-scale operation. This type of treatment is intended to replace activated carbon adsorption treatment of leachates containing hazardous chemical compounds. Instead, activated carbon would be used only as a "polishing" device after SBR treatment.

Conversion of a hazardous-classified waste to a delisted one can sometimes diminish the risk to the generator of environmental suits. This may amount to considerable expenditures of both time and money to demonstrate to the EPA that treatment residuals do not have the characteristics for which they were originally listed as hazardous. In addition, building and paying for the amortization and operation expenses

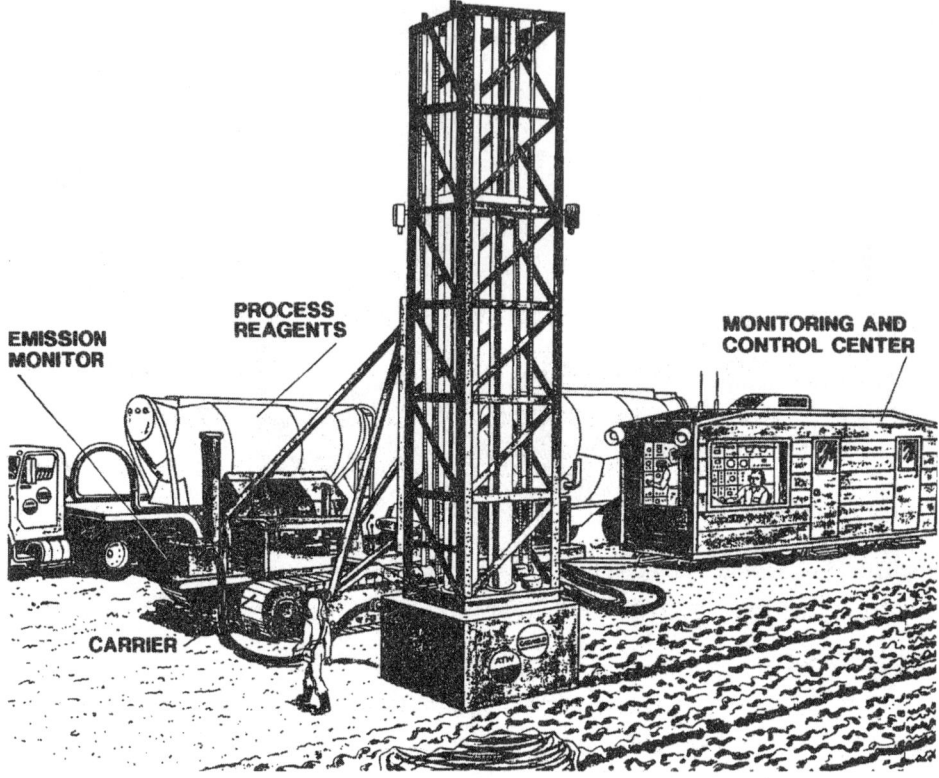

FIGURE 11.14. Detoxifier system. Used with permission from John Wiley and Sons.

of a treatment system can be costly. Also, sometime in the future, the generator's treated and delisted waste might—once again—be listed as hazardous by the EPA. In the long run, it may pay the industrial generator of a toxic waste to contract with an off-site disposal company for handling the waste. At least the industry will have a partner to share the liability of its environmental effects. Perhaps the off-site disposer can obtain delisting after treatment easier than the original industrial generator.

Regional Exchanges of Hazardous Wastes

The concept of regional exchanges or markets for transferring hazardous materials from one supplier to another user is a novel and intriguing potential solution to the dilemma facing society. A regional center enhances the exchanger rather than allowing the alternative disposal into the environment to occur. This process requires a great deal of communication, confidence, and consideration of all parties concerned. Advertisement of products, materials, and chemicals available for exchange must be extensive and discrete. Sometimes buyers and sellers must remain anonymous. Management of the exchange must be competent and trustworthy, somewhat like a stock market exchange, but without excessive product exposure to the general public. Certainly the objective of reuse without discharge is an admirable one. The procedure needs only refinement, practice, and more experience. There is a similarity in the objectives of these exchanges with those of zero-pollution attainment described in Chapter 15 of this book. The latter goal, however, is attained by all manufacturing of related products taking place in the same industrial complex.

LeRoy (1983) reports that these exchanges have been set up in France by local initiative. Their main purpose is to create and develop new outlets for using waste and saving raw materials.

The National Conference on Waste Exchange (1983) defines a waste exchange "as an operation that engages in transfer of either information concerning waste materials or the waste materials themselves." Waste exchanges were first organized in Europe where depletion of readily available natural resources and limited land-disposal areas forced manufacturers to find alternative sources of raw materials. All of the foreign exchanges are information clearinghouses and most of them are operated by trade organizations, primarily in the chemical industry. Material exchanges are the major innovation in the waste-exchange concept developed in the United States from the mid-1970s.

Surplus materials, equipment, scrap metals, and discontinued products, as well as traditional wastes, may enter the waste-exchange cycle. Industries are using the term "investment recovery" to describe the entire process of recycling and reuse within some plants. Some firms have already investigated or instituted process modifications designed to allow for or enhance the reuse potential of byproducts generated during normal manufacturing processes. Some modifications that have proven successful include (1) substitution of reclaimed acid for typical new electroplating acids, (2) source separation and segregation of various materials, (3) concentration and volume reduction, (4) altering raw materials specifications to allow substitution of lower quality inputs, (5) using intermediate reactions designed to modify waste stream components, (6) tightening process control to take advantage of byproduct (not waste) streams, and (7) educating plant management and workers on the benefits of resource reuse.

One deterrent to successful waste-exchange programs is that of exposing private information to persons outside a particular industrial plant. The program must maintain an agreement of "confidentiality" to protect the proprietary interests of generators and to limit the direct identification of specific firms that are generating a particular material. Industry identifies "potential liability" as the primary reason for nonparticipation in waste-transfer agreements.

Four requirements have been identified (National Conference on Waste Exchange 1983) for industrial resource reuse and waste transfer:

1. Participation in waste exchange must be uncomplicated and cost effective. The exchange itself must be reputable and reliable.
2. Alternatives to conventional treatment and disposal methods that are presented by a waste exchange must be ethical and cost effective.

3. The exchange should have as wide an audience of potential users as possible and should have extensive contacts in the waste management field to be aware of all waste management options.
4. The generator must know where and in what form the waste is being reused or disposed.

Many areas of cooperation between exchanges are possible including the following: (1) common database shared on regional or national level, (2) trading of listings between exchanges for catalogue distribution, (3) network of regional contacts for information referral, and (4) licensing of exchanges to ensure maintenance of quality of service. However, industrial managers hesitate to participate in cooperative exchanges because of the potential liability for mismanagement of waste. Unfavorable economics, especially when affected by high transportation costs for hazardous wastes, may be another deterrent to cooperative exchanges.

An Argonne National Laboratory (National Conference on Waste Exchange 1983) reports that the amount of energy saved by one waste exchange over 2½ years was 10×10^9 Btu. It calculated that a savings of 10^{12} Btu/yr (the equivalent of 100,000 barrels of oil at 10^7 Btu/barrel) would result if 50 exchanges as large or as effective as the one studied existed.

As with other liquid, gaseous, and solid wastes, there are generally no tax incentives, credits, or advantages for industries that recycle hazardous wastes. On the other hand, there are taxes levied for producing them. Exceptions are the states of New Jersey and California where tax advantages are given (or are being considered) for recycling waste.

One example of regional exchange of waste information is shown in Figures 11.15 and 11.16, which serves the southern region of the United States ("Southern Waste Information 1982). It publishes and distributes to interested firms listings of materials available, materials wanted, and services available and wanted. Examples of listing forms are included in Figures 11.15 and 11.16. Materials include acid and alkalis, inorganic chemicals, metals, organic chemicals, solvents, oils and paints, paper, wood plastic, rubber, glass, and miscellaneous. Services include recycling, equipment and supplies, consulting engineering, legal and health-related consulting, tank cleaning and lining, collection and transportation, storage, treatment and disposal, and miscellaneous.

Solar Evaporation

Some hazardous wastes, if sufficiently concentrated and of low volume, can be evaporated by natural sunlight to a very small volume for ultimate disposal or for permanent fixture at the same site. This method of treatment precludes ample sunlight, minimal rainfall and humidity, and adequate land for exposure to the sun's rays.

Sundaresan et al. (United Nations Environmental Protection 1983) report solar evaporation used to dispose of quinophos, a highly toxic pesticide. The semisolid sludge and the residue after solar evaporation are usually incinerated.

They also use solar evaporation ponds for concentrating urea plant spillages. The sludge after evaporation containing about 12% arsenic is stored in drums, sealed, and kept within the factory premises for final disposal (at an unknown destination).

Waste Management Services
Listing Form

A separate from is required for each each service listed.
Limit of two listings.

1. Company Name: _____
2. Mailing Address: _____
   _____
   _____

3. Company Contact: _____ 4. Title: _____
5. Signature: _____ 6. Date: _____
7. Phone Number: (____)_____ 8. SIC Code: _____
9. Fax Number: (____)_____

LISTING INFORMATION:

(1) CLASSIFICATION:
   (Review all first, then select the one that best describes your firm's service)
   ☐ Recycling
   ☐ Equipment and Supplies
   ☐ Environmental Consulting
   ☐ Legal and Health - Related Services
   ☐ Tank Management
   ☐ Collection and Transportation
   ☐ Storage, Treatment, Disposal
   ☐ Emergency Response/Clean-Up
   ☐ Laboratory Analysis
   ☐ Waste Minimization
   ☐ Miscellaneous

(2) DESCRIPTION OF SERVICE:
   (In 25 words or less, please describe your firm's service or product, keeping in mind what the reader of your listing may want to know. You may use the space provided below or type the listing on a separate piece of paper and attach it to this form.)

(3) LOCATION SERVED:
   (Give general area where service is available, e.g., Central Georgia, Southeastern U.S., etc)

Send completed form to: SWIX Clearinghouse
                               Post Office Box 960
                               Tallahassee, FL 32302

FIGURE 11.15. Waste Management Services listing form.

Landfilling with Leachate Treatment

The collection and subsequent treatment of leachates from hazardous waste landfills have become an acceptable alternative for disposing of these wastes. Collection systems must be designed to recover all the leachate, and treatment systems must be designed and operated to remove all the hazardous constituents from the leachates. Leachate collection systems require the use of subterranean impermeable barriers and drainage channels to central sump or reservoir sites. Treatments are concentrated mainly on removal of dissolved metals, minerals, and organic matter, as described in earlier sections of this chapter.

Material Available/Wanted Listing Form

A separate form is required for each item listed.
Limit of two listiage.

1. Company Name: _____ SIC Code #: _____
2. Mailing Address: _____
   _____
3. Company Contact: _____ 4. Title: _____
5. Signature: _____ 6. Date: _____ 7. Phone: ( ) _____
8. Fax Number: ( ) _____
9. Check One Only: ☐ MATERIAL AVAILABLE   ☐ MATERIAL WANTED

   Classifications (review all first; then select one that best describes your material):

   ☐ ACIDS   ☐ OTHER ORGANIC CHEMICALS   ☐ WOOD AND PAPER
   ☐ ALKALIS   ☐ OILS AND WAXES   ☐ METALS AND
   ☐ OTHER INORGANIC CHEMICALS   ☐ PLASTICS AND RUBBER   METAL SLUDGES
   ☐ SOLVENTS   ☐ TEXTILES AND LEATHER   ☐ MISCELLANEOUS

10. Material to be listed (Main usable constituent, generic name): _____
11. The industrial process that generates this waste: _____
12. Main constituent and percentage: _____
13. Other constituents (including contaminants): _____
14. Percent by (check one):   ☐ Volume   ☐ Wet Weight   ☐ Dry Weight
15. Physical State: ☐ Solution   ☐ Slurry   ☐ Sludge   ☐ Cake
                      ☐ Aggregate   ☐ Solid   ☐ Dust   ☐ Gas
16. Miscellaneous information (e.g. pH, toxicity, reactivity, color, particle size, flash point, total solids, purchase date, manufacturer): _____
    _____
17. Potential or intended use: _____
18. Packagine: ☐ Bulk   ☐ Drums   ☐ Pallets   ☐ Bales   ☐ Other: _____
19. Present Amount: _____   20. Frequency: ☐ Continuous ☐ Variable ☐ One Time
21. Quantity thereafter: _____   ☐ Pounds   ☐ Tons   ☐ Cubic Yards   ☐ Gallons
                        ☐ Kilograms ☐ Cubic Meters ☐ Liters   ☐ Other _____
    per: ☐ Day   ☐ Week   ☐ Month   ☐ Quarter   ☐ Year
22. Restrictions on amounts: ☐ None   ☐ Minimum   ☐ Maximum _____
23. Available to Interested parties: ☐ Sample   ☐ Lab Analysis   ☐ Independent Analysis
24. For material wanted, acceptable geographic area (i.e. states, regions, countries):
    _____
25. If necessary to speed communications, please check if your company's name, address and telephone number may by released.   ☐ YES   ☐ NO

        Send completed form to: SWIX Clearinghouse
                                          Post Office Box 960, Tallahassee, FL 32302

FIGURE 11.16. Material Available/wanted listing form.

Goldstein (1984) found that metal-plating leachates from landfills can sometimes be pumped to publicly operated treatment works (POTPs).

DuPont reports the use of powdered activated carbon for the treatment of leachates from toxic waste dumps to remove these contaminants (DuPont 1984).

Disposal into Publicly Owned Wastewater Treatment Plants

Hazardous wastes may be discharged to municipal sewer systems if and when they are properly pretreated to remove and/or detoxify the hazardous components. Because of the cost savings of sending large volumes of wastewater to sewage systems, this method of treatment is especially interesting to industries. However, the hazardous component must first be removed or sufficiently diluted to be acceptable to the receiving municipality.

In keeping with this, Coughlin et al. (United Nations Environmental Protection 1983; Toxic and Hazardous Wastes 1983) used a treatment technology for metallic wastes consisting of physicochemical processes that effectively change the hazardous soluble elements into recoverable nonsoluble solids. The solids are further concentrated by a low-pressure belt filter press that recovers 95% of the solids. The remaining solids are concentrated in an electroflotation system that results in a final treated effluent of 20 ppm of total suspended solids. The solid cakelike material is landfilled with no apparent hazard because of its relatively high pH. The water phase is then discharged to a publicly owned treatment plant. This system represents a typical one for pretreatment of hazardous wastes before admission to POTPs.

Membrane Technology

The use of semipermeable membranes as barriers for the removal of hazardous compounds or elements from reaching the external environment is gaining in popularity and acceptance. Membrane openings or pores are so minute that considerable pressure must be applied to the waste to drive the fluid through them. In normal filtration, microfiltration, the pore sizes are relatively large and the waste solution to be filtered approaches the polymeric membrane perpendicularly (Figure 11.17).

In another more common technique, known as "reverse osmosis" or sometimes "hyperfiltration," the pore sizes are many times smaller and require more liquid pressure to force the smaller molecules through the pores of the membrane. The solution is passed over the surface of a specific semipermeable membrane at a pressure in excess of the effective osmotic pressure of the feed solution.

In selecting the proper membrane to separate hazardous contaminants from wastewater, several environmental parameters must be considered:

1. *Pressure to which the membrane will be subjected:* This varies from as low as 3–50 psig with microfiltration to as high as 200–1,000 psig with reverse osmosis. At the higher pressures, certain membranes may deteriorate or deform because of the mechanical strain put on them.

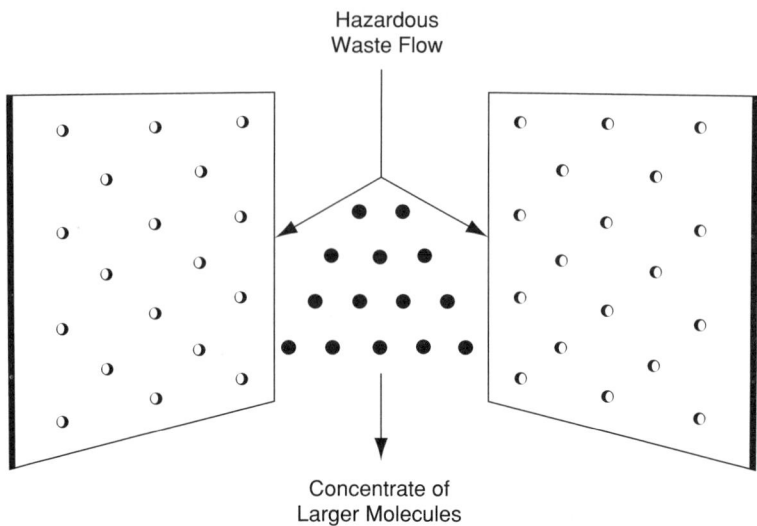

FIGURE 11.17. Separation of hazardous wastes using membrane filtration.

2. *Temperature to which the polymeric membrane will be subjected:* Most membranes will operate with varying efficiencies and lengths of life at temperatures between 0 and 85°C.
3. *pH range of operation of the membrane will affect its life:* The value will be influenced by the hazardous waste solution pH.
4. Chemical compatibility of each membrane to the specific chemical makeup of the waste solution is important in determining not only the membrane choice, but also its durability under continued operation.

Membrane separation has proven useful for treating and recovering metal salts from plating bath wastes, recycling lubricating oils and rinse waters from can-forming and other metal-working applications, concentrating wastes and conserving process waters in certain chemical processing plants, and in the future recovering pollutants from the large water users such as pulp and paper and textile industries. These applications will be enhanced by scarcity and increased prices of freshwater.

Electrochemical Treatment

One of the objections of chemical coagulation as a hazardous waste treatment device has been the difficulty in adjusting the exact dosages of coagulant required and pH needed to remove the metals. This results in higher operational costs for both chemicals and labor.

Electrochemical treatment eliminates or tends to overcome the objections of chemical coagulation. A direct current is conducted through a cell containing carbon steel electrodes. The current creates ferrous ions in the waste solution. The ferrous ion acts as a reducing agent and, in the case of the heavy metal chromium, results in the precipitation as the hydroxide. Other heavy metals such as Cu, Ni, Zn, Pb, and Sn are coprecipitated with the ferric hydroxide. It is claimed that no sulfates are formed, as in normal chemical coagulation, and that the effluent is acceptable for reuse in plating operations (Duffy 1983). When using this bipolar cell (carbon steel), one side is the anode and the other is the cathode. The ferrous ion produced at the anode and the hydroxide formed at the cathode react in the wastewater to form insoluble $Fe(OH)_3$. Although no chemicals are added, the carbon steel electrodes are consumed, usually 1–2 lb of iron per pound of heavy metal in the wastewater that is precipitated. Although the manufacturers admit that slightly more sludge is formed with conventional chemical coagulation, in the actual practice somewhat less sludge results. Presumably this is because the exact stoichiometric amount of iron required actually goes into solution in electrochemical treatment. The exact dosage in coagulation, on the other hand, is difficult to control.

A typical cell operating at about 25 amps of direct current (DC) requires about 5 kW of electricity per pound of heavy metal removed. Duffy (1983) reports the operational costs (which presumably include both iron metal and electricity) would be about $1.00/lb of heavy metal removed. He also claims that chelating compounds found in many plating wastewaters do not interfere with the treatment but are broken up by the iron. Although the effluent from electrochemical treatment may be nonhazardous and can be discharged safely to a receiving stream or sewage system, the sludge still remains to be disposed of (usually in landfills) and contains the toxic metals removed as hydroxides. A reduction of the pH of this sludge into the acid range will tend to liberate the metals into solution once again.

New Treatments

Since this field is so new and dynamic, many new treatment processes, in addition to those I have described, are being tested or researched. Randle (1989) summarizes many of these (Table 11.13).

Legal Aspects of Industrial Hazardous Wastes

The U.S. EPA (1998) has presented a tabular history of the U.S. laws regulating pollutant, hazardous, and toxic substance use and control of wastes. Of major concern are the Safe Drinking Water Act (1974, 1977), the Resource Conservation and Recovery Act (1976), the Toxic Substances Control Act (1976), and the Superfund Act and Reauthorization (1980, 1986).

Drinking-water standards were presented by this author in more specific terms and, in addition, Chapter 2 contains the 1996 standards. The list of hazardous and toxic substances continues to expand, while analytical capability, coupled with more refined risk analyses, has pressured the EPA to keep lowering the allowable levels of contaminant discharge. I believe that this trend will continue and, therefore, stress maximum utilization of waste products as the most economical approach to "zero pollution."

TABLE 11.13
Some New Treatments for Hazardous Waste, Many Still on the Drawing Board

Company	On Market Use/Method	Costs
BioTrol, Inc., Chaska, MN A. Dale Pflug (612) 448 2515	Organics Bacteria attack contaminant molecules introduced in water, working on soil cleanup	Processing and capital; groundwater (100 ppm contamination), $5–$2 cents/gal; soil, $50–$75 to $100–$125/ton
Groundwater Technology, Inc., Norwood, MA John Higley (800) 635 0053	Organics Small-diameter filter scavenger, part of Filter Scavenger series, removes floating hydrocarbons	Processing: electricity. Equipment: $10,000–$12,000/unit
Micro-Bac International, Inc., Austin, TX	Organics assimilate heavy metals Bacteria metabolize toxic materials	Cost NA
Bob Billingsley (512) 837 1145 Odgen Environmental Services, San Diego	Organics Mobile circulating-bed combustion incinerator	Processing and service by Ogden: $100–$200/ton of soil
Harold Diot (800) 876 4336 Permutit Co., Paramus, NJ Luis Rodguez (201) 967 6000	Heavy metals Sulfide precipitation process removes heavy metals from waste stream	Processing: for existing system removing metals as hydroxides, at 100 gpm, $10 to $30/day Equipment: small batch unit, $50,000; large, over $50 million
Resource Conservation Co., Bellevue, WA Douglas Austin (206) 828 2400	Organics, heavy metals Solvent extraction of contaminants from soil and sediment	Processing and capital: $50–$150/yd
Ultrox International, Santa Ana, CA Jerome Berich (714) 545 5557	Organics Waste or groundwater exposed to ultraviolet light plus ozone or hydrogen peroxide	Processing groundwater, low 15–20 cents/1,000 gal, high 5 cents/gal. Equipment: $60,000–$200,000 for 10 ppm TCE at 210 gpm

TABLE 11.13 *(continued)*

Company	On Market Use/Method	Costs
Westinghouse Environmental Services, Madison, PA C. Keith Paulson (412) 722 5447	Organics, heavy metals Plasma, fired cupola remelts and recovers metals Organics Pyroplasma destroys heat toxics in liquids Organics Infrared conveyor belt drives heat from soil	Service: $200–$300/ton Service: <$1/lb Service: $200–$300/ton
Zimpro/Passavant Rothschild, WI Bob Nicholson (715) 359 7211	Organics, some metals absorbed by carbon Biophysical system has powdered activated carbonic assist bacteria Wet air oxidation burns high-strength wastes	Processing: $1/1,000 gal Equipment: smallest unit $85,000; largest, $1 milion Processing: 6 cents/gal Equipment: 10 gpm skid mounted unit, $2 million
CF Systems Corp., Waltham, MA Tom Cody (617) 890 1200	Organics Liquified gas extraction uses gases as solvents to recover organic from refinery wastes including water, clay and oil	Processing, operating and equipment: sludge, $75/yd; water at 20 gpm, 20 cents/gal
Freeze Technologies, Raleigh, NC Ken Hunt (919) 850 0600	Liquid–liquid separation process based on freezing	Processing: NA Equipment: $1–2 million
EPA, Raritan Depot Edison, NJ Francine Everson (201) 548 8554	Organics Mobile carbon regenerator for water treatment systems Organics, heavy metals Mobile soils washer separates fine soils and contaminants from sands and gravels Organics, heavy metals	Costs NA Costs NA Costs NA

TABLE 11.13 (continued)

Company	On Market Use/Method	Costs
EPA, Cincinnati Alfred Kornel (513) 569 7421	Mobile *in situ* contaminant/treatment unit injects grout to contain spills; contaminants treated in place or removed for other treatment via flushing Organics Potassium-alkaline polyethylene glycolate reagent breaks halogen bond to chemically react with pollutants	Processing: pilot unit (1.75 tons/batch), $400–$800/ton. Equipment: NA
IIT Research Institute, Chicago Gug Stresty (312) 567 4232	Organics Radiofrequency heating, *in situ*, brings contaminants and water vapor to surface, where collected for treatment	Processing: $30–$60/ton. Equipment: 50–80 ton/day unit, $500,000
Solar Energy Research Institute, Golden, CO John Thornton (303) 231 1269	Organics Contaminated water pumped from ground; solar energy system catalyst forms free radicals to attack chemical bonds of pollutants	Costs NA
Vertech Treatment Systems, Denver Nathan Chesley (303) 452 8800	Organics Aqueous-phase oxidation of contaminated solution	Processing: 10 gpm at 50,000 mg/liter chemical oxygen demand, 30 cents/gal; 50 gpm at same COD, 13 cents/gal Equipment: 10 gpm unit, $3–$5.5 million
Westinghouse Environmental Services, Madison, PA C. Keith Paulson (412) 722 5547	Organics, heavy metals verified Electric pyrolizer burns solids and sludges at extremely high temperatures	Service: $200–$300/ton

Source: From Randle (1969).

Superfund

In 1980, the U.S. Congress enacted legislation creating the Superfund to provide for cleanup of abandoned dumpsites and dangerous spills of toxic materials and to facilitate compensation of victims. The law imposes a tax on chemical manufacturing firms, the revenues from which are placed in a fund ($1.6 billion) to be used solely for these cleanups. Congress intended to force producers and disposers of these hazardous wastes to accept responsibility for their own wastes and for the fund to pay for any residual costs involved.

According to Grossman (United Nations Environmental Protection 1983, 138), the Superfund legislation required the EPA to develop by June 1981 a National contingency plan to guide the search for and cleanup of dangerous sites and to prepare to respond to emergencies such as spills and explosions. A plan was finally proposed in March 1982. "The proposal is so vague as to provide no guarantee that Superfund resources and authority will be used to clean up any site. The plan implies that EPA cares more about saving money than cleaning up sites to protect human health."

Grossman reports that in its first use of Superfund authority, after a toxic dumpsite in Santa Fe Springs, $x^{1974,1977,}$ caught fire in July 1981, "top EPA officials quickly negotiated a private settlement with one of the responsible parties . . . The settlement limited the company's cleanup responsibility instead of requiring the cleanup to continue until the hazard was removed. It also committed EPA to testify on behalf of the company in any subsequent lawsuit against it arising from the dump and the fire."

It is alleged by EPA officials that the federal government will pay for 90% of the cleanup costs, while the individual states will pick up the remaining 10%. However, when the property is government owned, the federal share will be limited to 50%.

According to Skinner (United Nations Environmental Protection 1983, 69), the EPA published the list of national priorities sites, identifying 418 sites on December 20, 1982. Provisions were made to clean up these sites by any of the following options:

1. Remove drums from the site
2. Install a clay "cap" over the site
3. Construct ditches and dikes to control surface water
4. Construct drains, liners, and grout curtains to control groundwater
5. Provide an alternate water supply
6. Relocate residents on a temporary or permanent basis

An important part of the Superfund program (according to Skinner) is to encourage voluntary cleanup by private industries and individuals when they are responsible for their releases. Skinner (1983) reported that more than $121 million had been received from industry for cleanup. The EPA also encourages state government involvement in the Superfund program. States and the EPA may enter into a Cooperative Agreement in which a state takes a leading role in a remedial action. Federal money is transferred to the state and the state develops a work plan, schedule, and budget for the cleanup action. The state then contracts for any services it needs, and the EPA is responsible for monitoring the state's progress throughout the project.

In addition to the 10% state portion of the cleanup, the state must agree to maintain the site after remedial measures have been taken. Since CERCLA was passed in 1980 and up to September 1982, Superfund had agreed to allocations of $221 million (Skinner 1983).

Krog (1985), however, pleads for a revision of the Superfund Law as being lacking in both the original law, and the Superfund II being considered in September of 1985. She maintains that "toxic waste dumping doesn't generate money for anybody except attorneys." She points out that when an industry goes to court over a toxic waste dumpsite, the EPA must do likewise, which takes public funds out of the Superfund budget and transfers them to bank accounts for attorneys. And, according to her, this is not what the Superfund was intended for, but rather to clean up toxic dumpsites.

Former Congressman Robert Torricelli (1985), a member of the House Science and Technology Committee, points out the current problems and fallacies of the Superfund program. He laments that our "20,000 toxic waste sites are not being cleaned up. Hazardous dumps," he continues, "have merely been moved from one location to another, often creating new toxic sites in the process." He maintains that "the Superfund program promotes the storage of contaminated materials, not their permanent destruction." Although Torricelli does not say it, presumably this occurs because storage is still easier, less expensive, and less complicated than destruction of hazardous wastes. He contends that "despite the great intentions and accomplishments of the program, the result has been a lethal shell game that threatens to indefinitely prolong America's toxic legacy."

Some states have not been pleased by the progress made under Superfund. The New Jersey Department of Environmental Protection attempts to block the sale of portions of property containing hazardous wastes until an entire site is cleaned up.

Former Governor Bob Graham of Florida signed into state law, July 1, 1983 (*Miami Herald,* July 2, 1983), a comprehensive environmental bill. The bill included a "provision to impose a 2 cents/barrel tax on pollutants once a state hazardous waste cleanup fund drops to $3 million." Also included in the bill was "a provision to allow homeowners and businesses with small amounts of hazardous wastes to bring it to a state collection site for free disposal." The main part of the new law included a one-time transfer of $11 million plus interest from an existing trust fund to help clean up hazardous waste sites in Florida (estimated to number about 200).

As an example of government's intention to use new laws to prevent disasters from hazardous wastes, the premier of the Province of Ontario, Canada (statement by the Honorable David R. Peterson, July 2, 1985), said that "this government will ensure that environmental hazards do not eat away at the legacy we wish to leave our children. No longer will there be any question of who is responsible for preventing spills or cleaning them up. No longer will innocent victims be left without a route to compensation."

The U.S. Congress enacted the RCRA in 1976, which charged the EPA with implementation. Among its statutes was that of operating a hazardous waste treatment, storage, or disposal facility. An operator must complete Part A and Part B of this permit. The former is merely an application giving a bare minimum of vital information. Part B permit involves extensive study, survey, completion of forms, meetings, and approvals before a 10-year permit can be issued.

Thorsen and Petura (1983) give a schematic presentation of the steps needed to obtain a Part B permit (Figure 11.18). The entire review of the Part B permit application is expected to take up to 6 months and, when approved, contains very specific terms and conditions that the operator must follow.

In general, two types of applicant operator—each requiring a different approach and consideration for permit—are involved: (1) industrial and (2) regional. The former is an individual plant primarily in business to make a profit and produces a hazardous waste in the process. The latter is a collector, storer, and/or treater of hazardous wastes from a number of individual commercial plants. It, too, expects to make a profit, but as a direct result of the safe handling of the toxic waste. Thorsen and Petura (1983) visualize the decision of either type of applicant as a commitment of substantial time and money. They specify a master plan (not necessarily part of the application), which should describe the following:

1. Types and approximate quantities of wastes to be handled at the site
2. Activities and functions of the facility
3. The business-related aspects (for internal use)
4. Site-development sequence
5. Strategy and resources required for implementation, including schedule, budget, and permit application sequence required to secure all necessary regulatory approvals and permits

Because of the collective extensive hazardous waste management regulations, Thorsen and Petura (1983) recommend that "someone associated with the facility, with a substantial degree of regulatory familiarity, be actively involved with the team throughout the development of the permit application." They also point out that "the concerns and regulations of both federal and state agencies must be addressed."

In 1984, there were 10 regional offices of the EPA to handle these permit applications. Here, they are shown geographically for easy reference (Figure 11.19).

Whitescaver and Clarke (United Nations Environmental Protection 1983) present toxic concentration limits for eight toxic organics and illustrate the wide range of concentrations for the same toxic pollutant. This also shows that effluent guidelines were seldom used but that 14 other techniques were employed, including negotiation and "best practical judgment" (BPJ). Another process permit flowsheet is found in Figure 11.20.

On June 17, 1985, the Canadian Environmental Minister the new Regulation 309 under the Environmental Protection Act. This new regulation is aimed at controlling the process of handling industrial wastes from the generator to the receiver. It will clearly establish the responsibilities of the industries that create wastes, the haulers who carry them, and the site operators who treat and dispose of them.

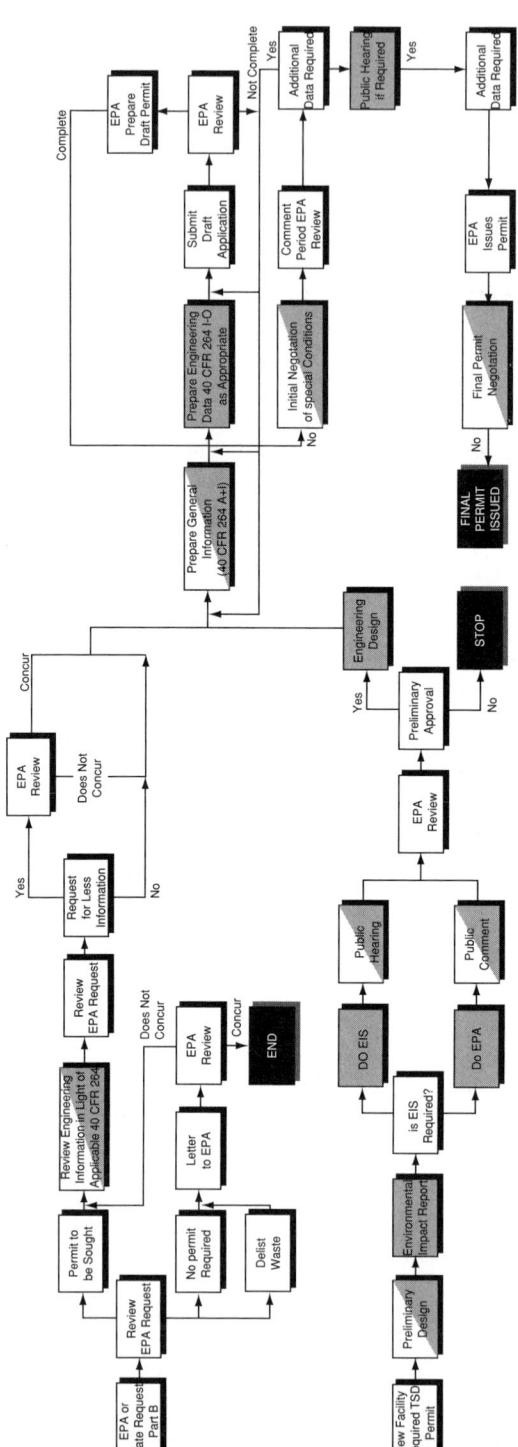

FIGURE 11.18. RCRA TSD Part B permit application.

Toxic limits for air and water are given in Tables 11.14 and 11.15. The reader can use these values to become familiar with the hazardous chemicals and to develop an idea of the concentrations likely to result in serious illnesses and death.

Even the Occupational Safety and Health Administration (OSHA) has become concerned and has pressured state labor departments to enact and enforce laws protecting plant workers from exposures to hazardous chemicals. In Florida, for example, as of October 1, 1985, employers who handle any one of 1,400 chemicals listed by the state must do the following:

1. Post notices informing employees of their rights to know about hazardous chemicals (and give a toll-free telephone number to call for information)
2. Keep for 30 years material safety sheets for each toxic chemical on the list, giving their health hazards and pertinent facts about them
3. Provide these information sheets to employees within 5 days of a request
4. Train employees in the handling, hazards, and emergency treatment for listed chemicals
5. Notify the local fire department of chemicals used and stored

A Water Pollution Control Federation workshop concerned with biomonitoring of toxic chemicals was summarized by Cairns (1985). He states that multispecies (rather than single test organism) are attractive in biomonitoring for toxics for the following reasons: they can give simultaneous data on toxicity and chemical fate; they are likely to use indigenous organisms; they can be cost effective; and they can be more representative of an ecosystem than single species tests. Only organisms can integrate all the factors involved in the toxicity of an effluent, and as Cairns reminds us, "one of the primary reasons for desiring clean water is to protect the aquatic ecosystems." Often we are more concerned with the quality of the water for external or "out-of-stream" uses.

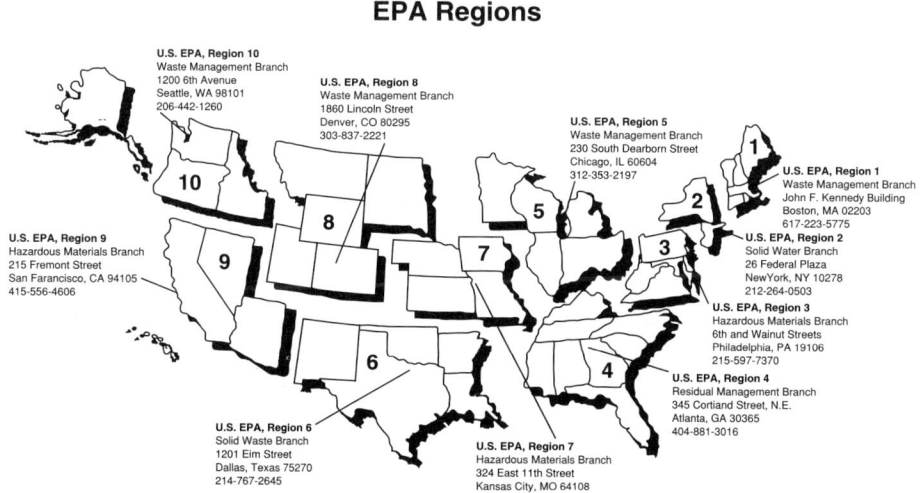

FIGURE 11.19. U.S. EPA office locations.

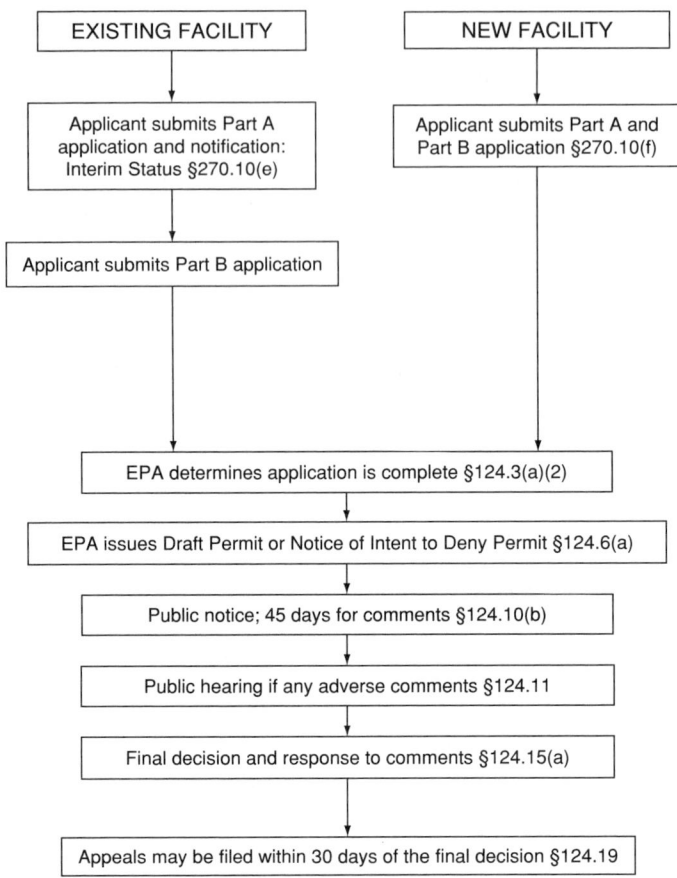

FIGURE 11.20. EPA process permit flowsheet.

On November 8, 1984, the Solid and Hazardous Waste Amendments were added to the RCRA as a reauthorization act (Weston Way 1985). This reauthorization mandates the EPA to selectively and aggressively prohibit the land disposal of specific hazardous wastes. It requires the EPA to require double liners and leachate collection systems for all landfill surface impoundments. In order to receive an RCRA permit, an owner of a landfill or impoundment will now be required to clean up or correct leaks of hazardous waste from the facility regardless of when the waste was managed. In 1986 the small-generator exclusion was lowered from 1 ton/mo to 220 lb/mo. More generators will be required to use RCRA disposal at costs of 15–30 times those of typical municipal landfill. When blending wastes with fuel oil and burning the mixture, the reauthorization requires the EPA to tighten up its rules and dates for complying, to abate abuses of the practice. Because of the need to use relatively large industrial boilers for blending and burning, some industries may decide to transport their wastes to central facilities with larger boilers. Governing and controlling by the EPA of underground storage tanks will now include feedstock products and fuels. The EPA is now required to develop new standards for existing and new underground storage tanks in 1987. The amendments state that one intention is to minimize the generation of hazardous wastes and the land disposal of hazardous waste by encouraging process substitution, materials recover, and properly conducted surveys.

Site Assessment
The Superfund law, CERCLA, establishes three classes of persons responsible for cleanup costs: (1) all owners and operators of facilities where hazardous wastes are located, (2) anyone who contracted for the disposal of hazardous waste (i.e., the generator), and (3) anyone who transported the hazardous waste. Note that waste may be present because of planned disposal, an accidental spill or release, or dumping by an unknown party (midnight dumping).

Superfund and its numerous state parallel statutes have created a wide net of liability that affects both businesses and individuals, even those who may never have had any involvement with the hazardous materials. These responsible parties may be the owner at the time of original disposal, the current property owner, intermediate property owners, successor corporations, or financial institutions that have foreclosed. Owners may be individual, corporations, or agencies of all levels of government. In the last case, many municipalities and counties today find themselves with major cleanup burdens caused by operation of public landfills. When the web of liability involves several responsible parties, there is always a scramble to decide who pays, and the losers are usually the parties with the greatest financial assets (i.e., the "deep pockets").

Superfund provides one escape clause from this open-ended liability: would-be property owners must attempt, before receiving title to the property, to identify the likelihood that hazardous waste may be present. This is the innocent landowner defense, to conduct "all appropriate inquiry into the previous ownership and uses of the property consistent with good commercial or customary practice." An environmental site assessment (ESA) is an investigation conducted for this purpose.

There are many sources of information that can be used for this purpose. The scope and level of detail can vary. Should a gas station receive the same level of scrutiny as a chemical plant? To bring standards to the preparation of ESAs, the American Society for Testing and Materials (ASTM), working with financial institutions, the real estate industry, and environmental consultants developed and, in 1993, issued "Standard Practice for Environmental Site Assessments: Phase 1 Environmental Site Assessment Process." The ASTM standard has been accepted as the minimum investigation to qualify for the innocent landowner defense.

The ASTM standard consists of four parts: records, review, site reconnaissance, interviews, and a report. Each part has detailed internal standards. The records review covers the widest range of information. It is necessary to review the records of environmental regulators regarding the presence of hazardous waste in soil or ground-

TABLE 11.14
Water Quality Limits for Toxic Pollutants in Three Uses

	Concentration of Toxic Material Considered Limit (ug/liter)				
	Freshwater Aquatic Life		Saltwater Aquatic Life		
Toxic Chemical	Acute	Chronic	Acute	Chronic	Human Health
1. Acenaphthene	1,700	—	970	710	20 (est.)
2. Acrolein	68	21	55	—	320
3. Acrylonitrile	7,550	—	Not available		0.58–0.006 lifetime
4. Aldrin-dieldrin	0.0019 (24-hr avg)	—	0.0019 (24-hr avg)		0.0071 ng/L–0.71 ng/L
	2.5 maximum conc	—	0.71 (max conc)		
4A. Aldrin	3.0	—	1.3	—	0.0074 ng/L– 0.74 ng/L
5. Antimony	9,000	1,600	Not available	—	146
6. Arsenic	440	40	508	—	0.22 ng/L–22 ng/L
7. Asbestos	Not available	—	Not available	—	3,000–300,000 fibers/L
8. Benzene	5,300	—	5,100	700	0.066–6.6
9. Benzidine	2,500	—	Not available	—	0.01 ng/L–1.2 ng/L
10. Beryllium	130	5.3	Not available	—	0.37 ng/L–37 ng/L
11. Cadmium	3.0 (100 ppm hardness) max		59 (maximum)	—	10
	0.025 (100 ppm hardness) avg		4.5 (avg)	—	
12. Carbon tetrachloride	35,200	—	500,000	—	0.04–4.0
13. Chlorodane	2.4 max	—	0.09 max	—	0.046–4.6
	0.0043 (24-hr avg)	—	0.0040 (24-hr avg)	—	
14. Chlorinated benzenes	250	50 (fish 7.5 days)	160	129	Hexachlorobenzene 0.072 ng/L–7.2 ng/L
					Tetrachlorobenzene 38 ug/L–48
					Pentachlorobenzene 74–85
					Monochlorobenzene 488

TABLE 11.14 (continued)

15. Chlorinated ethanes	118,000	20,000	113,000	—	0.094-9.4 (1,2 dichloro ethane)
	18,000	9,400	31,200	—	18.4 mg/L–1.03 g/L (two trichloroethane)
	9,320	2,400	9,020	261	017–1.7 ug/L (two tetra-chloroethanes)
	7,240	1,100	390	—	(pentachloroethane)
	980	540	940	7.5	0.19–19 ug/L (hexa-chloroethane)
16. Chlorinated naphthalenes	1,600	—	—	—	Not available
17. Chlorinated phenols	30–500,000	970	440–29,000	—	0.1 (3 monochlorophenol)
					0.1 (4 monochlorophenol)
					0.04 (2,3 dichlorophenol)
					0.5 (2,5 dichlorophenol)
					0.2 (2,6 dichlorophenol)
					0.3 (3,4 dichlorophenol)
					1.0 (2,3,4,6 tetrachloroph
					2.6 mg/L (2,4,5 trichlorop
					0.12–12 ug/L (2,4,5 trichlo
					1,800 ug/L (2 methyl 4 cl
					3,000 (3 methyl, 4 chlor
					20 (3 methyl, 6 chloroph
18. Chloroalkyl ethers	238,000	—	Not available	—	0.00038 ng/L to 0.038 ng/L (bischloromethyl ether)
					0.003 ug/L–0.3 ug/L (for bischloroethyl ether)
					34.7 ug/L (for bis-2-chlorophenol ether)

TABLE 11.14 (continued)

Concentration of Toxic Material Considered Limit (ug/liter)

	Freshwater Aquatic Life		Saltwater Aquatic Life		
Toxic Chemical	Acute	Chronic	Acute	Chronic	Human Health
19. Chloroform	28,900	1,240	Not available	—	0.019–1.9 ug/L
20. Chlorophenol	4,380	2,000 (one fish species)	Not available	—	0.01 mg/L
21. Chromium	21 (max) Hexavalent 0.29 (avg 24 hr) 4,700 ug/L (100 ppm trivalent hardness) 44 ug/L chronic toxicity		1,260 (max) Cr^{vi} 18 (24-hr avg) 10,300 (Cr^{+3}) chronic toxicity)	—	170 mg/L Cr^{iii} 50 ug/L (Cr^{vi})
22. Copper	5.6 (24-hr avg) 22 ug/L (100 ppm hardness) max	—	4.0 (24-hr avg) 23 (max)	—	1 mg/L (for taste and odor) no other available
23. Cyanide	32.5 (24-hr avg) 52 (max)	—	2–30 mg/L	—	200 ug/L
24. DDT and metabolites	1.1 max 0.001 (24-hr avg) 0.6 acute toxicity TDE 1,050 acute toxicity DDE	—	0.13 (max DDT) 0.001 (24-hr avg) 3.6 acute toxicity TDE 14 acute toxicity DDE	—	0.0024 mg/L to 0.24 ng/L for DDT
25. Dichlorobenzenes	1,120	763	1970	—	400
26. Dichlorobenzidines	Not available	—	Not available	—	0.00103–0.103
27. Dichloro ethylenes	11,600	—	224,000	—	0.0033–0.33 0.3 ng/L (for taste and odor)
28. 2,4 Dichlorophenol	2,020	365	Not available	—	3.09 mg/L (toxicity)

330

TABLE 11.14 (continued)

29. Dichloropropanes Dichloropropenes	23,000	5,700	10,300	3,040	87 ug/L
30. 2,4 Dimethyl phenol	2,210	—	Not available	—	400 ug/L (taste and odor)
31. 2,4 Dinitrotoluene	330	220	590	370	0.001–1.1 ug/L
32. 1,2-Diphenyl-hydrazine	270	—	not available	—	4-422 ng/L
33. Endosulfan	0.056 (24-hr avg) 0.22 (max)	—	0.0087 (24-hr avg) 0.034 (max)	—	74 ug/L
34. Endrin	0.0023 (24-hr avg) 0.18 (max)	—	0.0023 (24-hr avg) 0.037 (max)	—	1 ug/L
35. Ethylbenzene	32,000	—	430	—	1.4 mg/L
36. Fluoranthene	3,980	—	40	18	42 ug/L
37. Haloethers	360	122	Not available	—	Not available
38. Halomethanes	11,000	—	12,000	6,400	0.019–1.9 ug/L
39. Haptachlor	0.0038 (24-hr avg) 0.52 (max)	—	0.0036 (24-hr avg) 0.053 (max)	—	0.028–2.78 ng/L
40. Hexachloro butadiene	90	9.3	32	—	0.45–4.47 ug/L
41. Hexachlorocyclohexane (lindane) BHC	0.080 (24-hr avg) 2.0 (max)	—	0.16	—	0.92–9.2 ng/L
42. Hexachlorocyclopentadiene	100 7.0	— 5.2	0.34 7.0	—	1.63–163 ng/L 206 ug/L
43. Isophorone	117,000	—	12,900	—	1.0 µg/L (taste and odor)
44. Lead	3.8 ug/L (100 ppm hardness and 24-hr avg)				5.2 mg/L
45. Mercury	0.00057 ug/L (24-hr avg) 0.0017 ug/L (max)		0.025 (24-hr avg)	—	144 ng/L
46. Naphthalene	2,300	620	3.7 (max) 2,350	—	Not available

TABLE 11.14 (continued)

	Concentration of Toxic Material Considered Limit (ug/liter)				
	Freshwater Aquatic Life		Saltwater Aquatic Life		
Toxic Chemical	Acute	Chronic	Acute	Chronic	Human Health
47. Nickel	96 (100 ppm hardness and 24-hr avg)	—	7.1 (24-hr avg)	—	13.4
48. Nitrobenzene	1,800 (max) 27,000	—	140 (max) 6,680	—	19.8 mg/l 30 mg/l (taste and odor)
49. Nitrophenols	230	150	4,850	—	13.4 (for 2,4 dinitrocresol) 70 (for dinitrophenol)
50. Nitrosamines	5,850	—	3,300,000	—	0.14–14 ng/l (for n-nitrosodimethylamine) 0.08–8.0 ng/l (for n-nitrosodiethylamine) 0.064–64 ng/l (for n-nitrosobutylamine) 490–49,000 ng/l (for n-nitrosodiphenylamine) 1.60–160 ng/l (for n-nitrosopyrolidine)
51. Pentachloro-phenol	55	3.2	53	34	1.01 ng/l 30 ug/l (for taste and odor)
52. Phenol	10,200	2,560	5,800	—	3.5 mg/l 0.3 mg/l (for taste and odor)

TABLE 11.14 (continued)

53. Phthalate esters	940	3	2,944	3.4	313 mg/L (dimethyl phthalate) 350 mg/L (diethyl phthalate) 34 mg/L (dibutyl phthalate) 15 mg/L (di-2-ethyl-hexyl phthalate)
54. Polychlorinated biphenyls	0.014 (24-hr avg) 2.0 (max)	—	0.30 (24-hr avg) 10 (max)	—	0.0079–.79 ng/L
55. Polynuclear aromatic hydrocarbons (PAHs)	Not available	—	300	—	0.28–28 ng/L
56. Selenium	35 (24-hr avg) 260 (max) 760 (inorganic scienate)	—	54 (24-hr avg) 410 (max)	—	10 ug/L
57. Silver	4.1 (max 100 ppm hardness)	0.12 (average chronic)	2.3	—	50 ug/L
58. Tetrachloroethylene	5,280	840	10,200	450	0.08–8 ug/L
59. Thallium	1,400	40	2,130	—	13 ug/L
60. Toluene	17,500	—	6,300	5,000	14.3 mg/L
61. Toxaphene	0.013 (24-hr avg) 1.6 (max)	—	0.070 (max)	—	0.07–7.1 ng/L
62. Trichloroethylene	45,000	21,900	2,000	—	0.27–27 ug/L
63. Vinyl chloride	Not available	—	Not available	—	0.20–20 ug/L
64. Zinc	47 ug/L (24-hr avg)	—	58 (24-hr avg)	—	5 mg/L

TABLE 11.15
Allowable Concentrations for Air Contaminants Resulting from Hazardous Waste Treatment, Storage, and Disposal Emissions (EPA)

Substance	ppm^a	mg/m^{3b}
Acetaldehyde	20	40
Acetic acid	1	3
Acetic anhydride	0.5	2
Acetone	100	200
Acetonitrile	4	7
Acetylene dichloride (see *1,2-dichloroethylene*)		
Acetylene tetrabromide	0.1	1
Acrolein	0.001	0.003
Acrylamide—skin		0.03
Acrylonitrile—skin	2	5
Aldrin—skin		0.03
Allyl alcohol—skin	0.2	0.5
Allyl chloride	0.1	0.3
C allylglycidyl ether (AGE)	1	5
Allyl propyl disulfide	0.2	1
2-Aminoethanol (see *ethanolarine*)		
2-Aminopyridine	0.005	0.2
Ammonia	5	4
Ammonium sulfamate (ammate)		2
n-Amyl acetate	10	55
seo-Amyl acetate	13	65
Aniline—skin	0.5	2
Anisidine (o. p-isomers)-skin		0.005
Antimony and compounds (as SB)		0.005
ANTU (alpha naphthyl thiourea)		0.03
Arsenic and compounds (as AS)		0.05
Arsine	0.005	0.02
Azinphos-methyl—skin		0.02
Barium (soluble compounds)		0.05
Benzene	1	
p-Benzoquinone (see *quinone*)		
Benzoyl peroxide		0.5
Benzyl chloride	0.1	0.5
Biphenyl (see *diphenyl*)		
Bisphenol A (see *diglycidyl ether*)		
Boron oxide		2
C boron trifluoride	0.1	0.3
Bromine	0.01	0.07
Bromoform—skin	0.05	0.5
Butadiene (1,3-butadiene)	100	200
Butanethiol (see *butyl mercaptan*)		
2-Butanone	20	60
2-Butoxy-ethanol (butyl cellosolve)—skin	5	20

TABLE 11.15 *(continued)*

Substance	ppm[a]	mg/m³[b]
Butyl acetate (*n*-butyl acetate)	15	70
seo-butyl acetate	20	95
tert-butyl acetate	20	95
Butyl alcohol	10	30
sec-butyl alcohol	15	45
tert-butyl alcohol	10	30
C butylamine—skin	0.5	2
C tert-butyl chromate (as CrO_2)—skin		0.01
*n*Butyl glycidyl ether (BGE)	5	25
Butyl mercaptan	1	4
p-tert-butyltoluene	1	6
Cadmium fume		0.01
Cadmium dust		0.02
Calcium arsenate		0.1
Calcium oxide		0.5
Campher	0.2	
Carbaryl (Sevin®)		0.5
Carbon black		0.35
Carbon disulfide	2	
Carbon tetrachloride	1	
Chlordane—skin		0.05
Chlorinated camphene—skin		0.05
Chlorinated diphenyl oxide		0.05
Chlorine	0.1	0.3
Chlorine dioxide	0.01	0.02
C Chloride trifluoride	0.01	0.04
C Chloroacetaldehyde	0.1	0.3
α-Chloroacetophenone (phenacylchloride)	0.005	0.03
Chlorobenzene (monochlorobenzene)	10	35
o-Chlorobenzyliden malononitrile (OCBM)	20	100
2-Chloro 1,3-butadiene (see *chloroprene*)		
Chlorodiphenyl (42% chlorine)—skin		0.1
Chlorodiphenyl (54% chlorine)—skin		0.05
1-Chloro, 2,3-epoxypropane (see *epichlorhydrin*)		
2-Chloroethanol (see *ethylene chlorohydrin*)		
Chloroethylene (see *vinyl chloride*)		
C Chloroform (trichloromethane)	5	25
1-Chloro-1-nitropropane	2	10
Chloropicrin	0.01	0.07
Chloroprene (2-chloro-1,3-butadiene)—skin	3	9
Chromic acid and chromates		
Chromium, So. chromic		
Chromous salts as Cr		0.05
metal and insoluble salts		0.1

TABLE 11.15 (continued)

Substance	ppm[a]	mg/m³[b]
Coal tar pitch volatiles (benzene soluble fraction) anthracene, BaP, phenanthrene, acridine, chrysene, pyrene		0.01
Cooper fume		0.01
dusts and mists		0.1
Cotton dust (raw)		0.1
Crag herbicide		2
Cresol (all isomers)—skin	0.5	2
Crotonaldehyde	0.2	0.6
Cumene—skin	5	25
Cyanide (as CN)—skin		0.5
Cyclohexane	30	100
Cyclohexanol	5	20
Cyclohexanone	5	20
Cyclohexene	30	100
Cyclopentadiene	8	20
2,4-D		1
DDT—skin		0.1
DDVP (see *dichlorvos*)		
Decaberane—skin	0.005	0.03
Demetron—skin		0.01
Diacetone alcohol (4-hydroxy-4-mathyl-2-pentanone)	5	25
1,2-Diaminoethane (see *ethylenediamine*)		
Diazomethane	0.02	0.04
Diborane	0.01	0.01
Dibutylphthalate		0.5
C *o*-Dichlorobenzene	5	30
p-Dichlorobenzene	8	45
Dichlorodifluoromethane	100	500
1, 3-Dichloro-5,5-dimethyl hydantoin		0.02
1, 1-Dichloroethane	20	40
1, 2-Dichloroethylene	20	80
C Dichloroethyl ether—skin	2	9
Dichloromethane (see *methylene chloride*)		
Dichloromonofluoromethane	100	400
C 1, 1-Dichloro-1-nitroethane	1	6
1, 2-Dichloropropane (see *propylene dichloride*)		
Dichlorotetrafluoroethane	100	700
Dichlorvos (DDVP)—skin		0.1
Dieldrin—skin		0.02
Diethylamine	3	8
Diethylamin (see *ethyl ether*)		
Difluorodibromomethane	10	90
C Diglycidyl ether (DGE)	0.05	0.3
Dihydroxybenzene (see *hydroquinone*)		

TABLE 11.15 (continued)

Substance	ppm[a]	mg/m³[b]
Diisobutyl ketone	5	30
Diisopropylamine—skin	0.5	2
Dimethyl acetamide—skin	1	4
Dimethylamine	1	2
Dimethylaminobenzene (see *xylidene*)		
Dimethylaniline (*N*-dimethylaniline)—skin	0.5	3.0
Dimethylbenzene (see *xylene*)		
Dimethyl 1,2-dibromo-2,2-dichloroethyl phosphate (dibrom)		0.3
Dimethylformamide—skin	1	3
2,6-Dimethylheptanone (see *diisobutyl ketone*)		
1,1-Dimethylhydrazine—skin	0.05	0.1
Dimethylphthalate		0.5
Dimethylsulfate—skin	0.1	0.5
Dinitrobenzene (all isomers)—skin		0.1
Dinitro-*o*-cresol—skin		0.02
Dinitrotoluene—skin	1	0.2
Dioxane (diethylene dioxide)—skin	10	40
Diphenyl	0.02	0.1
Diphenylmethane diixocyanate (see *methylene bisphenyl isocyanate* [MBI])		
Diporopylene glycol methyl ether—skin	10	60
Di-sec, octyl Phthalate (Di-2-ethylhexylphthlate)		0.5
Endrin—skin		0.01
Epichlorohydrin—skin	0.5	2
EPN—skin		0.05
1,2-Epoxypropane (see *propyleneoxide*)		
1,3-Epoxy-1-propanol (see *glycidol*)		
Ethanethiol (see *ethylmercaptan*)		
Ethanolarine	0.3	0.6
2-Ethoxyethanol—skin	20	75
2-Ethoxyethylacetate (cellosolve acetate)—skin	10	53
Ethyl acetate	40	150
Ethyl acrylate—skin	3	10
Ethyl alcohol (ethanol)	100	200
Ethylamine	1	2
Ethyl seo-amyl ketone (5-heptanone)	5	25
Ethyl benzene	10	45
Ethyl bromide	20	90
Ethyl butyl ketone (3-heptanone)	5	25
Ethyl chloride	100	300
Ethyle ether	40	120
Ethyl formate	10	30
C Ethylmercaptan	1	3

TABLE 11.15 *(continued)*

Substance	ppm[a]	mg/m³[b]
Ethyl silicate	10	85
Ethylene chlorohydrin—skin	0.5	2
Ethylenediamine	1	3
Ethylene dibromide	2	
Ethylene dichloride	5	
C Ethylene glycol dinitrate and/or nitroglycerin—skin	0.02	0.1
Ethylene glycol monomethyl ether acetate (see *methyl cellosolve acetate*)		
Ethylene Imine—skin	0.05	0.1
Ethylene oxide	5	9
Ethylidine chloride (see 1,1-*dichloroethane*)		
N-Ethylmorpholine—skin	2	10
Ferbam		2
Ferrovanadium dust		0.1
Fluoride as dust		0.3
Fluoride (as F)		0.3
Fluorine	0.01	0.02
Fluorotrichloromethane	100	500
Formaldehyde	0.3	
Formic acid	0.5	0.9
Furfural—skin	0.5	2
Furfuryl alcohol	5	20
Glycidol (2,3-epoxy-1-propanol)	5	15
Glycol monoethyl ether (see 2-*ethoxyethanol*)		
Halnium		0.05
Heptachlor—skin		0.05
Heptane (*n*-heptane)	50	200
Hexachloroethane—skin	0.1	1
Hexachlorocaphthalene—skin		
Hexane (*n*-hexane)	50	180
2-Hexanone	10	40
Hexone (methyl isobutyl ketone)	10	40
	5	30
Hydrazine—skin	0.1	0.1
Hydrogen bromide	0.3	1
C Hydrogen chloride	0.5	0.7
Hydrogen cyanide—skin	1	1
Hydrogen fluoride	0.3	
Hydrogen peroxide (90%)	0.1	0.1
Hydrogen selenide	0.005	0.02
Hydrogen sulfide		
Hydroquinone		0.2
C Iodine	0.01	0.1
Iron oxide fume		1
Isoamyl acetate	10	55

TABLE 11.15 *(continued)*

Substance	ppm^a	mg/m^{3b}
Isoamyl alcohol	10	35
Isoamyl acetate	15	70
Isoamyl alcohol	10	30
Isophorone	3	14
Isopropyl acetate	25	95
Isopropyl alcohol	40	100
Isopropylamine	0.5	1
Isopropylether	50	210
Isopropyl glycidyl ether (IGE)	5	25
Ketene	0.05	0.09
Lead and its inorganic compounds		0.02
Lead arsenate		0.015
Lindane—skin		0.05
Lithium hydride		0.002
LPG (liquefied petroleum gas)	100	180
Magnesium oxide fume		2
Malathion—skin		2
Maleic anhydride	0.03	0.1
C Manganese		0.5
Mercury		
Mesityl oxide	3	10
Methanethiol (see *methyl mercaptan*)		
Methoxychlor		2
2-Methoxyethanol (see methyl cellosolve)		
Methyl acetate	20	60
Methyl acetylene (propyne)	100	165
Methyl acetylene-propadiene mixture (MAPP)	100	180
Methyl acrylate—skin	1	4
Methyl-a (dimethoxymethane)	100	300
Methyl alcohol (methanol)	20	25
Methylamine	1	1
Methyl amyl alcohol (see *methyl isobutyl carbinol*)		
Methyl (*n*-amyl) ketone (see *2-heptanone*)	10	50
C Methyl bromide—skin	2	8
Methyl butyl ketone (see *2-heptanone*)		
Methyl cellosolve—skin	3	8
Methyl cellosolve acetate—skin	3	12
Methyl chloride	10	
Methyl chloroform	35	190
Methylcyclohexane	50	200
Methylcyclohexanol	10	50
o-methylcyclohexanone—skin	10	45
Methyl ethyl ketone (MEK) (see *2-butanone*)		
Methyl formate	1-	25

TABLE 11.15 *(continued)*

Substance	ppm[a]	mg/m³[b]
Methyl iodide—skin	0.5	3
Methyl isobutyl ketone (see *hexone*)		
Methyl isocyanate—skin	0.002	0.005
C Methyl mercaptan	1	2
Methyl methacrylate	10	40
Methyl propyl ketone (see *2-pentanone*)		
C Methyl styrene	10	50
C Methylene bisphenyl isocyanate (MBI)	0.002	0.02
Methylene chloride	50	
Molybdenum		
Soluble compounds		0.5
Insoluble compounds		2
Monomethyl aniline—skin	0.2	0.9
C Monoethyl hydrazine—skin	0.02	0.04
Morpholine—skin	2	7
Naphtha (coal tar)	10	40
Naphthalene	1	5
Nickel carbonyl	0.0001	0.0007
Nickel, metal, and soluble compounds as Ni		0.1
Nicotine—skin		0.05
Nitric acid	0.2	0.5
Nitric oxide	3	3
p-Nitroaniline—skin	0.1	0.6
Nitrobenzene—skin	0.1	0.5
p-Nitrochlorobenzene—skin		0.1
Nitroethane	10	30
Nitrogen dioxide	0.5	0.9
Nitrogen trifluoride	1	3
Nitroglycerin—skin	0.02	0.2
Nitromethane	10	25
1-Nitropropane	3	9
2-Nitropropane	3	9
Nitrotoluene—skin	0.5	3
Nitrotrichloromethane (see *chloropicrin*)		
Octachloronaphthalene—skin		0.01
Octane	50	200
Oil mist, mineral		0.5
Organo (alkyl) mercury		0.001
Osmium tetroxide		0.000
Oxalic acid		0.1
Oxygen difluoride	0.005	0.01
Paraquat—skin		0.05
Parathion—skin		0.011
Pentaborane	0.0005	0.001
Pentachloronaphthalene—skin		0.05

TABLE 11.15 *(continued)*

Substance	ppm[a]	mg/m³[b]
Pentane	100	300
2-Pentanone	20	70
Perchloromethyl mercaptan	0.01	0.08
Perchloryl fluoride	0.3	1
Petchloryl fluoride	0.3	1
Petroleum distillates (naphtha)	30	200
Phenol—skin	0.5	2
p-Phenylene diamine—skin		0.01
Phenyl ether (vapor)	0.1	0.7
Phenyl ether-biphenyl mixture (vapor)	0.1	0.7
Phenylethylene (see *styrene*)		
Phenylglycidyl ether (PGE)	1	6
Phenylhydrazine—skin	0.5	2
Phosdrin (mevinphos)—skin		0.01
Phosgene (carbonyl chloride)	0.01	0.04
Phosphine	0.03	0.04
Phosphoric acid		0.1
Phosphorus (yellow)		0.01
Phosphorus pentachloride		0.1
Phosphorus pentasulfide		0.1
Phosphorus trichloride	0.05	0.3
Phthalic anhydride	0.2	1
Picric acid—skin		0.01
Pival (2-pivalyl-1,3-indandione)		0.01
Platinum (soluble salts) as Pt		0.000?
Propargyl alcohol—skin	0.1	
Propane	100	180
n-Propyl acetate	20	85
Propyl alcohol	20	50
n-Propyl nitrate	3	11
Propylene dichloride	8	35
Propylene imine—skin	0.2	0.5
Propylene oxide	10	25
Propyne (see *methyl acetylene*)		
Pyrethrum		0.5
Pyridine	0.5	2
Quinone	0.01	0.04
RDX—skin		0.2
Rhodium, metal fume, and dusts, as Rh-soluble salts		0.0001
Ronnel		1
Rocenone (commercial)		0.5
Selenium compounds (as Se)		0.02
Selenium hexafluoride	0.005	0.05
Silver, metal, and soluble compounds		0.001

TABLE 11.15 *(continued)*

Substance	ppm[a]	mg/m³[b]
Sodium fluoroacetate (1080)—skin		0.005
Sodium hydroxide		0.2
Stibine	0.01	0.05
Scoddard solvent	50	300
Strychnine		0.015
Styrene	10	
Sulfur hexafluoride	100	600
Sulfuric acid		0.1
Sulfur monochloride	0.1	0.6
Sulfur pentafluoride	0.003	0.03
Sulfuryl fluoride	0.5	2
Systox (see *Demetron*)		
2,4,5T		1
Tantalum		0.5
TEDP—skin		0.02
C Terphenyls	0.1	0.9
1.1.1, 2-Tetrachloro-2,2-difluoroethane	50	400
1.1.2, 2-Tetrachloroethane—skin	0.5	4
Tetrachloroethylene	10	
Tetrachloromethane (see *carbon tetrachloride*)		
Tetrachloronaphthalene—skin		0.2
Tetraethyl lead (as Pb)—skin		0.008
Tetrahydrofuran	20	60
Tetramethyl lead (as Pb)—skin		0.007
Tetramethyl succinonitrile—skin	0.05	0.3
Tetranitromethane	0.1	0.8
Tetryl (2,4,6-trinitrophenyl-methylinitramine)—skin		0.15
Thallium (soluble compounds)—skin as Tl		0.01
Thiram		0.5
Tin (inorganic compounds, except oxides)		0.2
Tin (organic compounds)		0.01
Toluene	20	
C Toluene-2,4-diisocyanate	0.002	0.001
o-Toluidine—skin	0.5	2
Toxaphene, (see *chlorinated camphene*)		
Tributyl phosphate		0.5
1,1,1-Trichloroethane (see *methyl chloroform*)		
1,1,2-Trichloroethane—skin	1	5
Titaniumdioxide		2
Trichloroethylene	10	
Trichloromethane (see *chloroform*)		
Trichloronaphthalene—skin		0.5

TABLE 11.15 (continued)

Substance	ppm[a]	mg/m³[b]
1,2,3-Trichloropropane	5	30
1,1,2-Trichloro 1,2,2-trifluoroethane	100	800
Triethylamine	3	10
Trifluoromonobromomethane	100	600
2,4,6-Trinitrophenol (see *picric acid*)		
2,4,6-Trinitrophenyl-methylnitramine (see *tetryl*)		
Trinitrotoluene—skin		0.2
Triorthocresyl phosphate		0.01
Triphenyl phosphate		0.3
Turpentine	10	60
Uranium (soluble compounds)		0.005
Uranium (insoluble compounds)		0.03
C Vanadium:		
V_2O_{25} dust		0.08
V_2O_5 fume		0.01
Vinyl benzene (see *styrene*)		
Vinylcyanide (see *acrylonitrile*)		
Vinyl toluene	10	50
Warfarin		0.01
Xylene (xylol)	10	45
Xylidine—skin	0.5	3
Yttrium		0.1
Zinc chloride fume		0.1
Zinc oxide fume		0.5
Zirconium compounds (as Zr)		0.5

[a]Parts of vapor or gas per million parts of contaminated air by volume at 25°C and 760 mmHg pressure.
[b]Approximate milligrams of particulate per cubic meter of air.

water on the property or on surrounding property. The history of use of the site is necessary. Numerous sources such as air photos, historical property-use maps, tax records, building permits, or any other type of historical information can be used.

The site reconnaissance is conducted by an environmental professional to observe the present and, to the extent possible, past use of chemicals and hazardous substances on the property and adjoining property. Areas to be observed include, for instance, storage facilities, process areas, and waste-disposal areas. Access to all parts of the facility is imperative during such a reconnaissance, but is often difficult.

The final information component is gathered from interviews with current and prior owners and occupants of the site. The site manager or superintendent is an invaluable resource. Facility records, particularly regarding environmental management and compliance, are useful. Interviews with local officials who deal with the facility on a day-to-day basis, such as fire department or environmental compliance personnel, are valuable.

The environmental professional then prepares a report following the recommended report format. This report includes documentation of the data compiled and supporting the conclusions of the report.

Who is qualified to prepare an ESA? Virtually any person who has an understanding of environmental processes or science, and who understands the purpose behind the methods specified by the ASTM, is qualified. Specific technical training in an environmental discipline may not be required; individuals with experience in the operation of chemical manufacturing or use, or of waste disposal, may be qualified. Some states have registration procedures for environmental professionals that include educational and experience criteria.

Bankruptcy apparently is no longer an escape route for an industry to avoid cleanup costs of hazardous waste dumps. The U.S. Supreme Court ("A clean victory" 1986) decided that bankrupt companies "retain an obligation to safeguard the public's safety after they go broke." In a five-to-four decision written by Justice Lewis F. Powell, Jr., "The Federal bankruptcy code is not meant to allow property to be abandoned in violation of state health laws." Although many more cases are expected to reach the court, the public's interest appears to be a major concern in any decision involving hazardous wastes whether old, abandoned, or current.

The federal government is not exempt from problems of disposing hazardous wastes legally. Federal agencies were found by a congressional study ("3 U.S. agencies" 1986) to have deposited tons of toxic wastes into a leaking California dump during 1985. Evidently the EPA banned the use of the facility for Superfund waste. The bulk of the 8,300 tons of toxic wastes came from the Department of Defense. The General Accounting Office maintained, however, "that although EPA can mandate how it handles toxic wastes under its own jurisdiction, it lacks authority to force others to follow its policies relating to licensed dumps that are having environmental problems. The only way EPA can prevent federal agencies from using a particular disposal facility is to close it."

The 104th U.S. Congress threatened to amend the Superfund cleanup law so that the taxpayers rather than the polluters would pay for the cleanup. The proposed bill, if it had passed, would have eliminated cleanup liability for pollution caused prior to 1980, when Superfund was passed, or 1986, when insurance companies stopped underwriting industrial liability. Sites contaminated earlier would have been cleaned up at taxpayers' expense or not cleaned up at all. Under CERCLA, liable polluters can obtain 30% of their cleanup costs paid by the pollution tax fund.

Industrial Insurance

Despite adequate precautions of waste planning, treatment, recovery, and reuse, industrial plants should protect themselves against eventual failure. Damages caused by accidental spills or unavoidable errors of discharge to the environment can be covered by adequate insurance.

In an editorial in the *Wall Street Journal* (1981), any company that stores, treats, or disposes of solvents, acids, alkalis, or other hazardous materials is advised to have some insurance. The RCRA of 1976 holds industry responsible for a "cradle to the grave" accounting of hazardous materials produced by themselves. The EPA was required by this act by January 1981 to ensure industry's coverage of $3 million for each case of long-term damage up to a maximum of $6 million a year, excluding legal defense costs. For sudden spills, the required coverage is $1 million for each case and up to $2 million a year. Many insurance brokers are of the opinion that the EPA's proposed limits are not sufficient to cover a major accident. In 1981, only three U.S. insurance companies were offering coverage for these damages. Therefore, many U.S. industries provide their own insurance for accidental spills.

Insurance problems have also carried over to the trucking industry, which is largely responsible for hauling away hazardous wastes. According to Giltenan (1985), "The reluctance of the insurance industry to cover pollution liability is causing near panic among handlers of hazardous materials, and truckers are being especially hard hit." As a result of insurance costs and reluctance to insure, many truckers—especially marginal ones—are getting out of the hazardous materials business. Under Section 30 of the Motor Carriers Act of 1981, truckers are required to carry a minimum of $5 million insurance coverage for bodily, property, and environmental-restoration (pollution) liability. The spill area of liability has also become too costly for insurance carriers to continue underwriting this coverage. Some large truckers are paying the higher premiums and carrying policies in excess of $5 million. Most of the higher insurance costs are blamed on the increased tendency of courts to rule in favor of environmental pollution victims. Some of the higher costs are being passed on to chemical shippers, making the costs of hazardous waste disposal even higher.

The Chemical Process Industry (CPI) notes the change in the insurance industry towards industrial environmental coverage ("CPI Scrambles 1985). It feels that " inexpensive (pollution) insurance may never again be available, certainly not in the short term." Without adequate pollution insurance, the CPI faces troublesome times. How can it protect itself against liabilities associated with sudden and accidental mishaps, such as gas leaks, and non-sudden and gradual instances, such as seepage from waste holding sites? Large companies are considering self-insurance to cover the dangers. Smaller companies are looking to trade-association and broker-assisted programs offering primary coverage for the major firm needs (worker's compensation, product liability, comprehensive general liability, and automobile liability/physical damage insurance). Insurance premiums, if obtainable, will be much more expensive, especially if pollution liability is clearly included. As a last resort, the federal government may enact legislation to assist the chemical industry. The exact nature of the assistance is not yet known, but it is likely that it will be similar to that offered the nuclear industry and for those needing flood insurance.

Federation (WEF, formerly WPCF) Committee Chairman Dan Hinricks revealed ("Hazardous waste conferences" 1985) that because of the complex liability issues involved in hazardous waste cleanup, the insurance market has almost totally dried up. He said, "The need is there, but if companies can't get insurance, they certainly aren't going to get into the business." He calls the insurance situation "the single, most important concern in the hazardous waste industry."

Illegal Dumping

Jaffe (1983) asserts that

> Despite tougher controls on the disposal of toxic wastes, illegal dumping has not been curbed. Instead, the law enforcement official say, waste haulers have found new ways to skirt the laws and new places to unload their poisonous cargo. States rely on a so-called "manifest system," which it forms for detailing the route and final location of all toxic wastes. In the manifest the kinds and quantity of waste chemicals are recorded. The hauler must include where the delivery is to be made and the disposer must tell the manner of final treatment. However, the illegal dumpers' answer to the manifest system has been "creative accounting and forgery."

Many illegal dumping suits have been filed and decisions have been rendered by the courts. It is not the purpose here to describe or elucidate the circumstances of each.

Storage of Hazardous Industrial Wastes

Hazardous wastes are generated by industry in small quantities generally. Whether they are exported for treatment and disposal or handled for treatment on the premises, they usually are stored temporarily. Storage is necessary in these cases to obtain sufficiently large quantities for efficient continual treatment. If these toxic wastes are collected and transported to some distant site for treatment, they often are also stored for some period at the distant site.

Once we accept storage as an integral step in the ultimate treatment of hazardous wastes, it becomes necessary to provide proper storage treatment. The following criteria should be integrated into the design and operation of hazardous waste storage:

1. Limited access: Entrance should be available only to previously certified and qualified plant personnel.
2. Proper ground location: Storage should be located at a proper distance from living and working people, with consideration for wind direction and velocity.
3. All storage ground area must be impermeable to spilled or leaked wastes. When spills or leaks occur, they cannot be allowed to permeate into the ground but must be drained off the impermeable surface to a collection sump from which they are pumped to treatment or transfer.
4. Containers or tanks must be corrosion proof and tight fitting.
5. Ample storage area: Provide several times the anticipated maximum storage area to account for breakdown in collection and treatment systems.

6. Monitoring devices: Leaks and fires into the surrounding air and on the ground out of the storage area should be monitored by continuously operated sensing equipment.
7. Storage area leak plan: An action plan for catastrophe-type leak occurrence should be available, publicized, and readily available if such a hazard takes place.

The Chemical Conservation Corporation (ChemCon) made available ("New transfer waste facility" 1985) a new hazardous waste transfer (storage) facility to generators in central Florida. The property is 1.2 acres, with undeveloped area adjacent for expansion. It is enclosed by a 6-ft chain-link fence, with a single security gate at the entrance. The truck area is graded concrete, which will contain spills and minimize chemical absorption. With a capacity of $5,500$ ft^2 for hazardous waste storage, the floor is heavily sealed concrete, divided into flammable and corrosive drum storage areas by a spill-containment curb. Each area slopes to a center drain, leading into one of two emerging 120-gallon waste containers located outside the building. Fire walls separate storage from personnel areas, and the storage electrical system is explosion proof. A hazardous waste sprinkler system protects the entire building with fire detection devices as well. These detection devices automatically signal the Orange County Fire Department and a portable telephone system ensures constant communication. Detection devices on all entries to the storage building are linked to the Orange County Sheriff's Department.

Hazardous wastes stored here are transferred to ultimate disposal facilities for incineration, reclamation, neutralization, chemical fixation, or sanitary landfill. Currently, this storage (transfer) facility is handling 1.5 million gallons of hazardous waste per year for 350 industrial customers in 1,500 drum quantities.

Safety Storage ("Chemical storage containers" 1985) has designed and constructed containers to comply with both NFPA standards and OSHA regulations. Many local hazardous material storage ordinances require hazardous chemicals to be stored in secondary containment structures to prevent spills or leaks from contaminating groundwater. Their containers are made of 10- to 14-gauge ASTM-A569 steel, possess secondary spill capacity of 500 gallons, and hold 30- to 55-gallon drums. The interiors are protected from chemical attack by a resistant epoxy coating. Outside dimensions are 8 by 8 by 22 ft. Each unit is static grounded to prevent ignition of flammable wastes by electrical discharge. Fire protection is provided with a water line and three fire sprinkler heads. Roofs are specially designed to handle explosions and floors are thick plywood underlain with subfloors of 12-gauge epoxy-lined steel. Many other optional features for hazardous waste storage safety are available on an optional basis. These containers may be leased to assist industry in conserving capital.

Stone (1985) advocates the use of selected existing mined space as technically and economically feasible for permanent storage of untreatable wastes or the toxic end products of hazardous waste treatment. Advantages of this type of storage include the following:

1. In deep mined space, the waste would be below drinking-water aquifers.
2. It is isolated from the public and the surface ecology.

3. If required, waste can be isolated from hydrological environment by encapsulation or containerization.
4. Security can be readily maintained.
5. In a sealed mine, no continuing maintenance will be required.
6. If retrievability is desired, the mine could be used as a long-term underground warehouse.

Such treatment must protect all parts of the environment during construction while adding waste, and for an indefinite period into the future, storage underground uses the minimum of existing land surface. One possibility in mines is that of abandoned salt mines, which are relatively free of people and industry. Although this practice has been used for storage of certain hazardous wastes for a number of years, one must be certain that the chemicals do not react with themselves or with the salt in the mine.

Storage of hazardous chemicals before and during use is a vital phase of material production. The reader can attest to this by reviewing some of the examples of accidents that have occurred from too little attention to this subject.

Transportation and Spill Prevention of Industrial Hazardous Wastes

Transporting hazardous wastes to an external site for storage, treatment, or disposal must be considered an integral part of its ultimate solution. It is usually done by truck or rail. Both modes of transportation involve hazards of accidental spills or wrecks that may release the wastes to an unsuspecting and unprepared environment. Trucks exhibit more flexibility in delivery and pickup. In the last decade, it was estimated also that rail accidents were two or three times more prevalent than truck accidents.

Although rail accidents are down, their severity is up (Cushman 1989). It was estimated that 4 billion tons of hazardous materials were shipped by rail annually, involving 250,000 shipments daily. The problem of moving hazardous chemicals by rail is not only that they may be explosive, poisonous, or corrosive, but also that trains move them through industrial areas and residential areas of cities. States and localities, along with the federal government, regulate chemical shipments by road and rail, but confusion and complex compliance have resulted in burdens for carriers. Congress is examining possibilities of legislation allowing larger cities to approve of transportation routes and the Department of Transportation to select the safest routes.

Several states have incurred 10 or more railroad accidents involving toxic wastes (Cushman 1989). In both modes of transportation, the waste generator must accept the responsibility of complete and correct disclosure of the chemical contents of the hazardous wastes. The more information about the composition and characteristics of the wastes to be transported, the safer will be the trip (in the case of an accident) to the disposal site. The transporter assumes the responsibility for safe delivery of the waste to the disposer. When the transporter accepts the waste, he or she should sign a verification statement with the generator that he or she is transporting a given volume of specific hazardous waste.

The same statement should be signed and given to the disposer once the wastes have arrived and been unloaded at the disposal site. Packaging of hazardous wastes should be done in conformance with Department of Transportation regulations. Usually, liquid wastes are packaged in 55-gallon drums, sealed, and surrounded by vermiculite for insulation against the shock of collisions or pumped into tank cars supplied for the purpose by truckers or railroad haulers.

The truck fleet used by ChemCon ("New transfer waste facility" 1985) travels more than 1 million miles per year and has been accident free for its 3 years of operation.

Nemerow (1987) discusses the scavenger system that hauls, treats, reclaims, and disposes of a variety of industrial wastes. These wastes are usually small in volume and difficult and/or hazardous to dispose of by the producer in normal ways. The expenses of transportation may be considerable and determine in the long run whether exportation/treatment is a better alternative to on-site treatment.

Spills

As an example of protection against spills, the Proform Company offers a system for eliminating spills in areas where rail cars are loaded, unloaded, washed, or fueled. It provides a structural fiberglass-reinforced plastic track collector pan system to collect and isolate spills for treatment and disposal. The system components include collector pans that attach to rails and cross drains that connect with preinstalled underground pipes leading to a sump for treatment (Proform, Inc. 1985).

Jim Bradley, the Minister of Environment for Ontario, Canada, announced on July 3, 1985, that "our initial priorities include regulation of the transportation of dangerous goods . . . " Bradley was new in this high position and recognized how vital it is to transport hazardous materials safely.

Example of Twentieth-Century Hazardous Waste Treatment

LaBranche and Collins (1996) conducted research experiments to determine whether hazardous wastes containing volatile organic petroleum products (VOCs) could be removed by air stripping and the optimum conditions for doing so. Full factorial experimental trials were conducted to determine the influence of inlet water flow rate and temperature on trichloroethylene (TCE), perchlorethylene (PCE), and total petroleum hydrocarbon (TPH) removal. Results indicated that economical air stripping of VOC and TPH compounds could be achieved using low liquid flow rates (20–75 L/min), high air/water ratios (225–898), and medium liquid temperatures (16–28°C) in tray-type air strippers.

References

3 U.S. agencies used banned toxic waste dump. *The Miami Herald,* January 1986.
A clean victory [Editorial]. *The Miami Herald,* February 3, 1986.
Agardy, F. J., J. W. Wilcox. 1990. Electrically heated high temperature incineration of air toxics. Proceedings of the 12th National Industrial Energy Technology Conference, Houston.
American Chemical Society, Department of Public Affairs. *Hazardous Waste Management.* Washington, DC: American Chemical Society.
The Amicus Journal, Spring 1985;6(4):25.
Areas in 7 states eyed for nuclear dump site. *The Miami Herald,* January 17, 1986.
ATW/Calweld, Inc. 1985. *Detoxifier-Hazardous Waste Clean-up and Compliance.* Santa Fe Springs, CA: ATW/Calweld, Inc.
Bahorsky, M. S. 1983. Land application treatment of textile knitting, dyeing, and finishing wastewater. In: *Toxic and Hazardous Wastes Proc. 15th Mid-Atlantic Industrial Waste Conf.,* p. 8. Woburn, ME: Butterworth Publishers.
Ban on asbestos is unnecessary, industry, Canadians advise EPA. *The Miami Herald,* January 24, 1986, p. 6A.
Bergenthal, J. *Textile Dyebath Reconstitution and Reuse. Future of Water Reuse Proc.,* Vol. 2, Water Reuse Symposium, Aug. 24–26, 1984, San Diego, CA, p. 840.
Boycem A. W. 1983. Reconditioning of dehydrated sludge with quicklime. *Toxic and Hazardous Wastes,* Washington, DC, p. 477.
Cartwright, P. S. 1984. Innovative technology to treat toxic wastes from a thin film head manufacturing facility: a case history. *Future of Water Reuse Proceedings,* Vol. 2, *Water Reuse Symposium III, Aug. 26–31, 1984, San Diego, CA,* p. 778.
Castleman, M. 1985. Toxics and male infertility. *Sierra Magazine,* March/April 1985, p. 49.
Cheremisinoff, P. N. 1988. Thermal treatment technologies for hazardous wastes. *Pollution Engineering* 20(8):50.
Civil Engineering. 1979. Waste processing firms to play critical role in hazardous waste management. Sept. 1979 (ASCE), p. 85.
Clean-up of the Gilson Road hazardous waste disposal site, Nashua, New Hampshire. *The Weston Way,* 1984;10 (Summer-Fall):3–7.
Cook, M. M., J. A., Lander, Sr., D. S. Littlehale. 1979. Case histories; reviewing the use of sodium borohydride for control of heavy metal discharge in industrial wastewaters. *Proc. 34th Purdue Univ. Industrial Waste Conf.,* May 8, 1979, p. 514.
CPI scrambles for pollution insurance. *Chemical Business,* November 1985, p. 41.
Cushman, J., Jr. 1989. Chemicals on rails: a growing peril. *New York Times,* Aug. 2, 1989, p. 8.
DeLatour, C. 1973. Magnetic separation in water pollution control. *E.E.E. Trans. Magnetics* 9(3):314.
Dotson, G. K., Dean, R. B., Cooke, W. B., et al. 1974. Yard Spreading: A Conserving and Non-Polluting Method of Disposing of Oily Wastes," Waste Oil Report to Congress. Environmental Protection Agency, April 1974, p. 151.
Doyle, B. W., D. A. Drum, J. D. Lauber. 1985. The smoldering question of hospital wastes. *Pollution Engineering,* July 1985, p. 85.

Duffy, J. G. 1983. Electrochemical removal of heavy metals from wastewater. *Products Finishing,* August 1983, p. 72.

Duke, D. D. 1994. Hazardous waste minimization: is it taking root in U.S. industry? *Waste Management* 14(1):49–59.

Duke, P. 1985. NASA roots out earthy cure for polluted air in space stations. *Wall Street Journal,* August 28, 1985.

DuPont. 1984. Dupont to enter waste treatment. *The Miami Herald,* June 5, 1984.

Eating raw shellfish health risk, MDs say. *The Miami Herald,* March 13, 1986, p. 12A.

ENELCO-Von Roll Environmental Corp. 1984. Thermal systems for waste and refuse. Baltimore, MD.

ENRECO, Inc. 1985. *Project Reports.* Amarillo, TX: ENRECO, Inc.

Environmental Protection Agency. 1978. *Land Cultivation of Industrial and Municipal Solid Wastes, A State of the Art Study.* EPA 600/2–78–1402. Washington, DC: EPA.

Environmental Protection Agency. 1980. *Everybody's Problem, Hazardous Waste.* Washington, DC: Office of Water and Waste Management.

Environmental Protection Agency. 1985. "Assessment of Incineration as a Treatment Method for Liquid organic Hazardous Wastes. Summary and Conclusions." Washington, DC: U.S. EPA, Office of Policy, Planning and Evaluation, March.

Environmental Protection Agency. Fungus eats away at toxins. *The Miami Herald,* June 15, 1985, p. 1.

Export plan for wastes stirs fears. *Miami Herald,* July 1, 1983, p. 1G.

Federal Register. 1980. Washington, DC. U.S.A. 261.21–261.24 EPA HW #D001-D003 and DO 17.

Findle, B. 1996. They're cleaning up. *San Diego Times Union,* June 26, 1996, p. C-1.

First plant in U.S. starts chemical destruction. *North County Times,* Washington, DC: AP, Aug. 21, 1996.

Fitzpatrick, D. T. 1983. Machine shop scrap processing and recycling. *Toxic and Hazardous Wastes,* p. 569.

Francis, G. Z. 1984. Landfilled wastes treated in place. *Pollution Engineering.* 16(Sept. 1984):37.

Freeman, H. 1981. Hazardous wastes incineration. In: Peirce and Vesiland, eds. *Hazardous Wastes Management,* Chapter 5, p. 59. Ann Arbor, MI: Ann Arbor Science Pub. Inc.

Giltenan, E. F. 1985. Pollution exclusions hit truckers hard. *Chemical Business,* October 1985, p. 38.

Graham, F. J. 1985. PCB's make a slow exit. *Chemical Business,* September 1985, p. 72.

Great Lakes study: chemicals imperil food for millions. *The Miami Herald,* December 12, 1985, p. 3C.

Greenhouse, S. 1980. Conference backs curbs on export of toxic waste. *New York Times,* March 23, 1989, p. 1.

Hanson, N. 1984. *Air Stripping Is Effective in Removing TCE from Groundwater,* p. 18. Water/Engineering and Management.

Hazardous waste conferences planned. *Highlights (WPCF)* 1985;22(12):5.

Hazardous waste landfill benefits from compost. *Biocycle,* Sept.-Oct. 1983, p. 28.

Herzbrun, P. A., R. L. Irvine, K. C. Malinowski. 1985. Biological treatment of hazardous waste in sequencing batch receptors. *J. Water Pollution Control Fed.* 57:1163.

Incinerator belches foul smoke. *The Miami Herald,* Section B, March 25, 1985, p. 1.

Irvine, R. L., A. W. Busch. 1979. Sequencing batch reactors—an overview. *J. Water Pollution Control Fed.* 51:235.

Irvine, R. L., et al. 1984. Enhanced biological treatment of leachates from industrial landfills. *Hazardous Wastes* 1:123.

Jacobs Engineering Group. 1988. *The EPA Manual for Waste Minimization Opportunity Assessments.* Pasadena, CA: Jacobs Engineering Group. EPA/600/2-88-025.

Jaffe, M. 1983. Authorities struggle to curb illegal toxic waste dumping. *The Miami Herald,* Dec. 1, 1983, p. 13C.

Jhaven, V., A. J. Mazzaca. 1983. Bio-reclamation of ground and groundwater: case history. Presented at: 4th National Conference on Management of Uncontrolled Hazardous Waste Sites, Oct. 31–Nov. 2., 1983.

Krishnan, S. V., N. L. Nemerow, et al. 1996. An Environmentally Balanced Complex of Fertilizer and Cement Plants, Master's Thesis, University of Miami, Mechanical Engineering Department, Nov. 1996.

Krog, K. 1985. Nation's toxic waste is a killer topic literally [Editorial]. *Miami Herald,* Sep. 1985, p. 22A.

LaBranche, D. F., M. R. Collins. 1996. Stripping volatile organic compounds and petroleum hydrocarbons from water. *Water Environment Fed.* 68(3)348.

Lagrega, M. D., Hendrian, L. K., eds. 1983. *Toxic and Hazardous Waste. Proceedings of the 15th Mid-Atlantic Industrial Waste Conference, June 26, 1983.* Boston, MA: Butterworth Publishers.

Land-poor but trash-rich, the Japanese take over Bay. *The Miami Herald,* February 4, 1986, p. 24.

LaPue, S. 1996. Concern voiced on cleanup of toxics. *San Diego Union Tribune,* Aug. 5, 1996, p. B-1.

LeRoy, E. 1983. Processing hazardous waste in France. *Industry and Environment-Industrial Hazardous Waste Management.* UNEP Special Issue No. 4, p. 46.

Make polluters pay [Editorial]. *The Miami Herald,* Oct. 14, 1995, p. 22A.

Metry, A. A., M. F. Coia, M. H. Corbin, et al. 1983. *In situ* closure of sludge lagoons. *Toxic and Hazardous Wastes,* p. 538.

Mims, B. 1996. Army starts destroying chemical arms. *San Diego Union Tribune,* Aug. 23, 1996, p. A-2.

National Conference on Waste Exchange. 1983. *Proceedings of the National Conference on Waste Exchange, Florida State University, Tallahassee, Florida, March 8–9, 1983.*

National Research Council. 1984. *Ocean Disposal Systems for Sewage Sludge and Effluent.* Washington, DC: National Academy Press.

Nemerow, N. L. *Industrial Solid Wastes,* Chapter 10. Ballinger Publishing Company.

Nemerow, N. L. 1952. Textile dye wastes. *Chemical Age.*

Nemerow, N. L. 1957. Color in industrial wastes. *J. San. Eng. Div. Am. Soc. Civ. Engrs.* Paper No. 1180. 83(SA1), Feb. 1957.

Nemerow, N. 1963. *Theories and Practices of Industrial Wastes Treatment,* p. 243. Addison Wesley Co.

Nemerow, N. L. 1974. *Scientific Stream Pollution Analysis.* New York: McGraw Hill.

Nemerow, N. L. 1982. *Industrial Water Pollution Control in Hong Kong.* UNIDO Report Vienna, Austria, July 5, 1982, DP/HOK/80/11-51/32.1J.

Nemerow, N. 1984. Industrial wastes. In: *Kirk-Othmer Encyclopedia of Chemical Technology, Vol. 24, Third Edition,* p. 228. New York: John Wiley & Sons.

Nemerow, N. L. 1985. *Stream, Lake, Estuary, and Ocean Pollution.* New York: Van Nostrand Reinhold Publishing Company.

Nemerow, N. L. 1987. *Industrial Water Pollution.* Malabar, FL: Robert Krieger Publishing Company.

Nemerow, N. L. 1988. *U.S. Pakistan Joint Research on Advanced Renovation of Small Industry Wastes.* Washington, DC: National Science Foundation. Project INT-85-20198.

Nemerow, N. L. 1995. *Zero Pollution of Industry.* New York: John Wiley & Sons.

New transfer waste facility built in central Florida. *The Florida Specifier,* April 1985, p. 20.

Noland, J. W. 1984. Incineration of explosives contaminated soils. *The Weston Way* 10(1) Winter: 3–6.

Nyer, E. K. 1985. *Groundwater Treatment Technology,* p. 149. New York: Van Nostrand Reinhold Pub. Co.

Oberteuffer, J. A. 1973. High gradient magnetic separation. *IEEE Trans. Magnetics* 9(3):303.

Oil Recovery Systems, Inc. *Air Stripping: A Technical Data Bulletin.* Norwood, ME: Oil Recovery Systems, Inc.

Piasecki, B. 1983. Unfouling the nest. *Science* Sep:78.

Pizzuto, J. S. 1981. Superfund: the U.S. response. Presented at: Symposium on Waste Disposal—The Challenge Madrid, Spain, International Association of Environmental Coordinators, Brussels, 1981.

Pro-form, Inc. 1985. Proform-SFRP Track Collector Pan System. Paducah, KY.

Prouty, M. F., J. Alleman, N. Berman. 1983. Sludge amended brick manufacture. *Toxic and Hazardous Wastes,* p. 492.

Public Service Electric and Gas Co. 1984. *Annual Report,* p. 14. Newark, NJ: PSE&G.

Raloff, J. 1984. Salt of the Earth. *Science News* 126(19):298.

Raeburn, P. 1985. Helping crops stand up to salt. *Industrial Technology,* May 1985, p. 88.

Randle, S. Some new treatments for hazardous waste. *Chemical Business,* February 1989, p. 36.

Reducing Hazardous Waste Generation. A Digest of the Report by National Academy Press. Washington, DC: National Academy Press, 1985.

Sachdev, D. R., et al. 1983. *Use of Fly Ash as a Liner Material for Utility Solid Waste Disposal Sites. 2nd Conference on Municipal, Hazardous and Coal Waste Management.* Coral Gable, FL: Pergamon Press.

Sax, I. 1979. *Dangerous Properties of Industrial Materials, A Text,* 5th edition, p. 824. New York: Van Nostrand Reinhold Publishing Co.

Shabecoff, P. 1985. EPA to permit waste burning off Jersey coast. *New York Times,* Nov. 27, 1985, p. 16.

Shabecoff, P. 1985. Tangled rules on chemical hazards hamper U.S. efforts to protect public. *New York Times,* November 27, 1985, p. 13.

Sludge extractor reduces quench oil use, waste load. *Pollution Equipment News,* October 1985, p. 93.
Statement by the Honorable David R. Peterson, The First Session of the Thirty-Third Parliament of the Province of Ontario, July 2, 1985, p. 29.
Stone, R. B. 1985. *Update: storage of hazardous waste in mined space. Disposal of Hazardous Waste Proceedings 11th Annual Research Symposium,* p. 339. Environmental Protection Agency. EPA/600/9-85/013.
Sunohio. New process to detoxify PCB's approved by U.S. *Miami Herald Newspaper,* May 29, 1981.
Taylor, R. EPA clears test on incinerating toxic waste at sea. *Wall Street Journal,* Nov. 27, 1985, p. 20.
The Southern Waste Information Exchange Catalog (SWIX), Vol. 11, No. 1, Oct. 1982.
Thorsen, J. W., J. C. Petura. 1983. Plan for your RCRA Part B permit. *Clearwaters,* Winter 1983:12–15.
Torricelli, R. G. Let's destroy toxic waste, not just move it around [Editorial]. *Wall Street Journal,* Sep. 26, 1985, p. 30.
Toxic sites close rather than comply. *The Miami Herald,* Dec. 7, 1985.
United Nations Environmental Protection. 1983. Industrial hazardous waste management. *Industrial and Environment Special Issue No. 4.* Paris, France: UNEP.
U.S. waste deadline likely to shut down many dumps. *New York Times,* Nov. 8, 1985, p. 11.
Van Gils, G. J., M. Pirbazari, S. H. Kim, et al. 1984. *Future of Water Reuse—Water Reuse Symposium,* Vol. 2, p. 911. San Diego, CA, August 26–31.
Van Noordwyck, H. J., et al. *Quantification of Municipal Disposal Methods for Industrially Generated Hazardous Wastes, Proceedings of the 6th Annual Research Symposium on Treatment of Hazardous Waste.* USEPA-MERL, Cincinnati, Ohio. EPA/9-80-011.
Weber, W., Jr. 1986. Organic contamination: whistling past the graveyard. *J. Water Pollution Control Fed.* 58(1):16.
Welch, K. Nuclear industry shouldn't escape liability fallout (U.S. Public Interest Research Group). *The Miami Herald,* January 9, 1986, p. 31A.
Wench, N. C., P. D. Josephson. 1984. Hazardous wastes sprayed away. *Pollution Engineering* 16(9):33.
Wilcox, J. B., F. J. Agardy. 1990. Thermal destruction of air toxic VOC's using packed bed technology. Proceedings of the 83rd Annual Meeting & Exhibition, Air & Waste Management Association, Pittsburgh.
Worthy, W. 1982. Hazardous waste: treatment technology grows. *Chemical Engineering News,* March 8 1982.
Wright, A. P., S. D. Caretsky. *In situ* treatment/containment and chemical fixation. In: *Hazardous Wastes Management,* Chapter 6. Ann Arbor, MI: Ann Arbor Science Pub., Inc.
Yakowitz, H. 1988. *Identifying, Classifying and Describing Hazardous Wastes.* UNEP Ind. & Env. Jan/Feb/Mar/1988, p. 6.
Zoltek, J. 1984. Bacteria digesting toxic wastes. In: *The Overflow.* Florida Water Pollution Control Operators Association, Nov-Dec. 1984, p. 13.

CHAPTER 12

Removal of Industrial Air Contaminants

Industries discharge two types of air contaminants: (1) objectionable solids dispersed in a gaseous waste and (2) objectionable gases dissolved in a gaseous discharge. It falls upon the environmental engineer—with his or her chemical engineering background—to understand these forms of contaminants and design systems for removing both types of contaminants.

Typical examples of the first type include dusts such as coke dust from steel mills, cement dust from cement plants, and fiber dust from pulp and paper mills. The second type includes ammonia as from coke mills, hydrogen sulfide, and sulfur oxides from food processing and paper mills, and organic vapors from paint and plastic manufacturing.

Dust particles not only foul the objects and land on which they are deposited, but also are breathed in by humans and animals causing irritations and consequential disease to respiration systems. Gaseous contaminants more directly interfere with respiratory systems and olfactory organs. Foul odors depict objectionable contaminants whether or not they are proven detrimental to human health.

Theories and Practices of Suspended Particle Removal

When contaminant particles are present in industrial waste gases, they require removal in order to protect the receiving environment. Treatment of these waste gases containing suspended solids is usually accomplished by passing them through solid bed contactors.

These solid beds can be either stationary or moving and can be in either concentrated or diluted form. Four examples of such bed contactors are described here briefly to provide you with basic background knowledge only.

1. *Dense fixed beds*, which depend on the densely packed solids in the contactor to remove by adsorption the waste gas particles. In this treatment system, the solids contactor is fixed in place and immobile (Figure 12.1A).
2. *Expanded beds*, which move the adsorbing solids about in some degree by either or both gravity or mechanical forces. Such contaminant particulates are typically removed by coming in contact with a slightly moving bed. An example of this treatment system is shown schematically in Figure 12.1B.
3. *Fluidized beds*, in which the adsorbing solids are moved by the impact of the incoming contaminated gases, are an effective and often used system. Such a treatment depends on the difference in forces of the incoming gas and the solid absorbent. Usually the gas flow is upward and the absorbent force is downward—the latter being greater, thus keeping the treatment unit intact. Such a system is shown schematically in Figure 12.1C.

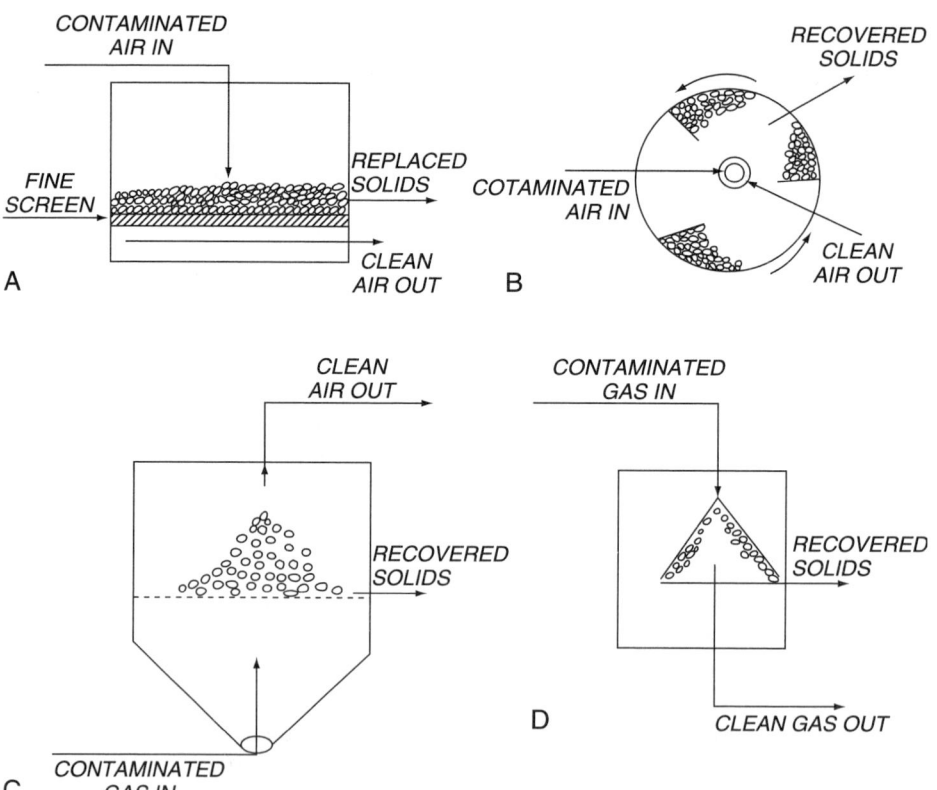

FIGURE 12.1. Treatment of waste gases.

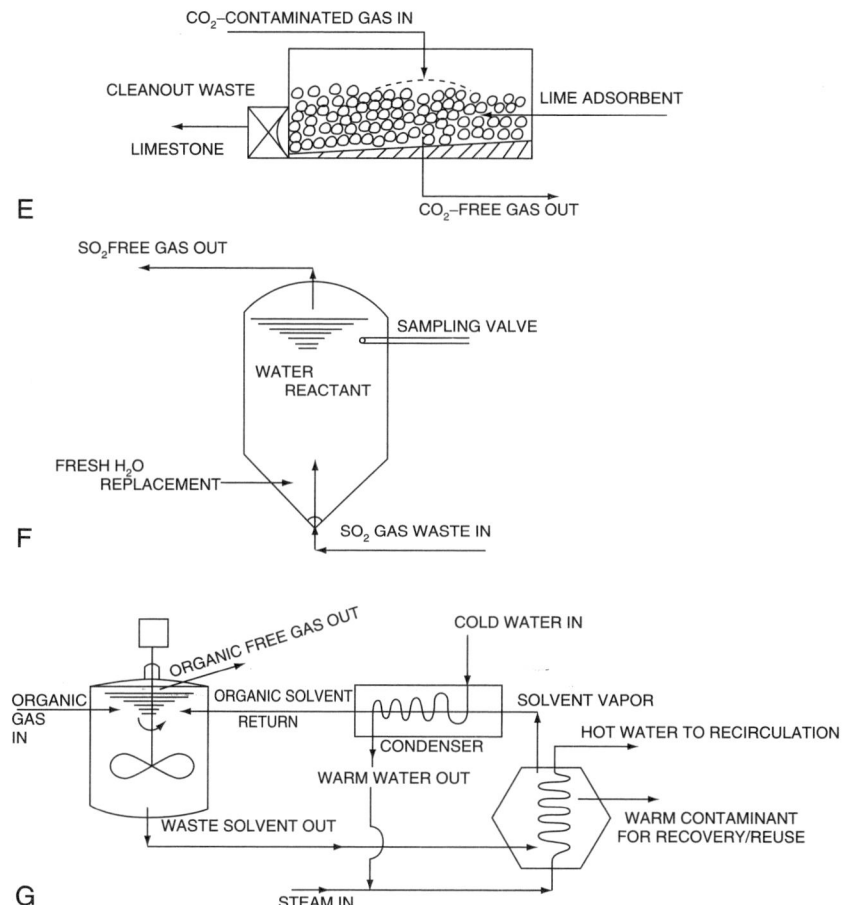

FIGURE 12.1. (*Continued*)

4. *Dispersed solid system*, in which the density of the suspended system is almost precisely that of the gas phase itself. This is such a dilute system that no bed of solids actually exists and the gas is in reality pressured into a chamber where its velocity decreases and the dispersed solids separate from the gas and are directed towards and collected along the tank or vessel walls. Such a system is presented schematically in Figure 12.1D.

Objectionable contaminant gases dissolved in gaseous discharges must also be treated and removed by the industrial waste environmental engineer. Typical of these types of industrial wastes are inorganic vapors such as ammonia (NH_3), carbon dioxide

(CO_2), sulfur dioxide (SO_2), and hydrogen sulfide (H_2S), and organic vapors such as trichloroethylene (C12C:CHCl), phenols (C6H5OH), and acetone or alcohols.

These gaseous contaminants are generally removed by contact with either solid or liquid reactants. For example, CO_2 is removed by passing the gas through a bed of lime reactant, and the resulting product is limestone ($CaCO_3$), while the remainder of the waste gas passes through. Periodically the product, $CaCO_3$, is removed and sold or reused and the reactor replaced with fresh lime. Such treatment systems are shown in Figure 12.1E.

Other gaseous wastes such as SOxs are removed by passing these gases through liquid reactants, such as water. When the reactant has been completely used in forming sulfuric acid (H_2SO_4), it is removed, sold, or reused, and freshwater replaced in the reactor. Such a treatment system is shown in Figure 12.F.

Ammonia gases can be reacted either as in Figure 12.1E and removed as (NH_4)$2CO_3$ or as in Figure 12.1F and removed as NH_4OH.

Organic vapors are retained in reactors with other organic solvents such as benzene or alcohol (depending on the type of organic contaminant). They are later removed from the organic reactor solvent by distillation for recovery and reuse of both the contaminant and the reactor solvent. Such systems are depicted schematically in Figure 12.1G.

In all these second-type systems, the removed and recovered contaminant products must be either sold or reused within the industrial plant to avoid the simple conversion of an air contaminant to a solid or liquid one. For more detailed designs and equipment specifications for all the industrial air contaminant reactors, see Nemerow (2003).

Example of Twentieth-Century Air Contaminant Removal

Heitbrink and McKinnery (1996) reported an evaluation by the National Institute for Occupational Health and Safety (NIOSH) of control techniques for airborne vapors and particulates in tire manufacturing. The mixers studied normally generate aerosols. They found that control of these air contaminants may be accomplished by using local exhaust ventilation or equipment configuration and material substitution.

Major Industrial Wastes

In 1998, I presented in *Strategies of Industrial and Hazardous Waste Treatment* a detailed description of the origin, character, and treatment of all of the major industrial wastes of the twentieth century. The reader is urged to consult this work for specific information concerning these wastes.

These wastes subdivided into eight categories: (1) apparel, (2) food, (3) materials, (4) chemical, (5) energy, (6) radioactive, (7) non-point, and (8) hazardous.

In the following section, I list these same wastes in convenient alphabetical order for your immediate reference. Keep in mind that many of these industries have modified their processes or eliminated certain products completely since 1998. In addition, new industries have begun production since that time. You can refer to the unnumbered table at the end of this chapter for some of these new industries, as well as Part B of this book.

Industrial Wastes Detailed in Nemerow and Agardy (1998)

Acid and alkali
Agriculture
Animal glue
Asbestos

Bakery
Beet sugar
Biological
Brewery

Cadmium
Cane sugar
Cannery
Caustic (spent)
Cement
Chloralkali
Coffee
Copper
Cornstarch
Corrosion hazardous

Dairy
Detergent
Distillery
Dry cleaning
Dye

Energy
E.P. toxic
Explosive

Feedlot
Fertilizer
Fish
Food
Formaldehyde
Fuel and lube oil
Fuel (radioactive)

Galvanizing and metal plating
Glass
Graphic and photographic

Hazardous
Hospital and laboratory

Ignitable
Iron foundry

Land farming
Land and non-point
Landfilling
Laundry
Leachate

Meat packing
Metal container
Metal plating
Mortuary
Motor

Naval storage
Nitrogen fertilizer
Nuclear power

Oils (waste)
Oil field
Olive

Paint and printing
Palm oil
Pesticide
Pharmaceutical
Phenolic
Phosphate
Photographic
Pickle
Pickling liquor
Plastic and resin
Plywood and glues
Polychlorinated biphenols
Porcelain enameling
Power plant
Pulp and paper

Radioactive
Reactive industry
Refinery
Rice
Rubber

Salts
Scrubber
Silver
Soap and detergent
Soft drink
Spent oil emulsion

Steam power plant
Spent solvents
Steel mill
Sulfides

Tannery
Textile
Toxics
Trihalomethane

Winery
Wood furniture
Wood preserving

I have added since 1998 the following four wastes:

	Origin	*Major Characteristics*	*Major Treatment*
Red bricks (building)	Cleaning of, grinding, mixing, screening and forming of equipment	Minerals	Return to original solids
Waxes carnuba	Semi-liquid filtered matter	Hazardous companies	Landfilled (safe) with solid wastes
Paraffin	Solvent extraction from lube oils	Complex organics	Evaporated for solvent recovery and landfilled
Beeswax	Boiled water discard	Complex organics	Filtered and landfilled

References

Heitbrink, W. A., W. N. McKinnery, Jr. Control of air contaminants at mixers and mills used in tire manufacturing. *Am. Industrial Hygiene Assoc.* June;47:312.

Nemerow, N. L. 2003. *Environmental Engineering*, 5th edition, Chapter 6. Hoboken, NJ: John Wiley Publisher.

Nemerow, N. L., F. J. Agardy. 1998. *Strategies of Industrial and Hazardous Waste Treatment, Part 4*, pp. 281–698. Hoboken, NJ: John Wiley Publishing Company.

Part B

Twenty-First Century

Foreword to the Twenty-First Century

After working in this field (what is now known as Environmental Engineering) for a half century, I feel the need to suggest a solution to the never-ending pollution problem. No solution is perfect, and certainly this one has its flaws. But not only is this the logical outcome of rational thinkers, but it is also ultimately economically and sociologically sound. If followed completely, it will require an entire restructuring of our industrial location and operation engineers, because they must consider other industry needs besides their own in making production choices. Society, however, will benefit from lower prices for goods and less pollution of the environment. What better goals to work for! The "waste utilization engineer" will replace the conventional "waste treatment engineer." Clean air and water will replace polluted environments. Lower costs with perhaps higher standards of living will replace more expensive products. All it takes is the strength of our conviction and an innovative mind. I've done my part by suggesting this primrose path. Now it's up to you to read, assimilate, and act on the contents of Part B of this book!

Preface to the Twenty-First Century

"When you look out from Mount Rainer, it is an awesome sight and it makes you realize yet again why it is so important to have clean air and clean water and save the global environment and I will fight to protect the environment with all my heart and soul."

—Al Gore
Everett, Washington
October 23, 2000

At the risk of sounding melodramatic, I believe that the answer to saving our environmental planet is in Part B of this book. I have been studying, observing, and analyzing the plight of industry as well as municipalities in attempting to abate their pollution for over 50 years. In our present era of merging forces for enhancing industrial economics, the time has come for industries and cities to use their wastes in ways that will result in zero pollution for the environment. Waste utilization must replace waste treatment in order to preserve our fragile environment.

Nelson Leonard Nemerow

CHAPTER 13

Prologue to the Twenty-First Century

In the 1940s, 1950s, and even into the 1960s, industrial waste treatments were patterned after those used to treat municipal wastes. In the 1960s and 1970s, we favored combining industrial wastes with municipal ones. During the 1970s and 1980s, we began to treat industrial wastes separately with more exotic and efficient methods and to seek exterior markets for them in other industries. All of these methods and systems from the 1940s to the 1980s were only partially successful and certainly inappropriate and generally outdated by the 1990s.

During the 1950–1980 period, when industry piped and even pumped industrial wastewaters into municipal sewers and sewage-treatment plants, representatives were aware of the accompanying problems and the inefficiencies that resulted. However, industry overlooked these and succumbed to the relative ease of disposal. Industrial costs were lower and legal responsibility for environmental degradation was avoided or at least jointly shared with the municipality.

Major requirements had to be met before industrial wastes could be treated along with municipal sewage. These requirements are valid even today and are more difficult to meet now than in earlier years (Nemerow and Agardy 1998). Such requirements include technical, economical, legal, and managerial ones.

Technical

Industrial wastes must be compatible (treatable) with the sewage. Industrial wastes must be equalized and proportioned to the flow and pollution load of the sewage. Industrial wastes should not contain any material toxic or detrimental to the operational objectives of the sewage-treatment plant. Industrial wastes should not contain any substances hazardous to the operating personnel or those of the environment near the effluent discharges. Industrial waste bypasses to the treatment plant must be provided if and when these wastes fail to meet the aforementioned four requirements. Proper treatment of the bypassed industrial waste must be employed to ensure environmental protection.

Economical

The cost to industry for combining its waste with city sewage must be low enough to provide sufficient incentive to use this system. Such provisions have usually been attained by the imposition of a sewer service charge. The charge often is based on the industrial pollution load that is being treated and/or removed by the treatment system. However, the variation in economic benefit to industry has been tremendous from one city to another. It has varied from the free use of the sewer and sewage-treatment plant or a nominal charge based on flow similar to householders to as prohibitive as a charge based on all extra pollutants. The free or nominal use of the municipal system has long been an unofficial "boon" to industry. Even the prohibitive charge to industry has often been economically acceptable because it avoided legal and/or managerial responsibilities on the part of the industry itself.

Legal

Once an agreement between the city and its industry has been reached, industry avoids at least part, if not all, of the lawful responsibility of any external environmental damages. Many times this alone has provided sufficient incentive to industry for this mode of solution to its waste problem. Industry often used this system when its wastes were difficult to treat or even hazardous when discharged separately to a watercourse.

Managerial

To industries of the 1940–1970 era, treatment and disposal of wastes represented a managerial burden. Because of this industrial attitude, combined treatment with municipal sewage was viewed as a welcome solution.

Conditions for Rejection

Despite its general use and acceptability, there were and still are situations in which combined treatment is not recommended. I recall the two most important ones here:

1. Perhaps most important of these in the long run is the lack of compatibility between municipal and industrial officials. As a result of a lack of a good understanding and cooperative relationship between them, technical problems will not be resolved satisfactorily.
2. Physical problems of industrial plant location, pumping, and the nature of hazardous waste may make this system unworkable.

All other adverse situations often can be overcome, if these two causes for rejection are absent.

Methods of Industrial Waste Reuse

Because of the problems and ineffectiveness of combined treatment and the situations that automatically call for its rejection, industry has sought other solutions. The most logical and most used solution is that of reusing its own wastes. The method first and generally used is reusing wastes within its own industrial plant. When possible, in-plant reuse is most economical and effective. However, such situations are ideal and seldom encountered in actual practice. As one plant manager put it, "Our wastes are wastes because they are not useful to us in any way."

The next potential for waste reuse is to contract with a "scavenger" collector to transport such wastes to a large central industrial waste-disposal plant. Such systems are costly to an industry for small volumes of wastes and impractical for larger volumes. They may, however, relieve the industrial plant of managing liability of disposal, which I discussed earlier.

The last waste-reuse technique involves the direct marketing of waste as a resource for another industrial plant. To use this method, industry must go through the familiar systems of selling its waste as it does its products. Selling is neither easy nor guaranteed and may involve giving more waste product characteristics information than it would like. Finding a suitable buyer for its waste, in fact, may even be more difficult than for its primary products. Waste exchanges, which publish industry waste needs and prices, assist plants to find buyers. Disclosure of waste character for sale remains a stumbling block for sellers using the direct method. A byproduct synergy system developed during the 1990s is an advanced method of direct reuse by another industry (see Chapter 9-D).

In replying to the question of whether the costs of recycling manufactured products is really less than the original cost of producing these same products Sherry Seethaler (2003) opines the following limited to aluminum cans, paper, plastic, and glass:

Aluminum: "Recycling aluminum is both economically and environmentally advantageous." She refers to rather well-known Environmental Protection Agency (EPA) data showing that it takes 95% less energy to make a can from recycled aluminum than from virgin bauxite. Also, the cans can be repeatedly recycled, thus saving valuable landfill space.
Paper: She admits that "recycling paper requires more water than reducing paper from wood but releases fewer toxic chemicals." Also, the recycling of paper eliminates the production of methane, a greenhouse gas, in landfills.
Plastic: She points out that plastic reuse (by melting and converting to other products) is "environmentally sensible because plastic is derived from crude oil." However, the wide variety of additives and other contaminants makes separation "arduous and expensive."
Glass: The original manufacturing of glass (like aluminum) requires very high production temperatures to melt and fuse the quartz sand, soda ash, limestone, and minerals, whereas recycled crushed glass melts at a much lower temperature, thus requiring less energy when it is added to the raw materials.

The Ultimate for Industrial Waste Reuse: EBIC

All of the previous descriptions of combined treatment lead to the main conclusion of this prologue. Because of the aforementioned inadequacies of combined treatment, a new solution is evolving. I have designated its terminology as the environmentally balanced industrial complex (EBIC). This new system has developed not only because the previous solutions were inefficient, but also because industry and society have grown exponentially so as to imperil our environmental resources beyond their capacity.

The EBIC can be described simply (Nemerow 1995) as a selective collection of compatible industrial plants located together in one area (complex) to minimize (or eliminate) both environmental impact and industrial production costs. These objectives are met by using the waste materials of one plant as the raw materials for another with minimal transportation, storage, and raw material preparation. When a manufacturing plant neither treats its waste nor stores or pretreats certain of its raw materials, its overall production costs must be reduced significantly.

Elimination of waste-treatment costs alone may be sufficient to influence industrial managements to continue to produce their products in the highly competitive world market. It should be our obligation to minimize waste-treatment costs and maximize protection of the environment.

Reuse costs within these complexes can be absorbed easier into production costs than end-of-the-line waste-treatment costs. Despite the advantages of the EBIC, many factors must be identified, clarified, and answered properly before a complex system can be accepted. For example, reasonable matches of waste quantities and raw materials must be established. In addition, the type of labor and worker numbers available in an area, as well as the marketing of products from the area, need clarifying. The key to feasibility for any complex lies finally in production economics and environmental protection.

Although we expect that during the twenty-first century many U.S. industries will transfer to overseas locations, certain ones are expected to thrive in the United States. Medical instrument manufacturing that requires careful monitoring and collaboration between hospitals, doctors, and manufacturers is one. Others include manufacturers that produce large equipment such as household items and furniture (difficult and costly to ship from afar), and producers of foodstuffs that are either frozen, perishable, or both. Hightech and innovative new industries will probably also be on the rise locally in the United States during the twenty-first century. Industries such as aluminum plants may also increase and return to mainland United States, if and when less expensive electrical energy becomes available. All of these types of industries are good candidates for an industrial complex system.

A realistic and optimistic suggested procedure for future industry was proposed on the op-ed page of the *Wall Street Journal* ("The greening of America, June 27, 2003, p. A12). "The key to future 'green' progress is maintaining (not stifling) the free market growth and innovation that can produce hydrogen cars or find a way to turn wind into cheap power." The op-ed writer acknowledges that free marketers admit that pollution is an "externality" that has yet to be internalized into normal production costs.

There are many potential EBICs that will yield several salable products and result in no wastes reaching the external environment (Nemerow 1995). Another important advantage that is readily apparent to the environmental engineer is that generally no waste treatment is necessary within the complex. When, in some cases, waste treatment is required to render it directly reusable, another product also results.

There is an opportunity for industry and municipality to combine in the future to produce industrial products directly from municipal wastes. In such cases, the solid contaminants contained in municipal sewage would be converted within the treatment plant to industrial products for sale instead of for disposal on the land.

For example, settling tank sludge can be rotary dried, pulverized, amended, bagged, and sold as fertilizer to the agricultural industry. Floating matter from this same settling basin can be skimmed and rendered by steam heat treatment to produce—with certain additions—animal feed for this same agriculture industry. We practice both these treatments today to some degree and in a few instances. However, a concerted effort needs to be made to design municipal treatment plants to include industrial production as an integral part of its operation. This also requires a closer collaboration of industrial and municipal services rather than a haphazard afterthought following municipal sewage treatment.

Another example is the fish-food industry and municipal sewage treatment. In this case, the sewage treatment plant would be designed not only to purify its wastes, but also to produce fish such as Tilappia for sale from this same treatment. Effluent from such a plant would be recirculated into algae ponds that serve as food for fish pond production. Excess effluent would be sold to the agricultural industry for irrigation water.

I challenge municipal, federal, industrial, and environmental engineers who read this book to "think" design and operation of truly combined treatment so that no effluents reach our fragile and disappearing environment. At the same time, we will be producing valuable industrial products at lower costs.

References

Nemerow, N. L., F. J. Agardy. 1998. *Strategies of Industrial and Hazardous Waste Treatment,* Chapter 16, pp. 231–242. New York: John Wiley Publishers.

Nemerow, N. L. 1995. *Zero Pollution for Industry,* p. 109. New York: Wiley-Interscience Publishing Company.

Seethaler, S. 1995. Questions answered. *The San Diego Union-Tribune* Section F.

CHAPTER 14

Rationale of Environmentally Balanced Industrial Complexes

Industry's contribution to the era of waste utilization is highlighted by its acceptance to use the ultimate in waste utilization: the environmentally balanced industrial complex (EBIC) system. The system comprises two or more compatible industrial plants located in close proximity to one another in one complex. Each plant uses the waste of another plant as part or all of its raw material. No wastes are discarded into the environment outside the complex. No wastes have to be transported or sold outside the complex. The environment is protected from environmental degradation. And people (consumers) benefit from lower-priced products. Industry also gains a competitive price advantage over other plants that choose to operate in separate locations and are encumbered with waste-treatment costs.

Although the real measurable cost of industrial environmental pollution control remains relatively small when compared to total production or value-added costs, it can be a significant amount when considered by itself. In fact, the amount may be enough to influence industry management to consider whether to produce or discontinue the manufacturing of specific goods. Although environmental engineers are usually not involved in that decision, the goal should be to reduce treatment costs to a minimum while protecting the environment to a maximum.

In conventional industrial solutions to waste problems, industry uses separate treatment plant units, such as physical, chemical, and biological systems. These separate treatment systems increase manufacturing costs. These costs are also easily identified and, even if relatively small when compared to other production costs, are opposed by industry. On the other hand, reuse costs, if any, in an EBIC will be difficult to identify and more easily absorbed into reasonable production costs.

Large, water-consuming and waste-producing industrial plants are ideally suited for location in such industrial complexes. Even though their wastes—if released to environment—might cause pollution, such wastes may be amenable to reuse by close association with satellite industrial plants using wastes and producing raw materials for others within the complex.

Examples of such major industries are steel mills, fertilizer plants, sugarcane refineries, pulp and paper mills, and tanneries. Cement plants may also produce the ideal product to allow a perfect match for the phosphate fertilizer plants in a balanced industrial complex.

One needs to choose the proper mix of industries of the appropriate size and locate them in a specific area isolated from other municipal, industrial, or commercial establishments. These choices will be highly influenced by marketing and socioeconomic factors.

Since 1977, I have proposed several typical complexes for tannery, pulp and paper, fertilizer, steel mill, sugarcane, and textile industries (Nemerow et al. 1978, 1980, 1987; Nemerow and Dasgupta 1981, 1984, 1985; Nemerow 1980a,b, 1984; Tewari and Nemerow 1982; Nemerow and Veziroglu 1988). Such complexes have the presumed advantages of minimizing production costs and adverse environmental impacts. Optimization of these advantages will meet the objectives of both industries and environmentalists.

Although the advantages of this type of complex are obvious, there are certain difficulties to overcome. One involves compatibility. There is no evidence that waste and product compatibility necessarily mean industrial working compatibility. Other plant operating requirements such as labor availability, marketing of products, and taxes may not mesh as easily.

Another involves optimal mass balances. Again, there is no evidence showing that all plants within such a complex can operate at or near their optimum production required for economic purposes. However, lack of evidence is no reason to discard the principle, but reason for more complete investigation and trials. In the middle to late 1990s, the field of "industrial ecology" became recognized as one step in promoting "waste utilization" rather than "waste treatment." An ardent supporter of this concept is Suren Erkman[1] who has surveyed this field and written extensively on the modern interpretation of industrial ecology. In one of his latest contributions to this concept (2001), he summarizes as follows:

> Industrial ecology aims at looking at the industrial system as a whole. Industrial ecology does not address just issues of pollution and environment, but considers as equally important, technologies, process economics, the inter-relationship of businesses, financing, overall government policy and the entire spectrum of issues that are involved in the management of commercial enterprises. As such, industrial ecology can provide a conceptual framework and an important tool for the process of planning economic development, particularly at the regional level. Also, industrial ecology may offer options, which are not only effective for protecting the environment but also for optimising the use of scarce resources.

---

[1] Suren Erkman, Institute for Communication and Analysis of Science and Technology, Geneva, Switzerland.

Thus, industrial ecology is especially relevant in the context of developing countries, where growing populations with increasing economic aspirations should make the best use of limited resources.

The reader should recognize the similar philosophies and goals of the "industrial ecology" concept described by Erkman and others and the EBIC concept proposed here. Both "beg" for implementation on a practical scale to verify and fortify their theoretical premises.

In the following chapters, I present justification for both potential and realistic industrial complexes that apply to the aforementioned principles and objectives.

References

Erkman, S. 2001. Industrial ecology: a new perspective on the future of the industrial system. *Swiss Medical Weekly* 131:531–538.

Nemerow, N. L. 1980a. "Preliminary Balanced Industrial Complex: A Three Stage Evolution." A Report to United States Environmental Protection Agency. Contract No. 68-02-3170 RTP, North Carolina, June.

Nemerow, N. L. 1980b. "Environmentally-Optimized Industrial Complexes." Lecture published in *Bound Proceedings of the National Environmental Engineering Research Institute Nagpur, India*.

Nemerow, N. L. 1984. *Environmentally-Balanced Industrial Complexes: The Biosphere: Problems and Solutions*, pp. 461–470. Amsterdam, The Netherlands: Elsevier Science.

Nemerow, N. L. 1995. *Zero Pollution for Industry*, Chapter 19, pp. 105–209. New York: John Wiley Publishing Company.

Nemerow, N. L., F. J. Agardy. 1998. *Strategies of Industrial and Hazardous Waste Management*. Hoboken, NJ: John Wiley & Company.

Nemerow, N. L., A. Dasgupta. 1981. Environmentally-Balanced Industrial Complexes. In: *36th Annual Purdue University Industrial Waste Conference Proceedings*, p. 416.

Nemerow, N. L., A. Dasgupta. 1984. Zero Pollution: A Sugarcane Refinery-Based Environmentally-Industrial Complex. Presented at the 57th Annual Conference of Water Pollution Control Federation, New Orleans, Louisiana, October 1984.

Nemerow, N. L., A. Dasgupta. 1985. Zero Pollution for Textile Wastes. Presented at the 7th Alternative Energy Conference, Miami, Florida, December 1985.

Nemerow, N. L., S. Farooq, S. Sengupta. 1978. Industrial Complexes and Their Relevance for Pulp and Papermills. Presented at the Seminar on Industrial Wastes, Calcutta, India, 1978.

Nemerow, N. L., S. Farooq, S. Sengupta. 1980. Industrial Complexes and Their Relevance for Pulp and Paper Mills. *J. Environ. Intl.* 3(1):133.

Nemerow, N. L., T. N. Veziroglu. 1988. *U.S.–India Joint Research on Industrial Complexing: A Solution to the Phosphogypsum Fertilizer Waste Problem.* Washington, DC: National Science Foundation.

Nemerow, N. L., Waite, T. D., T. Tekindur. 1987. Industrial Complexing and Ferrate Treatment for Reuse of Wastewater of Small Textile Mills. *Proceedings of 8th Miami International Conference on Alternative Energy Sources Session on Environmental Problems, Miami Beach, Florida, December 15, 1987.*

Tewari, R. N., N. L. Nemerow. 1982. Environmentally-Balanced and Resource Optimized Kraft Pulp and Papermill Complex. In: *37th Annual Purdue University Industrial Waste Conference Proceedings,* p. 353, May 12.

CHAPTER 15

Procedure for Industry in Attaining Zero Pollution

Planning for an EBIC

In planning for an environmentally balanced industrial complex (EBIC), several subjects and associated steps must be considered and taken prior to starting the system. They include, at a minimum, the following 11 subject steps:

1. Select and educate the EBIC developer
2. Location
3. Compatibility
4. Optimize production sizing of participating plants
5. Hold an informational meeting with participating plants and regulatory officials
6. Design the flow diagram for the complex
7. Develop a computer program of varied inputs and operating conditions
8. Perform a "dry run" (on paper) of the complex including all potential variations
9. Review architectural and engineering plans of the EBIC
10. Participate and observe construction of EBIC
11. Observe and consult with plants during startup of the EBIC

1. At the onset, the developer and/or purveyor of the land must be sought out and *indoctrinated* with the principles of an EBIC. Preferably, the developer should be shown schematic diagrams of industrial plants operating at separate locations with associated environmental damages. Then, he or she should be shown a schematic overlay of the same industrial plants located adjacent to each other in the EBIC, with no adverse environmental damages. At the very least, the developer should also be shown examples of EBICs with proven economic advantages of lowered real production costs (see Chapter 16B). The developer should be convinced to seek suitable industries to relate the economic and environmental advantages of his or her EBIC. The developer is then fortified as a seller of a more advantageous land-utilization system

2. The *location* of the proposed complex is an extremely vital component of the plan. The site must be acceptable to each of the industrial participants. Each will have its own preferences depending on many components such as source of raw materials, market for its products, and availability of economical utilities. Compromises by all participants will be necessary to arrive at a site agreeable to all of them. One thing that is certain, however, is that no longer will the managers of industrial plants need to be concerned with the effect of their wastes on the environment. This factor has been eliminated by the use of the EBIC. In fact, now the plants may be able to locate at a site more favorable to other production and marketing decisions. For example, a site adjacent to a metropolitan or residential area may be selected, because pollution of the city or homes is stopped by using an EBIC. A site may eventually be selected that will be less costly to purchase or operate on once concern for external pollution is relieved.
3. *Compatibility* between plants will be sought. The developer must select proper industrial plants that fill the needs for raw materials of each other. Their wastes must be reasonably suitable for raw materials for each other with little or no alteration. All plants should produce wastes that need reusing by others to avoid environmental contamination. The incentive—both moral and financial—would then exist for locating in an EBIC. In fact, economic incentives are major driving forces for such industries, with social environmental incentives secondary for most industries. It is important to remember that all industries exist to produce a useful product at a profit.
4. The developer must *search and obtain plants of the proper size* so as to optimize production quantities of the participants. It will not do to select an industry that produces more waste (as a normal operation) than can be used by another participant (also as a normal requirement). Naturally, there will be times when one participant will produce an excess of waste because of market demands for its product. The other industry should recognize this and be prepared to either increase its production or prepare to store some waste for future use. It is vital, however, that all of one plant's waste be used eventually by another plant in the complex.
5. An *informational meeting* between all complex parties and the proper state environmental regulatory authorities should be held before the decision to locate, build, and operate the EBIC. At this meeting, it should be made clear to the state that the concept of no wastes reaching the environment is new and binding upon all plants in the EBIC. Flow diagrams, mass balances, and even production data may have to be prepared ahead of time and explained to the state. The concept of waste utilization rather than waste treatment must be explained and defended to all regulatory authorities. These latter will facilitate receiving a state permit to operate.
6. It falls on the consulting engineer for the EBIC to design the *flow diagram* and to prepare and present each plant's flow diagram of its product, water, and wastes (air, water, and solids). The engineer will have to work first with each industry separately and then with personnel of all plants jointly in this venture. It must be explained to all participants that no wastes can be left out and all wastes must be reused by another plant. The consultant must substantiate that all wastes are reusable by another plant. Proof of these assertions may come from the literature,

the experiences of other plants, or even pilot plant studies, if necessary. The consulting engineer's new role is that of designing a mass balance flow diagram that guarantees full waste utilization within the complex. In some cases, it may be necessary for the engineer to design certain waste alteration or amelioration systems to provide a directly acceptable and reusable waste for one of the plants.

7. It is now necessary to *develop a computer program* that contains the various inputs and operating conditions of the EBIC. There are many possible computer programs that can optimize the efficiency of the EBIC, but all of them contain the objective of eliminating all unused wastes. A simple example can be expressed by the following equation:

$$(P1+P2)(W1+W2) = (P1+P2)W3,$$

where

P1 = production units of Plant 1,
P2 = production units of Plant 2,
W1 = wastes of Plant 1 used by Plant 2,
W2 = wastes of Plant 2 used by Plant 1, and
W3 = total wastes unused (not reused) from both plants, plus the reused wastes (W1+W2).

By increasing or decreasing P1 and P2, both W1 and W2 will increase or decrease accordingly. The goal of a perfect EBIC will be to produce no excess waste (W3) within the complex, regardless of the levels of P1 and P2. This can be accomplished only when W1 + W2 = W3.

For example, suppose Plant 1 when operating normally produces 100 units of goods and 20 units of wastes, while Plant 2 produces 50 units of goods and 10 units of wastes with no unused wastes. Then

$$(100+50)(20+10) = (100+50)(30)$$

or

$$150 \times 30 = 150 \times 30$$

and

$$4{,}500 = 4{,}500.$$

And now Plant 1 increases its production to 150 units and 30 units of waste. If Plant 2 does not also increase its production by 50%, then

$$(P1+P2) \times (W1+W2) \neq (P1+P2) \times W3$$

$$(150+50)(30+10) \neq (150+50)50$$

$$200 \times 40 \neq 200 \times 50$$

$$\text{and } 8{,}000 \neq 10{,}000.$$

Thus, Plant 2 must also increase its production to reuse all the waste from Plant 1. The program will reveal how much production each plant must have in order to completely reuse all of the wastes from both plants.

8. Before an agreement of the participants in the EBIC is made, they should engage in a "dry run" on paper including all potential variations of possibilities. Each plant participant would begin by revealing its planned starting production quantity and raw materials needed, along with amounts of liquid and solid wastes expected to occur from that production. At this level of operation, each plant would then agree on the amount of each plant's waste that would be acceptable as a replacement for a portion of its raw material. Ideally, all waste quantities could be used to replace part or all of the raw material requirements of other plants in the EBIC. This would then result in zero discharge of any waste external to the EBIC.

If and when the proportions of wastes to raw material is not an exact match, participating plants must make adjustments in production quantities so that all wastes are reused. If adjustments in productions cannot be made satisfactorily, participants have to agree on how to store or handle the wastes rather than discharge them externally.

Regardless, discussions should ensue concerning all potential variations in production possibilities. These should include no production situations, as well as greatly accelerated rate of production requirements. All potential situations should be discussed at this session so that no surprises or unusual events will occur under real conditions.

Emphasis should be placed on making certain that no situations arise in which unusable waste cannot be handled without discharging outside the complex.

9. *Architectural and engineering plans* for the EBIC are now in order. The engineering consultant for the EBIC should prepare preliminary plans for the location of the piping, plumbing, and mechanical and electrical equipment of the plants. These should be coordinated with those of the architect/engineer for each of the plant participants. The latter people have the responsibility to prepare the design and specifications and location for all production equipment of the plants. The engineering consultant, on the other hand, is responsible for all piping, pumping, and other treatment units for the water supplies and liquid and solid wastes arising from the production of manufactured goods in the complex. The two groups of consultants should work together closely, especially where production and wastes locations interface. In that manner, decisions involving responsibilities for design and rationale for location of equipment can be discussed and agreed upon. The final output from these two groups should provide a master plan containing the entire complex's layout of production and wastes reuse, piping, pumping, and equipment.

It is recommended that all plant engineers and/or managers be present at meetings with the two groups of consultants. After the final master plan has been presented and approved by all plant administrators, these plant managers would be familiar with reasons for and operations of all manufacturing units. This is extremely important if and when any unforeseen EBIC changes in operations are necessary.

The overall aim is to produce a team of EBIC plant managers capable of making and following through on all production and environmental decisions for all plants in the complex.

10. *The participation and observation* of the construction of the complex is of utmost importance. Once the master plan for the complex has been prepared and approved, construction permits must be obtained from local governments and state environmental agencies. The environmental consultant and engineers for all participating industries in the complex should supervise its overall construction.

Of prime importance during this phase of the project is that of making certain that the construction follows the plans accurately. The flow of all liquids and solids to and from all plants should exist as designed. Pipe and pump sizes should be ample to carry waters and wastes several times that of average flows. The consultant should pay special care that no liquids overflow or seep into the ground in the complex, and that no solid wastes be placed on bare ground while awaiting movement to be reused within the complex. Without any doubt, construction should ensure that neither liquid nor solid wastes arising from the participating plants can possibly escape the complex into the external air, land, or water environment.

11. The startup of operation should be *observed by all plant key operating personnel*. The environmental consultant should be on-site during the initial startup of the plants in the EBIC. Timing has to be such that all plants begin production simultaneously. Questions of waste reuse and utilization by the plants will be expected to occur. The consultant's presence provides assurance that these wastes, regardless of quality concerns, will be used entirely within the complex. Modifications may be necessary during the startup period to meet these requirements. Despite all previous planning, innovations may be deemed necessary. For example, some solid wastes may have to be altered or divided among production locations before acceptable reuse. Some liquid waste may require filtering, heating, disinfection, and so on so that production can proceed properly. Once again, the environmental consultant is on hand to answer questions and offer suggestions, but most of all to make sure that no external environments are affected by accidental or purposeful discharge of wastes.

Realistic Implementation of the EBIC System

Perhaps the most important step and most difficult to instigate is that of implementing the EBIC system. I propose that it is most likely to commence in any of the following manners:

1. State statute
2. Business acumen of a property developer
3. Either of the above, but with the aid of local governmental provision of tax-free land and/or services

There have been precedents for all of the above in environmental matters. However, some discussion of each of these procedures is in order.

1. The states have administered water-quality standards since the last half of the twentieth century. These include not only establishing the standards, but also advising and reviewing the planning and procedures, as well as enforcing the rules and regulations. In our case, states would have to be empowered to require all waste-producing industries to locate their plants in complexes so that their wastes are reused as raw materials by ancillary plants. The states would then be assured that no wastes from these industries would reach the environment outside the complex.
2. Ambitious entrepreneurs may acquire a large piece of property and seek and induce compatible industries to build and operate within the property complex. Such property owners/managers must understand and be champions of environmental protection. They must also be able to find and convince these industries that their manufacturing costs would be minimized within this complex, while eliminating their environmental pollution concerns. The support and encouragement of local and state governments would enhance land developers in obtaining industries for the complex. They might even join forces with environmental engineers to advise these industries on waste utilization and water quality. This would be a selling point for the developer and permit him or her to lease at a premium industrial property within the complex.
3. The enactment of state statutes and/or the enterprise of entrepreneurial land developers would be greatly beneficial as local inducements to industry. In the 1940s to 1960s, it was common practice among small governmental jurisdictions to set aside tax-free land to induce desirable industrial plants to build there. Often, they also provided utilities such as water and sewage at reduced rates. They also built pipelines to accept their wastes and roads or railroad spurs to facilitate transporting of raw materials and finished products. Obviously, the combination of 1 and 2 with 3 above would be the most desirable approach to implementing the EBIC concept.

From a practical standpoint, however, it will probably take a catastrophe of some kind before the logical use of EBICs will commence. *Industrial enlightenment* usually occurs as a result of critical situations such as dangerous environmental pollution or disastrous economic conditions. When these occur, and coupled with the three procedures listed earlier, we will see the day of the natural EBIC begin as "standard operating practice."

CHAPTER 16

Economic Justification for Industrial Complexes

Section One

Changing from a "waste-treatment" to "waste-utilization" culture involves overcoming many obstacles. The most important of these is proving to industry that by using an environmentally balanced industrial complex (EBIC), production dollars spent and society's dollars used will be lower. Even then, changes will come slowly. Human nature seems to be comfortable with tried and proven practice, while on the other hand, it abhors changing to an unknown one.

In the long run, history has shown that if we can "build a better mousetrap" at less cost, eventually industry will follow like sheep in the field. It is not within our ability to alter human nature, but we as engineers and scientists can show that the economics of production and conservation of natural resources favor the EBIC approach.

Industry is—and always was—in business to make a profit, and the more profit the better. This is not to say that industry does not recognize and is even sympathetic to the environmental cause, but in its eyes, the bottom line comes before any other consideration. Industry believes that it must concentrate on the bottom line to "stay in the game." And "staying in the game" is absolutely essential. It also has become increasingly aware that reusing water—a valuable diminishing and costly resource—is as important as reusing wastes.

I present one recent (1999) decision made by industry to substantiate the aforementioned principle. The College Retirement Equities Fund (CREF) was established and continues today to invest teachers' retirement money in stocks of industries to produce the greatest growth in profit (the bottom line). A group of CREF participants presented a proposal in November 1999 to the board that it divest itself of its holdings in a particular metals-producing industry stock. The participants' proposal was based on the fact that the industry "created and continues to pose unreasonable or major environmental, health, or safety hazards with respect to the rivers that are being impacted by the tailings, the surrounding terrestrial ecosystem and the local inhabitants."

CREF's board rejected the proposal, saying that "were we to divest from a specific company because some participants object to that company's environmental or social record, there would be no reason why a multitude of such types of requests could not be made." That was CREF's reasoning, but here is what lies behind that reasoning. It continued to state, "It would be difficult to fully consider those requests and run an effective investment program for participants who wish their investments to be based primarily upon financial analysis." In other words, money once again drives decisions and once again even at the expense of the environment, which belongs to everybody.

Manufacturing is and will continue to be a vital ingredient in the U.S. economy. Economist Joel Popklin ("Manufacturers warn of impact of plant closings" 2003) supports this conclusion by asserting that manufacturing spawns more additional economic activity and jobs than any other economic sector; each $1 of final demand for manufactured goods, for instance, generates an additional $0.67 in other manufactured products and $0.76 in products and services from nonfactory sectors.

The concept of EBICs was originally proposed for the pulp and paper industry by Nemerow et al. (1977). In the next 23 years, we published many papers describing potential industrial complexes for a number of other industries. Most of these are described in Nemerow (1995).

Rationale for EBics

The field of industrial waste treatment as practiced from the 1940s through the 1980s is now evolving from treatment to waste utilization. Society is calling for lower manufacturing costs along with less environmental degradation. The use of EBICs is not only the logical answer, but the only rational response to society's demands. This system reuses one plant's waste as another's raw material, thus simultaneously reducing raw feed costs and eliminating waste-treatment costs. However, the EBIC system depends on the inclusion of compatible industrial plants. Such a system completely changes our concept of industrial manufacturing. No longer should we locate industrial plants based solely on the economic marketability of our product, but we must consider the usefulness of wastes as raw materials for the ancillary plant. To discharge wastes untreated or partially treated into the environment is no longer an alternative. And, to completely treat the same waste before discharge is too costly for both the industry and society. Simple logic dictates that this waste be utilized directly by another manufacturer to save operating capital for the plants while improving the quality of the receiving environment for society. When industry also reuses water and wastes, it satisfies another of its objectives as well as that of environmentalists.

In the past, industry was more concerned with production problems and costs at a particular site rather than importation of raw material costs. This is no longer the case. When production costs are competitive, industry is concerned with importation costs of its raw materials. As an example of a decision involving these specific costs, Tyson Foods ("Plant closing shows" 2002) "company officials say they will shift production of its bacon brands, which include Thorn Apple Valley and Colonial, to more modern plants closer to its suppliers in the Midwest ... Tyson says it isn't sure how much money the move will save, but it is clear that the cost to this town (Holly Ridge, North Carolina) is huge."

The World Bank has indicated its worldwide support for industrial environmental protection—indirectly a stimulus for our environmental complex principle. A group of very large banks—such as Citigroup, ABN-AMRO, and Barclays—has announced that the banks will finance large industrial projects through the World Bank after applying strict (but voluntary) environmental standards (Phillips and Pacelle 2003). This policy should provide impetus for industries—especially in developing countries—to utilize concepts similar to the EBIC to economically avoid environmental pollution.

Fertilizer and Cement Production

In conventional practice, phosphate fertilizer and cement are manufactured as shown in the simplified schematic form in Figure 16.1. Three major wastes impose environmental damage on the surroundings. Two of these wastes originate in the fertilizer plant: (1) phosphate rock slime wash water and (2) phosphogypsum sludge waste. The first comes from washing the mined rock free from its varied impurities; the second arises from treating this washed and crushed rock with sulfuric acid, resulting in a calcium sulfate sludge (i.e., phosphogypsum). The third waste leaves the cement plant as dust, both raw material and kiln-type. The former comes from moving raw materials around the cement plant, whereas the latter results from burning the raw materials at very high temperatures.

Three distinct types of damage costs occur when industrial wastes such as these are discharged into the environment:

1. Primary costs to the industry itself, resulting in direct costs
2. Secondary costs to the people surrounding the plant, resulting in indirect costs
3. Intangible costs affecting society as a whole, which are more difficult to quantify, resulting in intangible costs

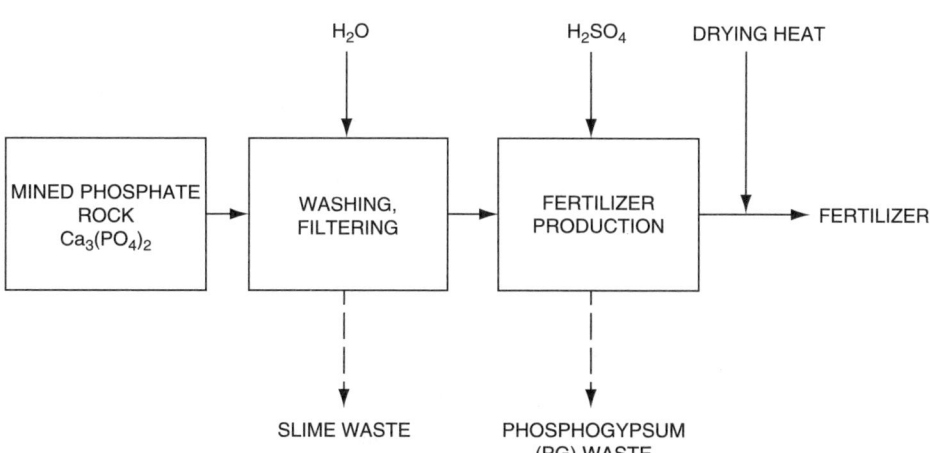

FIGURE 16.1. Mass flow diagram of a free-standing fertilizer plant.

TABLE 16.1
Comparison of Real Production Costs (1995 Dollars) of Free Standing Fertilizer and Cement Plants with EBIC Costs (1995 Dollars) per Ton of Fertilizer

Industry	Real Cost Includes 3 Damage Costs	EBIC Cost	Savings in Production Costs When Using EBIC	% Savings due to EBIC
Fertilizer	$245.76	$183.95	$61.81	25
Cement	$67.09	$34.00	$33.09	49

When these three costs are identified, totaled, and added to typical plant production costs, we obtain the real cost of manufacturing a product.

In a research project with these two industries, these costs were identified as shown in Table 16.1. More specifically, the damage costs identified for these two industries were attributed to the following:

1. Unsightly collection and storage of the voluminous sludges on increasingly valuable lands.
2. Leaching of contaminants from the sludge piles or stacks, which adversely affect drinking water and fish downstream.
3. Grinding and burning dusts from the cement plants affect all three environments.

Direct reuse of phosphogypsum and slime sludges in the cement plant within the complex and reuse of the dust by an adjacent fertilizer plant will lessen or eliminate all damage costs from these plants (as shown in column 3 of Table 16.1). A clean environment surrounding the plants will also be achieved (as depicted in Figure 16.2).

Case Study of Economic Proof of Industrial Complexes

Section Two
Fertilizer and Cement Plants: EBIC

In this section, I show that it is substantially less costly to produce two industrial products and reuse all wastes within one complex than to produce the same products at separate locations and discharge the untreated wastes to the surrounding environment. The original researchers of this project (Krishnan et al. 1996) have attempted to use real

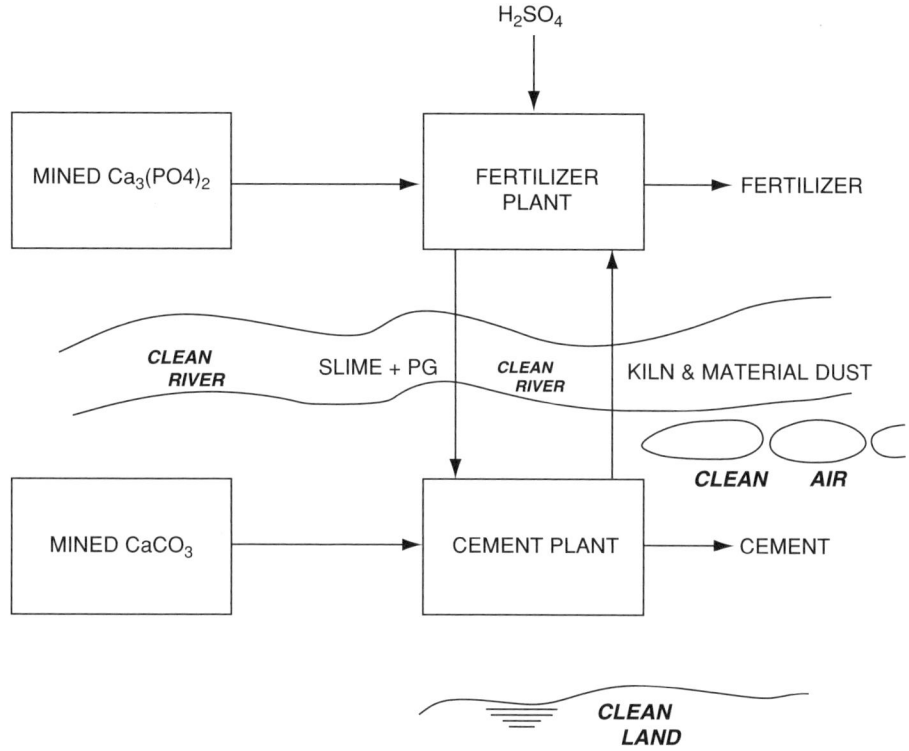

FIGURE 16.2. Clean environment surrounding the plants.

data to put a monetary value on the indiscriminate and wanton discharge of these plants' production wastes into the environment. When this value is added to the manufacturing costs of the two products, the real total production cost becomes 37% higher than that of the industrial complex.

Introduction

Industrial complexing is an innovative attempt to improve environmental quality—as I have argued many times in this text thus far—while lowering production costs. The ultimate goal of both environmental and production engineers is to attain zero pollution and minimum manufacturing cost. Before reaching this goal, industry must include all environmental damage costs—direct, indirect, and intangible—as part of the cost of manufacturing.

In this section, I attempt to measure all the environmental damage costs, add them to typical production costs, and obtain true and real costs of manufacturing a product. When armed with these true costs, industry can decide whether to cease damaging the environment by using EBICs.

Fertilizer and Cement Production

In conventional practice, phosphate fertilizer is manufactured as shown in Figure 16.1 and in the simplified schematic form in Figure 16.3. The two major wastes impose environmental damage on the surroundings. In a similar fashion, cement is manufactured as shown in Figures 16.3 and 16.4. Likewise, its two major wastes cause degradation of the air surrounding the plant.

Figures 16.2 and 16.5 present basic schematics of fertilizer and cement plants EBIC. All four wastes are reused. Some raw materials are substituted by these wastes. Transportation of some raw materials is eliminated, resulting in cost saving. This complexing system thereby reduces production costs even without considering the benefits of abating environmental damages.

Environmental Consequences of Complexing

Two significant types of damage caused by phosphate fertilizer plants' phosphogypsum wastes include:

1. Unsightly collection and storage of the voluminous sludges on increasingly valuable land
2. Leaching of contaminants from the sludge piles or stacks, which adversely affects drinking water and fish downstream

Direct reuse of phosphogypsum and slime sludges in the cement plant within the complex can eliminate the aforementioned damages.
In addition, the dust reaching air around cement plants—from both the grinding and the burning operations—damages the environment. Direct reuse of these dusts by an adjacent fertilizer plant will reduce damage costs from them (see the schematic configuration shown in Figure 16.3).

Environmental Damage Costs

Three types of environmental damage costs result from industrial wastes. The three categories of benefits accruing to society from using the environmental complex principle are presented here (Nemerow and Agardy 1998a).

1. Primary benefits: those affecting the industrial plants themselves, resulting in direct costs.
2. Secondary benefits: those affecting the people surrounding the plants, resulting in indirect costs.
3. Intangible benefits: those affecting society as a whole and those that are difficult to quantify, resulting in intangible costs.

When costs are tied to these benefits and added to normal plant production costs, one obtains the real cost of manufacturing a product.

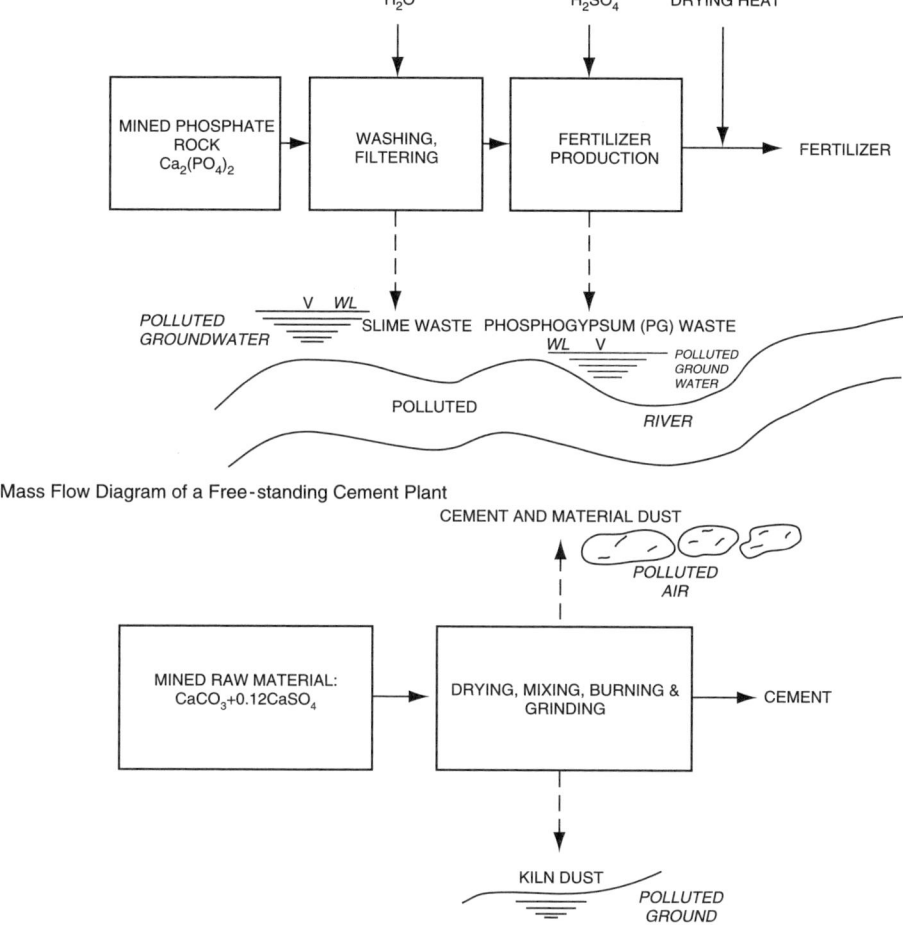

FIGURE 16.3. Free-standing cement plant.

Real Production Costs

When a phosphate fertilizer plant is located alone, its production costs include the usual operation costs: maintenance, materials, and labor (Fc), the direct costs of required waste treatment (Fwt), the indirect costs of environmental damage to nearby owners (Fnd), and the intangible costs of environmental damage away from the plant and to the public at large (Fxd). In summation, the real production cost (Fr) becomes

$$Fr = Fc + Fwt + Fnd + Fxd. \qquad (1)$$

Each of the quantities in Equation (1) will now be identified for the fertilizer plant.

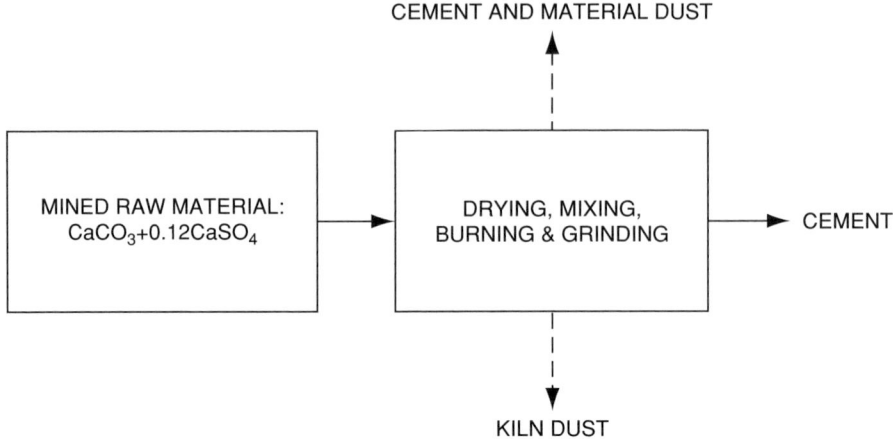

FIGURE 16.4. Mass flow diagram of a free-standing cement plant.

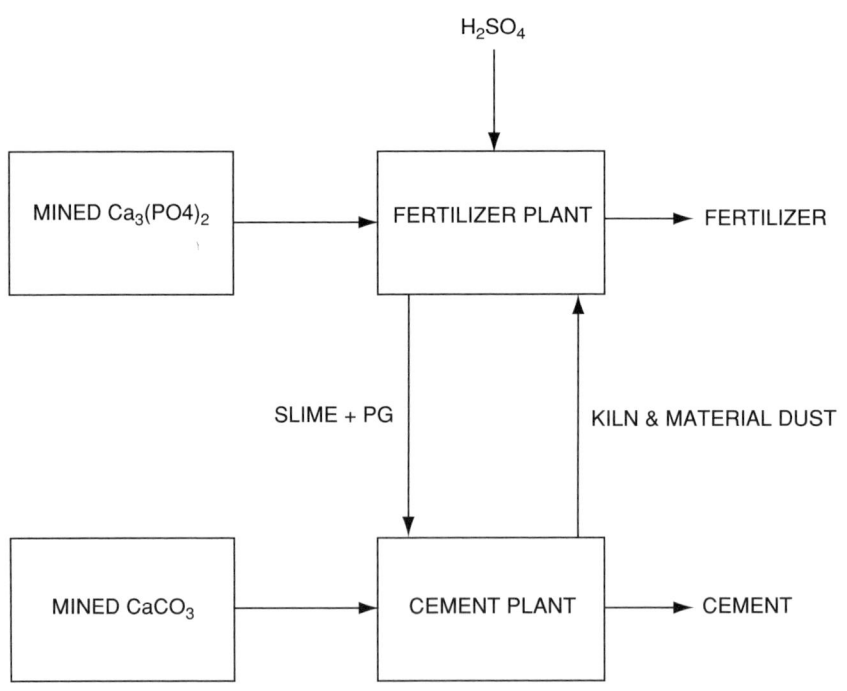

FIGURE 16.5. Mass flow diagram of a fertilizer-cement complex.

Case Study of Economic Proof of Industrial Complexes 387

Fc Is the Classic Production Cost
In 1995, the average U.S. fertilizer plant manufactured 831,607 tons of phosphate fertilizer. In addition, the average annual production cost in a fertilizer plant was $169.66 million. Therefore, the 1995 fertilizer production cost, Fc, was $169,660,000/831,607 or $204.02/ton.

Fwt Is the Direct Cost of Waste Treatment
In this example, this cost includes that spent in preventing phosphogypsum piles from reaching the environment; in storing and preventing the escape of tailing wastes; in abating air contaminants from the production of fertilizer from sulfuric acid reactors; and other types of waste treatment. Some real examples of plant expenditures for these waste treatments include the following:

> IMC-AGRICO spent $1 million in 1994 for water testing and new wells as a neighborly gesture but denies that its mining caused water pollution and subsidence problems (Satchell 1995). The company has voluntarily spent $6.8 million to plug a sinkhole and control the spread of contaminants to the groundwater. Assuming the life of a stack to be 10 years, this cost becomes $0.83/ton of fertilizer (Fwt2).
> CORGILL fertilizer placed at least 18 in. of compacted clay over a layer of at least 15 ft of natural clay at its Hillsborough County, Florida, mine (Newborn 1992). The company spent $22 million for lining the base of the new stack, plus another $5 million to close the existing gypsum stack. These expenditures for top coverings, when projected to 1995 costs, resulted in $3.73/ton of fertilizer (Fwt3).
> IMC completed lining a new stack at its New Wales mine in Polk County, Florida, with 20 million ft^2 of plastic for a cost of $70 million. This expenditure for bottom lining amounted to $9.67/ton (Fwt4).

> Therefore, Fwt = Fwt1 + Fwt2 + Fwt3 + Fwt4 = $15.46/ton.

Fnd Is the Indirect Cost of Environmental Damage to Nearby Owners
In 1987, the State of Connecticut established values of fish killed by acid leakage; this situation amounted to $1,082 (Anonymous 1988) (Fnd1) or $0.01/ton of fertilizer.

On December 21, 1994, IMC-AGRICO agreed to pay $1.1 million to settle a lawsuit filed by the Environmental Protection Agency (EPA), which had charged the company with violating water pollution limits at nine locations due to slime waste–contaminating groundwater. The cost of the legal proceedings came to $1.35/ton of fertilizer (Fnd2).

In 1994, the phosphate industry contributed $100,000 to some 25 "green ground societies." In 1988, the Audubon Society received $42,500 from IMC alone. These indirect costs amounted to $0.19/ton of fertilizer in 1995 dollars (Fnd3).

The industry also contributed to candidates for state and local offices. Total contributions were $160,000 for 1994. Donation indirect costs for election campaigns amounted to $0.20/ton of fertilizer (Fnd4).

Therefore, Fnd = Fnd1 + Fnd2 + Fnd3 + Fnd4 = $1.75/ton.

Fxd Is the Intangible Cost of Environmental Damage
These costs have been elusive to pin down and may very well be the greatest of the three types of damage costs. In the research effort described in this chapter, we established the following specific values.

Florida phosphate mining companies have paid about $1 billion in taxes to the state in the past 25 years. This amounts to a mean yearly value of $51/ton/yr, assuming that 50% of them are small and operate on a small margin of profit, thereby being adversely affected by state action to enforce pollution-control measures.

Assuming, again, that 50% of these plants are then estimated to have closed because of waste-treatment pressure, the public lost 25% of the $51 million of revenue taxes due to these plant closures. Hence, Fxd1 = $15.33/ton of fertilizer.

Land value in the Tampa, Florida, industrial area decreased from about $1,250/acre to $600/acre mainly because of fertilizer pollution over a period of 10 years, or $65/acre/yr (G. M. Lloyd, personal communication 1995).

Therefore, $65 times 1,000 acres times 17 industries times 5 gypsum stacks/industry divided by 831,607 tons/yr equals $6.64/ton of fertilizer equals Fxd2.

In 1996, the State of Florida purchased 38,251 acres of land along five riverbanks for a price of $21,270,000, or $547.77/acre (Browning 1996). This purchase was aimed at purifying these lands from the wastes mainly from fertilizer plants over a 10-year period. Therefore, Fxd3 = $2.56/ton of fertilizer. The total intangible costs we were able to identify were calculated to be

Fxd = Fxd1 + Fxd2 + Fxd3 = $15.33 + $6.64 + $2.56 = $24.53/ton of fertilizer.

Substituting our computed values in Equation (1), we obtained a real production cost of

$204.02 + $15.46 + $1.75 + $24.53 = $245.76,

when the major measurable environmental damages were considered.
Using the same procedure to compare cement plants, the cost relationship becomes

Creal = Cc + Cwt + Cnd + Cfd.

Cc Is the Conventional Cost of Cement Production
An average of 984,000 tons of cement was produced per cement plant in the United States in 1995 at an average production cost of $49.2 million.

Therefore, Cc = $49,200,000/984,000 tons = $50/ton of cement.

Cwt Is the Cement Waste Treatment Direct Cost
Cement kiln dust is a major waste to the air environment and needs collection and disposal to protect the surrounding air. Kessler (1995) reports that typically each percent of dust wasted increases the specific heat consumption by about 0.7% and decreases clinker production by 0.5%. He gives the dust losses costs as $Cwt1$ = \$4.08/ton of cement, because of the loss of raw material; $Cwt2$ = \$4.59/ton by feed crushing, conveying, drying, and grinding costs; $Cwt3$ = \$1.02/ton for transporting, conveying, handling, and de-dusting; and $Cwt4$ = \$3.06/ton for landfill maintenance, monitoring, pile maintenance, and closing.

Total waste-treatment costs for kiln dust becomes

$$Cwt = Cwt1 + Cwt2 + Cwt3 + Cwt4 = \$4.08 + \$4.59 + \$1.02 + \$3.06$$
$$= \$12.75/\text{ton of cement.}$$

CNN Is the Indirect Cost of Cement Plant Waste Environmental Damages
In 1992, the EPA fined Lafarge's Michigan and Alabama cement plants \$1.8 million for violating air emission operating rules (Ferguson 1993). The cost ($Cnd1$) was \$1.94/ton of cement.

Lafarge's switch to power-plant fly-ash from ground shale may have saved as much as \$600,000 in fines levied by the U.S. District Court (Anonymous 1994). The additional cost ($Cnd2$) resulting from environmental violations was \$0.62/ton of cement.

In 1992, a local customer sued Lafarge for \$1 million over improper disposal for chromium-tainted materials over a 5-year period. The cost of legal compensation ($Cnd3$) for affected victims was \$0.22/ton of cement.

These three indirect costs amounted to

$$CNN = Cnd1 + Cnd2 + Cnd3 = \$1.94 + \$0.62 + \$0.22 = \$2.78/\text{ton of cement.}$$

Cad Is the Intangible Cost of Cement Plant Environmental Damages
Since 1990, Hillary Clinton has been a director of Lafarge Cement Corporation, one of the largest operators of cement kilns fueled by burning hazardous wastes (Zweig 1992). In 1991 the former president's wife earned \$30,000 from Lafarge. It may have been inferred by some that this may indirectly influence state and federal policies on Lafarge. The cost of potential influence (Cad) in this case was \$0.03/ton of cement.

All cement damage costs equal

$$Cwt. Cnd + Cxd = \$12.75 + \$2.78 + \$0.03 = \$5.56/\text{ton of cement.}$$

The real cement plant production costs equal

$$Creal = Cc + Cwt + Cnd + Cxd = \$50 + \$12.75 + \$2.78 + \$.03 = \$65.56/\text{ton.}$$

EBIC of a Phosphate Fertilizer and Cement Plant

When these two plants are located in the same complex, all direct costs of waste treatment, indirect costs of environmental damages to nearby neighbors, and intangible costs to the general public are eliminated. In addition, transportation costs of replaced fertilizer raw material and cost of replaced fertilizer raw material decrease the real production costs.

The transportation cost of the replaced raw material (Fr) was computed to be $20/ton of fertilizer, and the cost of the replaced fertilizer (Fr) raw material was computed to be $0.07/ton of fertilizer. Likewise, corresponding values for the cement plant were

$$Car = \$6.00/\text{ton of cement}$$

and

$$Cram = \$10/\text{ton of cement}.$$

Calculations were made for the production of 1 ton (total) of EBIC product, as shown schematically in Figure 16.5. Because 3.5 tons of cement can be produced for each ton of fertilizer product, each ton of EBIC product would consist of 0.22 tons of fertilizer and 0.78 tons of cement.

To illustrate the truth of the economic value of transportation of wastes away from an industrial site, take the case of the American Waste Transport Corporation. This company is one of the major contract transporters of solid waste in southern California and southwest Arizona. In fiscal 1999, it had revenues in excess of $12 million according to a company statement (Clark 2000). It supplies materials to processors that convert waste into fertilizer and compost or burn it to produce energy. In our complex system, we would eliminate this massive transportation cost and perform the same or similar reuses within the complex. To continue, the CEO of U Biomes said that they currently transport 1,400 tons per day of "green wastes" and that "green waste recycling is expected to grow dramatically as California begins to comply with its state recycling law, which requires diversion of 50% of waste from landfills." This rush to transporting and reusing solid wastes is also occurring in many other states to comply with similar laws. The proposed system (EBIC) eliminates these unnecessary and burdensome transportation costs.

Conclusions[1]

It is seen that the real societal cost of products is more than the classic cost. In the case of fertilizer, it is 20% greater, and in the case of cement, it is 34% greater. By industrial

---

[1]The author acknowledges the financial assistance received from the National Science Foundation, in the form of partial funding for this project. Thanks, too, to G. Michael LLoyd of the Florida Institute of Phosphate Research and to Victor Turiel of Pennsuco Cement Corporation for assistance in obtaining significant data.

complexing, so that waste products of one industry become raw materials for the other, environmental damage costs can be eliminated and the raw material and transportation costs can be reduced. These result in a cleaner environment and lower product cost. In the case of fertilizer and cement plants—in a weighted sense—there is a 20% saving as compared with real societal cost and 37% saving as compared with real cost. The author wants these industries to become aware of the advantages of industrial complexing, both financially and environmentally, so that whenever feasible such two or more industry complexes can be achieved to the benefit of the manufacturer, the consumer, and the environment in which we live.

Tables 16.2 through 16.7 summarize the data previously described in this chapter. The real production costs of a ton of fertilizer and a ton of cement are $245.76 and $67.09, respectively, and are shown in Table 16.2.

Environmental damage accounts for 20% of the fertilizer production cost and 34% of the cement production cost (shown in Table 16.3). When credits for replacing some raw material and eliminating some transportation are taken into account, the classic fertilizer and cement production costs are $183.95 and $34.00/ton, respectively

TABLE 16.2
Real Production Costs[a] of Fertilizer and Cement Plants

Type of Cost	Cost per Ton of Fertilizer	Cost per Ton of Cement
Classical production cost	$204.02	$50.00
Direct environmental damage	$15.46	$14.28
Indirect environmental damage	$1.75	$2.78
Intangible environmental damage	$24.53	$0.03
Real production cost (Total production cost)	$245.76	$67.09

[a]All costs are projected to 1995 dollar values.

TABLE 16.3
Comparison of Classical Production Costs[a] of Free-Standing Fertilizer and Cement Plants with Real Production Costs[a] per Ton of Product

Industry	Classical Cost	Real Cost Including Damage Costs to Society	Cost of Environmental Damage	% Cost of Environmental Damage
Fertilizer	$204.02	$245.76	$41.74	20
Cement	$50.00	$67.09	$17.09	34

[a]All costs are projected to 1995 dollar values.

(shown in Table 16.4). These result in savings of 10% and 32%, respectively (shown in Table 16.5). The total savings to fertilizer and cement plants are $61.81 and $33.09/ton, respectively, when using the EBIC (Table 16.6). Table 16.7 presents a comparison of production costs of a ton of EBIC products and the corresponding classic and real costs for equivalent masses of products for free-standing plants. It can be seen that—in a

TABLE 16.4
EBIC Costs in a Fertilizer[a]-Cement Industrial Complex

Type of Cost	Cost per Ton of Fertilizer	Cost per Ton of Cement
Classical production cost	$204.02	$50.00
Credit for replaced raw material	−$0.07	−$10.00
Credit for transportation of replaced raw material	−$20.00	−$6.00
EBIC cost	$183.95	$34.00

[a]All costs are projected to 1995 dollar values.

TABLE 16.5
Comparison of Classical Production Costs[a] of Free-Standing Fertilizer and Cement Plants with EBIC Costs[a] per Ton of Product

Industry	Classical Cost	EBIC Cost	Savings Due to EBIC	% Savings Due to EBIC
Fertilizer	$204.02	$183.95	$20.07	10
Cement	$50.00	$34.00	$16.00	32

[a]All costs are projected to 1995 dollar values.

TABLE 16.6
Comparison of Real Production Costs[a] of Free-Standing Fertilizer and Cement Plants with EBIC Costs[a] per Ton of Product

Industry	Real Cost Including Damage Costs to Society	EBIC Cost	Savings Due to EBIC	% Savings Due to EBIC
Fertilizer	$245.76	$183.95	$61.81	25
Cement	$67.09	$34.00	$33.09	49

[a]All costs are projected to 1995 dollar values.

TABLE 16.7
Comparison of Production Cost[a] of a Ton of EBIC Product (0.22 Ton Fertilizer + 0.78 Ton of Cement) and Equivalent Masses of Free-Standing Plant Products[a]

Industry	Classical Cost	Real Cost	EBIC Cost	Saving Over Classical Cost	% Saving Over Classical Cost	Saving Over Real Cost	% Saving Over Real Cost
Fertilizer (per 0.22 ton)	$44.88	$54.07	$40.47	$4.41	10	$13.60	25
Cement (per 0.78 ton)	$39.00	$52.33	$26.52	$12.48	32	$25.81	49
Total (per ton of EBIC product)	$83.88	$106.40	$66.99	$16.89	20	$39.41	37

[a]All costs are projected to 1995 dollar values.

weighted sense—1 ton of EBIC product would be $16.89 cheaper than the classical cost and $39.41 less expensive than the real cost, which translates into 20% and 37% savings, respectively.

Marketing Unused Waste Resources

There may be instances in which the EBIC will not be able to reuse all the waste materials completely. These cases could occur despite the concerted efforts of the industrial plants to reuse all wastes as raw materials. In these rare situations, the EBIC participants should market these wastes in an efficient economical manner.

I recommend marketing these excess wastes using a system that I have been advocating officially since 1969 (Nemerow 1985). I maintained that the assimilative capacity of the environmental resource should play a major role in determining the price industry should pay for polluting it. The more assimilative capacity available, the lower the unit sale price of the wastes. As the assimilative capacity of the resource (air, water, or land) becomes limited, the higher the unit cost to discharge the waste. Instead of discharge or treatment, of course, I recommend that another external buyer purchase these excess units of waste at the price predetermined by its detrimental effect on the environment. In that way, the EBIC participants will not be forced to pay for polluting and the external environment will remain clean. If a buyer cannot be found, then the EBIC members have a choice either to pay a fair value for their untreated discharge into the environment or to pay for its treatment before discharge.

Lomborg's reviewer (Bailey 2001) points out that "clearly regulation has worked to improve these common areas (air and streams): our air and streams are cleaner than they were. But there is good evidence that assigning property rights and market mechanisms to such resources would have resulted in a faster and cheaper cleanup."

Further discussion of using this marketing system was also presented in another book (Nemerow and Agardy 1998b). As an example, you may determine that the beneficial damage of the industry's biochemical oxygen demand (BOD) pollutant was $10/lb, in which case, the industry would have the option to buy a certain number of BOD pounds "rights" to discharge at that price or treat its waste to remove that number of pounds. As described in the previous references in 1985 and 1998, as the available limit of BOD diminishes, the market price of a unit of BOD increases. The reason for this is to protect the water-quality level of the receiving water. Industry is discouraged from using the last available BOD units to preserve that water-quality level. Because the cost of buying BOD rights increases as the available resources decrease, industry must treat its waste, usually at a lower unit cost than buying rights. Of course, the possibility of buying lower cost BOD unit rights from another industry exists in the free market system.

The free market system could also be used for buying land unit rights for solid wastes and air-capacity rights for air pollutants. This system of buying and selling "pollution rights" should be based on the benefit lost of the resource by adding an incremental pollutant load. The environmental benefit lost becomes greater—and, hence, so does the pollutant right cost—as the available environmental resource gets used up. This method of pricing, though more difficult to compute unit costs, is preferable to an arbitrary price placed on the right by some overseeing agency. It is even preferable, in my opinion, than a truly free market pricing system because it is based on more tangible and measurable environmental damage costs.

Although Rinda E. Vas (2000), editor of *Environmental Technology,* found that "there are several kinds of market-like mechanisms that might be employed in environmental regulation," she does not quite include a *market charge based on damage costs* as one of them. In evaluating current emission trading systems, Bryner (1999) gets closer to my proposal of a system by concluding that "emission trading programs should lead to other, more powerful regulatory innovations that *will more effectively encourage ecologically sustainable activities* [the emphasis is mine]. Emission trading programs should be designed as a transition to a *system of emission fees or taxes and other efforts to reflect true costs in prices and to create more powerful incentives to reduce and prevent pollution* [again the emphasis is mine]." He goes on to conclude that "the ultimate test of an emission trading program is its contribution to a more fundamental shift in practices aimed at reducing pollution, improving efficiency, and conserving resources." All of these practices are incumbent in my market pricing system proposed in 1985 and 1998, and again in this book. Solomon and Lee (2000) write that "despite the success of these trading systems in affordably reducing emissions, they have been criticized by several environmental organizations for allegedly creating toxic hot spots (local areas with excessively high emissions or concentrations of a hazardous air pollutant."

The utility business is similar—in certain instances—to the environmental resource business. Units of power can be sold at a price based on real market value. The real market value can be reached by adding the existing kilowatt-hours charge to a unit local societal monetary loss to arrive at total real value.

Available power supplies are decreasing fast—especially during peak power demand periods—in San Diego, California, as an example (Rose 2000). Rose (2000) writes that "some suggest that power companies are holding back (construction of power plants), perhaps waiting for a crisis that would provide them with the financial incentive to build." Rose also quotes Edwin Guiles (president of San Diego Gas & Electric), "We are in favor of all solutions being considered—new generation, demand-side alternatives, distributed generation—but we have to make sure there is a solution that can deliver in the time period we have."

I suggest that the purchase of kilowatt-hours be based on benefit costs of not having power units above the basic level that exists. Such benefit losses include the following:

1. Lower standard of living from lack of adequate air conditioning and heating
2. Loss of industry production increase due to unavailability of power
3. Lack of municipal growth due to inadequate power, and so on

One can then put added values on each excess of these and sell kilowatt-hours to all consumers. The added dollar kilowatt-hour charge can be used by power companies as an incentive to build and produce more kilowatt-hour capacity. When extra kilowatt-hour capacity is met, the dollar extra kilowatt-hour charge can be dropped until demand exceeds supply again.

As recently as July 2001, the op-ed editor of *The Wall Street Journal* questioned whether pricing emissions is possible and advocated its use. The editor wrote that "thus, by providing flexibility and financial incentives, a cap-and-trade program [his term for selling resources] will result in more abatement from those firms who can do it at relatively lower cost. The net will be the same amount of overall pollution reduction, but achieved at lower cost than would be obtained under traditional regulation."

The editor referred to the Energy Information Administration as stating that the cost of power plant CO_2 reductions according to the requirements of the Kyoto Treaty agreement could be as much as 4% of the gross domestic product (GDP). However, "in a scenario offered back in 1998 by the Clinton administration's Council of Economic Advisors, if the U.S. buys permits for its excess emissions—so that it doesn't have to reduce by very much its own emissions—the cost would be only 0.1% of GDP." With these facts in mind, the editor recommends that the Bush administration propose a domestic cap-and-trade program for CO_2 that could, of course, be expanded to Canada and Mexico and later to Latin America and then the world.

Alternative Energy Sources to Reduce Resource Depletion, Costs, and Environmental Impacts

Alternative Energy Solutions

Because energy is so vital and integral to any industrial complex, it is appropriate to discuss ways and means of reducing its cost to a minimum before using some form of it in an industrial complex. We must consider all forms of reasonable alternative energy sources not only to minimize their production costs, but also to reduce depletion of natural resources and to eliminate any potential adverse effects on the environment.

Introduction

Since the current use of fossil fuels (coal and oil) and nuclear fuels is too costly, arises from nonrenewable fuel resources, or results in too great an adverse environmental impact, the search goes on for renewable, nonpolluting, and economical alternative energy sources. In this section, I present potential alternative solutions to the use of fossil and nuclear fuels and the adverse environmental effects that they create. You are urged to attempt to use any of these suggested alternatives in applications that are suitable to your particular situations.

Alternative Energy Sources

The following six fuel sources are suggested as alternatives to fossil and nuclear fuels for producing electricity with little or no adverse environmental impacts:

1. Hydrogen fuels
2. Wind energy
3. Solar energy
4. Geothermal energy
5. Wave energy
6. Other electrical energy sources

Each of the aforementioned sources of energy should be considered in solving environmental problems when it is desired to conserve and diminish the polluting effects of other nonrenewable fuels.

Hydrogen Fuels

Although hydrogen fuels are sufficiently important and currently in use to warrant a complete chapter, they are mentioned briefly here to make certain that their potential for use is considered to the fullest. Hydrogen is produced today primarily from the electrolysis of water into its separate constituents of hydrogen and oxygen. In this case, some conventional electric fossil fuel power is required, but theoretically, hydrogen can also be produced in a number of other ways. It can also be produced biologically from anaerobic decomposition of various agricultural wastes (see "Other Electrical Energy Sources" later in this section).

The production of hydrogen by electrical disassociation of water is shown in Figure 16.6. The most recent use of hydrogen fuel as an alternative to fossil (gasoline) is in automobiles. The hybrid cars—partly fueled by hydrogen—have been quite successful. Automobiles completely powered by hydrogen are also forecasted for the near future—within the next few years. One maker is even proposing a hydrogen-fueled auto with the hydrogen fuel tank installed in the rear seat area of the automobile to avoid the fuel delivery and loading problem. This would represent a highly desirable alternative solution to the combustion-gas air pollution problem caused by gasoline-fueled autos.

Wind-Generated Power

By the year 2004, more than 13,000 megawatts of wind power had been installed worldwide. California alone had 1,600 megawatts of wind power in use to provide enough electricity for over 750,000 homes. Wind farms—a collection of individual windmills at one location—have been increasing to a point where the U.S. Department of Energy predicts that wind power costs will drop to $0.2 cents/kW-hour from the current value of $0.3–0.6 cents.

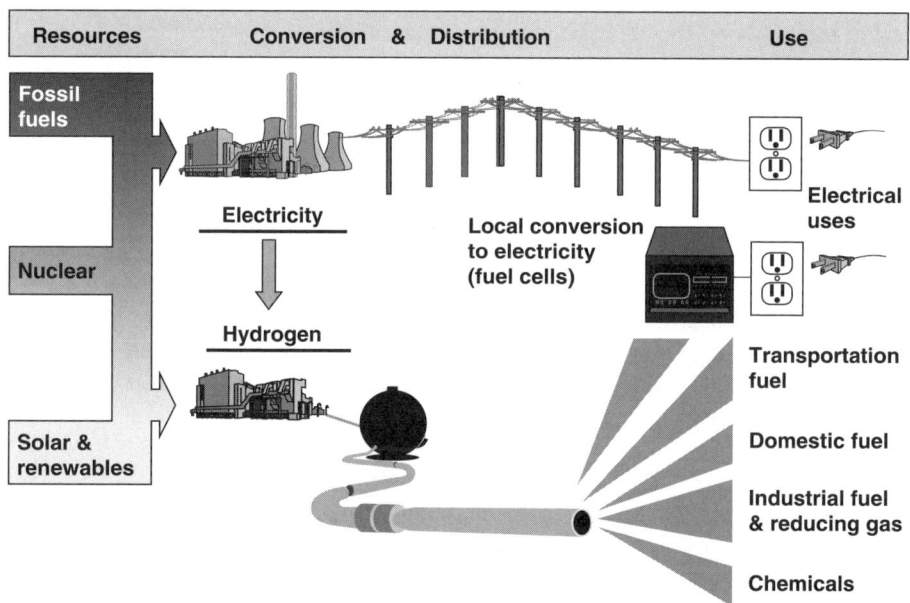

FIGURE 16.6. Production of hydrogen by electric disassociation of water. Used with permission from John Wiley and Sons.

Wind Production

Wind farms use large blades to catch the wind and turn rotors to produce electricity. A modern wind farm may contain as many as 500 wind turbines connected to a transmission grid. They produce electricity much the same as steam engines use steam to turn rotors of generators to produce electricity, except that in this case wind instead of steam does the work.

A wind speed production of at least 12 miles/hr is usually required to produce electricity. When the wind is not blowing at this speed, no electricity is produced, and the main transmission line is not being supplemented with wind-generated electricity. However, as soon as the wind speed picks up sufficiently, the wind electricity begins again to supplement that in the main transmission line. This creates no problems, because wind energy usually represents a small percentage (2–4%) of the total power being transmitted.

Environmental Impacts of Wind Power

Wind plants create no air pollution, do not use or waste any water, or despoil the land, but they may produce effects on vision, audio, and wildlife.

VISION

Because wind farms contain so many turbines that which are mounted on top of tall towers (some 350 ft high), they often are visible at a great distance from the farm. Some people object to the sight of large wind farms just as some people object to crowded buildings in a city. However, as these farms become more prevalent and as people get used to seeing them, they may become more acceptable to most people in the area. In fact, some people may even claim that their appearance is desirable from an artistic standpoint. Modern planners and architects face the challenge of designing these farms so that they are desirable for the surrounding humans.

AUDIO

Other people may be concerned about the noise created by the wind farms. That noise may be mechanical or aerodynamic in nature.

Mechanical noise is produced by parts rubbing against or hitting other parts, and has virtually disappeared in the newer designed rotors. The aerodynamic noise, which is that swishing sound emitted as the blades pass the tower, can be masked by proper use of sound-barrier construction. Once again, some people even claim that the swishing sound is rather soothing, similar to that generated by ocean waves hitting the shores.

WILDLIFE EFFECTS

Bird populations can be threatened by the wind farms. The ground below the mills is disturbed during construction, which in turn attracts mice, prairie dogs, and burrowing animals. These in turn attract raptors such as hawks and eagles that prey on them. These birds may perch on top of the wind generators for reasons of hunting for prey and often get caught in the spinning blades. Lately, farms are designed to contain tubular towers to prevent birds from perching on them. In addition, they turn more slowly than those of earlier design. When compared to other industries such as mining and coal combustion, the environmental effects on birds are much less.

The use of wind farms is greatly enhanced by the realization of farmers of the value of their land. They have found that by leasing the land for the wind farms, they can overcome increasing costs of fuel for cultivating crops on the land.

Solar Energy Power
The science of converting the sun's energy into electricity is often referred to as *photovoltaic (PV) science*. Although the concept was known since Edmond Becquerel discovered it in 1839, it wasn't until 1954 when the first photovoltaic cell was created in the Bell Labs. These cells are semiconductor devices that convert light directly into electricity. They are generally made of silicon with traces of other elements. Although PV cells are quite sophisticated in design, they are very simple to use. The PV cells are mainly low-voltage DC devices with no moving or wearing parts. Once they are installed, no maintenance outside of an occasional cleaning is required. However, most of these systems do include a battery (storage) and need some water just like automobile batteries.

PV Cells
PV cells (solar) consist of layers of semiconductor materials with various electronic properties. The bulk of the cell is usually silicon based, along with a minor amount of boron to lend a positive electrical charge to it. A thin layer on the front of the cell is painted with phosphorous to render the other part of the cell negatively. The union between the two cell layers will then contain an electric field at the junction. When the photons of daylight hit the solar cell, some of these photons are absorbed at the junction, which then frees the electrons of the silicon material. If and when the photons possess enough energy, the electrons will be able to move through the silicon and into an external circuit. They give up their energy as they flow through the external circuit and result in producing electricity to power all kinds of small and often larger electronic devices before returning to the solar cell. The entire process is solid state with no moving parts and no materials released or consumed in the process. Despite what many people believe, these solar cells work better in colder weather than in hot, mainly because they produce electricity from light rather than heat. As long as sunlight is reduced no more that 20% of full sun, sufficient photons will be released to operate under these partly cloudy conditions.

Benefits of Solar-Powered Electricity
Some of the many benefits of using solar energy to replace fossil- or nuclear-fueled sources include:

1. The fixed costs of operation remain constant for the life of the system.
2. The solar system is independent from any other source of fuel.
3. Several solar cells can be joined to increase the output capacity of the system.
4. No noise occurs from operating the system.

5. No so-called *greenhouse gases* are entitled, as is the case with fossil fuels.
6. The solar-power system possesses a long operating life.
7. Cost of operation and maintenance is competitive with other energy types.
8. Because of its nonpolluting nature, solar cells are widely known as "clean and green."

Some Environmentally Related Uses of Solar Cells

Because so many people in developing countries have little or no access to electricity, these solar cells are predicted to compete more favorably with conventional sources of power. Some of the uses that you might not normally consider for solar-derived electricity are the following, which also are environmentally enhancing:

1. Powering weather stations to provide dependable economical electricity
2. Powering pumps to transmit water from remote reservoirs and lakes or from rivers that are not readily accessible
3. Telecommunication of river water stages and even magnetic detection of earthquakes causing tsunami effects in oceans
4. Electricity to remote homes such as vacation or seldom-used buildings
5. Nuclear power radiation detection systems
6. Portable light and electric systems
7. Remote lagoon aeration systems for waste treatment
8. Disaster and all types of civil defense warning systems
9. Corrosion systems for pipes (cathodic protection)
10. Remote charging of batteries in hybrid and other autos
11. Powering difficult-to-access air pollution sampling and analysis stations

The reader is urged to think of his or her own uses of solar-generated electricity that may also benefit the environment and serve as an alternative to fossil-fueled electricity.

Geothermal Energy

Geothermal energy is obtained from heated water, steam, or soil that is derived from deep within certain land masses. There are two main uses for this energy: (1) hot water is used to create electricity or to provide hot water heating or warming, and (2) the thermal mass of the soil or groundwater is used to drive heat pumps that provide either heat or cooling. The first use is more widely known and used and is obtained from geothermal geysers that find their way to the earth's crust.

The aforementioned uses are not really from renewable resources; however, with properly calculated use, they can almost approach the "renewable" classification. The heated water, steam, or soil will gradually be depleted if overdrawn from the ground. This valuable ground resource will slowly regenerate itself over time so that if the withdrawal at the surface is timed to match the regeneration rate, the resource will be considered renewable. In any event, it will not deplete itself as fast as fossil or oil fuels are depleted by normal mining techniques. In addition, heat reservoirs are considered immense in magnitude compared to current or even projected use, thus rendering geothermal energy practically renewable.

In the United States, the production of electricity from the geothermal energy of the earth's interior heat is centered in northern California. These geothermal sources provided slightly more than 7% of California's electricity in the 15-year period ending the twentieth century. Geyser production has decreased from supplying about 2,000 megawatts in 1989 to 1,100 megawatts near the turn of the century. Unfortunately, because of the specific location of these geothermal fields, most individual households cannot use this energy. However, direct use of the heated water can save establishments as much as 80% in their fuel bills.

Geothermal Ground Source Heat Pumps for Residential Use
Heat pumps can reduce both air-conditioning peak loads and winter-heating loads. In addition, they are typically used to heat water (or as hot water) in households and buildings.

Economics of Geothermal Energy
Geothermal electricity can be produced practically and economically for about $0.5 cents/kW-hour, slightly higher than wind or solar energy. This higher cost is largely because it is necessary to drill deeper today to produce a given amount of power than in earlier years. It has been suggested (and even used) that the economics of geothermal power can be improved through co-production of other goods from high-temperature brine extracted from the depths of the ground. While geothermal power applications require more advances in exploration and drilling, heat pump direct uses require that the engineer and the consumer understand the technology. It may be more expensive to install geothermal energy systems at the start, but over the long term the benefits may make it economically and environmentally worthwhile.

Effects on the Environment
Air pollution relative to conventional fossil-fuel energy production will be minimized when selecting geothermal energy instead. It produces only about one-sixth of the CO_2 and none of the NOxs or sulfur gases that fossil-fuel plants emit. For these reasons alone, this method of energy production can be a very environmentally friendly alternative to fossil-fuel energy.

Amount of This Energy Already Being Produced
In 1998, geothermal energy provided 0.4% of the electricity generated in the United States. This amounted to 14.3 billion kW of electricity to more than 1,400,000 homes. At that time, it was growing at a rate of slightly less than 3% over an 8-year period. Worldwide, geothermal energy totaled slightly more than 8 million kW or about 3% of the 3,180 kW used worldwide.

Some Examples of Geothermal Energy Uses
The Oregon Institute of Technology has been heated by the direct geothermal energy since 1964. In Iceland, geothermal energy is used to provide the majority of households with heat. Tax neutrality, continued and increased federal funding, continued and expanded production tax credits, resource identification, renewable portfolio standards,

contractor education, and the issuing of air emission standards have been and are being used to encourage continued use of geothermal energy.

The reader is urged to consult the U.S. Department of Energy's web site for more information of geothermal energy. In addition, the Renewable Energy Policy Project maintains a rather detailed bibliography of the uses of this form of energy. It is located at 1612 K Street N.W., Suite 202, Washington, D.C.

Ocean Wave Energy
How Wave Electric Energy Is Created
The entire earth's surface including the ocean is heated by the sun. This creates wind that pushes against the surface of the ocean and forms waves. Waves can travel hundreds and thousands of miles from the beginning of their propagation. They are being continuously supplemented by new winds. These waves keep their energy long after the winds that created them have abated. These same waves represent one of the most concentrated and consistent sources of renewable energy. When compared to conventional fossil-fuel generation, wave energy provides the increased advantage of a limitless free supply of energy, along with a total lack of environmentally polluting emissions. However, even today there appears to be no agreement among professionals as to the most efficient technological approach to the use of wave energy.

Types of Wave-Energy Conversion Systems
The kinetic energy of waves may be converted into electrical energy mainly by four different systems:

1. *Tapered channel systems* that funnel incoming waves into shoreline reservoirs that raise the water above sea level. The head of water then is directed down through a turbine, which then drives a generator producing electric energy.
2. *Float systems* consist of buoys that sit on the ocean's surface. As the ocean rises and falls, the relative motion between the float and the ocean floor drives hydraulic pumps or pistons. This kinetic energy is also used to drive a turbine and a generator producing electricity.
3. *Oscillating water column systems* are fixed in place and are devices in which waves enter the column and force air up past a turbine. As the wave retreats, the air pressure drops, resulting in the turning of a turbine that once again drives a generator and produces electric energy. The first of this type of system was produced in Japan to power a light on a buoy used for navigation.
4. *Underwater turbines* collect and contain the movement of the ocean's currents and use this energy to drive slow-moving blades.

These, in turn, drive a generator directly—similar to an above-ground windmill—to produce electricity.

Advances
Many of today's professionals and equipment manufacturers think that the time has arrived for the era of wave energy usage to accelerate. Technological advances have

progressed sufficiently to make this form of energy cost effective when compared to fossil-fueled power. These advances include those of marine engineering that have come from the offshore drilling industry, which provide ocean-tested "off-the-shelf" components at reasonable prices. In addition, the cost of electronic control devices that optimize the efficiency of the technology has been reduced.

Environmental Considerations

Wave-generated electrical energy is a source of clean renewable energy and does not produce any objectionable greenhouse gases. When selecting this method to produce electrical energy, one must also be aware of certain disadvantages that may hamper their acceptance. First, the sea is unpredictable at best and devastating at worst (such as when the tsunami hit in the South Pacific). Under these conditions, the facilities must be designed to be able to withstand pressures many times the normal wave pressures. Second, these wave-generating systems may cause alterations to shore lines and local ecosystems. And thirdly, the electricity produced will vary because of the variability of the waves.

Generally, the average wave power level should be more than 15 kW/m to generate wave energy at competitive prices.

Other Electric Generated Systems

Agricultural residues, farm animal wastes, human sewage sludges, and other biomasses can be fermented to produce combustible gases such as methane. The gas can be burned directly in a boiler to convert water to steam, which then can drive a turbine connected to a generator to produce electrical energy.

When an ample supply of these biomasses is available, it is desirable to consider them as an alternative energy source. Not only does one produce a valuable energy resource at a competitive cost, but one also rids the environment of a source of waste causing adverse environmental effects. In these systems, bacteria do the work required in digesting the organic matter of these wastes to free CH_4 (methane). Bacteria require no compensation for this work, but they do require proper design and operation of equipment.

References

Anonymous. 1988. That sucker's going to cost you. *Discover Magazine,* May 1988, p. 6.
Anonymous. 1994. Cement maker finds cleaner product mix. *Engineering News Record,* Dec 1994;12:27.
Bailey, R. *The Skeptical Environmentalist* [by Lomborg]. *The Wall Street Journal* Oct. 2, 2001, p. A-17.
Browning, M. 1996. Glorious riverfronts, Crystal Waters, Florida. *The Miami Herald,* March 3, 1996, p. 6B.
Clark, B. E. USA biomes to buy American waste transport. *San Diego Union Tribune,* February 17, 2000, p. C-3.
Emissions impossible? *The Wall Street Journal,* July 23, 2001, Op-Ed, p. A-14.
Ferguson, J. 1993. Cement companies go toxic. *The Nation,* March 1993;8:307.

Kessler, G. R. 1995. Cement kiln dust (CKD): methods for reduction and control. In: *IEEE Transactions on Industry Applications,* March/April 1995, pp. 407–408.

Krishnan, S. V., N. L. Nemerow, T. N. Veziroglu, et al., under a National Science Foundation grant, and NEERI in India 1996 and University of Miami, Florida.

Nemerow, N. L., et al. 1977. *Industrial Complexes and Their Relevances for Pulp and Papermills,* Vol. 3, No. 1, p. 133. Calcutta, India (published also in 1980 by Pergamon Press, Oxford, England).

Nemerow, N. L. 1985. *Stream, Lake, Estuary and Ocean Pollution,* pp. 303–309. New York: Van Nostrand Reinhold Publishing Company.

Nemerow, N. L. 1995. *Zero Pollution for Industry,* pp. 105–209. New York: John Wiley Publishing Company.

Nemerow, N. L., F. J. Agardy. 1998a. *Strategies of Industrial and Hazardous Waste Management,* p. 81. New York: John Wiley & Company.

Nemerow, N. L., F. J. Agardy. 1998b. *Strategic Management of Industrial Wastes,* pp. 91–93. New York: John Wiley Publishing Company.

Newborn, S. 1992. Phosphate regulations proposed. *The Tampa, Florida Tribune,* September 30, l992.

Phillips, M., M. Pacelle. 2003. Banks accept environmental rule. *Wall Street Journal,* June 4, 2003, p. A3.

Plant closing shows intensity of cost cuts. *Wall Street Journal,* June 24, 2002, p. A2.

Rose, C. D. 2000. Power in peril. *San Diego Union Tribune,* February 13, 2000, p. I-1.

Satchell, M. 1995. Sinkholes and stacks. *U.S. News and World Report,* June 12, 1995, pp. 53–56.

Sherif, Veziroglu, Barbir. 2005. In: *Environmental Solutions,* Chapter 7. Elsevier Publishing Company.

Solomon, B., R. Lee. 2000. Why are some population segments more exposed to pollutants than others? *Environment,* October 2000, p. 34.

Vas, R. E. 2000. An assessment of market-based regulatory tools. *Environmental Technology* Jan-Feb 2000, pp. 15–17 [from an October 1999 report authored by Gary C. Bryner].

Zweig, J. 1992. Cement shoes. *Forbes Magazine,* 11 May 1992, p. 20.

CHAPTER 17

Realistic Industrial Complexes

In some cases, I have gathered more industrial production and waste data than others. These I am classifying as "Realistic Industrial Complexes" and presenting for illustration purposes in this chapter. In Chapter 18, I propose other possible industrial complexes about which little operating data have been amassed. These environmentally balanced industrial complexes (EBICs) are classified as "potential industrial complexes."

In this chapter, six EBICs are depicted as realistic. Some of these I have reported in earlier publications, some have been fortified with additional data, and one has been developed recently.

1. Phosphate fertilizer–ammonium sulfate–cement complex
2. Tannery–slaughterhouse–rendering complex
3. Sugarcane–power–alcohol complex
4. Textile mill complex
5. Pulp and paper mill complex
6. Sugarcane–briquette–fertilizer complex

Fertilizer–Cement Complex

Fertilizer Plant Wastes and Production

It has been reported and generally accepted that phosphate mining in central Florida accounts for about 75% of the U.S. needs and one-third of the world's supply. This alone makes it a vital industry not only to Florida, but also to the United States and the world.

After the rock is extracted, slurried, and separated from the clay and sand by screening and flotation, it is used to produce wet process phosphoric acid. The rock is digested by sulfuric acid to produce a slurry of contaminated gypsum ($CaSO_4 \cdot 2H_2O$) and phosphoric acid. The gypsum is pumped to holding ponds, where it represents a major disposal problem for the fertilizer industry. Because 4.5–5.0 tons of gypsum are formed during the production of each ton of phosphoric acid, the industry has a formidable volume of phosphogypsum (PG) waste with which to cope.

Any recovery and reuse system for PG will free up reclaimable land for productive purposes by the industry or by other private or public landowners. Further benefits can be derived from the elimination of adverse environmental consequences of leachates from the gypsum heaps. Leachates carry phosphate and other mineral nutrients that could contaminate drinking water supplies and cause algal blooms (red tide) in recreational waters. Direct reuse of PG presents the potential problem of incorporating radioactivity into building or road products. Sulfates also cause cement-contamination problems.

It is quite likely that the direct reuse of this PG within a closed industrial complex in making cement could eliminate all of the aforementioned problems. In addition, it anticipates a lowered production cost for both the fertilizer and the cement plants for reasons already mentioned in Chapter 16.

One other area of waste recovery that should be mentioned is "heat" energy. A phosphate complex generates and must dissipate large amounts of energy as waste heat. The recovery and utilization of much of this energy is a very real success story. Efforts are continuing to recover even more of the "waste" energy from the more difficult to recover sources, and there is no doubt that an even greater percentage of the available "waste heat" will be put to profitable use.

As early as 1968, it was reported that many firms in the United States had innovated processes for manufacturing useful products such as H_2SO_4 and cement from waste gypsum (*Chemical Week* 1968). Nothing would be gained by reporting here the numerous papers that have been published describing the potential or actual use of gypsum in cement making. However, I will report on a few representative ones. The British Sulfur Corporation, Ltd., described the MASAN product transformed from PG by the Brussels-based company Ultra International SA as a useful cement or plaster (*Chemical Week* 1968). It was reported to possess a compressive strength three to four times that of Portland cement (1,100 kg/cm^2 as compared to 300–400 for Portland cement). Moreover, the cost of cement from PG was $10/ton as compared to $30–40/ton for Portland cement at that time.

Ellwood (1969) describes a chemical process for converting PG into hemihydrate powder as a cement strong enough to compete with cement in applications such as sound-proofing dividing walls. Carmichael (1986) reported two Belgium plants that were using the Central-Prayon process for converting PG into the hemihydrate form of $CaSO_4$. The gypsum is then suitable for direct reuse in the plaster industry or as a cement retarder back in the cement plant.

Bhanumathidas and Kalida (1986) reported the conversion of anhydrite I grade of PG to calcine at 950°C to obtain a product similar to Portland cement. They claim that the product "has shown remarkable cementitious behaviors in parallel to those of white Portland cement."

Clur (1986) claims that the Fedmis (South Africa) fertilizer plant disposes about 25% of its PG production as soil conditioner, cement clinker, and cement retarder. "The quality of the cement compares favorably with that of local limestone-based cements, and is used in all classes of building construction and civil engineering." Clur (1986) also reports that "the technical problems of producing a good quality cement from phosphogypsum have largely been solved, the future of the process would seem to depend mainly on economic and environmental factors."

Cement Plant Raw Materials and Wastes

Portland cement is made by mixing and calcining calcereous and argillacerous materials in the proper ratio. Table 17.1 summarizes the raw materials consumed in 1972 (Shreve and Brink 1977).

One can observe in Table 17.1 that limestone represents the majority mass of cement raw materials. Replacement of some or all of this calcareous material with phosphogypsum would reduce the production cost of the cement as a result in savings of raw material.

Unit processes involved in cement manufacturing essentially include storage and mixing of raw materials, drying, grinding and crushing, calcining, clinker storage, finishing additives and ball milling, and packing for delivery. Although dry processing is practiced more than wet processing, both are shown in Figure 17.1A and B to provide a visual aid for cement production.

For each 376 barrels of finished cement by the dry process, 1,120,000 BTUs of fuel is required as well as 24.1 kWh of electricity, 30 gallons of water, and 0.17 hours of direct labor. Also required are 498 pounds of limestone, 124 pounds of shale, and 16 pounds of gypsum (Shreve and Brink 1977). I also mention here that as far back as 1945 I developed a wallboard for Johns Manville Corporation. This board was made of asbestos fibers and gypsum formed under high temperature and pressure.

TABLE 17.1
Raw Materials Consumed for Portland Cement in United States
3 (thousands of short tons)

Cement rock	23,799
Limestone	90,003
Marl	2,080
Clay and shale	12,158
Blast furnace slag	759
Gypsum	4,094
Sand and Sandstone	2,774
Iron Materials	839
Miscellaneous	414

Source: Shreve and Brinke (1977).

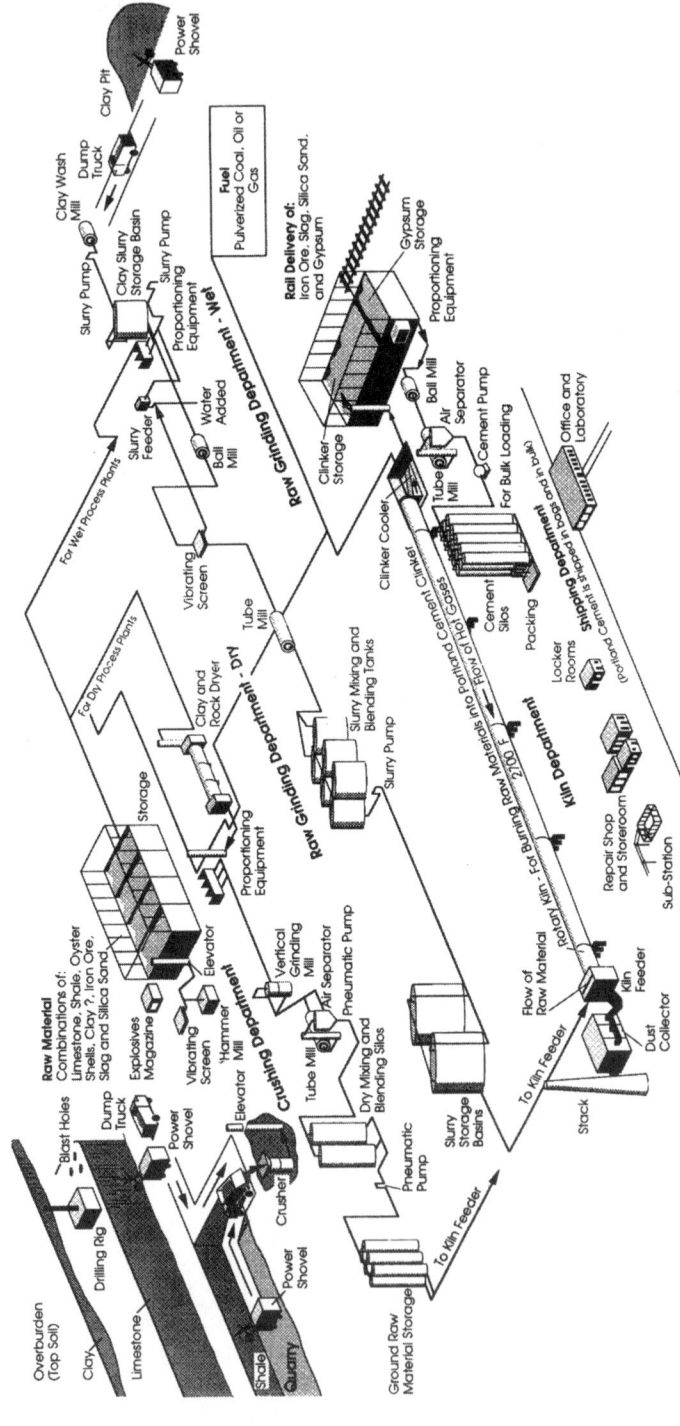

FIGURE 17.1A. Dry and wet processing. Used with permission from John Wiley and Sons.

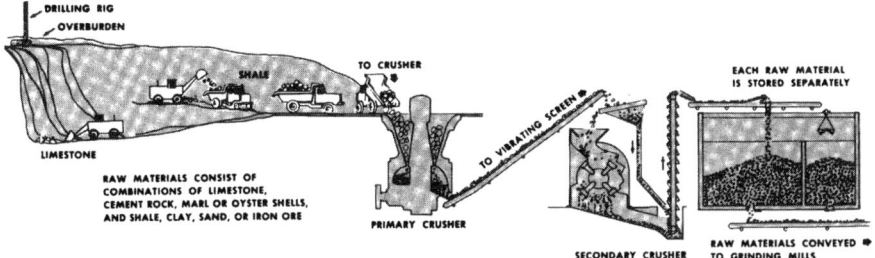

1. Stone is first reduced to 5-in. size, then to ¾ in., and stored.

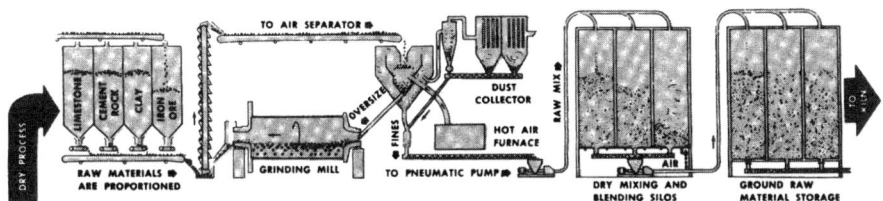

2. Raw materials are ground to powder and blended.

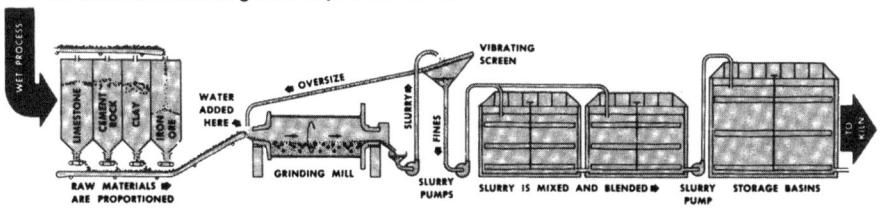

2. Raw materials are ground, mixed with water to form slurry, and blended.

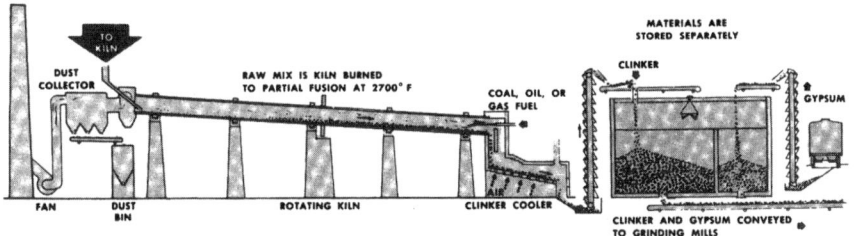

3. Burning changes raw mix chemically into cement clinker.

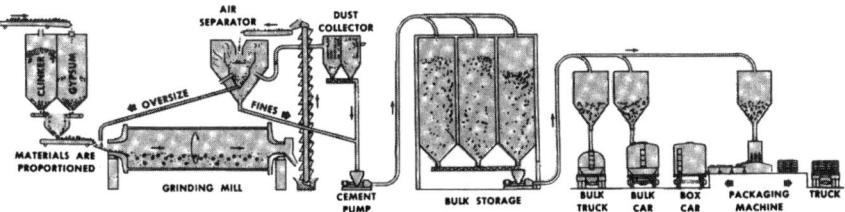

4. Clinker with gypsum is ground into portland cement and shipped.

FIGURE 17.1B. Dry and wet processing. Used with permission from John Wiley and Sons.

Statement of Problems and Objectives

The problems are twofold: (1) to lower production costs and (2) to eliminate adverse environmental impacts of industrial plants. These problems are especially severe or extensive when a particular industry is highly competitive, such as fertilizer and cement plants, and when these same plants produce significant wastes that pollute the environment (air, water, and land).

The overall objective was to determine the feasibility of location, building, and operating a two-industry complex, consisting of a phosphate fertilizer and a cement plant, within an EBIC at one site. The ultimate goal of this complex is to lower production costs at both plants while eliminating all adverse environmental impacts.

Further study (such as is presented on an economic basis in Chapter 16, Section two) should analyze and evaluate in depth the practicality of the complex that I presented earlier. Once again this complex as proposed is presented for your review in Figure 17.2A.

More precisely, it is necessary to determine: (1) the optimum size for each manufacturing plant included within the complex; (2) the suitability of the three products (wastes) for recovery and reuse as raw materials for ancillary adjacent plants within the complex (compatibility of plants); (3) the validity of total waste elimination from the two plants involved within the complex; and (4) the cost of production of the prime goods when manufactured at distinctly separated plants and compared to the same when manufactured within the complex (once again, as described in Chapter 16B). For convenience and further examination, this complex is shown in Figure 17.2B.

Next, the main effort concentrated on the extent of the economic gain by using the complex principle. This study included the economic cost of

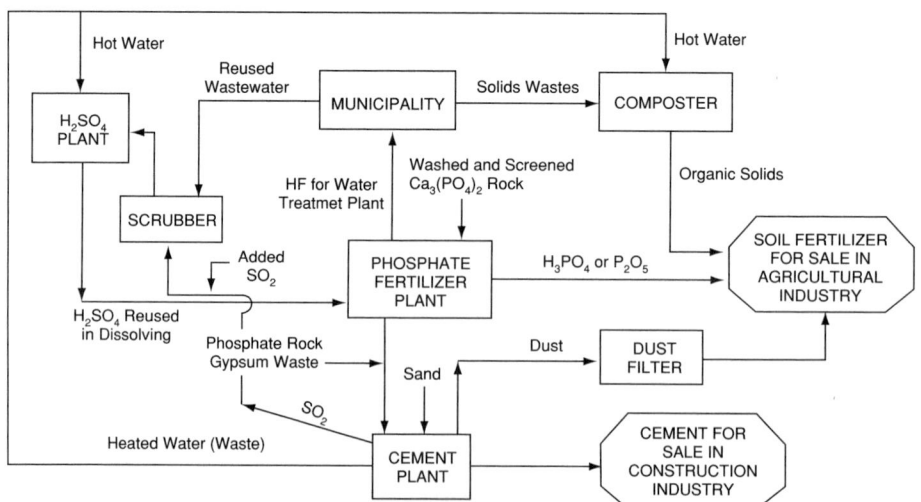

FIGURE 17.2A. Cement–fertilizer–municipal complex.

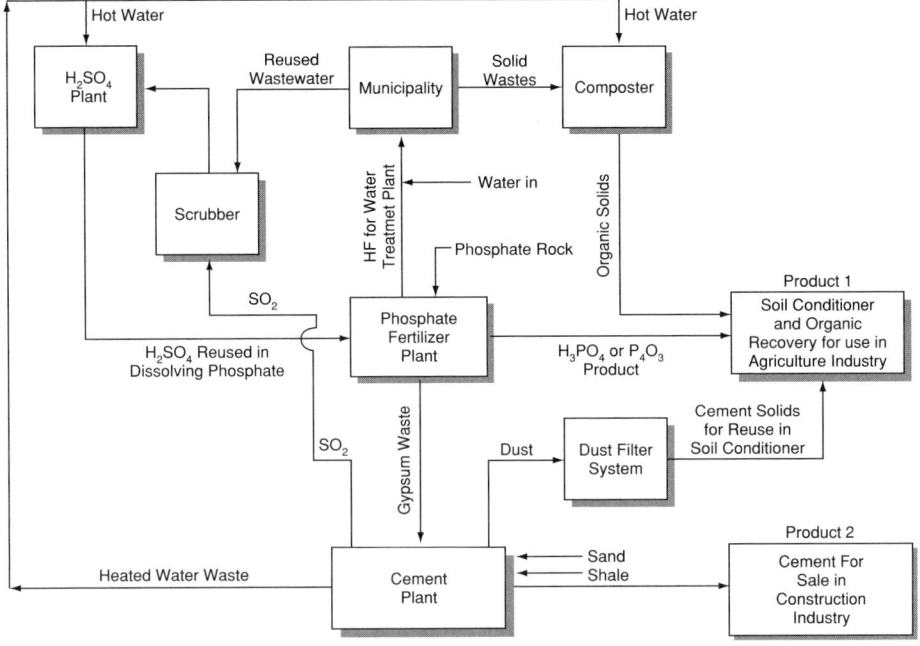

FIGURE 17.2B. Environmentally balanced fertilizer-cement plant complex phase.

environmental damage caused by wastes of all plants involved as part of the production costs.

This concept is not only "a gypsum for cement" idea, but also a totally new balanced industrial complex plan. The question is not whether this innovation is economical, but how much reduction in cost can be obtained by this complex principle when environmental costs are also included and all complex plant wastes are reused including excess heat.

Environmental costs include direct costs of wastewater treatment, indirect costs of environmental damages (to adjacent property owners), and intangible costs of environmental damages to "distant third parties," that is the public at large. These considerations are valid for both fertilizer and cement plants. The direct costs of fertilizer waste treatment (Fwt) included those for preventing PG piles from reaching groundwater or surface water, storage and preventing escape of mine tailings, air abatement from the production of fertilizer from sulfuric acid reactors, and other forms of waste treatment.

The *indirect costs of fertilizer waste environmental damages* (Fnd) include those for fish kills from pile or pond seepage, groundwater contamination from these seepages, and decontamination of company-owned and adjacent (third-party) property, as well as costs associated with the public relations aspects of these events (commonly referred to as "damage control"). It is much easier to maintain a "good corporate citizen image" than to win back that image after an environmental incident.

The *intangible costs of fertilizer environmental damages* (Frd) include costs such as those for loss in land value due to pollution and decrease in desirability of the property due to public reluctance to locate adjacent to contaminated facilities. Times Beach in Missouri, Love Canal in New York State, and the Stringfellow Acid Pits in southern California are examples to be studied and learned from.

One study (described in Chapter 16) (Krishnam et al. 1996) revealed the following values:

Fwt value of $15.24/ton of fertilizer
Fnd value of $1.74/ton of fertilizer
Frd value of $84.36/ton of fertilizer

When added to the conventional production cost of $204.02/ton, the real production cost became $305.36/ton, or about 50% greater than the conventional cost.

Similar analysis of the cement industry led to the following:

Cwt value of $14.28/ton of cement
Cnd value of $2.90/ton of cement
Crd value of $0.03/ton of cement

When added to the conventional product cost of $50/ton, the production cost becomes $67.21/ton, or about a 34% increase.

When combining these two plants in the complex described earlier, these environmental costs can be significantly reduced or avoided altogether. The result would be an overall savings resulting from facility complexing of $154.62/ton of product or 42%. In addition, the costs of mining, transporting (of raw materials), associated transportation spill damage, material handling, and storage are reduced or eliminated using the complexing principle.

Prevention rather than pollution appears more often than not to be the most profitable alternative. In some instances, when the PG is unsuitable for direct reuse in making cement because of its sulfur content, intermediate treatment with ammonia and CO_2 may be necessary. This results in another fertilizer product, $(NH_4)_2SO_4$. Phosphochalk is also produced, which can be used directly in making cement. Such a potential EBIC is presented in Figure 17.3 for a typical 600-ton/day phosphoric acid fertilizer plant.

Two-Tannery Complex

Tannery wastes from upper sole chrome tanning mills contribute to a significant pollution problem in the United States. The wastes are hot, highly alkaline, odorous, highly colored, and contain elevated quantities of dissolved organic matter, biochemical oxygen demand (BOD), total suspended solids, lime, sulfides, and chromium. The

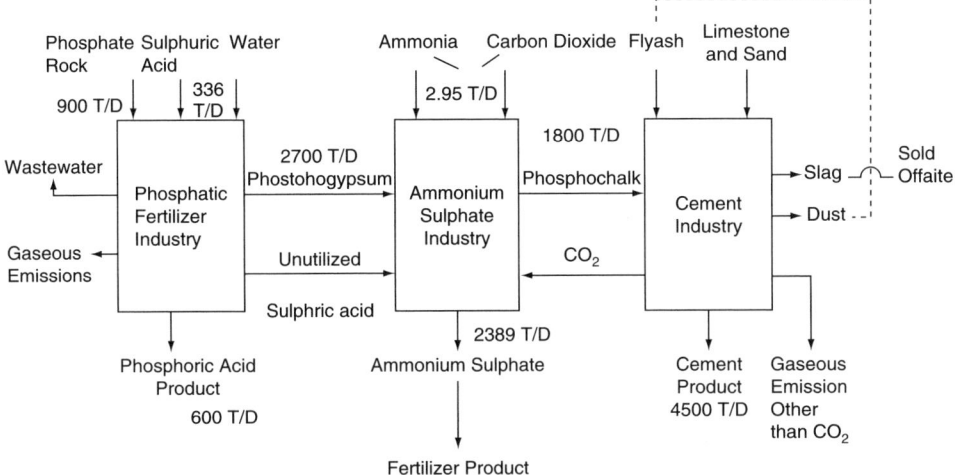

FIGURE 17.3. Schematic diagram of environmentally balanced phosphate–fertilizer–cement industrial complex.

treatment of such wastes has been difficult because of the conflicting pollutional parameters of pH, organic matter, and potential toxic compounds. Most successful treatment plants use some form of biological treatment to reduce the oxygen demand on receiving waters. This necessitates the use of well-designed and operated preliminary treatments to ensure safe and efficient biodegradation. High sludge quantities result from these treatments. Therefore, properly designed and operated tannery waste-treatment systems may be costly to build and operate, whereas the lack of these facilities will cause excessive stream pollution. Placing the tannery in an environmentally optimized industrial complex eliminates both of these negatives. One such complex involves three separate industries and is described next.

The Slaughterhouse–Tannery–Rendering Complex

I have presented two formal papers at technical meetings on this subject (Nemerow 1980a; Nemerow and Dasgupta 1981) and a report (Nemerow 1980b). These represented a first attempt at providing a complete mass balance of reference-validated inputs and outputs of plants within an industrial complex.

The fulcrum industrial plant of this complex is a tannery. Supporting industries include slaughterhouse and rendering plants. The three-industry complex is also expanded to consist of an animal grazing and feedlot facility, as well as a residential area for homes of all personnel working in the complex. As the complex is expanded to include the feedlot and residences and biogas and power plant services, the complex becomes more self-sustaining. Outside service requirements are minimized by the expansion. All power is generated within the complex—in the expanded third-stage version. Excess products of leather, meat, meal, soap, and even electricity are sold to consumers outside the complex. Chemicals, water, cattle, and animal feed are imported to the complex. Wastewater, blood and bone meal, hide and leather trimmings, cattle dung, and residential solid wastes are recovered and reused within (internally) the expanded complex. The complex can be constructed as shown in the first stage, second stage, or fully expanded to the third stage (Figure 17.4). Criteria for decision making will be based on area requirements and individual local objectives.

Stage 1
This is the first of the three-stage industrial complex, which is balanced internally so that little or no adverse environmental impact results from any of the industrial plants' production activities. Each subsequent stage represents a totally balanced and individual industrial complex. This first stage consists of a three-industry plant complex: (1) a slaughterhouse, (2) a tannery, and (3) a rendering plant (Nemerow 1980b).

Stage 2
The second of the three-stage industrial complex is also balanced internally so that little or no adverse environmental impact results from any of the industrial plants' production activities. It differs from the first stage in that it provides a more complete and self-sufficient complex. It also provides more reuse potential for the three industrial effluents than the first stage. In addition, it provides living space in the complex for employees of the industrial plants and feedlot and grazing area for raising the animals to the required weight. Whenever feasible, the second-stage complex is recommended in preference to the first stage only (Nemerow 1980b).

Stage 3
The third stage of the three-stage industrial complex enlarges the smaller complex and is more balanced internally so that little or no adverse environmental impact results from any of the industrial plants' productive activities. Agriculture and municipal residence services are provided in this phase of the complex. Residential solid wastes from both industrial and municipal facilities are fermented to methane gas, which is used subsequently to produce electrical energy for use in the complex. Waste sludge from the fermenter is incinerated to produce additional electrical energy for use in the complex. The schematic arrangement of the third phase of the complex is shown along with the mass balances of each unit in Figure 17.4. External raw materials and manufactured products for external sale are given in Table 17.2.

General Discussion
As we proceed with the three-industry complex by adding stages, some potential problems arise. For example, when we add stage 2 to the complex, we compute that a cattle grazing and feedlot area of 620 acres is required for the 135,000 cattle. This vast acreage may be difficult to obtain. In addition, 1,350/tons/day of feed must be supplied from internal and external sources.

In the third stage of the complex, we are proposing to produce methane gas from solid waste residues. This gas will subsequently be used for power production.

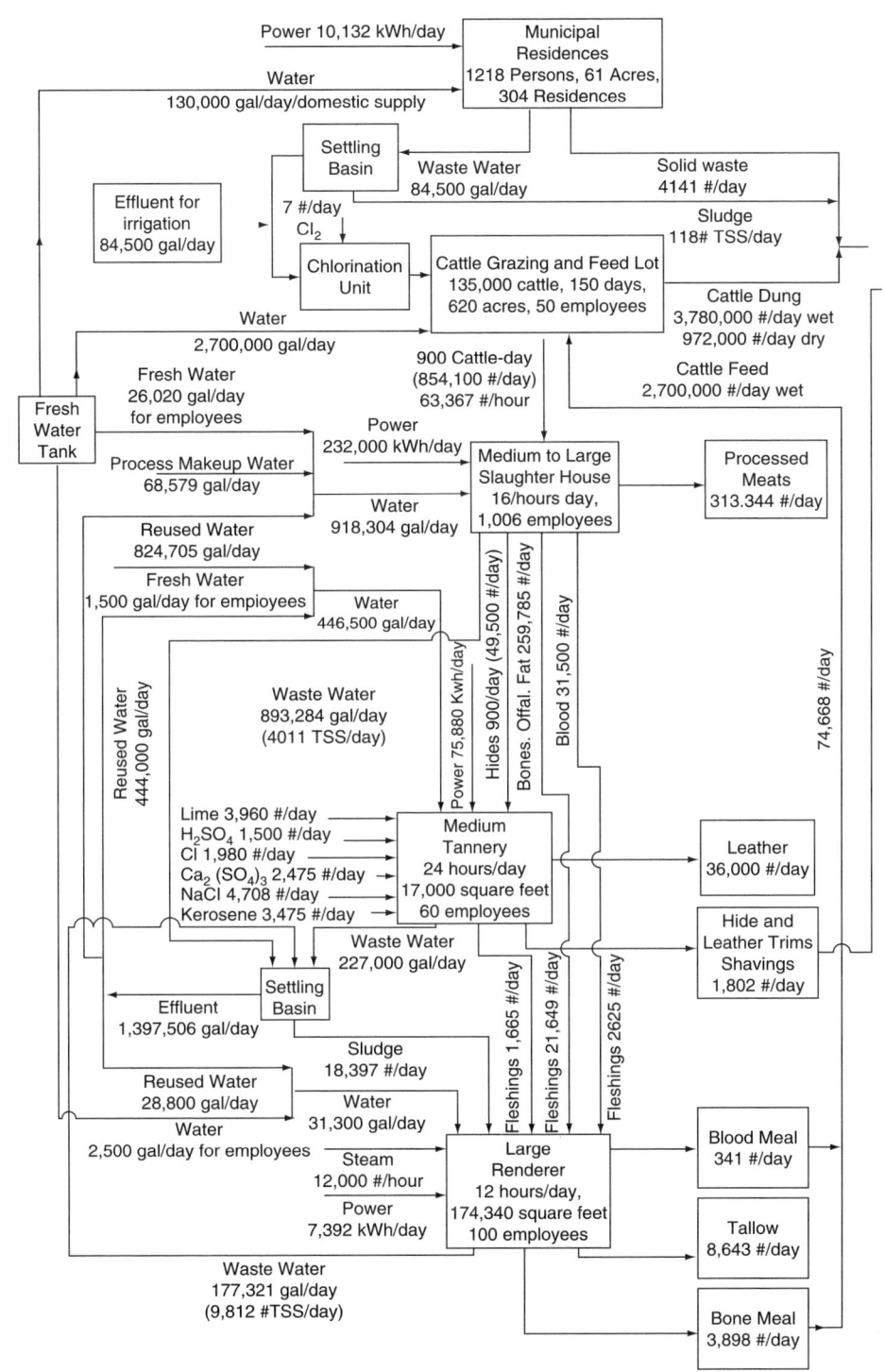

FIGURE 17.4A. Three-industry complex: tannery–slaughterhouse–rendering.

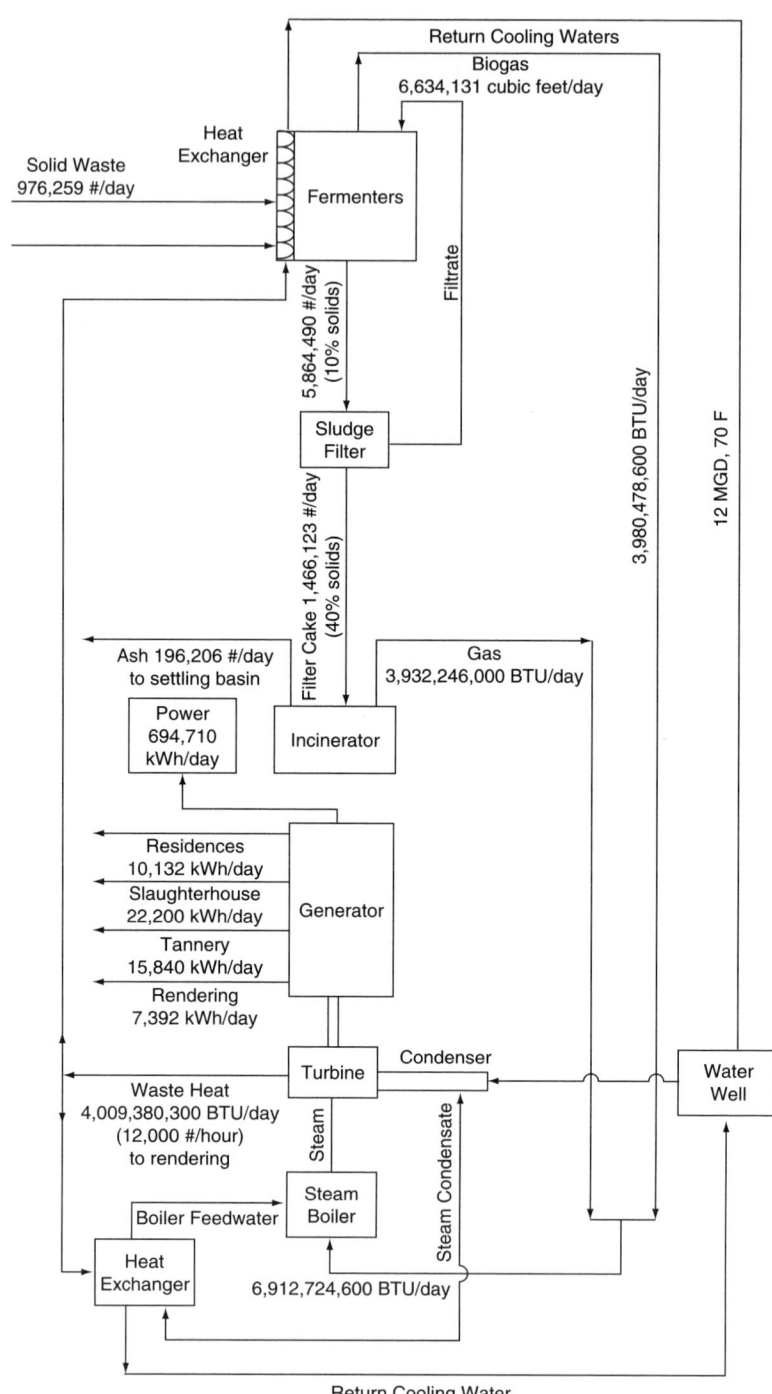

FIGURE 17.4B. Three-industry complex: slaughterhouse–tannery–rendering.

TABLE 17.2
External Raw Materials and Manufactured Products in Three-Industry Complex (Stage 3)

Raw Material Required from Outside the Complex		*Manufactured Products for Outside Sale*	
Material	Amount	Material	Amount
1. Fresh Makeup water	2,927,599 gal/d	1. Meat products	513.341 #/d
1A. Well water (one time only)	12 mgd	2. Tanned leather	36,000 sq.ft/d
2. Calves	900/d (150 days)	3. Tallow	79,740 #/d
	540,00 #/d	4. Energy	694,710 kwh/d
3. Chemicals	495 #/d Na_2S		
	3960 #/d $Ca(OH)_2$		
	1500 #/d H_2SO_4		
	2475 gal/d kerosene		
	1980 #/d oil or wax		
	2475 #/d $Cr_2(SO_4)_3$		
	4208 #/d NaCl		
	7 #/d Cl_2		
4. Cattlefeed	2,625,000 #/d		

Source: Nemerow and Dasgupta (1981), p. 209.

An excess of power within the complex results from this sequence of operations. An alternative to exporting power for sale outside the complex would be the production of other valuable intermediate products such as alcohol from the ferments. This can be determined from market conditions at the time of establishment of the complex.

This three-stage complex analysis is the deepest study of the new concept. As shown in Table 17.2, the managers of the three-stage complex still must import four basic materials: water, calves, chemicals, and cattle feed. About 3 million gallons of water, 2.6 million pounds of feed, 900 cattle, and about 6 tons of chemicals are needed each and every production day. This complex also will produce for external sale about 250 tons of meat, 36,000 ft² of leather, 40 tons of tallow, and almost 700,000 kW of energy each production day. Although complete economic analysis of such a system has not been made, it appears at least self-sustaining and probably will show a considerable net profit. The implications of such complexes are obvious. However, if the complex is able to produce a profit and protect the environment from any degradation, its major goals will have been achieved.

Conclusion
A three-stage environmentally balanced complex has been designed. Mass balances of all plant inputs and outputs have been computed based on the most recent published industrial data. From an analytical standpoint, an industrial complex consisting of a slaughterhouse,

tannery, and rendering plant is technically feasible. This complex is also technically feasible when expanded to include animal grazing and feedlots, as well as municipal residences (second stage). The expanded version (third stage) of the complex is more self-sustaining as far as reused products and electrical energy generation are concerned.

Sugarcane Complexes

The Cane Sugar Industry

The cane sugar manufacturing industry is essential to the production of many varieties of foods. In the United States, there are about 6,400 sugarcane plantations, 94 sugar mills, and 24 sugar refineries, mostly located in Florida, Louisiana, and Hawaii.

Because of recent dietary recommendations, alternative sweeteners have entered the market. Competition from the lower prices of other sweeteners has caused a reduction in refined sugar prices. This is true even though there has been a deficit in sugar produced in the United States. Florida, the largest sugar-producing state in the nation, grows about one-fifth of all sugar consumed in the United States. It is imperative to the Florida mills, as well as sugar refineries elsewhere, that production costs be kept to a minimum to keep the industry healthy.

Brief Outline of Sugar Manufacturing Process

In the manufacturing of sugar, the sugarcane stalks are chopped into small pieces by rotary knives, and the cane juice is extracted from these pieces by crushing them through one or more roller mills. The solid residual material from this operation, consisting of fibrous residue of the cane sugar stalks, is termed "bagasse" and is a solid waste of the cane sugar industry. After the juice is extracted from the stalks, it goes to the boiler room where lime is added to precipitate insoluble sugars. The precipitate, in the form of thick slurry, is vacuum-filtered to produce a filter cake often termed "cachaza" and constitutes the second type of solid waste from sugarcane manufacturing operations. Then, the clarified juice is thickened in evaporators, and the resulting syrup containing sugar and molasses is boiled in vacuum pans to form raw sugar crystals. The sugar crystals are separated from molasses by centrifugation, and the molasses is sometimes further evaporated to recover more sugar. The final products are coarse, crystalline brown raw sugar and molasses. The raw sugar is transported for further processing in sugar refineries to produce the various forms of white refined sugar. The bulk of the molasses is used for production of various types of fermentation products and a small portion is used for animal feed. A schematic diagram of a sugar mill operation is shown in Figure 17.5.

The Solid Waste Problem

The two forms of solid wastes generated in the manufacturing of cane sugar are bagasse and cachaza. Every 1,000 tons of processed sugarcane generates about 270 tons of bagasse and 34 tons of cachaza.

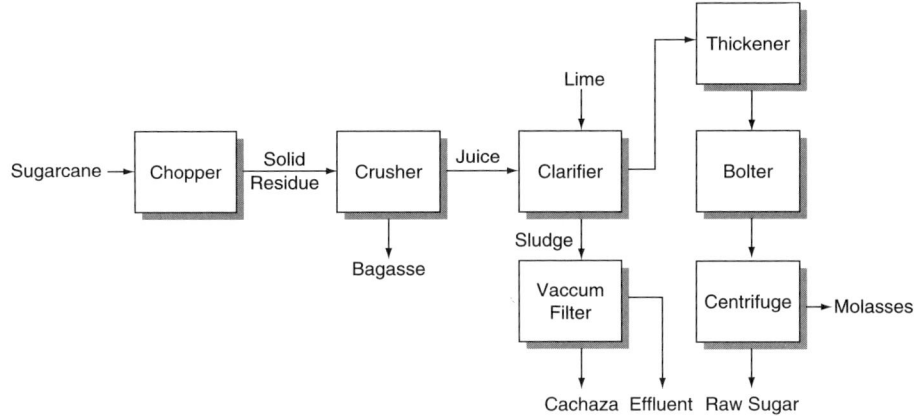

FIGURE 17.5. Raw sugar manufacturer-flow diagram.

The sugar industry is faced with the problem of proper and economical disposal of large quantities of bagasse and cachaza. The most common bagasse disposal method has involved burning as much as possible in boilers operated at sugar mills. Burning bagasse presents problems of its own. It is not a particularly clean fuel, and mills require installation and maintenance of stack scrubbers to clean the emissions. Moreover, utilization of bagasse as a boiler fuel is impaired by the high degree of moisture (45–60%). In addition, its bulkiness requires the construction of special furnaces to operate efficiently.

The other type of solid waste generated, namely cachaza, generally is slurried for disposal by lagooning or disposed as landfill, resulting in land and water pollution. Even if a large portion (usually 70%) of the bagasse generated is burned directly in boilers, a considerable amount of bagasse (30%) remains to be disposed of with the entire quantity of cachaza.

Considering the high cellulose content of bagasse and the organic matter in cachaza, these are potential renewable sources of biomass for biochemical conversion to methane by anaerobic fermentation. In addition, the residual digested sludge can have beneficial uses as fertilizer/soil conditioner.

Environmentally Balanced Industrial Complex Solution

Anaerobic digestion of a 2.4:1 mixture of bagasse to cachaza was demonstrated to be effective in producing methane gas and reducing organic solids (Nemerow and Dasgupta 1984). Despite this development, residual wastes remain to be considered.

An evaluation of the sugarcane refinery based on products and wastes after digestion suggested that a "closed-loop" complex would result in the discharge of little or no final residual wastes. Figure 17.6 presents a schematic diagram of a sugarcane refinery–based

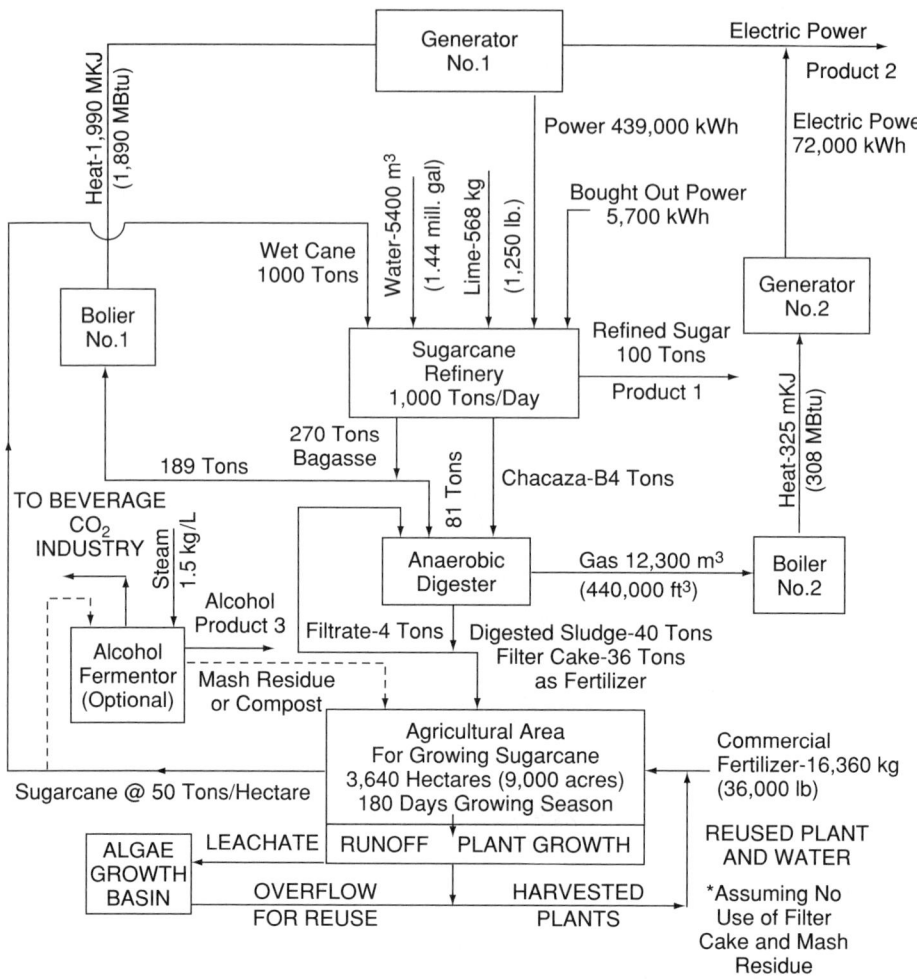

FIGURE 17.6. Sugarcane refinery-based EBIC (sugarcane-power-alcohol complex).

EBIC. For purposes of this evaluation, the mass balances are estimated based on the refining of 1,000 tons of sugarcane, resulting in a generation of about 270 tons of bagasse and 34 tons of cachaza (Nemerow and Dasgupta 1984). In many mills, these wastes are discharged to the environment with a variety of adverse impacts.

Sugarcane–Briquette–Fertilizer Complex

Introduction
Sugarcane is cultivated in tropical and subtropical areas as it requires at least 60 in. of irrigation or rainfall per year. The major producing countries of sugarcane are Brazil,

India, and China, as shown in Table 17.3. Their combined total production exceeds half of global production. The composition of sugar cane is shown in Table 17.4.

Sugar Production Process

The sugar production process is very energy intensive, it requires steam and electricity at many stages. Initially, the sugarcane stalks are chopped into smaller pieces by means of a chopper, as shown in Figure 17.7. The chopped stalks then pass through a crusher, which has one or more roller mills for the extraction of the juice. The resulting fibrous residue is called *bagasse,* a solid waste. The juice then passes through a clarifier where lime is added to precipitate the insoluble sugars. The precipitate, a thick slurry, is vacuum filtered to produce filter mud cake (cachaza), another solid waste of the sugarcane industry. The clarified juice is thickened and then heated in a boiler to form raw

TABLE 17.3
World Production

	Area Harvested (ha)	Yield (tons/ha)	Production (tons)
Brazil	5,303,560	73.83	386,232,000
India	4,300,000	67.44	290,000,000
China	1,328,000	70.71	93,900,000
Thailand	970,000	76.36	74,071,952
Pakistan	1,086,000	47.93	52,055,800
Mexico	639,061	70.61	45,126,500
Colombia	435,000	84.14	36,600,000
Australia	423,000	85.13	36,012,000
Cuba	1,041,200	33.33	34,700,000
USA	403,390	77.29	31,178,130
Philippines	385,000	67.10	25,835,000
Other	4,091,132		244,581,738
Total	**20,405,343**		**1,350,293,120**

TABLE 17.4
Sugarcane Composition

Water	69–75%
Sucrose	8–16%
Reducing sugars (dextrose and levulose)	0.5–2%
Organic matter other than sugar	0.5–1%
Inorganic compounds (phosphates, chlorides, nitrates, silicates, sodium, potassium, etc.)	0.2–0.6%
Nitrogenous bodies (albuminoids, amides, amino acids, ammonia, etc.)	0.5–1%
Ash	0.3–0.8%

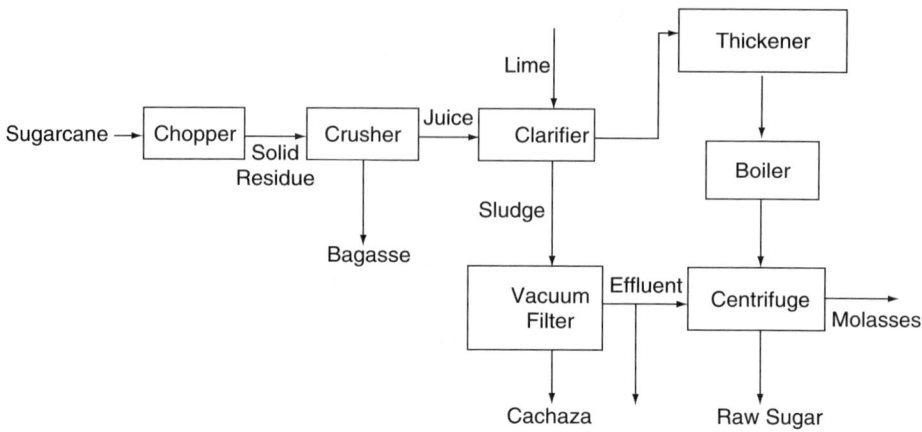

FIGURE 17.7. Sugarcane production.

sugar crystals. These crystals are a mixture of sugar and molasses. The sugar is separated from the molasses by centrifugation. The raw sugar is transported to sugar refineries and the molasses is used in the production of animal feed.

Solid Wastes of Sugarcane Industry

As seen in the cane sugar manufacturing process, there are three byproducts, namely, bagasse, filter mud cake, and molasses. Molasses is widely reused for animal feed in ethanol production or in other molasses-based products and, thus, is not considered an environmental waste. Bagasse and cachaza have limited applications and their disposal causes environmental problems; therefore, they are the main solid wastes in the cane sugar industry.

Every 1,000 tons of processed sugarcane generates 270 tons of bagasse and 34 tons of cachaza (Nemerow 1995). Proper and economical disposal of these solid wastes is a major problem to the sugarcane industry because of their significant quantities and adverse effects on the environment.

Bagasse

Historically, bagasse was burned in boilers at the sugar mill without any treatment as a boiler fuel, although it has a high moisture content (45–60%) (Nemerow 1995), which reduces its efficiency. Bagasse has a gross calorific value of 19,250 kJ/kg at zero moisture and 9,950 kJ/kg at 48% moisture. The net calorific value at 48% moisture is around 8,000 kJ/kg (Deepchand 2001).

Direct burning of bagasse is not an efficient disposal method because the emissions are harmful to the environment and requires the installation of filters to clean the smoke. In addition, the bulkiness of bagasse causes it to have low energy value per unit volume and requires the construction of special furnaces for efficient

TABLE 17.5
Chemical and Physical Composition of Bagasse and Mud Cake (4)

Constituent	Bagasse (%)	Mud Cake (%)
Cellulose	46	8.9
Hemicellulose	24.5	2.4
Lignin	19.9	1.2
Fats and wax carbon	3.5	9.5
	48.7	32.5
Hydrogen	4.9	2.2
Nitrogen	1.3	2.2
Phosphorous	1.1	2.4
Silica	—	7.0
Ash	2.4	14.5
Fiber	40.8	15.0

Source: Dusgupta (1963).

operation. Because of the inefficient burning of bagasse, usually only 70% is burned and the remaining 30% is disposed with cachaza in a landfill or sold at a very low price.

The high cellulose content of bagasse and the high organic content of mud cake qualify them as potential energy sources (Adekeke 2003). The chemical and physical composition of bagasse and mud cake is presented in Table 17.5.

Proposed Solution to Solid Waste Problem in Cane Sugar Industry

Energy consumption throughout the world is gradually increasing. Most of today's energy sources are carbon-based nonrenewable sources. Consumption of these sources needs to be reduced because of their significant negative impacts on the environment. Attention should be focused on the utilization of renewable energy sources. Biomass, organic matter derived from plants, is one of the renewable sources of energy being used in the production of liquid and gaseous fuels such as ethanol, methanol, hydrogen, and biogas. Efficient use of biomass is crucial for higher energy gains. Direct burning is inefficient because the energy value per unit volume is low and maintaining a steady fire becomes problematic because of difficulty in controlling the combustion process. In addition, collection, transportation, storage, and handling are tedious. One method for the efficient utilization of agricultural residues is their densification into solid fuel pellets called *briquettes*.

Briquette Quality Considerations

Briquette quality aspects are combustion, environmental concern, durability, and stability. *Combustion* is the energy value from the solid fuel as well as ease of the briquette getting ignited. *Environmental concern,* the level of toxic emissions during burning; *durability and stability,* how long the briquette can remain in its compact solid form for its intended purpose. Properties of the agricultural waste used and the efficiency of the briquetting process determine briquette quality. Other parameters against which briquette quality is measured include calorific value, compressive strength, porosity, and density.

Bagasse Briquetting
Adverse Environmental Impacts

In sugar mills that do not preserve the environment, 70% of the resulting bagasse is inefficiently burned in boilers to generate steam, which is reused in the sugar production process. A proportion of the ash from bagasse burning is lost to the surrounding air as fly-ash, thus polluting the environment. Consequently, sugar mills are required to install costly stack scrubbers to clean emissions. The remaining 30% of bagasse is mixed with cachaza and disposed nearby. Figure 17.8 illustrates the current situation in sugar mills; the values of the Komombo sugar cane factory in Aswan, Egypt, are used for illustration.

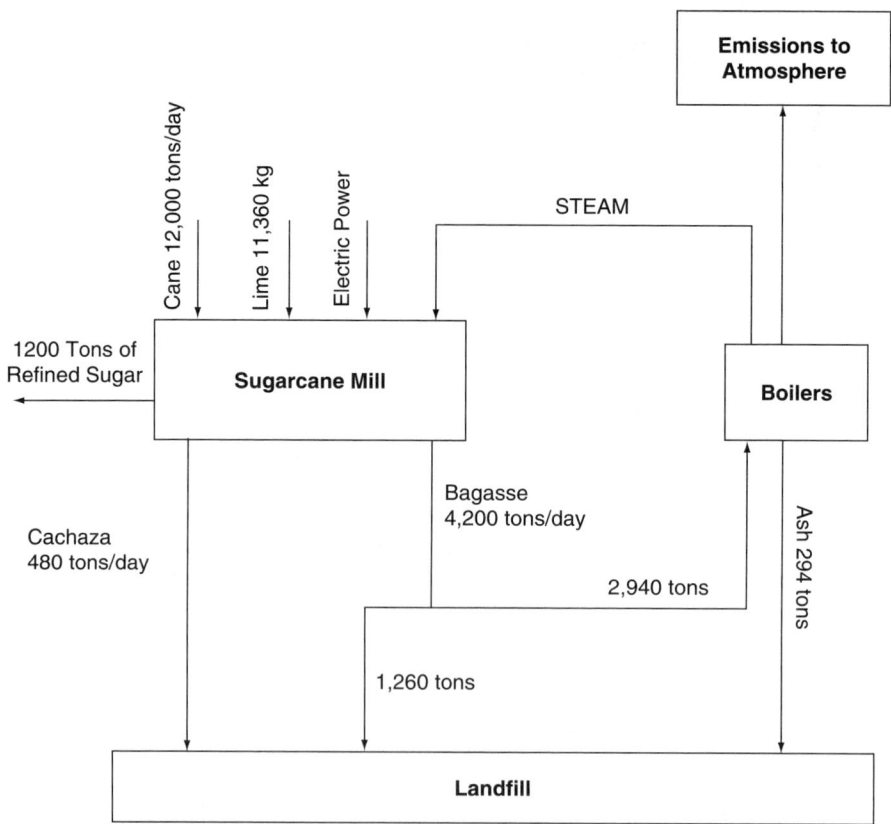

FIGURE 17.8. Current sugar-mill situation.

The previously described system is an inefficient system that needs to be modified. Two modifications are suggested as follows: introducing a briquetting unit for the bagasse and cachaza and developing an EBIC in which the briquetting unit is combined with the sugar mill at a single site. Briquetting of bagasse and cachaza has many advantages; however, more advantages will arise when both types of production are combined into a complex.

Advantages of Briquetting Unit

- When briquettes are burned in boilers, the compact form allows the ash to precipitate, thereby significantly reducing emissions. The precipitating ash, which is very rich in nutrients, can be retrieved and used as a fertilizer. Table 17.6 shows the chemical analysis of the ash generated from burning bagasse and cachaza.
- A briquette has a higher combustion efficiency, 80% on average as opposed to bagasse in its loose bulky form, which has a combustion efficiency less than 60% on average. The difference in combustion efficiencies is primarily related to the lower moisture content and improved properties of the briquettes compared to their loose bulky form.
- The amount of briquettes required for generating sufficient heat for the boiler can be determined with accuracy; therefore, increasing efficiency as opposed to using bagasse in loose bulky form.
- The resulting compact form of the briquettes allows them to have high energy per unit volume and high specific weight and makes them easier to store and transport. These properties make them an attractive fuel for home and industry use, so excess briquettes can be sold by the sugar factory.
- The cachaza and excess bagasse that were being disposed will be used in the production of briquettes.

TABLE 17.6
Composition of Ash from Cachaza and Bagasse [Dasgupta 1983]

	Ash of	
Composition	Cachaza (%)	Bagasse (%)
Organic Matter	9.77	17.13
Total Potassium	9.97	12.5
Total Phosphorus	0.67	1.24
Iron	0.5885	0.181
Manganese	0.0863	0.006
Copper	0.0353	0.0121
Zinc	0.0314	0.009
Calcium	3.6289	0.4133
Magnesium	0.0173	0.0127

Source: Dasgupta (1983).

Briquetting Unit

In bagasse–cachaza briquettes, most of the mixture is bagasse and the cachaza acts as a binder because of its fat and wax content. Several parameters determine briquette quality: residue size, moisture content, cachaza content, and compression temperature and pressure. Experimentation showed that bagasse–cachaza briquettes can be compressed with or without heating (Ishaq 2003). The briquettes from both processes almost have the same calorific value, which is 15,000 kJ/kg; however, they differ in compressive strength. The heat-pressed briquettes have a higher compressive strength, which allows better storage and handling (Ishaq 2003). Because the briquettes will be used as boiler fuel in the sugar mill, production without heat is more economical.

Briquetting at Varied Pressure without Heat

Increasing the pressure increases the density of the briquette; a higher density is desirable as it contributes to a higher energy per unit volume and improved handling and storage characteristics.

Different densities are obtained at different residue sizes; however, high pressure yields briquettes having densities equal to 0.7 g/cm^3 for all residue sizes, which is high enough to give sufficient energy per unit volume. The optimum process conditions are as follows (Ishaq 2003):

- Applied pressure: 100–120 MPa.
- Residue moisture content should range between 9 and 12%.
- Cachaza inclusion should not exceed 10%.

Theory of Complexing Technology

As shown earlier, briquetting is an appropriate method for the utilization of bagasse and cachaza; however, the sugar mill will realize more benefits by combining the briquetting unit in the same site to form an EBIC. The economic production objectives (in addition to the environmental benefits) of developing an EBIC include the following:

- *Saving in transportation costs:* Transportation costs would otherwise be high because of the large amounts of bagasse and cachaza that would have to be transported from the mill to the briquetting unit and back as briquettes.
- *No time delay:* No time will be lost in transporting the bagasse and cachaza to the briquetting unit and back to the mill to be used as boiler fuel. Time efficiency is essential for sugar mills because they operate only during the cane growing and harvesting season, which is usually 5 mo/yr. During this period, sugar mills usually operate 24 hours a day, 7 days a week because of time constraints.
- *High durability and efficiency:* The briquettes will be subjected to less handling such as loading and unloading on/from trucks, which maintains high durability and energy per unit volume for the briquettes.

- *Maintaining moisture content:* There will not be a significant change in moisture content of the briquettes during the period between their production and usage in the mill as boiler fuel.
- *Space utilization:* There will be no need to store the solid wastes until they are transported or store the briquettes when they arrive, thus reducing storage costs and storage areas. The bagasse–cachaza mixtures will be continuously fed to the briquetting unit and the produced briquettes will be used as boiler fuel right away. In the case of excess briquettes, the storage area required will be less than that needed for loose bulky bagasse.

The Environmentally Balanced Industrial Complex

Sugarcane–Briquettes–Fertilizer Plants

For illustrative purposes, the concept of developing an EBIC will be theoretically applied to the Komombo sugar mill in Aswan, Egypt. One potential complex is shown in Figure 17.9. The Komombo sugar mill produces 180,000 tons of refined sugar per year. The mill operates 24 hours a day, 7 days a week during the cane growing and harvesting season, which lasts for 5 months. The average daily production of the mill is 1,200 tons. The resulting amounts of bagasse and cachaza are 4,200 and 480 tons/day, respectively.

In the production of briquettes, the proportion of cachaza should not exceed 10% by weight, so 420 tons/day is used in production of briquettes and 60 tons/day is used in the organic fertilizer. The corresponding number of briquettes produced per day, assuming a briquette weighs 100 g per briquette, is 46.2 million.

Burning of the briquettes generates 55,440 GJ of steam for energy production. The steam can be used to drive turbines to produce low-pressure steam for heating purposes and 175,575 kW of electric power, both of which can be used to supply the refinery; in addition, excess electric power can be sold to the national grid.

Due to burning briquettes in the boilers, 462 tons/day of ash precipitates. The resulting ash is then transported to the organic fertilizer unit to be mixed with the excess cachaza and used in the production of fertilizer. The fertilizer produced is sold to cane growers and other consumers.

First Alternative: Establishing an EBIC for Each Sugar Mill

The complex proposed is illustrated in Figure 17.9, with a theoretical application to the Komombo sugar mill in Aswan. Comparing the complex in Figure 17.9 with the typical sugar mill in Figure 17.8 shows the additional investments necessary for the establishment of the EBIC. Although these investments will require a large capital outlay, the overall benefits will outweigh the large initial investment. A more complete economic analysis of this industrial concept solution is presented later in this chapter.

From a practical standpoint, we have verified that it is feasible to develop an EBIC utilizing the sugarcane mill, the briquetting plant, and a fertilizer plant to allow no liquid or solid wastes to enter the air, water, or land and cause adverse environmental

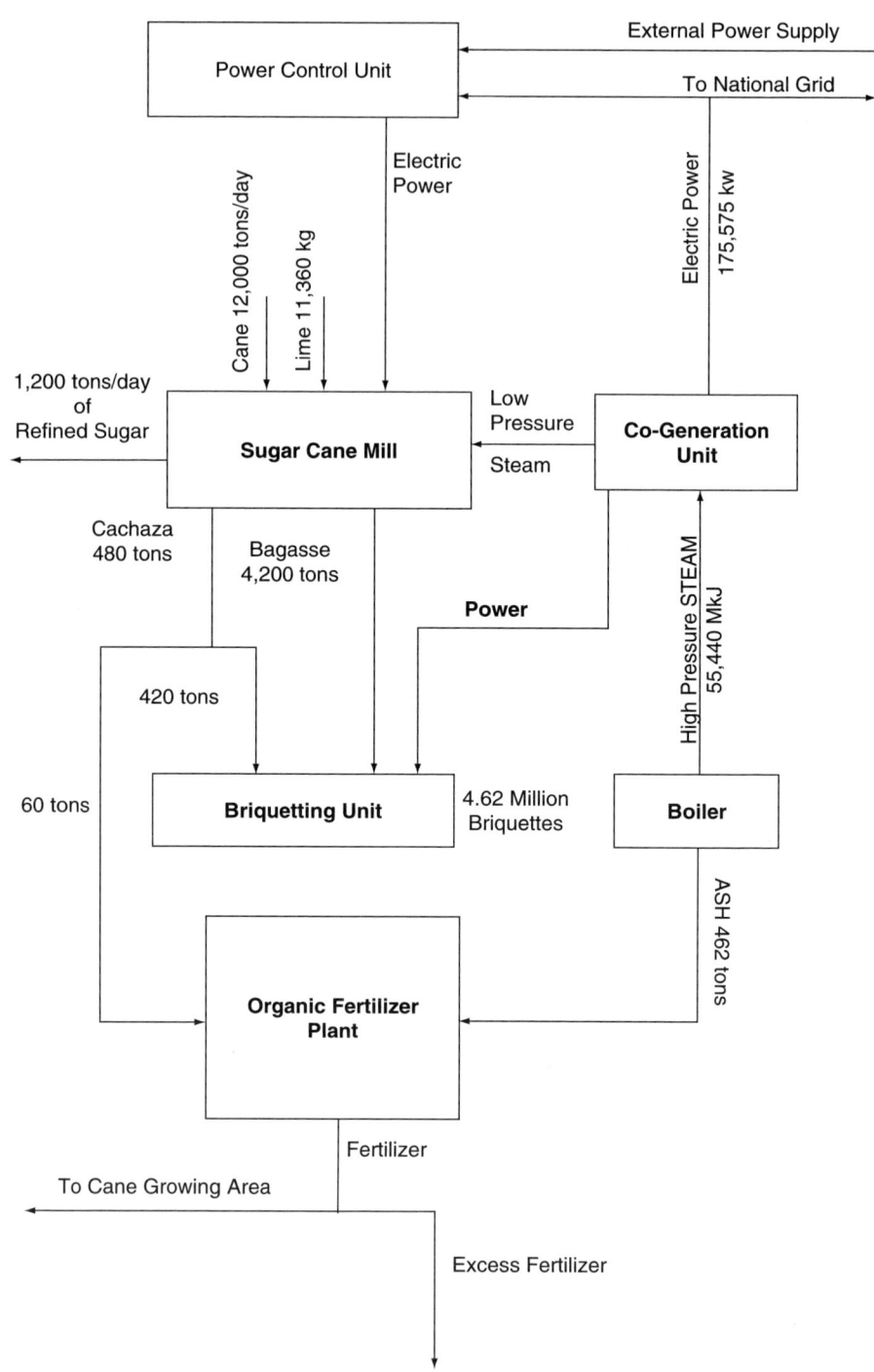

FIGURE 17.9. Environmentally balanced sugarcane complex.

effects. We also show in this book that it is economically feasible as well to develop this EBIC for all plants, especially for the sugarcane mill.

Acknowledgments

The author gratefully acknowledges the support of the U.S.–Egypt Science & Technology Joint Fund Program and the U.S. National Science Foundation and its Senior Program Director for Egypt, Dr. Osman A. Shinaishin.

Textile Mill Complex

Plight of Small Textile Mills

The textile industry represents one of the most competitive fields of production worldwide. Each plant attempts to reduce its cost to compete with other similar plants within its region and those in other countries. One answer to competition has been to increase production, sometimes by merging with other plants and sometimes merely by expanding one plant's capacity. Lower unit costs generally result from increased production in accordance with accepted economic principles. However, some mills for one reason or another cannot increase production to reduce costs. These smaller plants are vital to local economies but are finding it difficult to compete with other larger mills.

In addition, these small textile mills are often located on small watercourses where their waste exerts an unusually high pollutional demand on the environment. Pressure is being applied by water pollution control agencies to avoid and avert this pollution. Treatment of these wastes may also increase production costs.

When one couples the economic size and environmental pollution problems with the reality of dwindling supplies of fresh, raw process water, the small textile mill is currently being squeezed either out of business or to disproportionately increase its product cost or both. Larger mills are usually located where process water is more abundant and hence cheaper and where receiving streams or domestic wastewater treatment plants are more able to handle the pollutional load.

The ultimate survival of small textile mills–indeed small water-using industrial plants of all types–depends on solving both economic and the environmental resource problems. This section contains an innovative, potential solution to the plight of these small mills.

Cost of Raw Water

Although process water cost generally represents a minor portion of total manufacturing cost, it is significant because it is becoming an increasing percentage of that cost. Process water is also becoming a scarcer raw material. Little information is published on the actual cost of raw water. In general, municipal water utilities charge from $0.50 to $1.50/100 ft^3 (or 1,000 gallons of water). In fact, our survey of textile mills using public water supplies showed that they pay $0.44–1.43/1,000 gallons.

For a typical small mill finishing woven fabric in a series of complex processes that uses 600,000 gallons/day, its daily cost would be $264–858. Even these charges may be misleading because they occurred only where this amount of water was available for sale, as reported by the mills.

Conventional Wastewater Treatment

Wastewater treatment from small textile finishing mills has been either (1) separate treatment and reuse of dye wastes only or (2) complete treatment of the whole finishing mill waste (Nemerow and Dasgupta 1985). The first has been accomplished mainly by hyperfiltration and/or dye bath reconstitution, while the second has been done mainly by chemical coagulation and/or biological aeration.

Both methods have produced certain amounts of reusable water. However, economic considerations and government environmental regulations play major roles in the decision to produce reusable wastewater. Costs of producing acceptable-quality reusable wastewater to the small mill will need further definition and reduced to a minimum before reuse becomes standard practice, regardless of receiving water quality degradation.

Costs of Conventional Wastewater Treatment

In our search of the literature thus far, we have found that capital and operating costs of small textile mill wastewater treatment depend largely on the type and extensiveness of the treatment used. The capital costs range from as low as $31,500 for simple dye bath reconstitution to as high as $303,000–492,000 for chemical coagulation, filtration, and activated-sludge treatment. Operational costs for similar treatments range from $40,000 to $328,000 per year.

A typical small mill produces about 25,000 lb/day with average capital costs of $500,000 for complete treatment and annual operating costs of $150,000. This results in capital costs of $20/lb of production per day and annual operating costs of $0.02/lb of production per day (assuming 6 days per week and 50 weeks per year of production). These are very approximate costs for presumed average small mills. The range of true costs may vary greatly from. However, it is apparent that both capital and operating costs to these small mills represent a very significant expenditure.

Minimization of these costs by subtracting them from the benefits of wastewater reuse would constitute a real boon to the small mills.

Alternate Solutions to the Dilemma

There are two potential methods for reducing waste treatment costs of the small textile plant and, at the same time, producing reusable wastewater to replace or replenish the mills' costly water supply. These are (1) industrial complexing and (2) chemical coagulation. Other methods reported in the literature may reduce waste-treatment costs or produce a partial supply of raw water but will not accomplish both of these objectives. For example, dispersed growth aeration as suggested by Nemerow and Dasgupta (1985) will treat the wastewater at reduced costs but will not, by itself, produce acceptable reusable water. Also, Brandon and Porter (1976) hyperfiltered dye wastes through membranes to produce both recyclable water and dyes but failed to treat a sufficient portion of the plants' total waste at a lowered cost to result in satisfactory overall waste treatment. In order to be cost effective for the small textile manufacturer, the solution must satisfy both environmental and production concerns.

Industrial Complexing

Water-consuming and waste-producing textile finishing mills are ideally suited in these industrial complexes. Although wastes may pollute the fragile environment, they may be amenable to reuse by close association with satellite industrial plants, that in turn, produce raw materials for others within the complex.

An ideal illustrative EBIC for small textile finishing mills is shown in Figure 17.10. This complex contains five manufacturing plants producing 12,000 lb of woven fabric

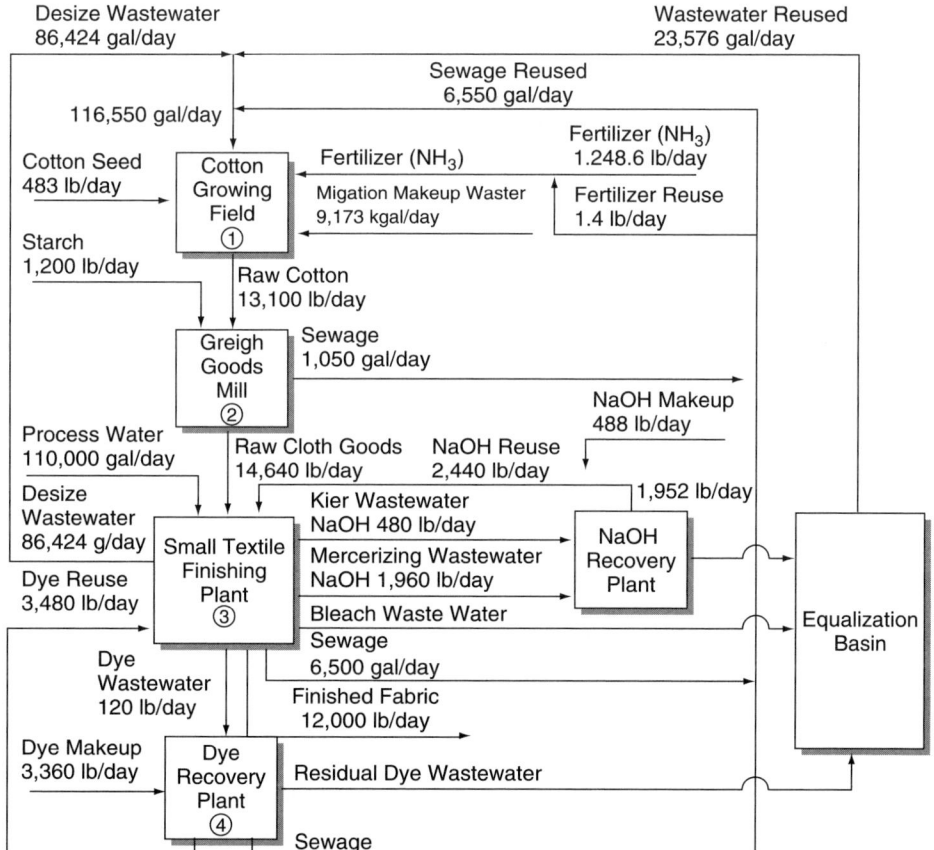

FIGURE 17.10. Diagram of the integrated five-plant industrial complex.

for sale outside the complex and 13,200 lb of cotton, 14,640 lb of greigh goods, 1,952 lb of NaOH, and 120 lb of dyes for reuse within the complex.

In addition, all sewages and wastewaters are reused without treatment within the complex. Of notable interest and importance is the reuse of 86,424 gallons of untreated finishing mill desize waste that contains 732 lb of BOD.

In Table 17.7, a raw material balance justification is given, which compares the raw material quantities and costs for the five separate plants manufacturing at distant locations and manufacturing within the EBIC. From Table 17.7, it is apparent that the cost savings of the industrial complex from a material balance perspective alone is $6,726 less $6,093.75 or $623.25 per day, which represents a savings of $52.69/1,000 lb of finished cotton fabric.

In addition, environmental costs would have to be considered. Within the industrial complex, we are presuming no external environmental costs are needed. As separate plants operating at distant locations from each other, environmental costs would include both domestic and industrial waste treatment charges as well as any measurable adverse environmental impact costs of the residual effluent wastes. These additional costs are currently being assessed by your author.

From a preliminary study, total environmental savings appear to be greater than $1,248/day, which represents the average costs of treatment for separate greigh goods and finishing mill wastes. To both savings the savings from transportation of raw cotton and sized woven goods must be added. Presuming transportation cost is $0.026/lb of cotton transported, we can estimate the additional cost for transportation as $343.20/day.

Pulp and Paper Mill Complex

The products of pulp and paper mills, the fifth largest in the U.S. economy, are consumed at the annual rate of about 400 lb/person. The pulping of the wood and the formation of the paper product produce wastes containing considerable quantities of sulfates, fine pulp solids, bleaching chemicals, mercaptans, sodium sulfides, carbonates and hydroxides, sizing casein, clay ink, dyes, waxes, grease, oils, and other small fibers. The overall wastes can be high or low in pH, and contain high-color, suspended, colloidal, and dissolved solids and inorganic fibers. Because of its high water consumption and wastewater discharge of 20,000–60,000 gal/ton of product, the wastes contain large total quantities of organic oxygen-demanding matter.

The high water use and wastewater production usually preclude the possibility of joint treatment with municipal sewage. These wastes also create considerable environmental impacts because of their concentrated loads of air, water, and land pollutants. The siting of new pulp and paper mills has become a major endeavor. They must be located near vast quantities of relatively clean water, as well as receiving water resources downwind and at a distance from residential habitation (because of common air pollutants such as SO_2 and mercaptans), usually on a rail line and near major highways for shipping, and near adequate land area for waste treatment and sludge disposal. Such sites are also difficult to find. For these and other reasons previously given,

TABLE 17.7
Raw Material Balance as Part of Total Production Cost

Mass Balance of Small Textile Complex Raw Material

Raw Materials, Amount Needed	Plant Type Amount Needed	By the Complex Quantities	Cost ($)	When Plants Are Separate Quantities	Cost ($)	Cost/Unit of (1982–1983)
	Agricultural growing field					
Irrigation water		9.173 mgd	4,036.00	9.29 mgd	4,087.00	$0.44–1.43/1,000 gal
Cotton seed		483 lb/day	48.30	483 lb/day	48.30	0.10 lb
Fertilizer		1,248.6 lb/day	81.16	1,250 lb/day	81.25	0.065 lb (1987)
Greigh goods manufacturing						
Starch sizing	Textile mill finishing	1200 lb/day	180.00	1,200 lb/day	180.00	$0.15
NaOH for fabrics		488 lb/day	131.80	2,440 lb/day	658.80	$0.27 lb
Dyes for fabrics		3,360 lb/day	1,512.00	3,480 lb/day	1566.00	0.45 lb
Process water		0.11 mgd	104.50	0.11 mgd	104.50	0.44–1.43/1,000 gal
Total material cost			$6,093.75/day		$6,726/day	

[a]Assume typical small textile finishing plant producing 12,000 lb/day of woven finished cloth product.

I recommend consideration of a pulp and paper mill complex, with little or no adverse environmental effect. Figure 17.11 describes one possible complex centered about an average-sized paper mill producing 1,000 tons of paper product per day

In the first publication (Nemerow et al. 1977), a balanced industrial complex centered about a pulp and paper mill was presented and is produced here as Figure 17.11 for further clarification. Eight separate industrial plants were included as part of this complex, five of which would produce products to be used within the complex.

The Pulp and Paper-Mill Environmental Problem

Pulp and paper-mills are among the top five industrial water users in the United States (Nemerow 1978). They use about two times 10–12 gal/day (Gould 1976). The largest percentage of this water is discharged into the environment as wastewater from pulp washing and paper making or as steam from the drying plant. The pulp and paper industry is also the ninth largest in the United States, accounting for nearly 4% of the value of all manufacturing, producing $7,866 million of our gross national product in 1971 (American Paper Institute 1973). In the cooking and bleaching of the pulp, these mills may also contaminate the air surrounding the plant. Approximately one-half of all weight of the wood entering the pulp mill leaves the mill as product paper. The greatest percentage of the loss in weight ends up as solid material to be disposed of in the environment. Bark, waste pulp, and paper mill fines constitute most of these solids and potentially end up on the land or in the air. Therefore, this industry—because of its great volume of water, wood, and chemical intake per mill—has an adverse impact on the air, water, and land environments. The large quantity of wastes also represents a potential supply of valuable resources for ancillary and compatible industries.

With that in mind, we derived, from our general knowledge and from theory, a pulp and paper-mill industrial complex for further investigation. As mentioned earlier, this complex is depicted in Figure 17.11 and produces 1,000 tons/day of fine paper.

Timber is brought into the complex to the pulp mill (no. 1 in Figure 17.11) where it is converted into pulp for use by the paper mill (no. 2). Major wastes from (no. 1) are bark, which is burned subsequently in the steam plant, and sulfate waste liquor, which is used in three internal complex plants—road binder (no. 3), vanillin (no. 4), and sulfate concentrating (no. 8). Products from no. 3 and 4 can be sold locally or internationally, while those from no. 8 are reused in the complex by no. 1 or by the hardboard manufacturing plant (no. 7). Fine paper product from no. 2 can be sold in the world market. Wastes from no. 2 include heat, fillers, and fines, which can be used internally in the ground-wood pulp mill (no. 5), which also uses a percentage of used newspaper stock.

The pulp product from no. 5 will be used partially in the complex by no. 1 and sold as paperboard externally. The plant (no. 5) produces waste suspended solids, which are used internally by the wrapping paper plant (no. 6 and 7). The products of no. 6 and 7 can be sold regionally. In total, this pulp and paper mill complex produces six products for external sale (fine paper, wrapping paper, hardboard, vanillin, paperboard, and road binder) and four products for internal use (concentrated sulfate, wood pulp, wrapping paper, and ground wood pulp). In addition, all major wastes of suspended

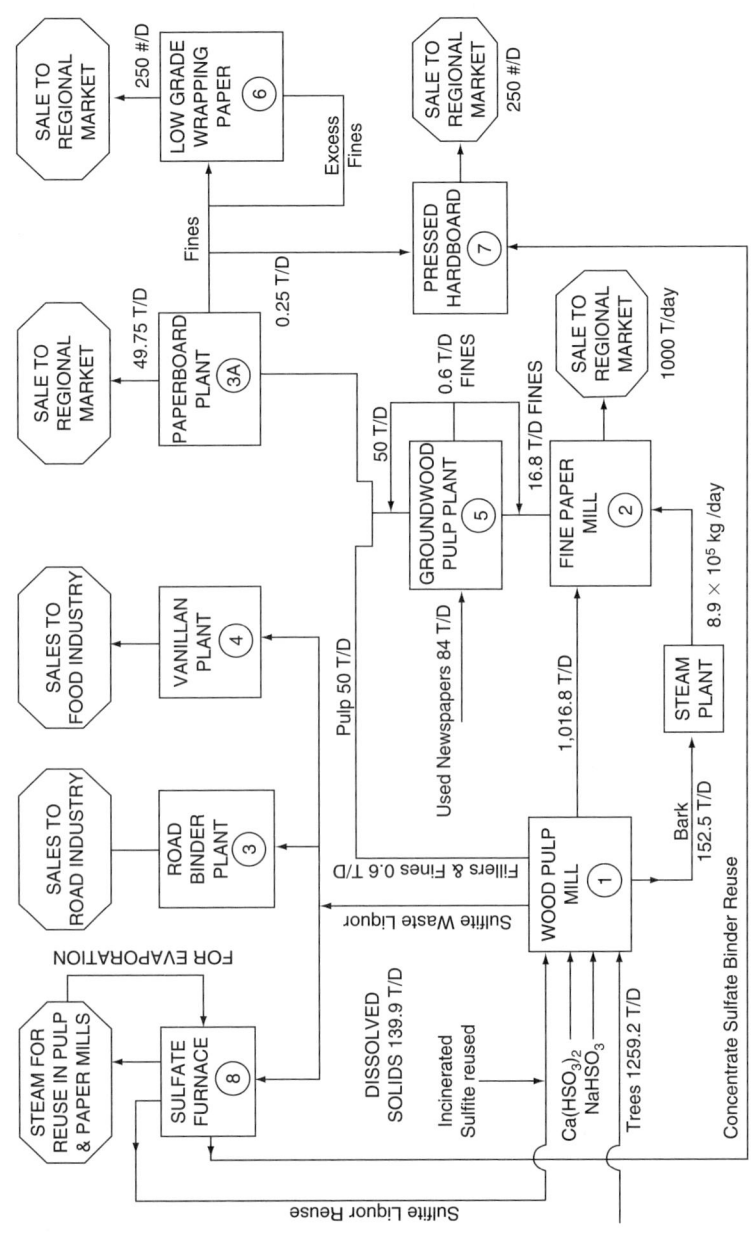

FIGURE 17.11. Pulp and paper mill complex.

solids, cooking liquor, fillers, heat, and bark are reused within the complex in the manufacturing of these products.

Mass Balance of Products

Based on a literature review evolved typical concentrations of recoverable suspended solids in various process effluents were calculated. A mass balance was prepared assuming that the total production of fine paper is 1,000 tons/day. The remaining quantities are calculated based on this production.

Computation of Trees Required at the Complex

Production of fine paper: 1,000 tons/day (907.2 kg/day × 1,000). Fiber losses from the paper mill = 1.68% of production. Therefore, suspended solids going into waste streams from the paper mill = 1.68/100 × 1,000 = 16.8 tons/day (15.24 kg/day × 1,000). Total wood pulp produced per day = 1,000 + 16.8 = 1,016 tons (922.44 kg × 1,000). Quantity of sulfite liquor generated in wood pulp mill = 300 gal/ton, while concentration of dissolved solids in sulfite liquor = 11%.

Thus, dissolved solids going into sulfite liquor = 110,000 × 8.34–6 × 300 × 10 = 275.22 lb/ton (0.1376 kg/kg) of pulp. Total sulfite wastewater dissolved solids produced per day = 275.22 lb/ton × 1,016.8 tons/day × tons/2,000 lb = 139.9 tons/day = 0.00001269 kg/day. On an assumption that the amount of bark produced is generally 15% (by weight) of the pulp production, bark production = 15/100 × 1,016.8 = 152.5 tons/day (138,000 kg/day). Total tonnage of trees used in the complex = 1,016.8 + 139.9 + 152.5 = 1,309.2 tons/day (1,187,000 kg/day).

Ground Wood Pulp Production

Recovery of suspended solids from paper mill = 16.8 tons/day = (15,200 kg/day). Assume that 100 tons (90,720 kg) of ground pulp is required for production every day. Fiber loss in the ground wood pulp plant = 0.6 tons/100 tons of the ground wood pulp (Nemerow 1978) = 0.6/100 = 0.6 tons/day (544.3 kg/day). Total ground wood pulp produced and lost per day = 100 + 0.6 = 100.6 tons/day (91,260 kg/day). Therefore, used newspaper required = 100.6 – 16.8 = 83.6 tons/day (75,840 kg/day assuming 50% of the ground wood pulp is recycled as shown in Figure 17.11 and the remainder is used in the production of paperboard.

Paperboard Production

Loss of fines from ground wood pulp production is about 0.5% of production (15,1978). Let us say that paperboard production = x tons/day and $X + 15/100X = 50$ tons of pulp/day (54,359 kg/day) and $1.005 X = 50$

$$X = 49.75 \text{ tons paperboard/day } (45{,}132 \text{ kg/day}).$$

Fines recovered from paperboard waste = 49.75 = 0.25 tons/day (226.79 kg/day). A total of 0.25 tons/day of fines can be used to produce low-grade wrapping paper and pressed hardboard. With no loss of fines and with a 50–50 product production split, 1.25 tons (113.64 kg) of each product can be manufactured.

Sulfite Recovery

The solids concentration of spent sulfite liquor drawn from the digesters may vary from 6 to 16%, with an average value of 11%. These solids may contain as much as 68% lignosulfonic acid, 20% reducing sugars, and 6.7% calcium (Nemerow 1978). Complete evaporation of the sulfite waste liquor produces both a fuel, which can be burned without an additional outside fuel supply, and a salable byproduct such as synthetic vanillin and road binder.

An overall mass balance regarding the production of different quality of papers is given in Figure 17.11. No attempt is made here to correlate the effects of the complete recovery of suspended solids on the reduction of final BOD of the wastewater. Similarly, no detailed information is given about the recycling of the wastewater effluent. However, it is reported in the literature that 90% of the effluent can be recycled wastewater; it must be presumed that a considerable portion of the dissolved and colloidal organic matter is being reincorporated into the various products. This is especially true in the case of the sulfite waste liquor, which is completely reused or recovered and contains the major portion of BOD in the complex.

Energy Management

Integrated production complexes have a significant advantage over conventional plants from the energy management standpoint. Waste heat from one section of the complex can be used as process heat for another section, the concept being minimization of waste heat. The environmental problems associated with waste heat discharged to ecosystems have been well documented. It is accepted that thermal discharges may result in anomalous stratification in the receiving basin, lowering of capacity to hold oxygen, and increased reaction rates and metabolism. These effects vary significantly with the chemical and meteorological conditions associated with the water body. The lethal effects of thermal pollution are sometimes obvious, whereas the sublethal effects on food chains and waste assimilative capacities are not easy to foresee without careful study.

The present industrial complex outlined can reduce waste heat discharged to the hydrosphere and atmosphere. The two significant areas of concern follow:

1. Utilization of solid wastes from the plant to achieve energy efficiency
2. Utilization of low-grade heat from one section in another suitable section

The first area has two possible applications. Bark from the shredding plant is used to provide heat from the steam plant. The estimated bark production for the plant is 152.5 tons/day (138,345 kg/day). Because the heating value of bark varies considerably with the type of tree and aging, an average heating value of 4,000 cal/g is used.

The heat available by combustion of bark is

$$= 152.5 \times 1,000 \times 454 \times 4 \text{ kcal}$$
$$= 550,000,000 \ (75+54) \text{ kg} = 890,000 \text{ kg/day.}$$

Assuming incoming water temperature at 25°C., the total steam production = 550,000,000/(75+54) kg = 890,000 kg/day.

The second solid waste to energy recovery application lies in evaporation and burning of sulfite liquor. Sulfite liquor is evaporated to enough of a solid content suitable for burning. Difficulties with scaling, corrosion, and fly-ash may result. However, this burning procedure can also be justified because this will eliminate the need to discharge sulfite wastes to the environment.

Utilization of low-grade waste heat from the proposed complex is somewhat difficult, because details of process thermodynamics are needed. Further, research on this concept provides the needed data for such analysis. However, some conceptual comments can be made regarding the proposed complex. Low-grade heat from fine paper mills can be used in the ground wood pulp plant. In cooler regions of the world, waste heat from any of the effluents in the complex could be used for space heating and providing hot water for use by plant personnel.

Economy of Complex or Residual Environmental Impact

Little or no air or water pollution results from this complex. In addition, it is anticipated that no expensive wastewater-treatment plant would be required for the pulp and paper mill complex. This, in itself, represents a savings not only in capital equipment costs, but also in operating or production costs equal to 1–5% of production costs. Additional operating costs will be reduced by the following practices:

1. Reusing burned sulfite waste liquor to replace a portion of calcium or sodium bisulfite cooking liquor
2. Reusing ground wood pulp (50 tons/day) or (43,359 kg/day) to replace a similar weight of trees
3. Burning bark to generate steam for use in the fine paper mill
4. Reusing concentrated sulfite waste liquor in making pressed hardboard
5. Reusing fillers, fines, and heat from the paper mill in making ground wood
6. Reusing ground wood pulp mill fines (0.6 tons/day or 544.31 kg/day) to make additional pulp or paperboard
7. Reusing paperboard mill fines (0.25 tons/day or 226.80 kg/day) to make both low-grade wrapping paper and pressed hardboard
8. The sale of additional products as follows:
    a. Low-grade wrapping paper: 250 lb/day (113.64 kg/day)
    b. Pressed hardboard: 250 lb/day (113.64 kg/day)
    c. Paperboard: 49.75 tons/day (45,132 kg/day)
    d. Vanillin
    e. Road binder
9. Combustion of concentrated liquor for heat recovery and pollution abatement
10. Use of low-grade waste heat for space and water heating

An exact and more detailed economic analysis of this complex will be made in any continuing study. It will be necessary to obtain more precise data on the production requirements of the small service industrial plants in the complex. In the meantime, I propose this analysis as a beginning to what may develop into a revolutionary new system of industrial plant design.

A major producer of the present-day pulp and paper uses the Kraft process. Under normal circumstances, a variety of liquid, solid, and air wastes are created in this process. One possibility for integrating the typical Kraft mill into an EBIC is shown in Figure 17.12.

The approach adopted for developing this complex follows:

1. *Single-system concept of water, power, raw materials, and wastes management:* The operation of utilities system, so critical to the production process, requires efficient performance. Centralized utilities and management allows the achievement of self-containment.
2. *Waste utilization/byproduct recovery:* Industries based on chemicals from the other half of the tree and wastes have been included in the complex. This step

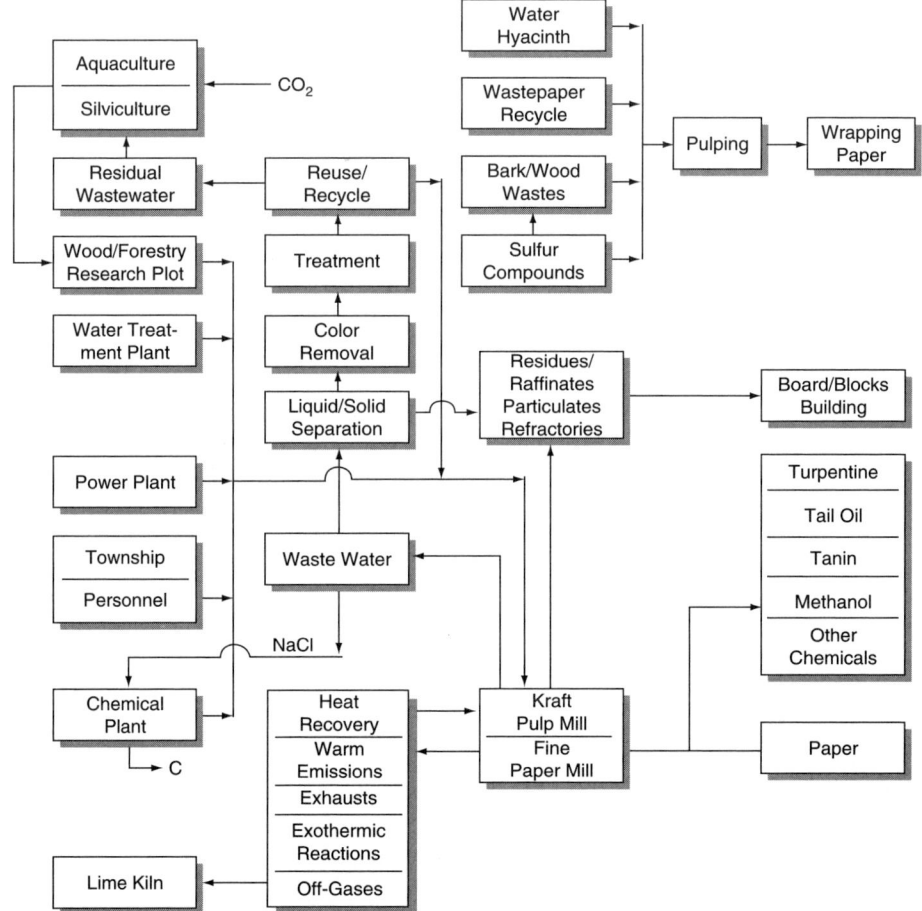

FIGURE 17.12. Paper mill complex.

made of pollution control is of significant merit since it allows twofold benefits of minimizing production and treatment costs.
3. Grouping of other compatible and complementary industrial plants such as chemical, forestry research, aquaculture (water hyacinth), and so on.
4. *Integrated development in stages and overall management of all operations within the complex:* The central starting point is the pulp industry, based on which the complex develops in stages by inclusion of other industries and utilities. The gradual development in this way approaches self-sustainment.

Although complete analysis with respect to criteria for acceptability has not been made, primary technoeconomic considerations make this complex extremely attractive. As resources become increasingly scarce and environmental regulations become more stringent, multiple benefits will be more than apparent.

References

Adekeke. I. O. 2003. *Durability and Stability of Briquetting from Solid Agricultural Wastes* [MS thesis]. The American University of Cairo, Egypt, Fall.

American Paper Institute. 1973. General Statistics for the United States Pulp and Paper Industry. Washington, DC.

Bhanumathidas, N., N. Kalida. 1986. *Anhydrite: A Possible Solution for White Portland Cement.* 2nd International Symposium on Phosphogypsum FIPR, p. 65. University of Miami, Coral Gables, December 10–12, 1986.

Brandon, C. A., J. J. Porter. 1976. *Hyperfiltration for Renovation of Textile Finishing Plant Wastewater.* EPA 600-2-76-060. Environmental Protection Agency. Washington, DC, p. 157.

Carmichael, J. B. 1986. *World-Wide Production and Utilization of Phosphogypsum.* 2nd International Symposium on Phosphogypsum FIPR. University of Miami, Coral Gables, Florida, p. 34, December 10–12, l986.

Chemical Week, August 3, 1968.

Clur, D. A. 1986. *Fedmis Sulphuric Acid/Cement from Phosphogypsum.* 2nd International Symposium on Phosphogypsum FIPR. University of Miami, Coral Gables, Florida, p. 141, December 10–12, 1986.

Dasgupta, A. 1983. *Anaerobic Digestion of Sugarcane Wastes* [Master' thesis]. University of Miami Coral Gables, Florida.

Deepchand, K. 2001. Commercial scale cogeneration Bagasse Energy in Mauritius. *Energy Sustainable Development,* 1(1).

Ellwood, P. 1969. Turning by-product gypsum into a valuable asset. *Chemical Engineering,* March 24, 1969.

Gould, M. 1976. *Water Pollution Control in the Paper and Allied Products Industry.* New York: McGraw-Hill Publishing Company.

Krishnam, S. V., et al. 1996. *11th Annual Conference of Phosphate Research Institute,* Chemical Processing Session, Lakeland, Florida, October 18.

Nemerow, N. L. 1978. *Industrial Water Pollution Theories, Characteristics and Treatment.* Reading, MA: Addison-Wesley Publishing Company.

Nemerow, N. L. 1980a. Environmentally Optimized Industrial Complexes. In: *Bound Proceedings of the National Environmental Engineering Research Institute, Nagpur, India.*

Nemerow, N. L. 1980b. *Preliminary Assessment of Environmentally Balanced Industrial Complex Three Stage Evolution,* Report to U.S. E.P.A. Contract No. 68 02-3170 RTP North Carolina, June.

Nemerow, N. L. 1995. *Zero Pollution for Industry.* New York: John Wiley & Sons.

Nemerow, N. L, A. Dasgupta. 1981. Environmentally-balanced industrial complexes. In: *36th Annual Industrial Waste Conference Purdue University,* p. 416.

Nemerow, N. L., A. Dasgupta. 1984. *Zero Pollution: A Sugarcane Refinery-Based Environmentally-Balanced Industrial Complex.* 57th Annual Conference of Water Pollution Control Federation. New Orleans, Louisiana, October 1984.

Nemerow, N. L., A. Dasgupta. 1985. Zero Pollution for Textile Waste. In: *Proceedings of the 7th Alternative Energy Sources Conference Miami, Florida, December 1985,* Vol. 6, p. 499. Hemisphere Publishing Company.

Nemerow, N. L., S. Farooq, S. Sengupta. 1977. Industrial Complexes and Their Relevances to the Pulp and Papermills. Presented at Seminar on Industrial Wastes Bombay, India.

Shreve, R. N., J. A. Brink, Jr. 1977. *Chemical Process Industries,* p. 162, 4th. New York: McGraw-Hill Publishing Company.

CHAPTER 18

Potential Industrial Complexes

INTRODUCTION

There are many possible examples of potential industrial complexes that may accomplish the objectives set forth in Chapter 15. I list some here and provide brief descriptions of their configuration mostly in schematic ways. The reader may also be aware of others may also meet requirements for workable environmentally balanced industrial complexes (EBICs). Those in this chapter do not contain mass balances or any other detailed operating data. As the latter become available, I intend to transfer these to Chapter 17 in updated editions. Those already in Chapter 17 will be updated and, hopefully, include actual production and operation data, efficiencies, and problems, if they exist.

I am not so presumptuous as to assert positively that the complexes in this chapter are (or ever will be) feasible. I present them here only to stimulate your thinking about these possibilities. If, in fact, it turns out that these potential complexes actually become reality, then I can look back and take some comfort in the fact that practice sometimes results from theory. If, on the other hand, they never become reality, I hope they at least provided you with the incentive to innovate. The principle of industrial complexing is an art of (1) finding a troubled waste-producing industry and (2) matching it with another industry that can alleviate the trouble by consuming its contaminating waste. Once you develop the technique of mastering this principle, you will have used successfully the concept of EBIC.

In Chapter 20, I include some examples of complexes that were not designed as EBICs originally but that developed over time from "industrial estates."

The following are some of the potential EBICs known to the author and proposed at this time:

A   Wood–Papermill Complex
B   Steel Mill–Coke and Gas and Fertilizer Plants
C   Finished Metals–Plastic Plant Complex
D   Organic Chemical–Wood Processing Plant Complex
E   Wastewaters–Power Plant Complex

F   Steel Mill–Fertilizer–Cement Plants Complex
G   Coal Power Plant–Cement–Concrete Block Plant Complex
H   Plastic Plant Complex
I   Cement–Lime–Power Plant Complex
J   Wood–Lumber Mill Complex
K   Power Plant–Agriculture Complex
L   Cannery–Agriculture Complex
M   Nuclear Power–Glass Block Complex
N   Animal Feedlot–Plant Food Complex
O   Coke (Steel Mill)–Tar–Benzol Plant Complex
P   Wood–Ethanol Plant Complex
Q   Water, Electricity, Chlorine, Lye Plant Complex
R   Aluminum, Electricity, Red Brick Plant Complex
S   Corn Growing, Alcohol Producing Plants Complex
T   Restaurant, Paint Manufacturing Complex
U   Oil Drilling Offshore–Seashore Recreation
V   Metal Plants–Dry Cleaning Plants Complex
W   Electrical Storing and/or Converting Voltage Wax Manufacturing Complex
X   Nuclear Power Plant, Waste Reprocessing, Cannery Complex
Y   Electric Power, Drinking Water Plant Complex
Z   Vegetable Pickling Cannery, Inorganic Chemical and Chlorine Plant Complex
AA Sugar, Ethanol, Gasoline Plants Complex
BB Reclaimed Cell Phones, Cement Plant and Concrete Products Complex
CC Sugarcane–Fuel Briquettes Complex
DD Hog Production–Animal Feed–Energy Production Complex
EE Seawater Desalination Plant–Boric Acid Plant Complex
FF Animal Feedlot–Power Plant–Fertilizer Complex
GG Used Plastic–Textile Manufacturing Complex
HH Lumber–Textile–Corn-Growing–Alcohol-Producing Complex

A. Wood–Paper Mill Complex

A "natural" complex exists with a lumber mill that manufactures lumber from trees, a chipboard mill that uses the waste sawdust and bark to produce paperboard, and a fine paper mill that uses the paper pulp to make fine paper. Each of these three plants discharges wastes along with the valuable products it manufactures. These wastes can be reused as raw materials by another of the complex's plants to convert into its product. These systems are shown schematically in Figure 18.1.

In this figure, sawdust, wood chips, and bark solid wastes from the lumber mill are reused directly in the chipboard mill instead of being incinerated or landfilled to cause environmental degradation and costs. The wood pulp mill's waste sulfite liquor is concentrated and reused back in the pulp mill instead of causing immense pollution on

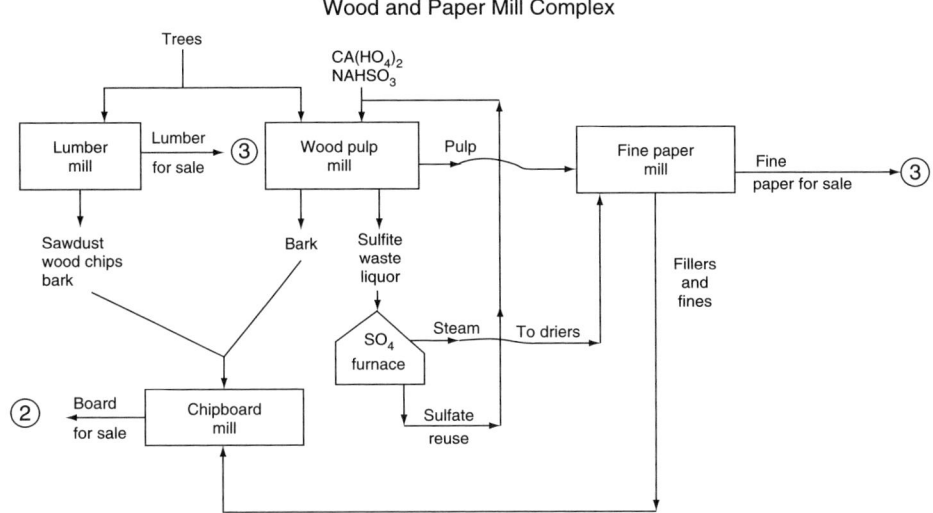

FIGURE 18.1. Wood–Paper Mill complex.

surrounding waterways. Its steam is also captured and reused by the paper mill to aid in drying the fine paper.

Three major products evolve from this complex: lumber for external sale, chipboard for external sale, and fine paper also for external sale. No wastes from these three plants would leave the complex to impose environmental costs on society.

Wood represents an excellent and renewable resource that is not only very valuable but also very preserving if reused. The waste that exists in this industry is enormous. Myerly et al. (1981) suggested that biomass such as cellulose must be handled like any other valuable product: everything must be sold except the sound of the tree falling.

Austin in 1984 admits that underuse of the wastes from wood production comes from the complexity of wood itself and lack of integration of chemical, pulp, and lumber companies. He adds to those the disinterest of processing companies in producing and selling byproducts, lack of chemical knowledge or interest, and the dilute form in which many of the byproducts are available. But he counters with the fact that because environmental laws have made the stream dumping of pulp mill waste products impossible, some real interest in waste use has developed, but most such envisioned uses are as a fuel.

Austin also notes that because one of the major problems in wood use is its collection and transportation out of difficult terrain to the place in which it is desired, use of waste wood is particularly attractive at lumber and pulp mills, where these chores are already done. This agreement makes our proposed complex even more substantiated.

B. Steel Mill–Coke and Gas–Fertilizer Plants Complex

Another "ideal" marriage of compatible industrial plants includes a steel mill with its ancillary coke and gas plant, as well as an ammonia fertilizer plant. With this mix of plants, the waste pickle liquor, $FeSO_4$ from the steel mill, is reacted with ammonia gas waste from the coke plant to make ammonium sulfate, another fertilizer component in the fertilizer plant. Such a combination produces steel, coke, and ammonium sulfate for external complex sale. When these plants operate in separate locations, they also produce extremely pollutional pickle liquor and ammonia gas wastes.

It takes 2.5 tons of raw materials (coal, iron ore, scrap, limestone) to produce 1 ton of liquid steel (Nemerow 1976). Of these raw materials, 1 ton ends up as liquid steel and 0.385 tons as slag (2.5–1.385) leaving 1.115 tons of residual air, water, and solid wastes, assuming that all of the slag is normally ground and reused in road building and/or landfill.

If we assume that all of these wastes in air, water, and solids, except for the coke plant, end up as solid wastes after collection, this amounts to 0.923 tons of solid contaminants resulting from the production of 1 ton of liquid steel. More solid wastes of an unpredictable quantity must be added to this in the finishing of steel unless 100% of the waste is returned to the furnaces as scrap. At least 50% and usually 75% of the solid matter is iron.

In addition, the waste ammonia gas from the coke plant is reacted with steam and caustic soda to remove the contaminant phenol. The resulting sodium phenolate is recovered and sold to a chemical company (either within or outside the complex), which purifies the phenol for use as a raw material, usually for the manufacturing of plastics or wood preservatives. The coke plant ammonia gas waste can also be reacted with sulfuric acid to produce additional ammonium sulfate fertilizer product.

Thus, in this complex, two potentially contaminating wastes—pickling liquor ferric sulfate and coke gas plant ammonia—are recovered as ammonium sulfate and sodium phenolate for subsequent sale to and use by fertilizer and chemical plants, respectively. No waste will leave the complex to enter the surrounding environment. A schematic diagram of this potential EBIC is shown in Figure 18.2.

The air emissions from model coke plants have been given by Codd (1973) and reported by Nemerow (1976). It is repeated here as Table 18.1 to give a fair idea of what to expect from coke plant wastes.

I recommended as far back as 1976 that the production of iron and steel results in the discharge of numerous contaminants but need not be constrained by environmental factors because the control technology is known and abatement costs are relatively insignificant when related to other production costs. I continued that these plants should be located primarily but not solely in developing countries at sites that minimize the sum of the production costs, including environmental control costs. I believe I recognized 25 years ago that developing countries represented the ideal place for industrial complexes. This is where we could expect new plants and new construction in the wet industries to take place.

I further recommended then that the plants could be located in integrated industrial complexes to minimize these production costs while minimizing the total environmental damage costs and maximizing production values. I did qualify my recommendations of

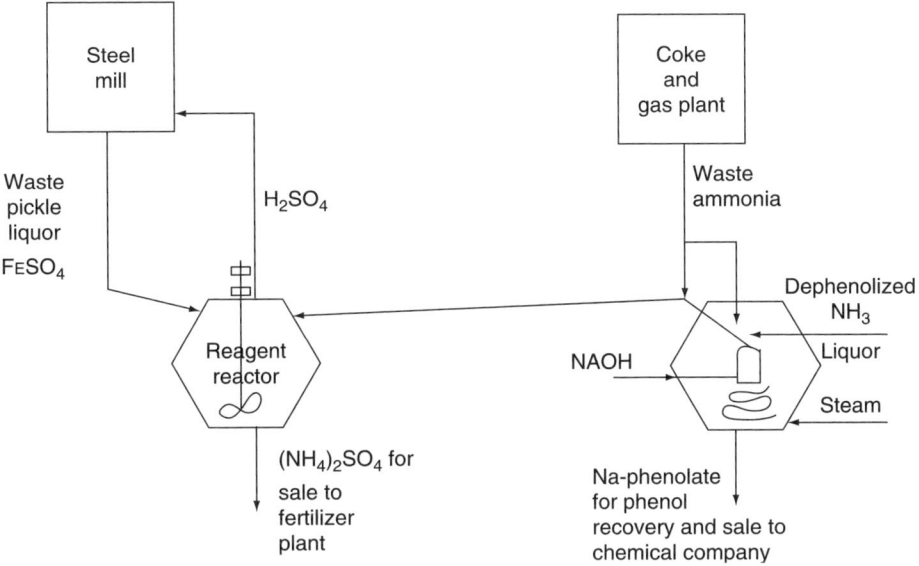

FIGURE 18.2. Steel mill–coke and gas plant complex.

TABLE 18.1
Air Emissions from Model Coke Plants

Pollutant	Emission Level (kg/ton dry coke)
Coke and Coke dust	2.0
Coke oven gas	0.7 (approximately 0.4 vol % of total gas made-300 m³/ton coke)
SO_2	0.63
H_2S	0.12
Phenols	0.13
Aromatics	0.21
HCn	0.07
NH_3	0.14
Pyridine bases	0.02

1976 by stating that considerable research, evaluation, and pilot experimentation are necessary to optimize these industrial complexes (which may contain a different mix of industries for each site in each country). Further, wherever possible steel product should be manufactured by the direct reduction, electric arc furnace, continuous casting sequence of production for minimizing environmental damage.

C. Finished Metals–Plastic Plant Complex

One unusual combination of plants that offers a potential balanced complex consists of a metal parts plant and a plastic production plant.

After metal parts are plated and protected from corrosion with a coating of grease, they are shipped or transferred to a "finished metals parts" plant. Here the grease is removed and the part further fabricated into a finished product. A commonly used solvent is trichloroethylene (TCE), a highly toxic organic compound. When released to the environment, it often exceeds both air- and water-quality standards, thus contaminating groundwater and receiving streams as depicted schematically in Figure 18.3.

At the same time, a polyvinyl plastic manufacturing plant uses as a raw material vinyl chloride monomer to produce polyvinyl chloride (PVC). Various suspension agents and catalysts are added to the monomers. Polymerization takes place in a heated reactor tank, as shown schematically also in Figure 18.3. The polymerized mass is then centrifuged and sold to the plastic industry as PVC. Hot wash wastewaters and centrifugates from this polymerization process, when released to the environment, contaminate groundwater and receiving streams, once again shown in Figure 18.3.

When these two plants are located adjacent to each other in an EBIC (Figure 18.4), it is possible to maintain clean groundwater and rivers. In this case, all that is needed to make this possible is an anaerobic digester reactor to receive the waste TCE and convert it to vinyl chloride, which is then used as raw material for the plastics plant. The metal parts plant's sewage is used as seed for the digester reactor, the polymer plant's sewage is used as seed for the digester reactor, and the polymer plant's centrifugate and TCE are used as feed for the digester reactor.

Hot water from the digester is used to heat the polymerization tank and returned to the digester in a closed system. No waste from either plant is discharged to the air,

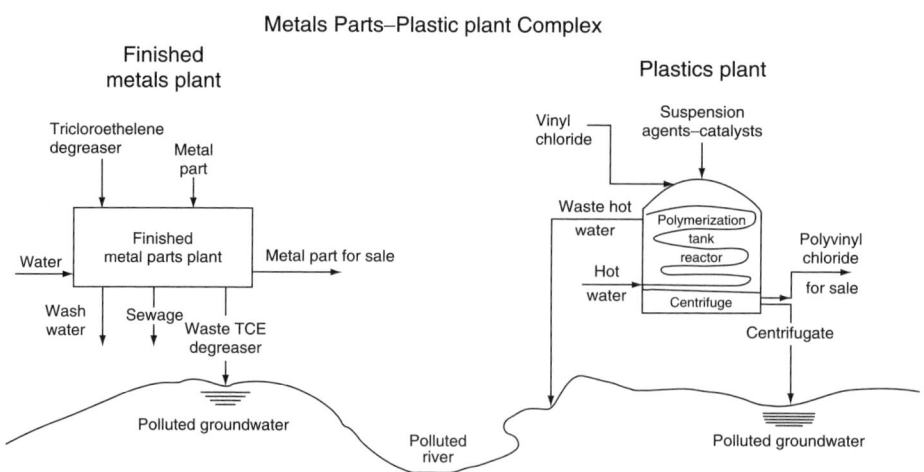

FIGURE 18.3. Metal parts–plastic plant complex.

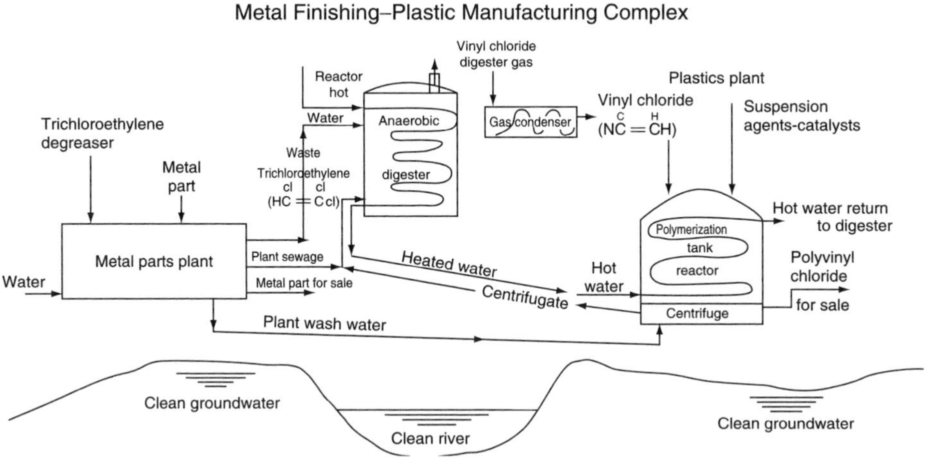

Metal Finishing–Plastic Manufacturing Complex

FIGURE 18.4. Metal finishing–plastic manufacturing complex.

ground, or river, as depicted in Figure 18.4. By placing the schematic of Figure 18.3 directly over Figure 18.4, one can observe the value and efficiency of using an EBIC to produce clean metal parts and PVC for external sale.

Of course, not all plastic manufacturing plants produce only PVC. Other plastic manufacturing plants might not fit into this particular environmental complex. Because of this, some facts are presented here about plastic plants in general. The common basic raw materials for plastics are petrochemicals, coal, carbohydrates such as wood or cotton, and gas and saltwater.

Each plastic possesses properties that warrant its manufacture above all others. In general, they are all tough, water and corrosion resistant, many colored, and rather easily manufactured. They started to be produced in 1868 (cellulose nitrate) and continued with different ones through the middle of the twentieth century. The vinyls, which we use as an example in this complex, were started in about 1927.

Shreve (1984) presents an interesting breakdown of plastics into three groups: thermosetting resins, thermoplastic resins, and polymer resins.

D. Organic Chemical–Wood Processing Plant Complex

It is common practice among some organic chemical manufacturing plants to begin with a purchased raw material such as benzene and alter it by reactions to produce other organics such as chlorinated phenols. One typical product is polychlorinated biphenol (PCP). Wastes from the manufacturing of PCP may include mainly phenolic compounds from the original raw material, as well as formaldehyde, hydrochloric acid, and various other organic compounds. Wastes are, therefore, high in phenols and biochemical oxygen demand (BOD) and low in pH. They contaminate the groundwater and receiving streams by contributing odorous matter and oxygen-depleting organic matter and cause potential toxicity to fish and fauna, as depicted schematically in Figure 18.5.

At the same time, wood-preserving plants use creosote and pentachlorophenol to impregnate telephone poles, fence posts, railroad ties, and lumber of all types to preserve them against degradation when put into service. These plants produce, in addition to the aforementioned products, various phenolic wastes, which, when entering the environment, contaminate groundwater and flowing streams. These effects are also shown schematically in Figure 18.5.

The conditioning of wood and its impregnation with preservative vary in materials, procedures, and mainly chemicals. The specific variation depends on the requirement for the use of the final product. In Table 18.2, I have prepared a description of the processes involved in both conditioning and impregnation of the wood. Also included in this table are the wastes resulting from the process, the character of the wastes, and their recommended treatment (if treatment rather than industrial complexing with an organic chemical plant is used).

In a potential EBIC containing both plants adjacent to the other, no wastes reach the groundwater or receiving stream. Phenolic wastes from the wood-preserving plant are extracted, the phenols are returned to the organic chemical plant as supplemental raw material, and the remaining organics are incinerated. Waste heat in the form of steam from the incinerator is reused in the wood-processing plant for curing wood. Polychlorinated biphenols in the organic chemical plants' waste are fed directly into the wood-preserving plant to supplement product PCP raw material for wood preserving.

An EBIC showing complete reuse of all wastes by both plants resulting in an uncontaminated water environment is shown in Figure 18.6. Here, waste phenols and acids from the chemical plant are used as feed supplements for the wood-preserving plant, as is a portion of the PCP product. The wood-preserving plant sends its wastes to

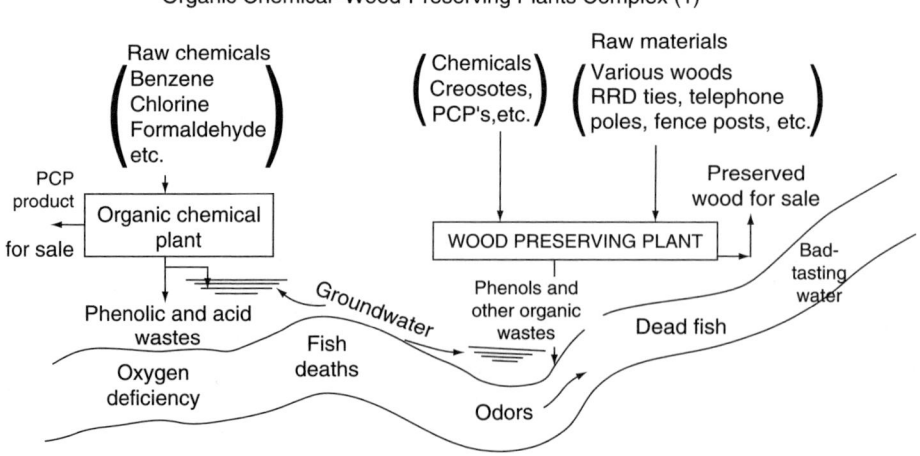

FIGURE 18.5. Organic chemical–wood-preserving plants complex.

TABLE 18.2
Wood-Preserving Processes

Wood Conditioning	Wood Impregnation with Preservative
To reduce moisture and increase permeability	To provide toxicity to all forms of wood destroyers
Drying usually by air seasoning, sometimes by kiln seasoning	A. With or without air pressure, always with pressure-closed tanks
	B. Sometimes without pressure and superficial application of oils
Conditioning usually by steaming under pressure	A. Waterborne preservatives use surface retention primarily
Green Wood Products	
	B. Oil-borne (creosotes) use deep penetration (pentachlorophenols),
Wastes resulting mainly steam condensates crankcase oil drainings boiler blowdowns	mainly separated vacuum or pressure water condensates from retorts and drippings from withdrawn treated wood and charge
Wastes character lignous and cellulosic, sugar organic matter	$ZnCl_2$, As, Cu, Cr, F, and B from use of waterborne preservatives, phenols, coal tar, creosote, petroleum, aldehydes, PCP, and copper naphthenate from use of oil-borne preservatives toxicity, refractory-
Objectionable qualities and recommended treatment Color, odor, and oxygen demand keep out of ground biological treatment concentrate and burn	nature, color, odor, acid pH keep out of ground collect all unusable waste contain all in impermeable facility for reuse, incineration for exportation for recovery or disposal

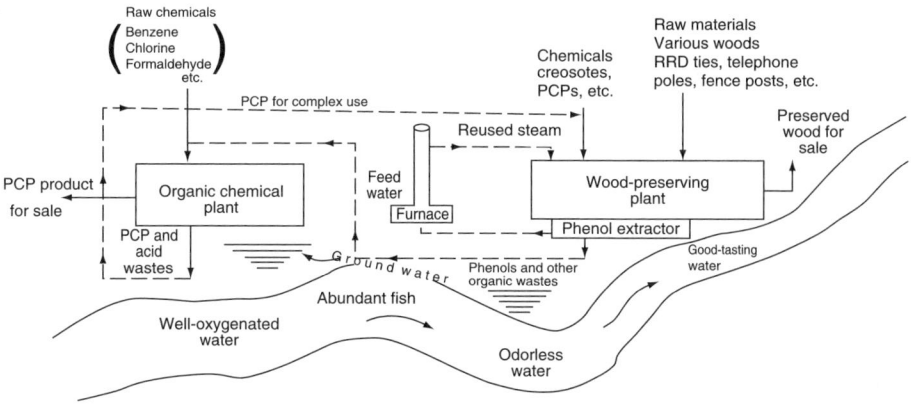

FIGURE 18.6. Organic chemical–Wood-preserving plants process.

a phenol extractor and the purified phenol back to the chemical plant for raw chemical supplement, while the other organic wastes from the extractor are burned in an incinerator. Steam from the incinerator is fed into the wood plant for heating the soak drums. No wastes are discharged into the groundwater or river and raw materials of both plants are supplemented by reused wastes.

E. Biomass Power Plant–Municipal–Forestry–Agriculture Complex

When power plants generate electricity by burning fossil fuels such as coal or oil, they also produce noxious, odorous, and greenhouse gases, as well as potential toxic residual unburned ash.

At the same time, when cities treat their wastes to produce sludges, foresters grow trees that result in waste trimmings, barks, and sawdust, farmers grow sugarcane and leave stalks, and other farmers graze animals that deposit various defecated sludge solids, pollution of the air, land, and water also results. Some of these solids are burned in the field creating air pollution, and some contaminate the land and groundwater and even surface water supplies during runoff periods.

When both the power plant facility and the biomass producers are located in one complex, many of these adverse environmental effects are eliminated. One potential configuration of this complex is presented in Figure 18.7.

Reicher (1999), an assistant secretary at the U.S. Department of Energy, believes that biomass can be used as an alternative to the traditional fossil fuels for producing electricity and other products.

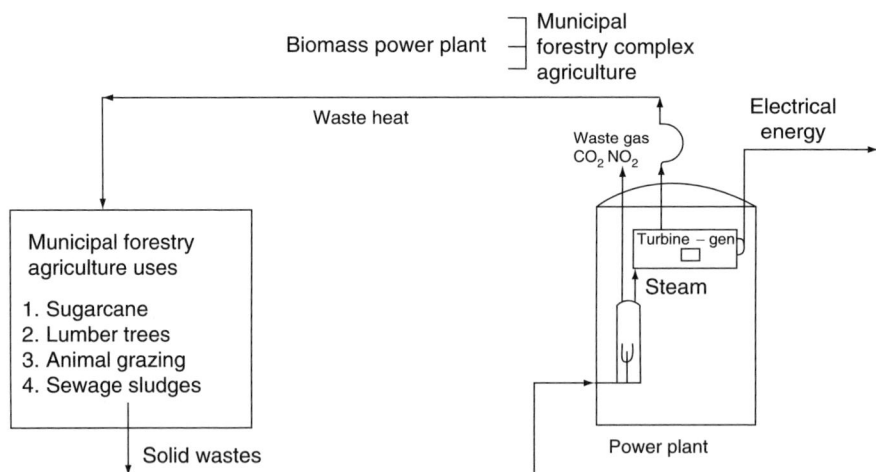

FIGURE 18.7. Biomass power plant–municipal–forestry–agriculture complex.

Reicher (1999) also believes that we will soon see federal legislation that encourages the use of biomass, perhaps as part of a larger electricity restructuring initiative. This indeed is good news for our EBIC concept, which includes the power plant in this case. Biomass, rather than fossil fuel, has the advantages of being diverse in origin and of releasing substantially less carbon dioxide into the air. Biomass also possesses much less, if any, mercury and sulfur, as do fossil fuel, and releases less (~20% less) nitrogen oxide when burned than fossil fuels.

Because biomass represents a renewable resource, environmental engineers should make every effort to use it to produce electrical energy even when not incorporating it into a complex. But using it in a complex, such as suggested in Figure 18.7, appears to be even more environmentally preserving and economically advantageous.

Waste heat contained in the turbine condensate of the power plant can also be reused to heat the facilities that produce the biomasses in the first place.

The city of Fallbrook, California, is considering building a large power plant fueled by burning green waste produced by agricultural and home wastes, turn it into electricity, and sell the power to San Diego Gas and Electric Company (SDGE) for distribution to its customers (McCormac 2003). A contract between SDGE and the developer of the complex calls for the plant to produce 40 MW of electricity per day to supply about 40,000 homes (1,000 MW/household) in 2006.

F. Steel Mill–Fertilizer–Cement Complex

Steel mills are actually five separate industrial plants in one, consisting of the following: (1) coke plant, (2) iron ore reduction plant, (3) steel production, (4) hot rolling mill, and (5) cold rolling mill. Predominant wastes originate from the coke and steel plants, although certain dusts, slag, and iron also come from the other plants.

Troublesome waste products include ammonia, cyanide, phenol, heat, and acidic ferrous sulphate or chloride pickle liquor. Steel mills also use huge volumes of water, mostly for cooling and quenching the steel ingots, and produce large volumes of air, water, and solid contaminants. They have developed a worldwide reputation as one of the most polluting industries of modern times. At the time of this writing, about 20 U.S. steelmakers had filed for Chapter 11 bankruptcy law protection since 1998, in part due to a glut of steel that sent prices to 20-year lows (Mathews 2001). Domestic steelmakers have tried to reduce their debts by increasing prices for specialty steel. However, domestic steel makers are also hurting because of poor investments, inefficient capacity, and failure to be competitive (Mathews 2003). By using this industrial complex, the steel will be more competitive because production costs will be lowered.

As already mentioned, I described this industry and recommended solutions to its waste-treatment dilemma in 1976. These plants require so much land area and employ so many people that their location in a separate industrial complex would be a natural development. Fertilizer and building material plants are likely candidates to join the complex as auxiliary industries. I have proposed and presented such a complex in Figure 18.8.

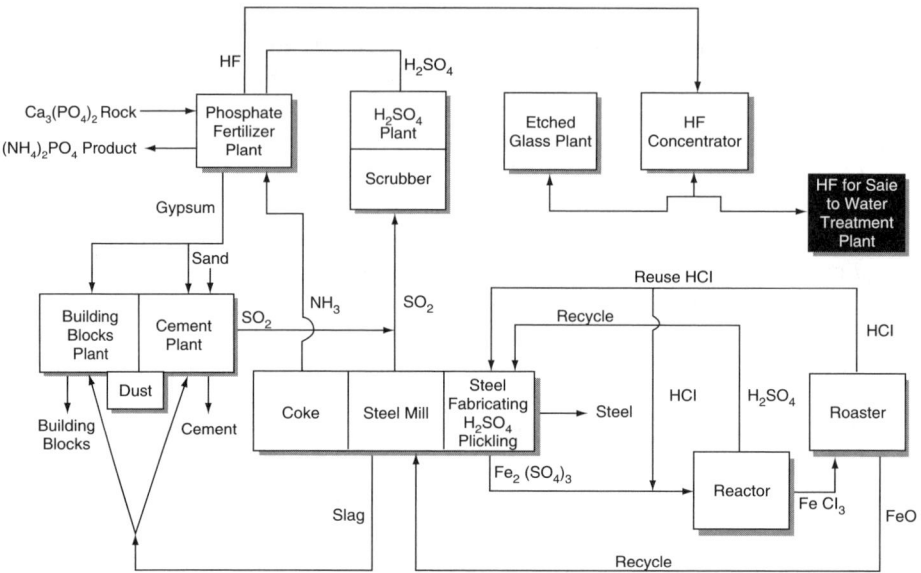

FIGURE 18.8. Steel mill–fertilizer–cement complex.

G. Fossil-Fueled Power Plant Complexes

Electric power plants face the task of producing more electricity at lower cost while minimizing external damage to the surrounding environment. This is increasingly more difficult to accomplish because of the problem of obtaining permits for producing nuclear power, the polluting characteristics of both oil and coal fossil fuels, and the untried utilization of more sophisticated wind-, solar-, and hydrogen-generated power. Because fossil-fueled power plants are currently cost effective and generally acceptable to the public, it is reasonable to use this type of fuel and attempt to ameliorate or abate adverse environmental consequences. The challenge is to accomplish this effectively, that is, at a minimum production cost and with little or no adverse external environmental consequences. To do this, one needs a more complete knowledge of the background of coal-fired power plants.

Coal-fired power plants generate the majority of electricity in the United States. For example, in 1986 these electric utilities produced 2,487.3 billion kilowatt-hours (kWh) of electricity, about 56% of the nation's total production for that year ("Monthly power plant report" 1986).

The U.S. Department of Energy encouraged the use of coal as a principal fuel (in lieu of gas or oil) by the electric utility and industrial sectors. Combustion residues from coal-fired power plants—fly-ash, bottom ash, boiler slag, and fuel gas desulfurization (FGD) sludge—are currently exempted from the Resource Conservation and Recovery

Act (RCRA), which requires the Environmental Protection Agency (EPA) to promulgate regulations for the disposal of hazardous and nonhazardous wastes ("Impacts of proposed RCRA" 1987).

On the basis of chemical origin, the EPA categorizes wet waste streams for the steam electric power-generating point source category as follows:

1. Once-through cooling water
2. Recirculating cooling system blowdown
3. Fly-ash transport discharge
4. Bottom-ash transport discharge
5. Metal-cleaning wastes (air preheater, fireside wash, etc.)
6. Low-volume wastes (boiler, evaporator, softener blowdowns, sanitary wastes, drains, etc.)
7. Ash pile runoff
8. Coal pile runoff
9. Wet flue gas cleaning blowdown

It should be noted that cooling water and wet ash handling systems and flue gas scrubber processes constitute the major discharge volume. Mean discharge flow rate per installed capacity from once-through cooling water systems is approximately 900 times as much as that from recirculating cooling water systems.

However, discharge from one-through systems is relatively clean and does not need treatment before discharge. General description of systems handling major wastewater streams are as follows:

Cooling water systems: In a steam electric power plant, cooling water absorbs the heat that is liberated from the steam when it is condensed to water in the condensers. Depending on the size, location of the power plant, and availability of a water body into which it is being discharged into, there may be either of the cooling water systems, once-through and recirculating, described as follows:

Once-through cooling water systems: In a once-through system, the cooling water is withdrawn from the water source, passed through the system, and returned directly to the water source. Discharge flow rates from such a system in coal-fired power plants may reach up to 55.4 million gallons per day (mgd)/MW. In the United States, about 65% of all power plants have once-through cooling water systems.

Recirculating cooling water system: In a recirculating cooling water system, the water is withdrawn from the water source and passed through the condensers several times before being discharged to the receiving water. After each pass, the heat is removed from the water by three major methods: cooling ponds or cooling canals, mechanical draft evaporative cooling towers, and natural draft evaporative cooling towers. Discharge from such a system in coal-fired power plants may reach up to 63,057 gal/day/MW.

Ash-handling systems: The chemical compositions of both types of bottom ash (i.e., dry or slag) are quite similar. The major species present in bottom ash are silica (20–60% as SiO_2), alumina (10–35% as Al_2O_3), ferric oxide (5–35% as Fe_2O_3), calcium oxide (1–20% as CaO), and small amounts of metal oxides. Fly-ash generally consists of very fine particles. The major species present in fly-ash are silica (30–50% as SiO_2), alumina (20–30% as Al_2O_3), and others including sulfur trioxide, carbon, boron, and so on. Distribution between bottom ash and fly-ash varies depending on the type of boiler bottom. Typically, bottom ash to fly-ash ratio is 35:65 for wet bottom boilers, and it is 15:85 for dry bottom boilers (Costle 1980). Wet ash handling (sluicing) systems produce wastewaters that are currently either discharged as blowdown from recycle systems or discharged directly to receiving streams in a once-through manner. In a coal-fired power plant, wet ash handling system discharges may change reaching up to 16,387 gal/day/MW for fly-ash ponds and 38,333 gal/day/MW for bottom ash ponds.

Flue-gas desulfurization processes: In the lime or limestone flue-gas desulfurization processes, SO_2 is removed from the flue gas by wet scrubbing with slurry of calcium oxide (lime) or calcium carbonate (limestone). The principal reactions for absorption of SO_2 by slurry are

$$\text{lime: } SO_2 + CaO + \tfrac{1}{2}H_2O = CaSO_4 \cdot \tfrac{1}{2}H_2O$$

$$\text{limestone: } SO_2 + CaCO_3 + \tfrac{1}{2}H_2O = CaSO_3 + \tfrac{1}{2}H_2O + CO_2.$$

Oxygen absorbed from the flue gas or surrounding atmosphere causes the oxidation of absorbed SO_2. The calcium sulfite formed in the principal reaction and calcium sulfate formed through oxidation are precipitated as crystals in a holding tank. The potential exists to use calcium sulfite in manufacturing cement within the proposed complex.

Background on Cement Manufacturing Plants

Portland cement is made by mixing and calcining calcerous and argillaceous materials in the proper ratio. Table 18.3 summarizes the relative amount of raw materials consumed for the production of Portland cement (Shreve and Brink 1977).

It can be observed that limestone represents the major amount of the raw material consumed.

Unit processes involved in cement manufacturing include storage and mixing of raw materials, drying, grinding and crushing, calcining, clinker storage, finishing additives and ball milling, and packaging for delivery. A schematic diagram of a typical rotary steam kiln boiler and isometric flowchart for the manufacturing of Portland cement is shown in Figure 18.9.

For each 100 barrels of finished cement by the dry process, 297,872 Btu of fuel as well as 6.4 kWh of electricity and 8 gall of water are required. Also required are 132 lb of limestone, 33 lb of shale, and 4.3 lb of gypsum (Shreve and Brink 1977).

TABLE 18.3
Amount of Raw Materials Consumed for Production of Portland Cement

Raw Materials	Amount Consumed in 1981 (thousands of metric tons)
Cement rock	24,204
Limestone	66,380
Clay and shale	8,536
Sand and sandstone	2,298
Blast furnace slag	86
Gypsum	3,272
Iron materials	1,040
Fly-ash	688
Miscellaneous	3,149

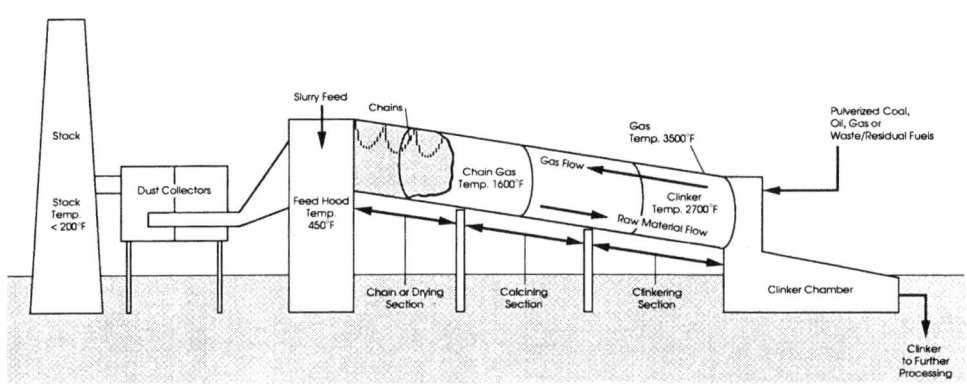

FIGURE 18.9. Schematic diagram of a typical rotary steam kiln boiler.

Background on Concrete Block Manufacturing Plants

The aggregates used for mortars and concretes can be conveniently divided into dense and lightweight types. The former class includes all the aggregates typically used in mass and reinforced concrete, such as sand, gravel, crushed rock, and slag. The lightweight class includes pumice, furnace clinker (or cinders in the United States), foamed slag, expanded clay, shale, and slate.

Environmentally Balanced Industrial Complex

An EBIC has been proposed for a coal-fired power plant in the schematic flow diagram shown in Figure 18.10A. The complex consists of three industrial plants, including a coal-fired power plant, and two ancillary plants: cement and concrete block manufacturing plants. The coal-fired power plant has the following waste streams:

1. Recirculating system cooling water blowdown
2. Boiler/evaporator blowdown
3. Fly-ash discharge
4. Bottom-ash discharge
5. Flue-gas discharge

Cooling water blowdown, boiler blowdown, and evaporator blowdown are determined to be the major wet waste streams from a coal-fired power plant. These streams will be directed to the kiln steam boiler in the cement manufacturing plant, as shown in Figure 18.10A.

Sulfur dioxide, which is released during the combustion of coal, will be scrubbed with lime/limestone slurry. The calcium sulfate formed after oxidation will then be utilized in cement manufacturing plant as a cement additive.

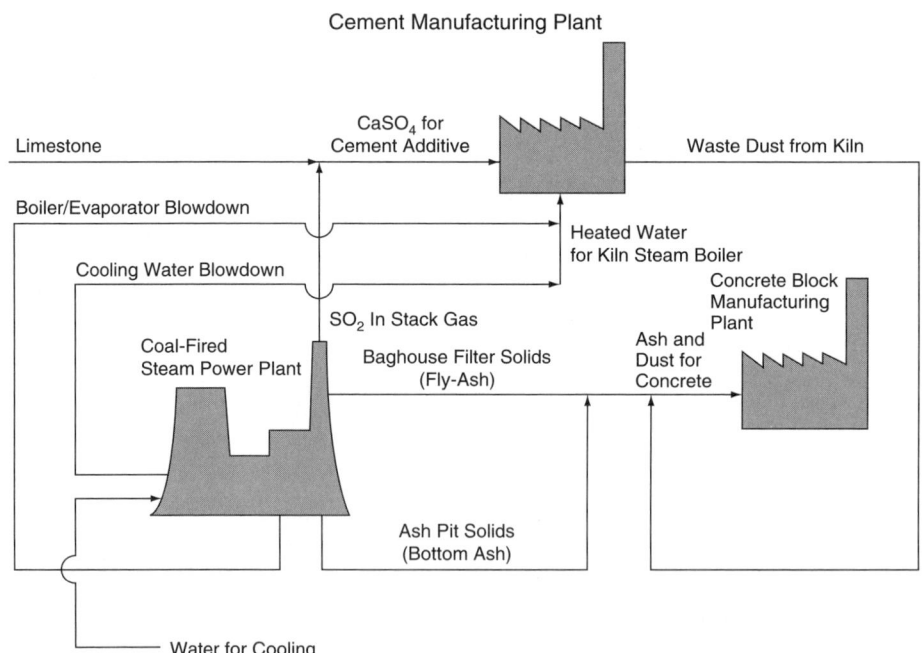

FIGURE 18.10A. Schematic flow diagram of the EBIC for power plant industry.

Waste dust from the kiln steam boiler in the cement manufacturing plant, as well as fly-ash and bottom ash formed during the combustion of coal in the power plant, will be transported to the concrete block manufacturing plant. These solid wastes will be used in the production of concrete blocks.

Reduction in Production and Environmental Costs

According to the proposed theory and obvious implementation of this zero pollution solution, production costs must decrease. Likewise, by eliminating industrial waste treatment, environmental costs must also decrease. The latter is especially valid when we include the benefits of eliminating adverse environmental impacts caused by the plant wastes. The sum of these two positive cost savings should be significant.

Nemerow (1987) reported as an example, that savings alone within a textile mill complex was $52.69/1,000 lb of cotton fabric from a typical small mill (12,000 lb of cotton fabric per day). From a preliminary study, the total environmental savings (from eliminating waste treatment) appeared to be greater than $104/1,000 lb of fabric. To both of these savings, we must add the savings from transportation of raw cotton and sized woven goods. In one instance and presuming a typical transportation cost in the southeastern United States, if $0.026/lb of cotton shipped, we can add $343/1,000 lb of fabric produced. The total savings of these three costs are almost $500/1,000 lb of cotton fabric produced. To allow the reader to grasp the significance of this, these savings represent 77% of the cost of cotton required ($590/1,000 lb of cotton) (*New York Times,* April 12, 1988, p. 48).

Simple Coal-Fired Power Plant Complexes

Almost half of all energy produced in the United States originates from burning coal. We still have an abundant supply of coal yet to be mined in the United States. It stands to reason that any future energy plan we use will highlight the use of coal. With that in mind, we should take a look at the environmental problems arising from the use of coal-derived power.

Some coals contain high sulfur contents that when burned give off excessive amounts of sulfur dioxide, an undesirable air contaminant. All coals, when burned, emit great quantities of carbon dioxide from the combustion of carbon. In addition, all coals, even when burned optimally, release both fly-ash into the smokestacks and surrounding air as well as boiler ash as solid waste usually to the land.

Huge capital and operating expenses are needed by these utilities to protect the environment from the detrimental effects of these gases and solid wastes. The utilities claim that they are not able to pass such costs on to the consumer without exceeding the limit of the consumers' ability to pay.

Without debating this last assertion by the utilities, it falls on the environmental engineer to provide some assistance. I have already proposed a rather complicated three-industry environmental complex in Figure 18.10, which may not be suitable in all instances where an isolated coal power plant is located. What is needed here is a simpler, easily applied form of complex that can be used in most cases with an isolated coal-fired power plant.

The advent of the problem of global warming (caused by an excess of combustion gases such as carbon dioxide) and a country-wide energy shortage (demanding more coal-fired power facilities along with other forms of energy) cause our current dilemma: how to produce more energy in an environmentally friendly way!

In 1999, for example, 2.265 billion metric tons of CO_2 resulted from all electric power generation, about one-third of the CO_2 released from all sources that year (Bradsher and Revkin 2001). American Electric Power is mentioned as "accepting the idea that government limits on carbon dioxide and other greenhouse gases are inevitable."

I propose here two rather uncomplicated and readily adaptable types of industrial complexes to contain both the coal-fired power plant and another compatible industrial plant. These are shown in Figure 18.10B. In case one, that plant would simply

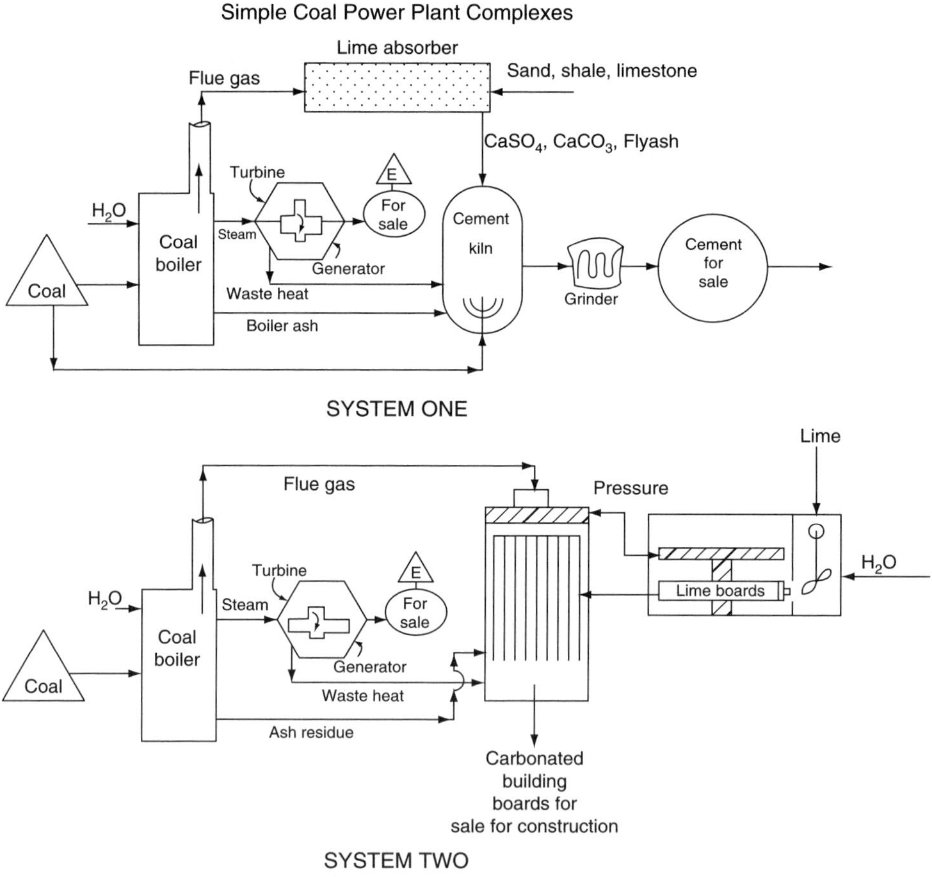

FIGURE 18.10B. Simple coal power plant complexes.

manufacture cement that would incorporate the three contaminants SO_2, CO_2, and fly and boiler ashes. The product cement may not be as high quality as Portland-type cement but would be usable, generate a profit, and eliminate the environmental contaminants of the coal plant.

The other complex, case two, would include a low-cost carbonated building board, which would be structurally sound for a limited number of construction purposes. It would contain the dusts and ashes from the combustion of coal and the chemically converted $CaCO_3$ and $CaSO_4$. The dusts and ashes may even lend a decorative appearance to the board, which will be strengthened by these solids and the carbonates and sulfates.

The key fact here is that no costly waste treatment would be required by the electric utility. The steel industry has been in desperate need of such cost-saving systems to survive in the global market. As proof of this cost-saving need by the steel industry, President Bush put tariff restrictions on foreign steel imports. "The administration expects the industry to use the tariff cushion to consolidate and rationalize operations, reduce costs, enhance efficiency, increase productivity, improve quality and service, and develop new products and markets" ("Steelmakers get Bush" 2002). In addition, the profitability of the ancillary cement and/or board plants would be ensured by the free supply of a portion of their raw materials and perhaps a partial subsidization of their operation by the owners of the industrial complex or by a government entity.

H. Plastic Industry Complexes

Plastics and resins are chainlike structures known chemically as *polymers*. Polymers are synthesized mainly by adding a free radical initiator and a modifier to the monomer (the building block of the polymer). Although not a great deal of wastewater arises from the polymerization process, more results from synthesizing the original monomer.

As long ago as 1967, the U.S. Department of Interior classified plastics in nine categories, but we are primarily interested in only two predominant types: polystyrene and polyolefins. In 1967, these types made up 46% of the total plastics and resins produced (Nemerow and Dasgupta 1991).

Polystyrene's combination of physical properties and ease of injection molding and extruding makes its use desirable. The crystal-clear product has excellent thermal and dimensional stability, high flexural and tensile strength, and good electrical properties. Styrene monomers (or mixtures) are purified by distillation or caustic washing to remove inhibitors. The purified raw materials, together with an initiator, are fed to a stainless-steel or aluminum polymerization vessel, jacketed for heating and cooling and contain an agitator. Polymerization of the monomer is carried out at about 90°C to approximately 30% conversion to create a syrupy mass. Water is used during this stage as a heat-exchange medium and is recirculated without contamination. Then the syrupy mass is transferred to suspension-polymerization reactors containing water and proprietary suspending and dispersing agents. These reactors are usually jacketed, and the contents stirred in stainless-steel vessels.

The syrupy mass is broken into droplets by means of the stirrer and held in suspension in the aqueous phase. Temperature is a critical variable in further polymerization of the product. The polymer is then sent to a blowdown tank where any unreacted monomer is stripped. The stripped batch is then centrifuged, and the polymer product is filtered, washed, and de-watered. A flowchart of this manufacturing process is presented in Figure 18.11 to aid you in following the manufacturing process.

Reaction water (suspension medium) and wash water are the two significant sources of wastewater from the manufacturing of polystyrene. They are shown in figure 18.11 as wastes A and B. About 1.5 gallons of water, excluding cooling water, is used and wasted for each pound of polystyrene produced. These wastes are not very polluting; they contain small amounts of catalyst and suspending agents used in suspension polymerization and heated water (120–180°F).

The catalysts are generally of the peroxide type, and the suspending agents may be methyl or ethyl cellulose, polyacrylic acids, polyvinyl alcohol, or miscellaneous compounds such as gelatin, starch, gum, casein, zein, and alginate. Inorganic materials such as calcium carbonate, calcium phosphate, talc, clay, and silicate may also be present in effluent reaction water. No plants employing typical technology have waste-treatment units; instead, 90% discharge these wastes into municipal sewers.

Polyolefins (polyethylenes) are composed of many different molecular weights from waxes of a few thousand to molecular weights of several million. Polyethylene is used for film and sheet, injection molding, blow-molded bottles, cable insulation, coatings, and other products.

There are two processes for manufacturing polyethylene: one for a low-density and the other for a high-density product. Both start with ethylene as the raw stock material. Heat, pressure, catalysts, and solvents are reacted with the ethylene, and then the product is purified by various unit operations, as shown schematically in Figures 18.12 and 18.13.

Both processes produce little wastewater. However, potential hazards that may generate water-borne waste are improper operation, spills, and wash-down of equipment and facilities. Typical process wastewaters contain a BOD of less than 10 ppm (U.S. Department of Interior 1967).

The Ultimate Plastic Problem

According to Sherman (1989), the United States produces 60 billion pounds of plastics annually, and sales of plastic products exceed $150 billion a year. By this year, 2000, the U.S. output may reach 76 billion pounds. Thayer (1989) expects plastic waste to reach 38 billion lb/yr by 2000 and to account for 10% of total municipal solid waste.

Because plastic production wastes are minimal as far as contaminants are concerned, little or no waste treatment is done at plastics manufacturing plants. This industry is rather unique in that the major wastes and environmental costs occur during and following the use of plastic products, which makes plastic waste treatment very difficult.

Until the late 1980s, most plastic products were disposed of in municipal solid wastes and were given the same treatment as these wastes. Landfilling and incineration were the predominant systems for plastic waste treatment for solid wastes.

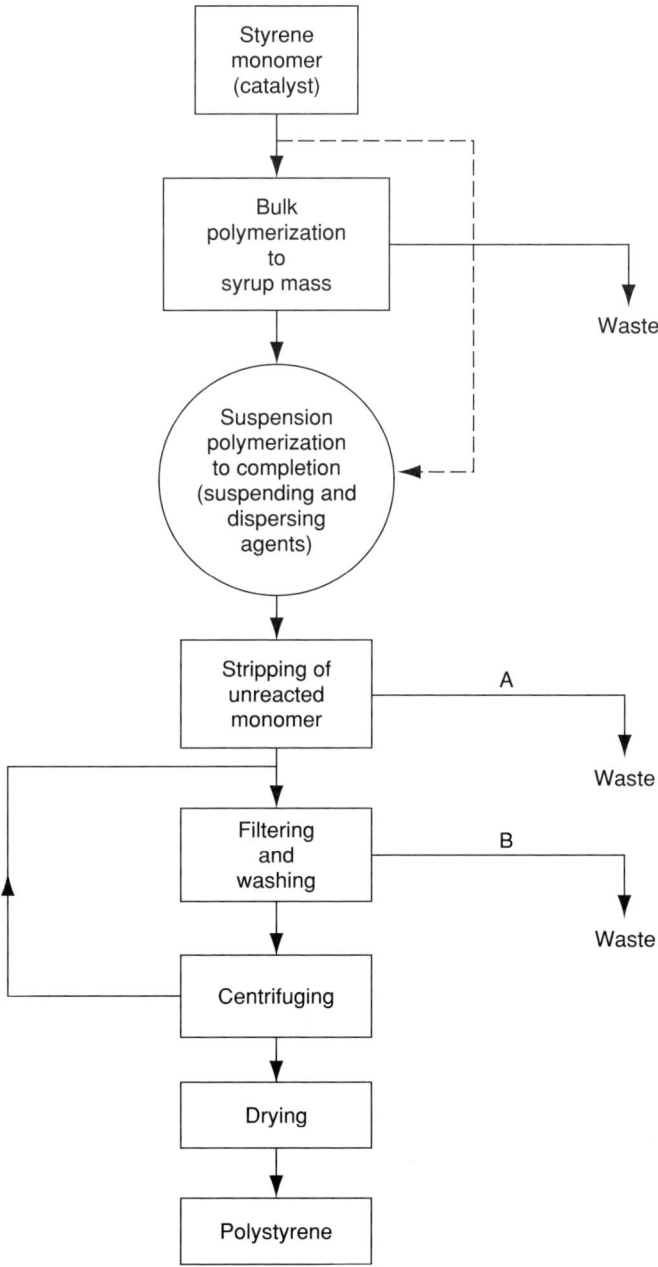

FIGURE 18.11. Flowchart for polystyrene production.

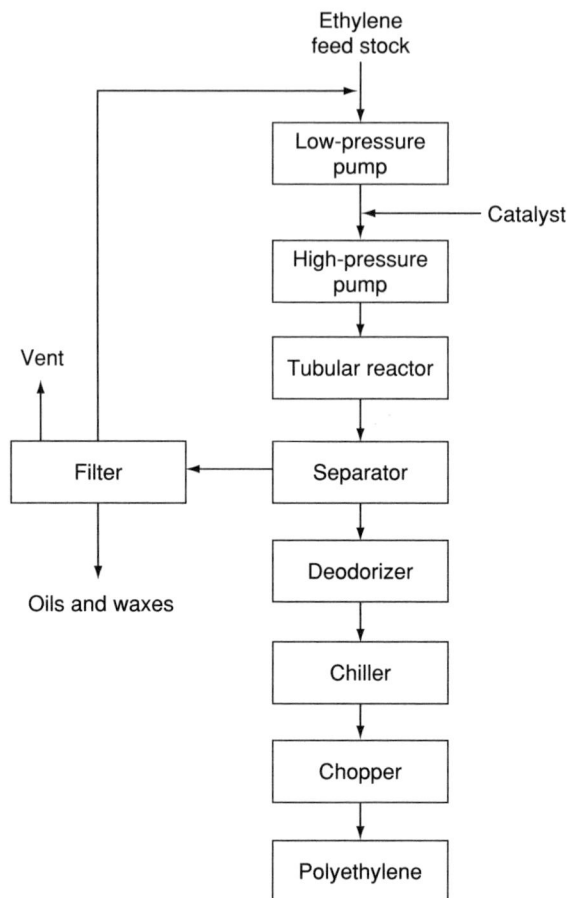

FIGURE 18.12. Tubular reactor process for low-density polyethylene production.

Cheremismoff and Cheremismoff (1989) report that "the option of landfill as a disposal method is rapidly diminishing." For example, the number of legal landfills declined from a 1976 figure of 18,000 to 9,000 in 1989. Not only are landfill sites diminishing, but costs and limits for disposal of plastic materials are increasing. Plastics degrade slowly in soils relative to other ingredients in municipal solid wastes. Because we are rapidly approaching a crisis in the treatment/disposal of plastics, we must seek alternative and new methods.

The best choice, according to Sherman (1989), is waste-to-energy incineration. Most plastics ignite as easily as natural gas and emit CO_2, NOxs, and H_2O vapor. However, some plastics, mainly PVCs, can adversely affect the air environment. These PVCs, unless burned at more than 1,200–1,600°F, will evolve hydrochloric acid, which

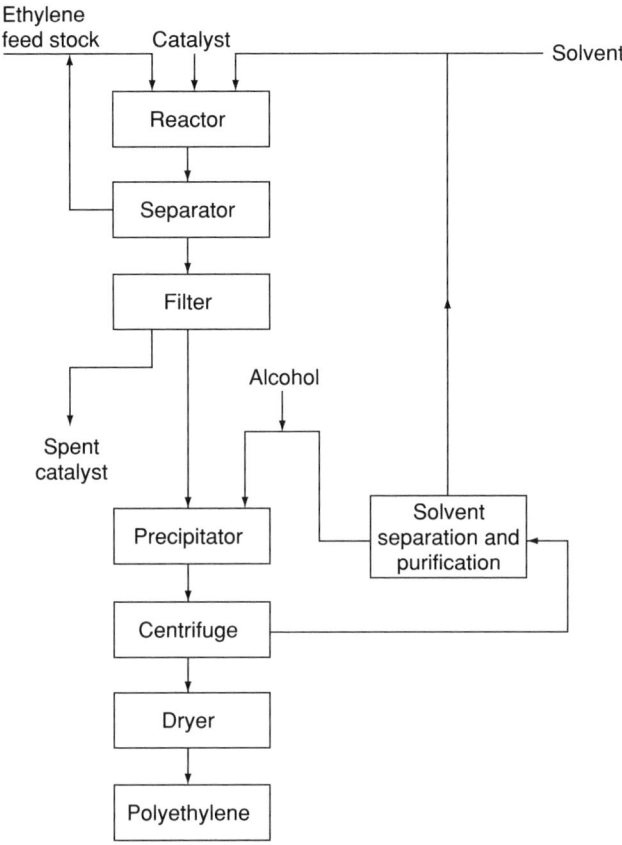

FIGURE 18.13. Philips process for high-density polyethylene production.

is corrosive to metals and antagonistic to humans who come in contact with the acid. Some other plastics contain cadmium or other heavy metals that remain unburned in the incinerator ash. Such metals are toxic when they enter the water environment through landfill leachates. One advantage of incinerating plastics is the high heat energy released from the burning: about 16,000 Btu/lb. Some waste-to-energy type of plants, which are rather expensive to build (~$100,000/ton/day) are described in *Industrial Solid Wastes* (Nemerow 1984).

Other potential problems with incineration as a method of treatment/disposal for plastics include enhancement of the "greenhouse effect" and ash disposal. The burning emits CO_2, which keeps the heat energy rays (infrared) from escaping the earth. The subsequent heating may alter the earth's climate, affecting agricultural output and ocean levels. However, there is still much controversy about any true CO_2 increase in the atmosphere. Even so, the small increase in CO_2 may also stimulate increased tree and plant growth. These green growths will evolve oxygen into the atmosphere and oceans to aid in their purification.

The ash resulting from incineration may constitute about 10% of the waste burned, contain heavy metals, and be expensive to dispose of by normal processes. Some progress has been made in using ash as an additive in concrete products meeting utilization specifications.

Another possibility for eliminating the plastic disposal problem is using waste plastics to make other useful products. Rubbermaid Company, for example, makes trash containers in Winchester, Virginia, out of reused plastic chips (Sherman 1984). Lehrman owns BTW, a company that reprocesses used plastics to make woodlike fence posts, car stops, and picnic tables (Kollin 1991). Main reprocessing units include pulverizers, extruders, and other pressure from machinery. The cost of reused plastic varies from $0.23 to $0.40/lb, almost as much as virgin resins ("Profitability problems plague" 1990). Prices, however, do not present the problem that collection and separation of the different plastics cause. Recycled polyethylene can also be used for carpet fibers, and polystyrene is used for a variety of durable goods such as office supplies, hair accessories, cafeteria trays, license plate holders, and loose-fill packing material.

At this point in the solution to the problem, I suggest, once again, the use of a plastic industry environmentally balanced complex. However, in this case, I propose merging only two facets of the plastic industry: virgin plastic manufacturing and recycling plastic collectors. Such mergers would result in recycled plastic products similar to the originals. An example of such a complex is shown in Figure 18.14.

Some limits on quantity and quality of recycled plastics and additions of processing equipment undoubtedly will be necessary to optimize the complex. However, the merger appears worthwhile to both industries. The plastic manufacturer will have lower raw material costs and recyclers will not have typical marketing problems with the final product. In addition, as usual no wastes result from the complex and a considerable plastic solid waste problem will have been removed from the environment.

I. Cement–Lime–Power Plant Complexes

In many cases, it is possible to combine more than two industries in a complex to more fully utilize all waste products and optimize production costs. Such a situation exists in the manufacturing of cement, lime, and power.

The separate production of cement and coal-fired power production has already been presented and shown in Figure 18.10. I refer you to this earlier material for review of their manufacturing materials processes and wastes. Here, I present the background of lime manufacturing and illustrate how it would work into the complex concept.

Lime production is an ancient industry. Today, lime and limestone are employed in more industries than any other natural substances (Shreve and Brink 1977).

Lime itself is used for medicinal purposes, plant and animal feed, insecticides, gas absorption, precipitation, dehydration, and caustisizing. It is used as a reactant in paper making, de-hairing hides, manufacturing high-grade steel and cement, water softening, recovery of byproduct ammonia, and the manufacture of soap, rubber, varnish,

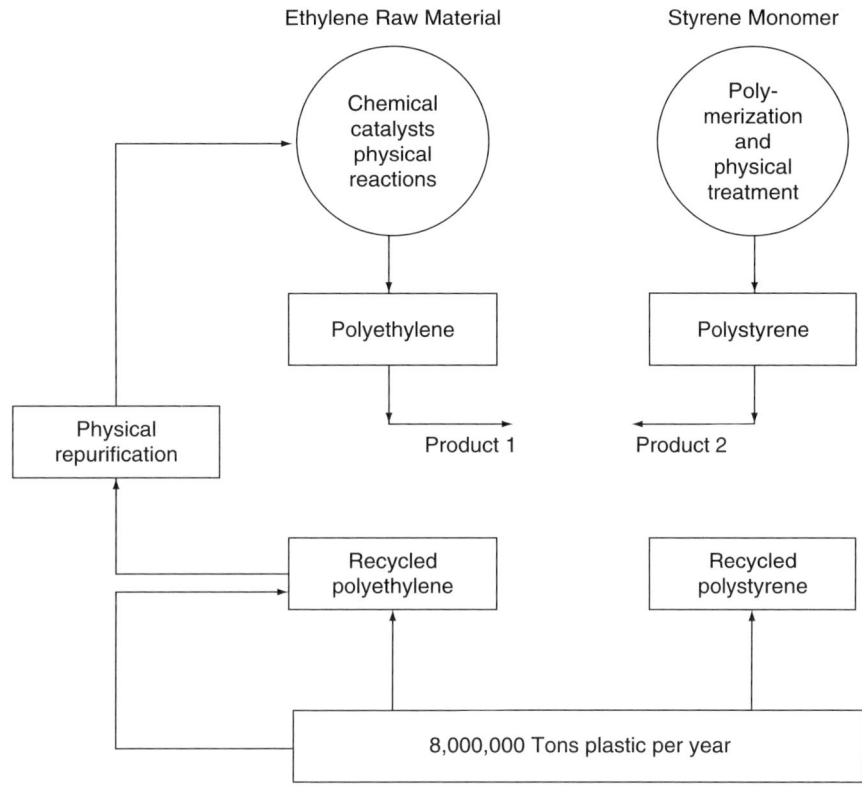

FIGURE 18.14. Plastic manufacturing industrial complex.

refractories, and sand-lime brick. Lime is also an essential ingredient in mortar, plaster, and soil additives.

There are several types of lime, varying with the lime content, water content, and specific use requirements. All types originate with limestone, so lime manufacturing is usually done near limestone mining deposits.

The carbonates of calcium and magnesium are obtained from deposits of limestone, marble, chalk, dolomite, and oyster shells. Quarries are selected that yield a rock consisting of low amounts of silica, clay, or iron and a naturally high concentration of calcium, magnesium, or both.

Impurities can adversely affect the desired hydraulic qualities of the resulting lime. Both overburned and underburned limestone leave undesirable lumps in the product lime.

Considerable energy (power) is required in this industry for blasting limestone out of the mines, for transporting and sizing of rock, and for burning (calcining) it. Calcining requires 4.25 million Btus per ton of lime produced, and subsequent hydration liberates 15.9 kcal. The volume of rock declines during calcining and swells during hydration. The amount of coal required for calcining varies from 1 to 3⅓ pounds (depending on the kiln type) per 3¼ pounds of lime produced. Calcining takes place at 1,200–1,300°C.

The sequence of actual operations in manufacturing lime is: (1) blasting the limestone in the quarry, (2) transporting the stone to the plant, (3) crushing and sizing of stones, (4) screening to remove small (<4 in.) and large (>8 in.) stones, (5) moving the uniform-size stones to a vertical kiln (large size), (6) taking fines to a pulverizer to produce powdered limestone for agriculture uses, (7) burning the limestones in vertical kilns to give lump lime or in a horizontal rotary kiln to make fine lime, (8) packaging of finished lime in barrels or drums or sending it to a hydrator to make hydrated lime, and (9) packaging of slaked lime in bags.

A typical schematic flow sheet is presented in Figure 18.15. An existing complex of this type is already in operation in Brooksville, Florida, as shown schematically in Figure 18.16. The complex is claimed to be "the world's first combination of pulverized coal/fluidized bed combustion boiler producing lime and electric power, integrated with a cement plant. Nowhere else in the state (Florida) is there a facility that's using the waste material from the mining, in the case of limestone fines which could not be sold otherwise, and making major products like Portland cement and lime" (Lawhome 1989).

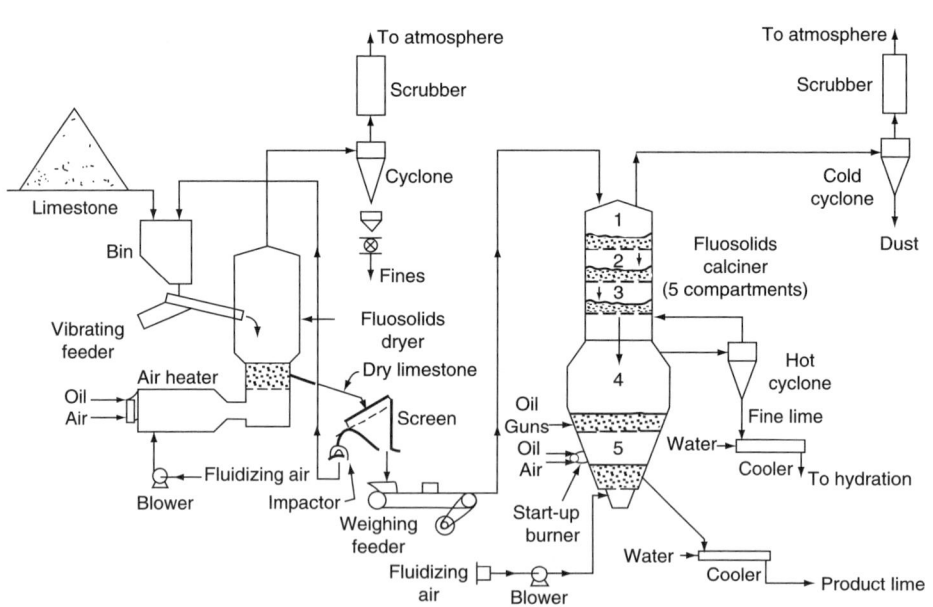

FIGURE 18.15. Fluosolids system (Dorr-Oliver, Inc.).

Potential Industrial Complexes 469

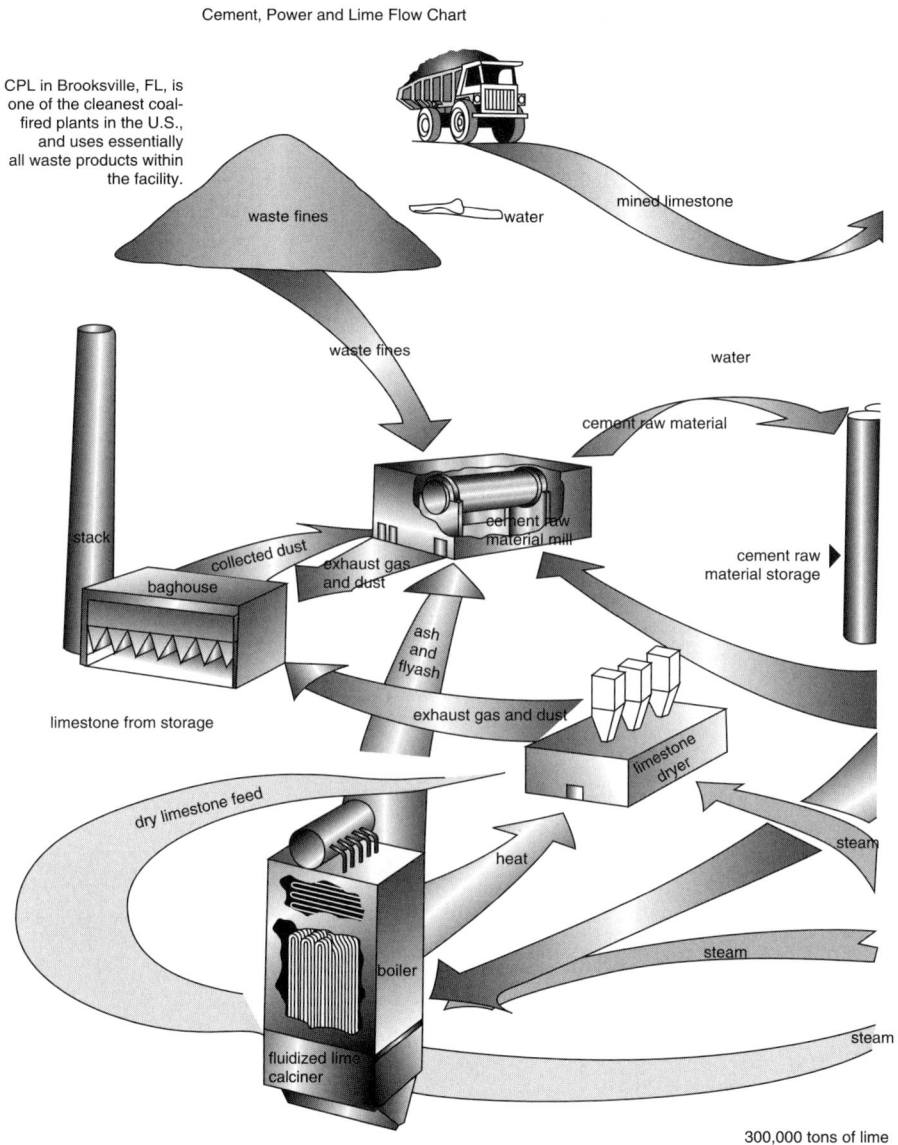

FIGURE 18.16. Cement, power, and lime flowchart. Used with permission from John Wiley and Sons.

The Brooksville plant will produce about 600,000 tons of cement a year and 330,000 tons/yr of chemical lime. The lime or CaO will be used for gas desulfurization, acid pond neutralization, agriculture, or in building products. According to Lawhome (1989), the limestone quarry plant generates about a ton of waste product for each ton of usable stone. The limestone fines are not suitable as an aggregate or as a stable fill material, but do contain large quantities of calcium and silica.

This composition is required in raw material in the manufacture of cement. The plant's energy efficient co-generative design allows the simultaneous manufacture of lime and cement, which, in turn, reduces operational costs of the electrical power plant. Hot air, which comes from the cooling plant cement clinker, is used as combustion air for the power plant boiler. Waste heat in the form of cement produced by the generator is used to dry the limestone at the cement and lime operations. In addition to lower power costs, ash from the combustion of coal in the power plant provides additional iron and aluminum—two cement raw materials that are not present in adequate quantities in the limestone fines.

A state-of-the-art computerized control room, manned 24 hours a day, supervises every aspect of the plant's production facilities. Highly trained personnel monitor production quality to ensure that finished products meet or exceed strict specifications. A chemical analysis department repeatedly tests lime and cement samples for mineral content and consistency.

Initial engineering and design for the cement power and lime cogeneration project got under way in 1982, and production began in 1984. Various components of the facility, including the cement plant, have been in operation since 1988.

Design for the 125-megawatt coal-fired plant was provided by Larramore, Douglass, and Popham, Inc. of Chicago, Illinois. The power produced at this facility will be sold under a long-term contract to Florida Power and Light Company.

J. Wood (Lumber) Mill Complexes

Sawmill and planing mills (Standard Industrial Classification Code 2421) produced more than 162 tons of solid waste per employee as far back as the 1980s (Nemerow 1984). Small producers of lumber and wood products also generated 16,083 yd^3 of solid waste per firm per week, or 836.33 yd^3/firm/yr. This was contributed by 17,247 employees per firm, for an annual discharge of 48,492 yd^3/employee.

Rough lumber from trees is brought to the lumber mill and sawed and planed to appropriate lengths and widths, largely for sale in the housing market. Planing removes the tree bark, and saws produce sawdust in addition to finished lumber product. Most sawmill waste (sawdust and bark) has typically been burned or reclaimed as soil conditioner (Nemerow 1984).

The burning of these solid wastes, though not requiring an outside source of energy, contributes potential air pollutants of unburned carbon and ash. Soil conditioning with these same wastes requires preparation, transporting, and locating a suitable market, all of which cost lumber mill owners time and money.

Potential Industrial Complexes

A potential solution to the problem of sawdust and bark wastes originating from lumber mills is an EBIC in which these wastes are used directly to produce other products. Such a complex is shown schematically in Figure 18.17.

In this complex, the plant will pyrolyze the wastes (sawdust and wood chips) into oil vapors. The vapors are condensed to make fuel oil or other chemicals (product 3). The remaining gases are burned to generate electricity (product 2). Finished lumber, naturally, is the prime product of the complex (product 1). Nonburnable gases such as CO_2, CO, and NOs are also obtained from the condenser/pyrolyzer and used to stimulate the growth of algae in a separate unit operation. This takes place in a greenhouse to use natural sunlight. The algae serves as food for small seafood such as shrimp, which, along with some algae/herbs, are sold as food products (product 4).

Such a complex has been suggested ("Process converts wood waste" 1991) to convert 70–90% of the wood waste into salable products; at the same time, virtually no

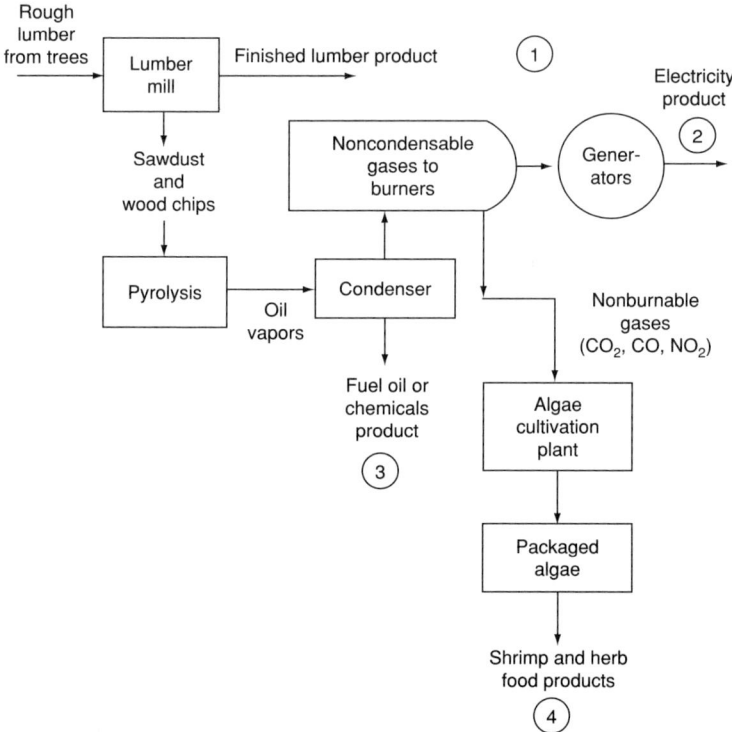

FIGURE 18.17. Lumber mill complex—four products.

pollutants are emitted. There are also possibilities for converting bark and sawdust into other useful products, such as resins and tanning agents from bark and product fillers from sawdust. These would need exploration in individual cases.

K. Power Plant–Agriculture Complexes

Electric power and food are two of the vital necessities for human life. It is imperative that we preserve both industries and produce their products as efficiently as possible. By combining operations in a single complex, we can accomplish two objectives: efficient production and a pollution-free environment.

One possible configuration of combining electric power production and food production is presented in Figure 18.18. In this complex, a low-sulfur fossil-fuel power plant produces three main wastes: (1) heated water, (2) flue-gas fly-ash, and (3) boiler residue ash. These wastes are described in detail in Section G, earlier in this chapter. The heated cooling water is reused within the complex to enhance the growing of fish. Fish will metabolize food faster and thereby grow at a higher rate if the temperature is raised, which is especially significant for colder climates. Aquaculture has more than tripled in size in the last 10 years, with annual U.S. sales of farm-raised bass, salmon, and catfish at $1 billion, and worldwide sales at $40 billion ("A farm for fish" 2000).

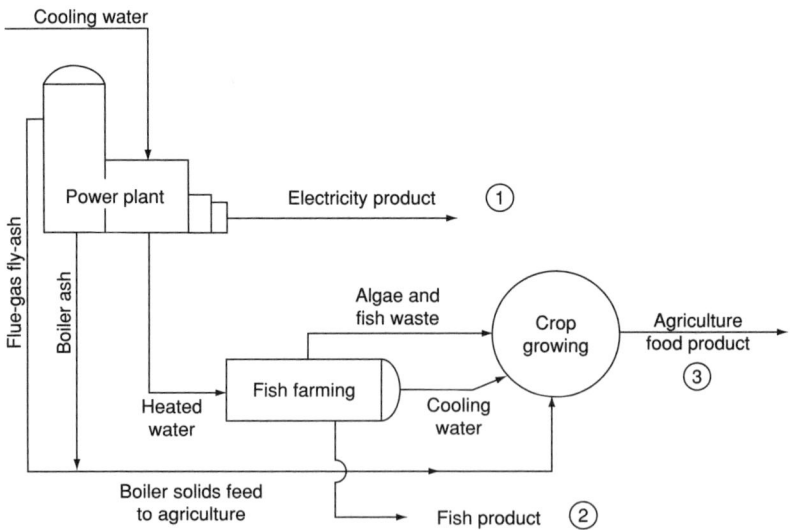

FIGURE 18.18. Power plant-agriculture complex.

Both ashes from the flue-gas and boiler residue are fed to the agriculture growing area to enhance the soil. The minerals in these ashes will increase crop growth and yield more food per acre. At the same time, no ash wastes will reach the outside environment.

As shown in Figure 18.18, three products will be sold outside the complex: (1) electric power, (2) fish, and (3) food crops. Algae and fish wastes, produced in the fish-farming waters, will also be reused as fertilizer for the crops. No pollution will reach the air, water, or land environments outside this complex, especially if the power plant uses low- or no-sulfur fuel. This would be possible by using only low-sulfur coal or oil or natural gas as fuel.

The Electric Power Research Institute (EPRI) (2000) announced a new filtration technology that could make aquaculture more efficient. Its improved-recirculation aquaculture systems cost less and provide better filtration than traditional systems, ensuring healthier fish stocks and improved productivity. The EPRI says its filters use waste heat from power plants to harness energy that otherwise would be lost. In addition, the aquaculture systems are relatively small and can be located near urban areas. These systems are claimed to offer a higher degree of environmental control than traditional fishing methods.

L. Cannery–Agriculture Complexes

Canneries are in the business of producing food products for direct consumption by the purchaser. Under most existing conditions, they import fresh fruits and vegetables from considerable distances. This not only results in added costs of transportation but also increases spoilage and waste of raw materials during the travel and extra handling.

It would make good economic sense for canneries to grow their own raw materials within a complex. But because farming is an entirely different enterprise, coordination of farming with the cannery operation would be required. The land areas and their usage required in agriculture must be integrated with the relatively smaller areas of canning. Moreover, the two enterprises require workers with different backgrounds and abilities.

However, if the two plants' processing operations could be combined in one complex, the entire system would be more cost effective. In addition, the normal wastes produced by each plant would be used within the complex to eliminate any adverse environmental impact to the surrounding area. A typical configuration of such a complex is shown in Figure 18.19.

In this complex, the farming industry would grow fruits such as tomatoes and peaches and vegetables such as beans and carrots. These crops would be cut and/or picked in the field and mechanically conveyed directly to the adjacent canning plant for processing. Waste cuttings, rot, and extraneous organic residues remaining after harvesting would be collected and trucked directly to the fermenter for digestion to methane gas.

Runoff from the farming area carrying excess or unused fertilizers and pesticides would be collected in drainage ditches surrounding the growing area. Instead of polluting

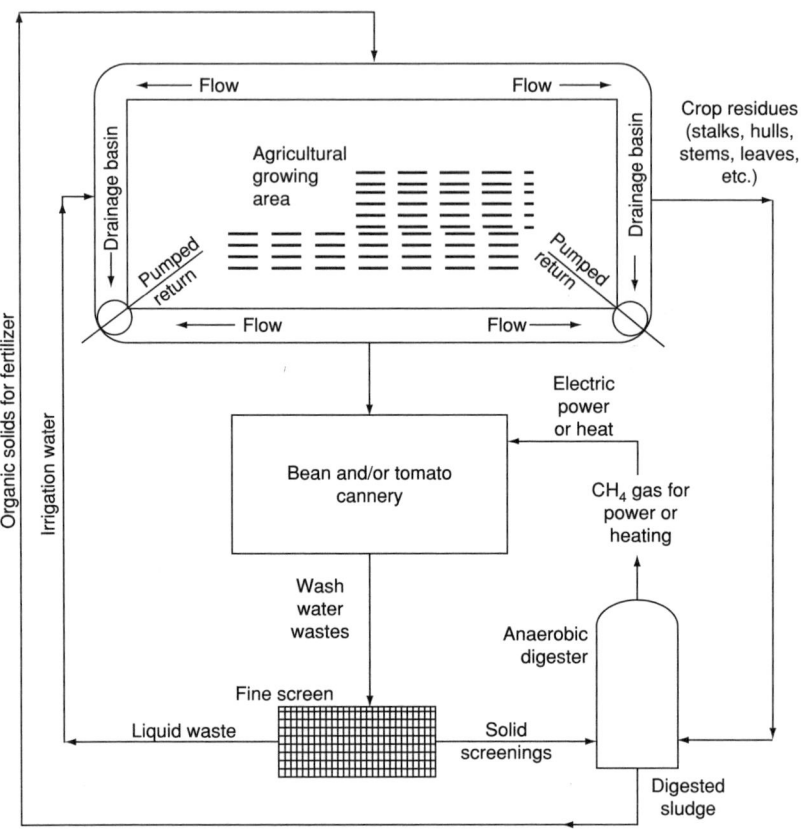

FIGURE 18.19. Cannery-agriculture complex.

the area water environment, this runoff is collected and pumped back to the growing area through perforated aluminum pipelines. In this way, water is conserved, as well as potentially contaminating chemical phosphates and insecticides.

The wastewater from the cannery is screened and returned to the farm growing area as a source of valuable water. The screenings are delivered to the farm to enhance the production of methane gas.

The digester sludge waste, which is distributed on the farmland crop-growing area, increases the fruit and/or vegetable yield because of its value as an organic fertilizer. The methane gas (CH_4) produced by the digester is used internally to heat the cannery buildings, sold externally to heat the cannery buildings, or sold externally to local power plants to produce power for the cannery or local homes of industrial workers.

No wastes leave the complex to cause adverse effects on the environment. Furthermore, canned fruits and vegetables are produced at a minimum cost.

M. Nuclear Power Plant–Glass Block Complexes

One of the biggest environmental concerns facing the world today is the problem of waste disposal from nuclear power plants. Safety hazards appear to be waning as a result of better plant design, operation, and supervision. Low-level wastes have never caused serious environmental problems. Only high-level radioactivity resulting from replacing spent fuel rods has been of serious and ongoing concern to the Atomic Energy Commission (AEC) and to environmentalists. Nuclear energy as a viable alternative to the use of coal or oil fossil fuel is an attractive possibility for the future.

If the environmental consequences of nuclear energy production can be overcome, its acceptance and use will solve an enormous energy dilemma. This involves devising solutions for both the excessive heat releases in the cooling water and the dangerous radioactivity continued in the high-level fuel reprocessing wastes.

A possibility for abating both waste problems is the use of a balanced environmental complex, such as that shown in Figure 18.20. This example complex contains

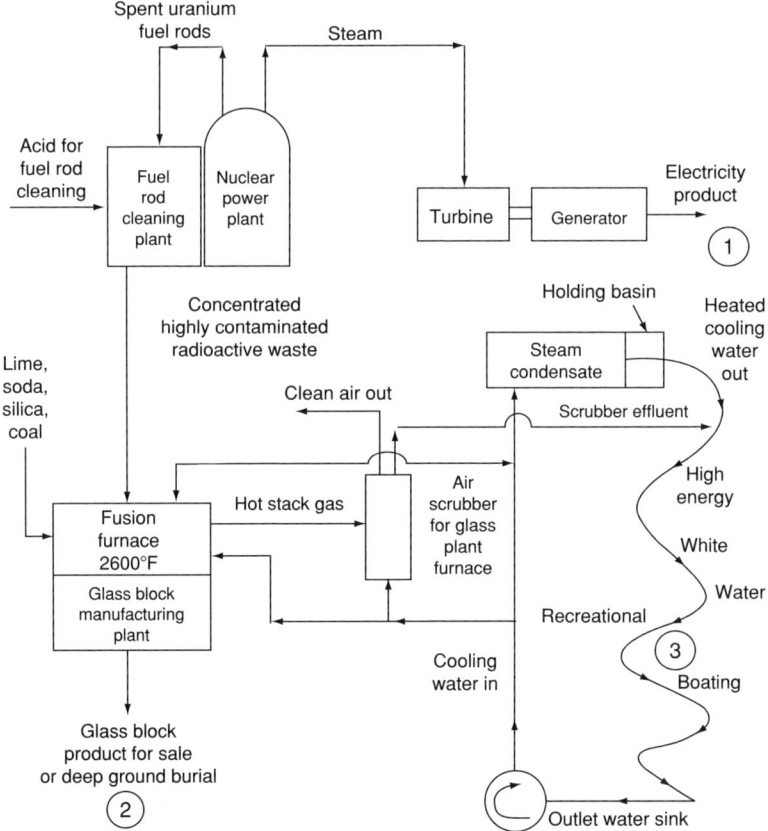

FIGURE 18.20. Nuclear power–glass block complex.

plants producing three major products: (1) electricity, (2) decorative glass blocks, and (3) a recreational boating facility.

Electricity is produced by the nuclear power plant. The cooling water used to condense steam exhausting from the turbine contains excessive heat and high energy. It is used as a source of water for a white-water recreational boating river. This river meanders toward an outlet water sink (lake, river, or ocean), losing its heat and potential energy as it flows downhill and around bends to the outlet. Boating obstacles are placed along the recreational path to enhance the challenge to its users. The outlet sink also serves as the source of cooling water intake for the condenser. Any potential low-level radioactivity leakage from the uranium fuel rods into the stream, and subsequently into the condenser, would be retained in the holding basin, whose effluent is continuously monitored before discharge to the recreational river. Recreation, naturally, would be discontinued during periods in which any release of radiation was detected.

The uranium fuel rods, when reprocessed, are removed to the adjacent cleaning plant where nitric acid and other cleaning agents are added to refurbish the rods. The high-level wastes are concentrated by evaporation before being sent to the block manufacturing plant. Here, lime, silica, soda, and some coal are added, and the mixture is sintered at 2,600°F to form a molten glass product. The liquid glass is poured into blocks, cooled, and removed to be sold after complete monitoring for radiation and testing for physical durability as a decorative building material or buried in deep, dry underground locations for environmental security. The hot furnace gases are precooled by using some of the outlet sink water and then sent to a scrubber to remove any excess sulfur and carbon dioxide.

The scrubber effluent is also returned to the white-water recreational boating river. Thus, no radioactivity or heated water escapes from the complex. Furthermore, a recreational industry is supported, and electricity and glass block products are manufactured at lower total costs than if located in separate places.

In 2000, the federal government reported that it will spend $265 million over at least 9 years to remove safely and store a leaking radioactive waste dump ("Radioactive waste cleanup" 2000). What a waste of both time and money! It could have considered constructing an environmental complex on the dumpsite. This solution could have been done for less than $265 million, taken a shorter time than 9 years, and have been more environmentally safe. The radioactive waste could have been fused and fabricated into decorative glass blocks in an adjacent block manufacturing plant. The block plant could produce a product for sale outside the complex, the costs of which would diminish (or eliminate) the disposal costs that would have been incurred by the federal government. Further, it is much safer to reuse the environmentally secure glass blocks in building construction than placing the waste pile in a new lined dump area.

In fact, designing and operating an EBIC on an existing hazardous waste site might be considered an economically superior alternative solution to waste treatment at the same site. Such sites would have to be made secure by excavating and diking before reusing the waste contents in manufacturing a compatible product.

By the year 2001, the nuclear industry and the federal government's Department of Energy had been working on a "dump" for its wastes for almost 20 years instead of reusing them in a safe product. Millions of dollars have already been spent on the Yucca

Mountain U-shaped tunnel "dump" and no radioactive debris has yet been stored there. The major deterrents to this solution have been political, court decisions, and public outrage (the NIMBY reaction). Conceivably, if we had put that much time and money into the EBIC system, we would have operating complexes by now. In addition, it would have been easier to increase production of nuclear energy to alleviate the existing electric energy crisis in the United States.

It is true that the reuse of nuclear wastes represents a formidable problem. Most of these wastes are bulky and dangerous to handle. Low-level radioactive wastes—most of which are sent to a depository in South Carolina until 2008—consist of clothing, tools, soils, and construction materials used in and about the reactor. High-level wastes, which consist of filters, resins, and equipment parts, come from the reactor core. These are mainly stored nearby the 103 reactors of origin in each state in either very deep (>40 ft) or oblong-shaped steel-lined concrete vessels. It is claimed that "on average, each reactor produces enough low-level nuclear waste each year to fill one sports utility vehicle" (Wilkie 2001). The Department of Defense admits that these storage vessels are secure for only about 100 years and represent a danger of explosion if water seeps in.

However, the Bush administration appears content to allow the 10 millions of tons of uranium waste piled up near Moab, Utah, to sit exposed to rain and seepage into the Colorado River. Instead, the administration proposes to cap the pile in place despite the chance of groundwater leaching ("Toxins and Water" 2001). Another objection to underground storage such as the Yucca site was brought out by W. Kenneth Davis, a former Department of Energy official, in a protest against shipping long distances the dangerous radioactive material ("Ex-official ends support" 2001).

However, the Yucca Mountain project was finally approved by the Senate in 2002. "Every year the nation's commercial nuclear power plants generate 2,000 tons of spent reactor fuel, and the accumulation of highly radioactive waste has grown to 45,000 tons" ("Yucca Mountain" 2002). We have been paying a tax for this disposal for years (added onto our electric bills). Americans have been taxed $14.1 billion for this nuclear dump—and the government has earned an additional $6 billion in interest ("$20 billion sits in nuclear dump" 2002). "Utilities have been collecting the fee—about $0.1 cent for every 10 kWh of nuclear-based electricity used—since 1982." The dump could begin taking waste as soon as 2010 ("$20 billion sits in nuclear dump" 2002).

Most of these bulky low- and high-level radioactive wastes would have to be reduced in volume by careful incineration before they could be used in an EBIC such as suggested here in Figure 18.20.

N. Animal Feedlot–Plant Food Complexes

Large-scale livestock operations have removed animals from pasturage and now handle large numbers in small confinement areas (feedlots) where feed and water are brought to the livestock (Nemerow and Dasgupta 1991a). Loehr (1967) reported that treatment and disposal of animal wastes collected in the feedlots are complicated by the nature of the wastes, the volume of wastes to be handled, the lack of interest by the livestock producer in waste treatment, and the proximity of a suburban population. These wastes are high in organic solids—both suspended and dissolved BOD, and nutrients such as ammonia. Land disposal and anaerobic lagoons have been used to treat the wastes. As a result, production costs have been increased to include waste treatment. It makes sense to eliminate these treatment costs by substituting ancillary industries to use the wastes.

Figure 18.21 shows a potential combination of plants to handle all feedlot wastes and, at the same time, produce other useful products. In this complex, the animal feedlot is the major plant and the hyacinth food plant and digester are ancillary plants. The feedlot produces mature animals ready for sale in the open market, and hyacinth production makes plant food, partially for reuse as animal food and partially for feed for the fermenter. The fermenter makes methane (CH_4) for outside sale or for heating any of the complex buildings. A secondary hyacinth basin recovers settled sludge from the first basin and makes more hyacinth plants for similar usage.

All feedlot wastes are used in the complex to grow hyacinth plants in shallow-growing basins in natural or artificial sunlight. Hyacinths are crushed and returned to

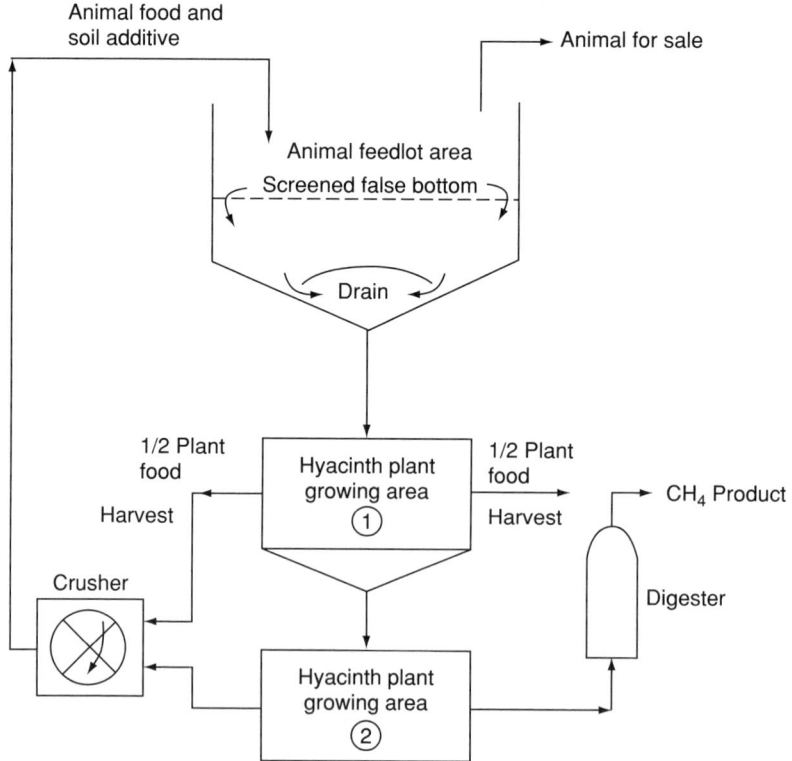

FIGURE 18.21. Feedlot–food production complex.

the feedlot area as food for animals. The fermenter produces methane gas for digesting one-half of the hyacinth plants from the first growing area. The methane becomes the third product made in the complex.

A typical application of this complex could be used for the recovery of hog wastes from hog farms, which generate great quantities of manure. Inefficient disposal by lagooning of these wastes leads to all kinds of adverse environmental effects including seepage into the groundwater. Such a situation was reported in *The Wall Street Journal* (Kilman 2001b). By using a three-product industrial complex, we have provided animals with a feedlot area for a given period without creating any adverse outside environmental effects.

O. Coke (Steel Mill)–Tar–Benzol Plants Complex

At the turn of the twentieth century–and before–coal was burned in what was known as byproduct coke ovens to produce gas fuel and sometimes high-grade tar. The gas was sold and used to provide lamp illumination and to heat local residences and factories. Today this gas has been replaced largely by natural gas piping and transportation systems. However, in some places on the world, byproduct coke ovens are still used for a variety of products.

Where these ovens are still used for gas production in individual locations, many detrimental wastes also are evolved. These wastes include coke and coke breeze, tar, gas liquors and sludges, benzol, ammonia, and various quench liquors, cooling water, and scrubber chemical contaminants.

Present-day coking is carried out by a series of manufacturing processes: (1) coal is brought into the plant, crushed, and screened; (2) this screened coal is then charged into a hot, empty oven; (3) the process of pyrolysis converts the coal into coke and gases; (4) the residual hot coke is discharged from the oven and quenched; (5) condensible gases are cooled and liquified and collected separately; (6) (foul) residual gases are cooled and tar extracted from them; (7) ammonia is separated from the gas with sulfuric acid; (8) further gas cooling can yield other organics such as benzene and toluene; and (9) contaminating hydrogen sulfide can then be removed from the gas leaving it purified sufficiently for sale for lighting and/or heating.

The whole process starting with coking of coal and ending with purified fuel gas involves and includes several small industrial operations. These can be combined in one efficient and effective industrial complex in which all byproducts are turned into valuable products for sale outside the complex or reused within the complex to aid in manufacturing.

The products for sale would include tar, sold for roofing fuel, or road surfacing; gas for illumination and/or heating, coke for steel making, benzol sold to chemical companies, and ammonium sulfate for sale to fertilizer manufacturers. Other wastes and byproducts such as coke breeze, filtered quench water, ammonia cooler water, and gas liquor sludge can be reused within the complex.

One potential complex for this industry is schematically presented in Figure 18.22.

480 Twenty-First Century

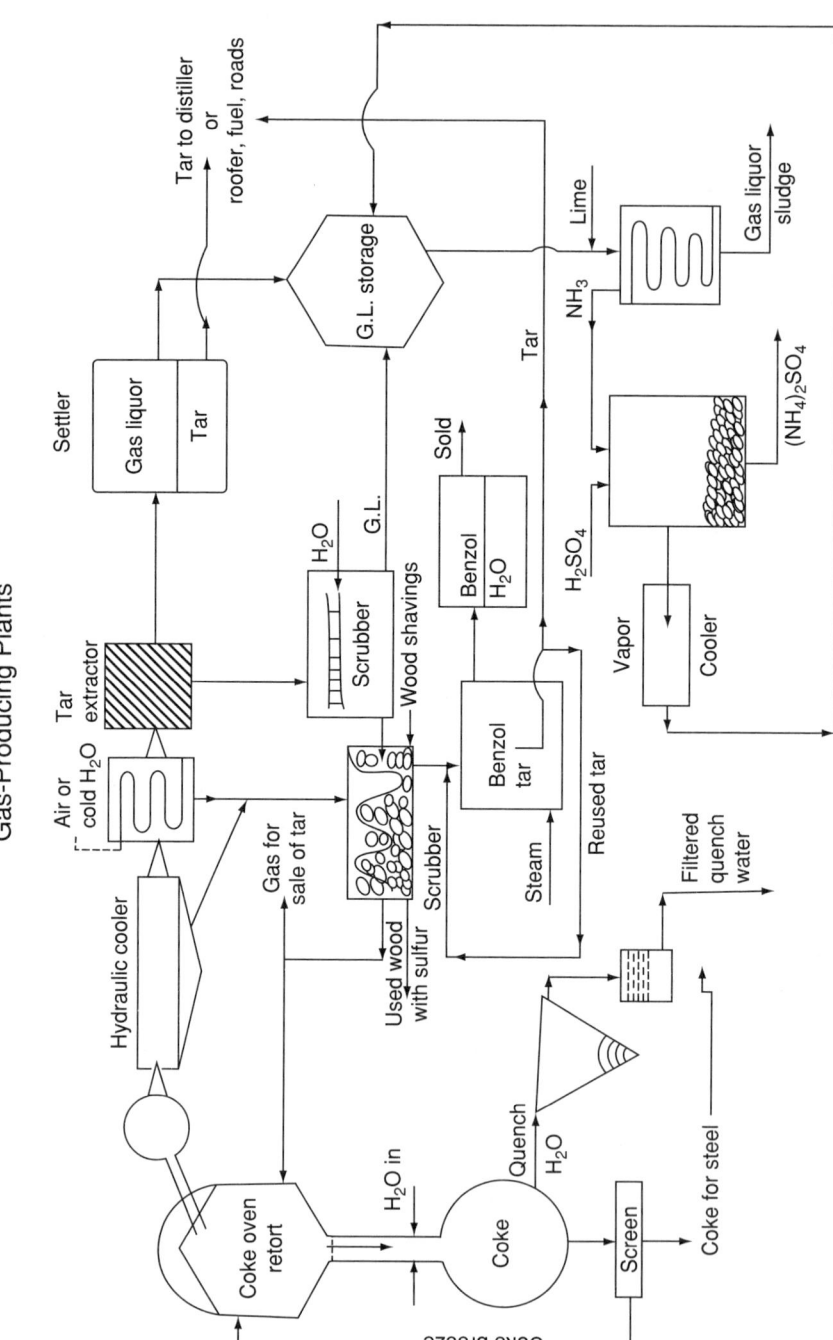

FIGURE 18.22. Gas-producing plants.

P. Wood–Ethanol Complex

It is becoming increasingly evident that we must devise an economical, suitable, and less polluting fuel for automobiles. One potential fuel is ethanol–already used to some degree as an additive to gasoline fuel. Ethanol can be produced by fermentation of many cellulosic organics. Wood waste is an ideal source of the cellulose. Wood cellulose has been shown to yield ethanol effectively by a number of researchers since at least the last quarter of the twentieth century (Shreve 1984).

Wood waste contains cellulose, which can be converted to sugar, which can then be fermented to yield ethanol, a source of fuel for motor vehicles. In Figure 18.23 (adapted from Shreve 1984), I show a flowchart for a process that has been used to make a dilute alcohol.

The cellulose wood waste must first be hydrolyzed. The hydrolyzed vapor is then neutralized with limestone and the resulting gypsum separated before feeding the partially decomposed cellulosic material (mainly hemicelluloses and sugars) to the fermenter.

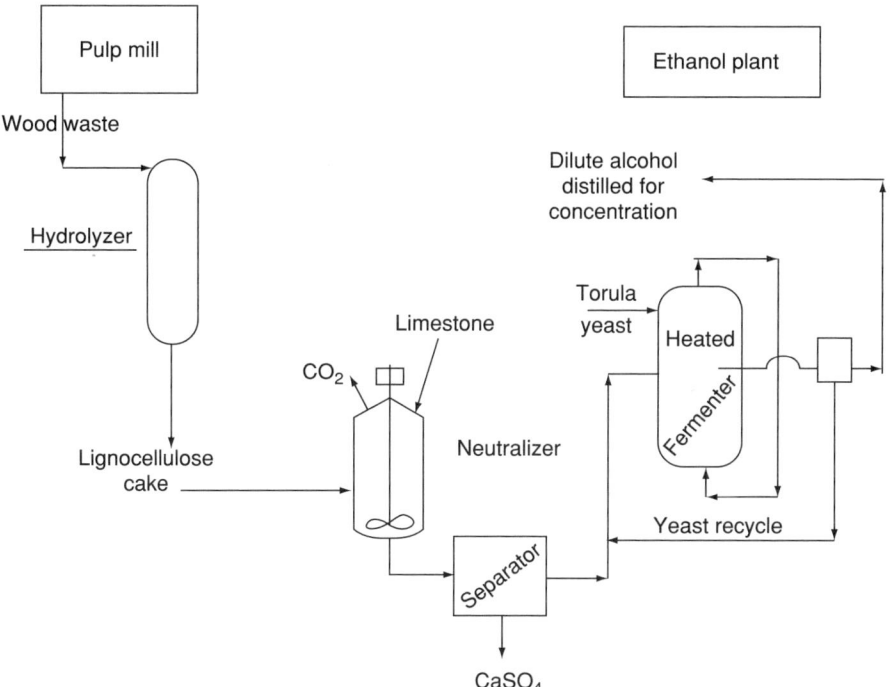

FIGURE 18.23. Flow chart of dilute alcohol production.

In the fermenter, the sugar-like residual of the cellulosic waste is converted—with the aid of a starter yeast culture—to a dilute alcohol product. The latter is distilled to concentrate it for sale as a fuel.

Pulp mills using hardwoods to make paper by the sulfite process evolve waste liquors containing up to 3% sugar. These waste liquors are often steam-stripped to remove their sulfites after which diammonium acid phosphate is added to enhance the growth of yeast, pH adjusted, and the mixture added to continuous fermenters inoculated with yeast cultures. Aerobic fermentation follow at about 35°C, after which the torula yeast obtained contains about 47% protein along with high vitamin content. The yeast product is used as animal feed supplement quite comparable to meat and milk. This offers an alternative to ethanol production.

Q. Water, Electricity, Chlorine, and Lye Plant Complex

Besides food and habitat, water and electricity are vital commodities. This complex produces both commodities as well as chlorine and sodium hydroxide for chemical suppliers (Figure 18.24). Because more than one-third of the U.S. population lives within a few miles of the oceans, industrial plants, which use ocean raw materials and serve the near-ocean inhabitants, ideally should be located there as well.

Ocean waters contain about 3% salt (30,000 ppm) and are full of seaweed or kelp, especially near the shoreline and even on the beach. Today freshwater can be produced from seawater more effectively and economically than ever before. The increased population of the coastal areas is hungry for this water to support life. The seaweed and kelp can be dried and burned and the heat used to boil water to produce steam to drive a turbine tied to a generator, which will result in electric energy for this same coastal population. Some will also be reused within the complex to electrolyze the brine and to provide pressure to the reverse-osmosis system. More burnable organic matter in the form of algae can be grown in ponds through which the boiler flue gas is sent. The carbon dioxide in the flue gas would stimulate the algae growth, aided by natural sunlight usually abundant in seacoast areas.

Newer and better designed reverse-osmosis seawater treatment systems are now available for use. More durable and effective membranes, which can withstand up to 1,000 psi pressures, will remove larger molecules from the seawater and permeate pure water for coastal societal use. The impurities, largely NaCl molecules, build up in the rejected membrane material and become a brine solution, which is very high in NaCl concentration. Usually this brine is discharged back to the sea (LaRue 2001). However, in our complex this brine would serve as a raw material to manufacture both chlorine and sodium hydroxide. This would be accomplished by electrolysis of the brine to liberate chlorine at one pole and sodium ions at the other. Chlorine would be reused at the water plant in the complex to disinfect the water while sodium dissolved in water would be sold to chemical companies to manufacture products such as soaps.

Here we have a four-product industrial complex with little or no wastes entering the environment outside its confines. Two of these products are vital for human survival, whereas the other two provide necessary and useful chemicals with a real monetary value.

Water, Electricity, Chlorine, and Lye Plants Complex

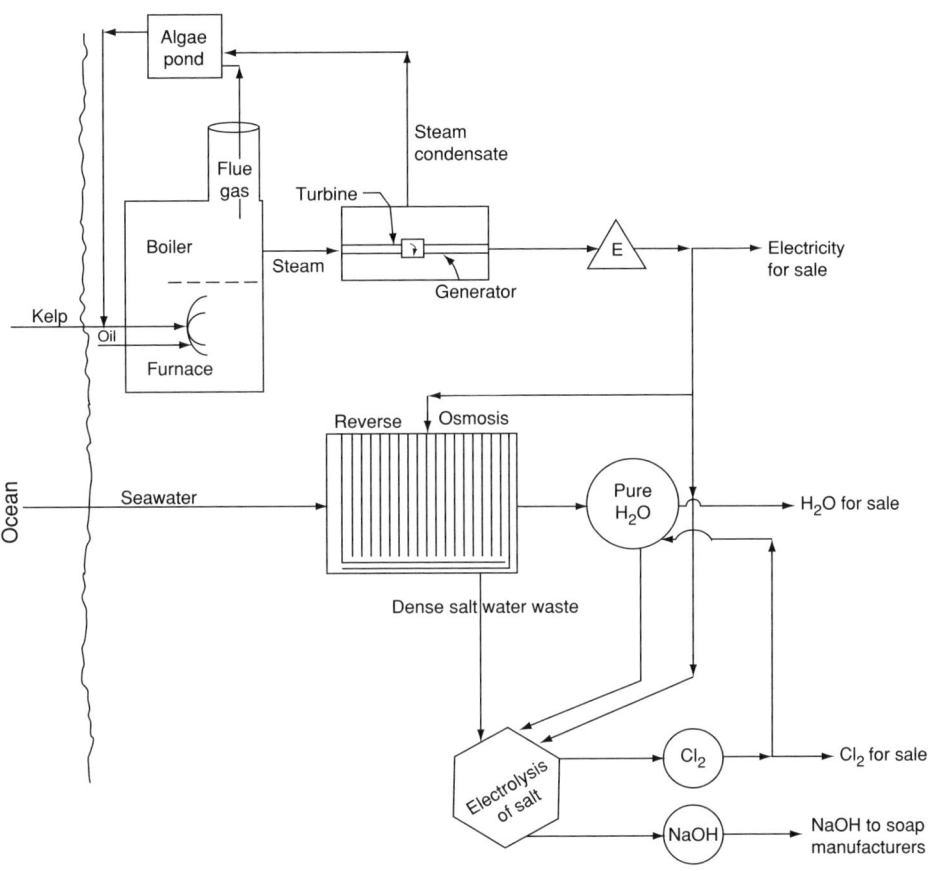

FIGURE 18.24. Wood, electricity, chlorine, and lye plants complex.

R. Aluminum, Electricity, Red Brick Plant Complex

The production of aluminum requires a great quantity of power, 51.4 kWh/ton of aluminum (Nemerow 1984). This electrical demand represents about 70% of the total aluminum production cost. These plants are usually located, therefore, in places with inexpensive power costs. In the past, this meant near or as a part of a hydroelectric power facility, the lowest-cost source of electrical power. Because most of these sites are already taken, or the competitive demand for elevated water stored in the reservoir for power is so great, other sources and types of low-cost power must be found for new aluminum plants. Such is the case proposed in Figure 18.25, in which Canadian power is imported from long distances through undersea electric cables to the refinery complex. Electricity is generated in Canada by burning local natural gas and the

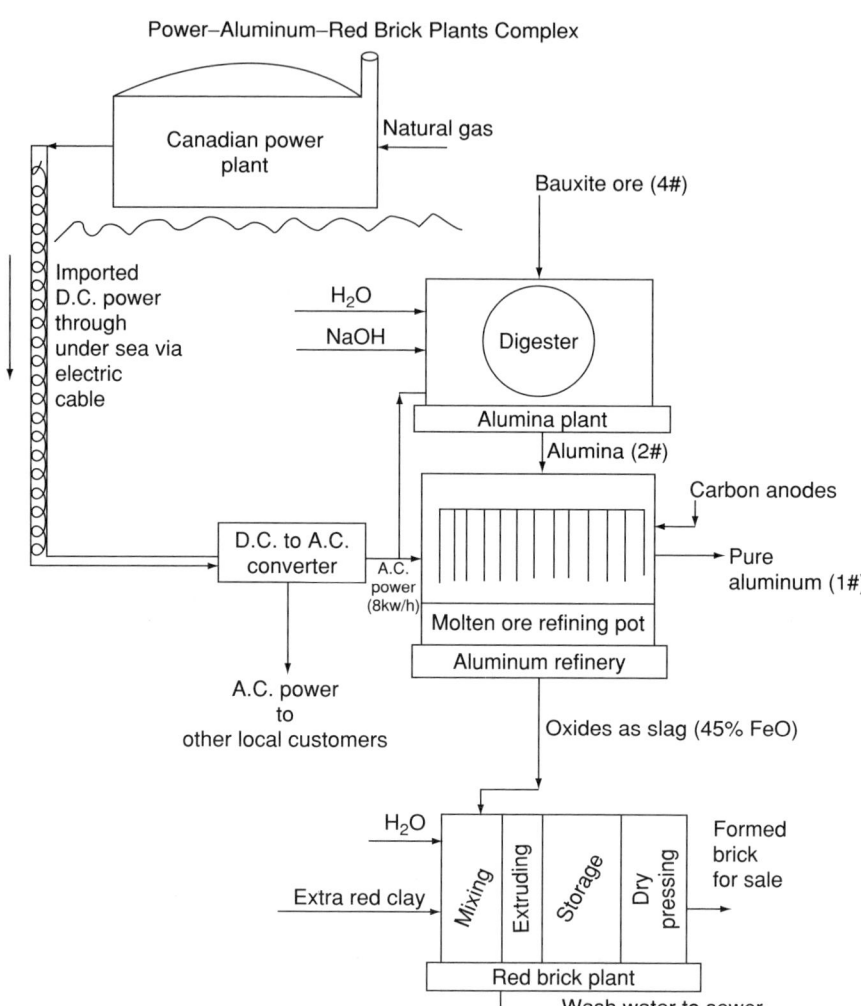

FIGURE 18.25. Power–aluminium–red brick plants complex.

manufactured DC electricity transmitted through deep-water cables to the refinery (Caffrey 2001). Naturally, a local hydroelectric plant's electricity could also be used (rather than importation of Canadian power) if such were available. If electricity is imported, the Canadian facility would shoulder the environmental burden of treating flue gas and heated water emissions, as well as boiler ashes. Because the capital cost of the cable transmission system is considerable, excess AC power can be sold at the converter site to other local customers.

The importance of electricity in the production of aluminum is evidenced in the closing of Alcoa's Maryland smelter ("High costs weigh" 2006), which reported an average aluminum price in the fourth quarter of 2005 rose 12% to $2,177 per metric ton, largely because of high prices in electricity, which can account for one-third of total production cost.

Figure 18.25 depicts a complex in which power is imported to operate the aluminum plant's refinery. The predominantly iron oxide, red mud sludge is reused as the main raw material for the red brick plant. Some imported red clay is used to supplement the refinery sludge. Pure aluminum is sold in various forms (usually in 5-lb ingots) to outside customers.

The red mud sludge is mixed with a very small amount of water and red clay and extruded into storage bins. From there, the red clay mixture is pressed and formed into bricks for sale to local building contractors.

In the refining of bauxite ore, two sequential steps are necessary: (1) reduction of the bauxite ore to alumina (an oxygen alumina compound) and (2) aluminum production by smelting at the refinery. Capital costs are high (in one case, $200 for mining the bauxite, $750 for alumina production, and $1,630 for alumina smelting, for each ton of aluminum metal produced) (Nemerow and Agardy 1998). These costs are 1979 ones and must be updated for current values; however, the relative proportional power capital costs remain about the same.

In this complex, aluminum is produced at a reasonable manufacturing cost. Wash waters are reused within the plant and waste sludges are sent to the brick plant as raw material. No wastes of any consequence leave the complex to contaminate the environment.

S. Corn Growing and Processing–Alcohol Producing Complex

The manufacture of industrial alcohol by fermentation has always been an important production in the United States. At the beginning of the twenty-first century, it is becoming an even more vital product. This is because federal regulations require cities subjected to air pollution to put oxygen additives in gasoline (Kilman 2001a). Oxygen additives aid in burning gasoline more completely to leave less reduced carbon (such as CO) in the air. California had been using a petroleum byproduct, MTBE, to provide the oxygen. However, after finding excessive MTBE exhausted by autos in groundwater supplies, the State of California will force drivers to put corn-derived ethanol (mixed with gasoline) in their gas tanks. The industry predicted that it would need to expand by about 30% just to produce the ethanol that California required by 2003. Further, the U.S. annual market for ethanol has doubled to 3.5 billion gallons over the last few years. All this means that probably new corn fermentation plants will be built in the near future. An opportunity exists to include such new industrial alcohol plants as part of a more efficient industrial complex. A logical system would include corn growing and processing as a component of the complex. In this way, the agricultural wastes resulting from the corn growing and the contaminated liquid wastes from the alcohol plant can all be reused within the complex. I have depicted such a potential complex in Figure 18.26.

In this system, corn is grown in an agricultural area within the complex. The cornfield is fortified with fermenter slops and waste condensates from the alcohol plant.

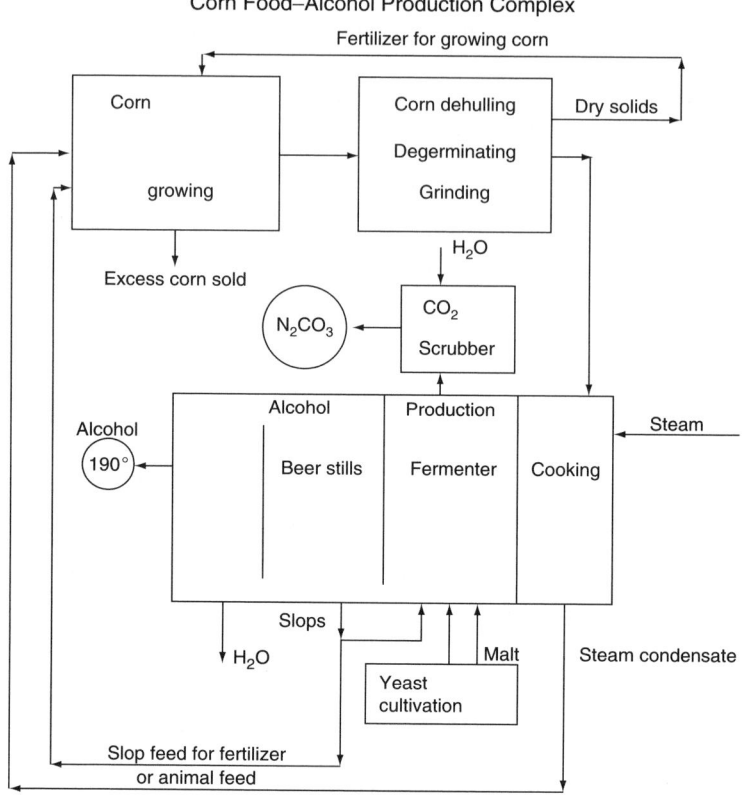

FIGURE 18.26. Corn food–alcohol production complex.

The corn dehulling, degerminating, and grinding of solid wastes are also distributed evenly over the corn field between plantings for soil improvement. The processed corn will be fed to the alcohol plant as its major raw material. Excess corn not required in the production of alcohol will be sold in the open market for human and/or animal consumption.

The ground corn will be hydrolyzed with malt or acid, mixed with yeast cultures, and fermented to a "beer." Further distillation of the "beer" by steam will produce industrial alcohol for sale to the refineries to make "modified gasoline." Slops, contaminated condensates, and wash waters will be fed to the cornfield to increase agricultural product yield.

The two-industry complex will not discharge any contaminants outside the complex.

T. Restaurant–Paint Manufacturing Complex

The fast-food restaurant business is plagued with the dilemma of how to dispose of its used vegetable oils from deep-frying potatoes and meats. The usual practice is to dump waste oils into a grease trap type of treatment. Most of the oil floats to the surface of the grease trap and is usually skimmed off into vats to be picked up by grease scavengers for transportation and sale to processors such as renderers. This practice is not only inefficient, but also costly and prone to operational problems. For example, George Markowvics, an employee of the Department of Environmental Protection (New York City), experiences many of these problems (Newman 2001). "America's sewers are in a bad way. Three quarters are so bunged up that they work at half capacity, causing 40,000 illegal spews a year into open water. Local governments already spend $25 billion a year to keep sewers running." Although many things such as roots, corrosion, and bottles will clog sewers, now "blockages are almost all wrapped in fat" (Newman 2001).

Grease traps, if and when used in restaurants, are often accessible for cleaning and pumping with difficulty. The grease collectors often complain that the grease that they collect, as is, is too watery for reuse by renderers. If grease traps are too small (MGM Hotel in Las Vegas required five 15,000 gallon traps) or are designed or operated inefficiently, grease will overflow into the municipal sewer. In any event, grease recovery from individual grease traps is neither the easiest nor the most desirable task. And the reward for proper operation and collection is usually insufficient to result in a problem-free system.

A solution to the restaurant grease problem is to make the recovery of reusable oils cost effective. One way to do this is to combine all the waste fats of two or three restaurants in one "strip mall" area into one large well-designed and well-operated grease trap and in a complex in which the oil recovered can be reused on-site by an industry as part of its raw material.

One such industrial plant is a paint manufacturer. Usually paint producers require certain amounts of oils as thinners, extenders, and antifoam agents in their raw paint mixture. Even water-based or alkyd paints use some oils. As pure raw materials, these oils add to the cost of the finished paint product. Potentially, grease-trap oil could be used for this purpose after some refinement (probably acid treatment and filtration).

In Figure 18.27, one potential complex of restaurants and paint manufacturing is proposed. In this complex, three fast-food restaurants (perhaps one Chinese, one ribs, and one chicken special) collect their waste cooking oils in a single large easily accessible central grease trap. The trap is maintained by an EBIC employee to ensure that the floated oil is piped continually to the paint manufacturer. At the same time, the grease trap underflow is led directly to the municipal sewer for normal treatment by the city. The recovered oil is acidified, filtered, and added to the paint mix of pigments, resins, and so on.

Various paint plant cleanup wastes are collected, stored, and reused in the proper amounts and types of ensuing mixes.

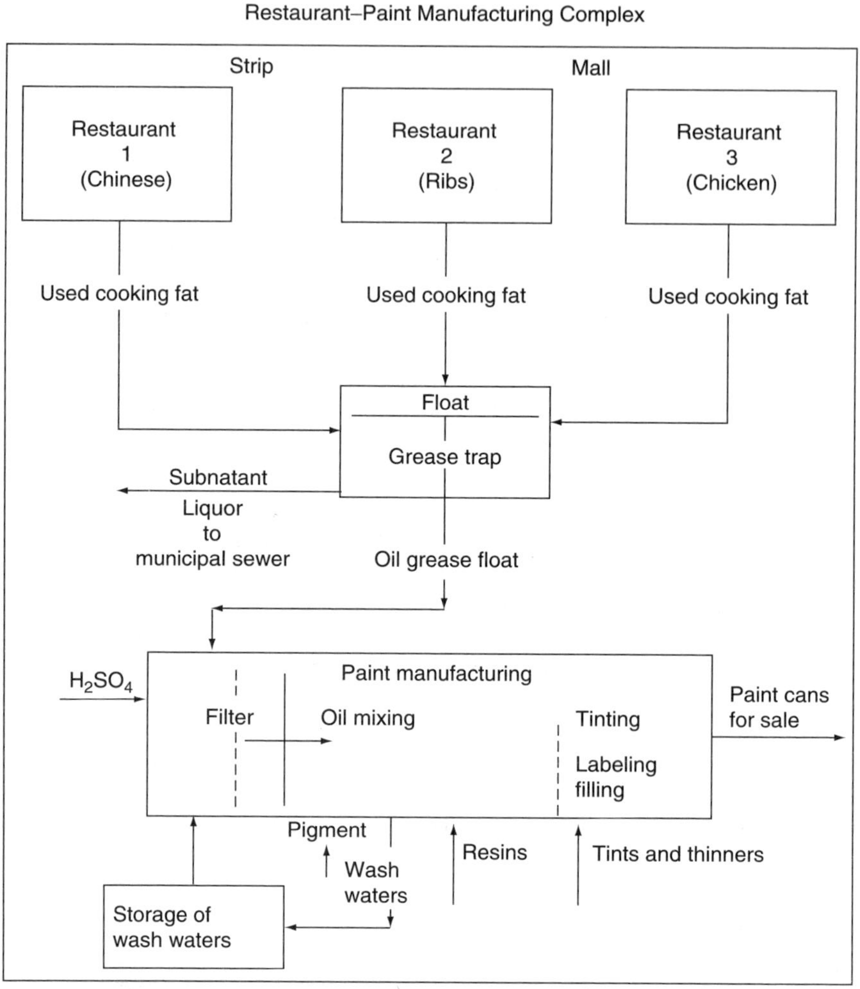

Restaurant–Paint Manufacturing Complex

FIGURE 18.27. Restaurant–paint manufacturing complex.

In this way, neither restaurant nor paint liquid wastes reach the environment outside the complex. Instead they are all reused within the complex to avoid contamination and reduce costs of products for sale.

U. Offshore Oil Drilling –Seashore Recreation Industrial Complex

This complex offers a unique opportunity to unite the wastes of one offshore oil-drilling industry with the needs and uses of another nearby ocean-using industry, ocean beach recreation. It is unusual because the two industries do not have to be relocated; they already exist together. They operate by using the same offshore ocean water and will both be enhanced by an even closer symbiotic relationship proposed in this complex.

The offshore oil-drilling industry produces not only the dangers of oil spills, toxicity, and explosions, but also the final disposal of its oil-drilling platform and drill-cutting piles following its useful life. Oil drillers use a sometimes modified drilling fluid—usually water—to penetrate the depths of the well to drive the oil out to the surface for recovery and sale. The drilling water is often contaminated to some degree and represents a problem of proper disposal. In addition, oil spills, though infrequent and unpredictable at the onset of operation, also represent environmental concern.

Beach recreation often supports an offshore fishing operation that can be improved by any techniques that result in increased fish life. The beach sand itself is usually eroded annually by excessive storms causing higher peak tides and winds. Any practice that diminishes these wind and tidal effects will protect the beach sand, reducing the inconvenience and the cost of replacement. Ocean beach users require water for various purposes such as washing sand from their bodies and flushing toilets and watering adjacent plants and trees. These uses do not normally require water of pristine purity. In fact, the use of potable water for these purposes has long been claimed as both a waste of a valuable and diminishing resource, and an unnecessary expense for the user.

The suggestion of leaving the oil drill-cutting piles in place was offered by the Phillips Petroleum Company of Norway in 1999 (Alastair 2000). There "is no proven technology that could remove large amounts of heterogeneous sediments from the deep water of the northern North Sea." The inference, then, is that despite its potential toxicity to marine life for up to 20 years, it is best left *in situ* after finishing the drilling.

Another offshore drilling problem occurs in the canals used by the oil industry to transport oil (and wastes), which may contaminate wetlands with saltwater. Pipelines could add to the problem (Pflum 2000). As a result, any practice such as proposed here should eliminate this problem.

Herrick (2001) reports on platforms and rigs by admitting that when the work (of the oil drillers) is done, "an enormous mass of metal remains to be cleaned up." He confirms that federal lease agreements require that the oil drillers leave the ocean floor as they found it. But he also confirms that "tearing down the multi-million dollar platforms and hauling them to shore to sell for scrap can cost as much as tens of millions of dollars." He reports as well that "the leading alternative is converting rigs into artificial reefs." He found that first coral and colorful sponges cling to the underwater jacket, then shrimp and crab follow, and soon fish appear.

Vickie Chachere (2001) points out the current practices and objections to offshore drilling. She also reports on modern more environmentally safe types of equipment such as "the sleek and modern, $1.2 billion structure 130 miles southeast of New Orleans—twice as tall as the Sears Tower and known in the drilling business as a

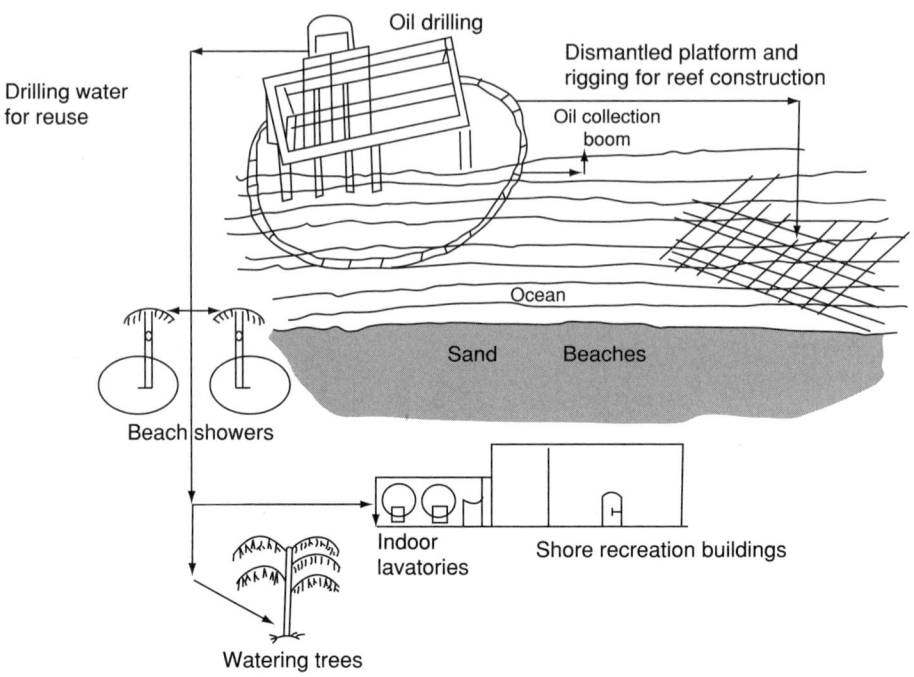

FIGURE 18.28. Oil drilling offshore–seashore recreation complex.

tension-leg platform. It can produce up to 200,000 barrels of oil a day or 2 percent of current domestic consumption."

In our proposed complex, waste oil-drilling waters will be pumped to shore to be used by the shore-side recreation facility for sand washing, toilet flushing, and plant watering. The used platform and rig will be dismantled and used near shore to form reefs for fish propagation and sand erosion protection. Oil leaks will be minimized by modern oil-drilling operations and, if and when they occur, will be collected by booms and recovered and reused for heating purposes onshore. All these are depicted in Figure 18.28.

V. Metals Plants–Dry-Cleaning Coffee-Decaffeination–Plants Complex

Trichloroethylene (TCE, $CL-C=CH$ with Cl substituents) is another very toxic and hazardous industrial waste that "begs" for safe and economical recovery and reuse. Fortunately, the major industries requiring and using this organic chemical solvent are now well known to the environmental engineer. The metal parts plants use TCE to clean the grease preservative off its incoming metal parts. The dry cleaners use TCE to dissolve and remove the organic clothing contaminants from the garments. Discharge of either of these plants' TCE wastes into the environment has been known to me to cause pollution. It makes good common sense to recover the wastes of dry-cleaning plants for potential reuse in coffee manufacturing. In the latter, TCE is used to remove the caffein from the coffee bean. The slightly contaminated TCE waste from the coffee plant can be reused by the metal parts plant to remove its grease and then by the dry cleaner to remove more grease from clothing.

Eventually excessively contaminated TCE waste must be distilled before reuse. The still bottoms can be landfilled periodically or reused as grease lubricants within the metal parts plant.

Such a three-plant complex is shown in Figure 18.29, in which no hazardous TCE wastes reach the outside environment and a valuable (useful) raw material (TCE) is provided internally.

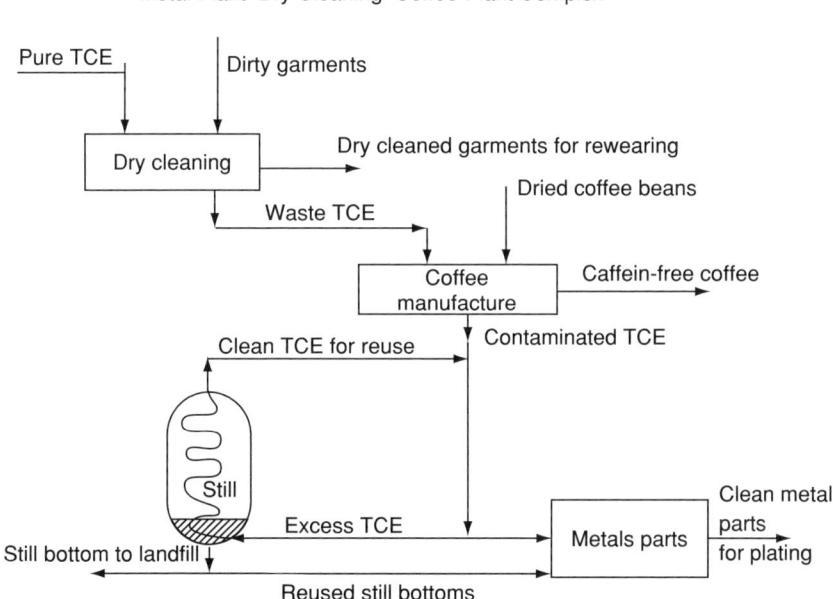

FIGURE 18.29. Metal plant–dry cleaning-coffee plant complex.

W. Electrical Storing and/or Converting Voltage–Wax Manufacturing Complex

Polychlorinated biphenols (PCBs) comprise a unique group of toxic chemicals that can be represented by the following chemical structure:

$$\text{Cl}-\text{C}_6\text{H}_3\text{Cl}-\text{C}_6\text{H}_3\text{Cl}-\text{Cl}$$

They are used in small quantities in the manufacturing of lubricants, duplicating paper, printing inks, paints and coatings, adhesives, plastics, and so on. When paper wastes are reused and de-inked, some small amounts of PCBs are also found in the mills' wastes. Therefore, they may be found in small concentrations in wastes from these product manufacturing plants. However, their concentration is usually too small for economical recovery and reuse, but large enough to cause environmental problems because of their resistance to decomposition and perseverance in water and soil.

In one industry, they are used and discharged in large enough amounts to be hazardous to the environment and to warrant their collection, recovery, and reuse in an industrial complex. Such an industry stores power and converts the power to various voltages (using capacitors and transformers). PCBs are used there as both coolants and lubricants. When they become used to a point of excess contamination or degradation, they are wasted. If wasted to a flowing river such as General Electric did in the Hudson River, they can and did cause decades of problems mainly with bottom-feeding fish and the utilization of these fish for human food.

When attempts are made to dredge river bottoms, problems arise as to what to do with the contaminated sludge, environmental safety, and the huge cost of such procedures. One possibility that appears feasible is to mix the sludge with cement kiln dust, layer it in the atmosphere to neutralize the chlorine, and then safely place the treated sludge in a chemically safe landfill (Sell 1992). However, direct reuse of the PCB waste in an industrial complex would eliminate all these intermediate treatment costs, hazards, and problems with recovery after contaminating the river environment.

Direct PCB reuse in the wax manufacturing industry appears to present the best opportunity for successful environmental protection at the least cost. PCBs are used as "extenders" in the manufacturing of wax, especially carnauba and paraffin types of wax. Product compatibility and mass balances of PCB wastes and wax needs will play important roles in the acceptance of this system. Such a complex is shown schematically

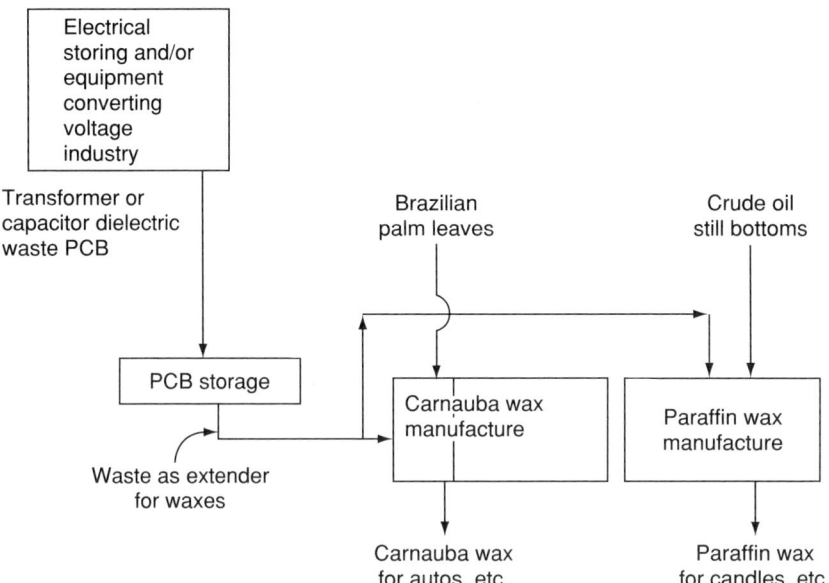

FIGURE 18.30. Electrical storing/converting voltage–wax plant complex.

in Figure 18.30. The figure is self-explanatory and only requires the import of Brazilian palm leaves and/or crude oil still bottoms. If the PCB wastes can be stored and fed to the wax plants properly, no hazardous PCB wastes should enter the outside environment.

X. Nuclear Power Plant Waste Processing–Cannery Complex

Another potential industrial complex arises out of the need to protect not only our normal environment, but also in emergency situations. Simply stated, nuclear power plants produce a hazardous waste from the reprocessing of spent uranium fuel rods. In usual practice, these wastes are stored (before or after dangerous transportation in "safe" facilities, hopefully until after many generations they someday can be released safely into the environment). These operations can and do result in wastes that pollute the environment.

In a similar mode, but in an entirely different industry, when we process fresh fruits and vegetables, the produce must be preserved and protected from bacterial contamination until it is ingested by the consumer. Once again, in usual practice, this is accomplished by washing, heating, and/or steam sterilization. These operations, likewise, result in wastes that pollute the environment.

Although we would not likely think of the "marriage" of these two dissimilar industries in one complex, I propose we consider co-locating them together. The primary purpose of this proposed merger is to protect the environment from radiation and microbiological-type wastes. In these days when the heretofore unthinkable "intentional" contamination is possible, both nuclear and bacterial contamination may be prevented by industrial complexing.

At this juncture, I suggest that you refer to the schematic layout shown in Figure 18.31. Here, we are merging, in reality, three separate industrial operations in one complex: nuclear power production, nuclear waste reprocessing, and cannery production.

The reader should be familiar with the basic dangers of ionizing radiation, which can be released from these wastes if they are left unshielded from the outside environment. Ionizing radiation is simply radiation with sufficient energy to remove electrons from atoms. It can cause changes in the chemical balance of cells that may result in cancer and/or harmful genetic mutations, which can be passed on to future generations. Current regulations permit a maximum of 5,000 millirems of this radiation annually.

Storage of nuclear wastes on-site in water presents certain environmental hazards. This is one reason for treating and reusing the rods and wastes as soon as possible. If a breach of a waste pool occurs from an earthquake or sabotage, the leak would release radiation, as well as exposing the stored rods or pellets to the air and overheating, leading to further release of radiation. Without water to cover the fuel elements, a possible fire or meltdown can occur.

Nuclear power, which represents about one-fifth of all power produced in the United States, is obtained from the heat energy released from uranium rods. As a result of this operation, these rods when eventually "spent" must be reprocessed, mainly by acid cleaning. The wastes contain small amounts of radioactive elements dissolved in strong acids. If these wastes can be concentrated and separated into their two major components, acids and radioactive elements, the radioactive component could be made into an "enclosed beam radiator," while the acids can be reused in further reprocessing operations.

Such a radiation beam is already a reality and manufactured by Surebeam in San Diego, California (Calbreath 2001), and used in Hawaii to irradiate papayas before shipment around the world. Similar devices are proposed for and may already be in use in U.S. Post Office processing buildings to prevent anthrax contamination of the mail. The radiation source produces gamma rays, electrons, or x-rays, which kill disease-causing bacteria. This treatment, however, is not without some drawbacks cited by its critics (Hauter 2001). These critics claim that "we oppose the use of ionizing radiation to 'treat' food because this process destroys vitamins and other nutrients, and can form chemicals known or suspected to cause cancer and birth defects." When radiation beam treatment was proposed for sterilizing mail in post offices, it was pointed out that it ruins unprocessed and unused film and damages some electronics as well. This should not be a problem in this example, however.

As shown in Figure 18.31, the spent fuel rod wastes are concentrated and acids separated from the uranium component, the latter being formulated and housed in a storage vessel to be used by the cannery to irradiate fresh produce. The waste acid portion is recirculated and reused in cleaning the next batch of spent fuel rods.

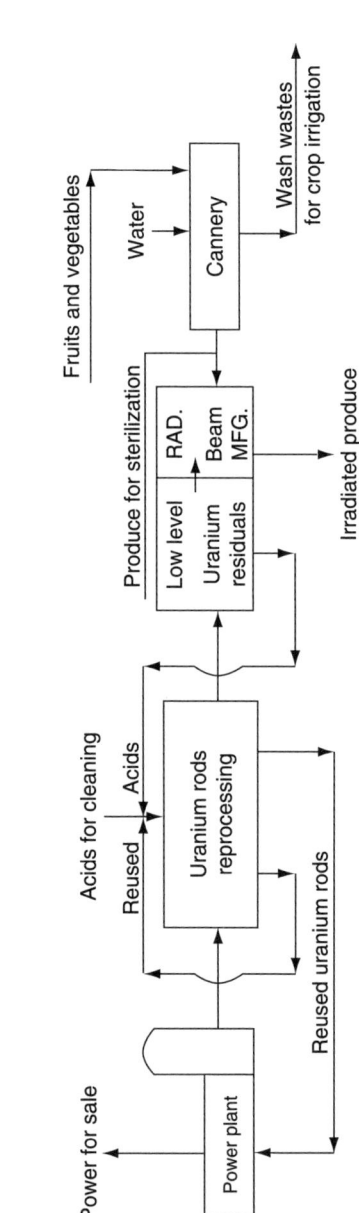

FIGURE 18.31. Nuclear power–waste recovery–cannery complex.

The only wastes emanating from the cannery are the wash waters from cleaning the produce. These wash waters containing mainly soil, leaves, and ground detritus are also recirculated and reused in irrigating the adjoining fruits and vegetables growing in the agricultural fields nearby. This is vital because water scarcity has a direct effect on food production. About two-thirds of all the freshwater consumed each year is used to irrigate crops.

As usual, with these examples of balanced industrial environmental complexes, no wastes leave the complex. Instead, radiation and suspended solid wastes are reused within the complex in making ancillary products for sale. The products for sale are nuclear power, radiation beams, and fruit and vegetable produce.

Y. Electric Power–Drinking Water Plant Complex

More and more situations are arising in which people in or near an expanding metropolitan area require both additional electrical energy and water. If we examine these two industries—electrical energy and drinking water—they appear to possess some interdependency.

The power plant needs cooling water for its steam condensers and clean water for steam generation. The water desalination plant produces both the excess clean water for the power plant and salty cooling water for condensing the steam. The water plant is able to send most of its pure water to people for drinking water and the rest to the power plant for steam production. Its saltwater residue (i.e., rejected impure water) because of its low (seawater) temperature is beneficial to the power plant for condensing the turbine steam, which can then be reused as a source of clean water.

The desalination plant will have no unusable wastes when made a part of the two-plant complex shown in Figure 18.32. The electrical energy plant will have no liquid wastes if its cooling waters are stored to lower its temperature and then reused. Also, if its fly-ash is collected by electrostatic precipitation and its flue gases passed through sodium hydroxide (to make a soda ash product) before discharging to the air, no air contaminants will arise. The soda ash is then used in the desalination plant to soften the drinking water before municipal use.

Two current examples of this potential complex have been made public. In 2002, Marathon Oil Company (Lindquist 2002). Announced that it would be the "lead partner in an energy complex near Tijuana (Mexico) that would include a liquefied natural gas regasification plant and an electrical power plant." Plans for Marathon's LNG facility are reported to include a marine terminal to accommodate tankers transporting the LNG from worldwide locations, an off-loading terminal, and an onshore plant that can re-gasify 750 mcf of LNG daily and a pipeline to deliver the gas to customers. The company proposes to cool the power plant with Tijuana wastewater and may also build and use a desalination plant to use Pacific Ocean water to provide 20 mgd of additional cooling water for the electric turbines.

The second (Roletti 2002) is planned by the City of Carlsbad, California, which is studying the feasibility of building a seawater desalination plant next to the Encino power plant. The city would like to "extract salt from ocean water and convert the water

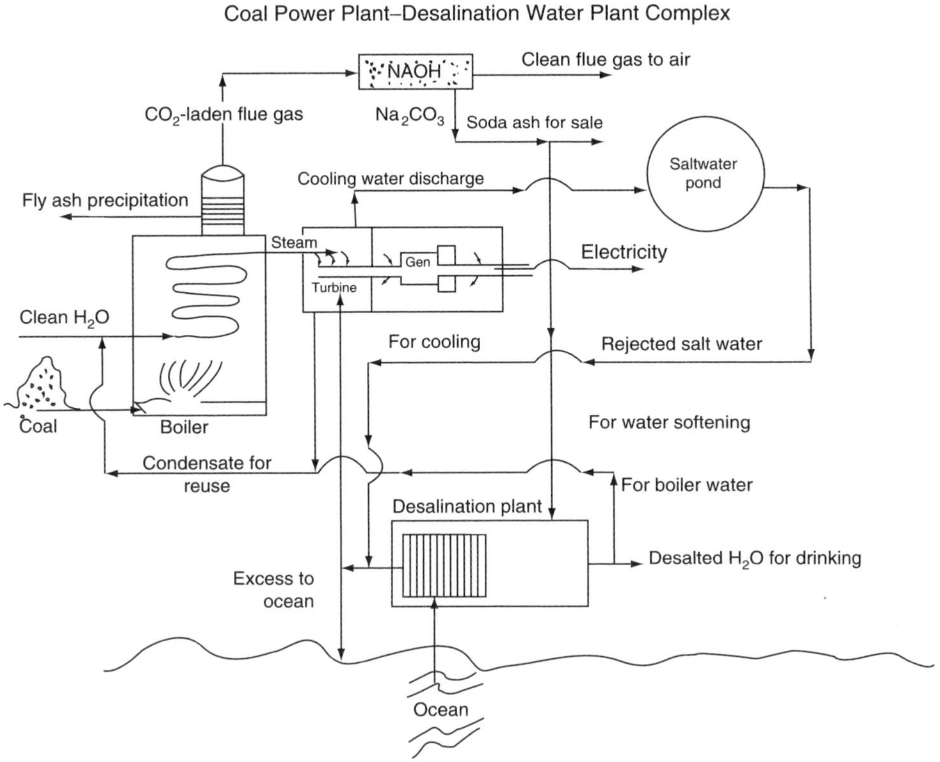

Coal Power Plant–Desalination Water Plant Complex

FIGURE 18.32. Coal power plant–desalination water plant complex.

into 50 mgd of drinking water. Building the plant next to the power plant is considered feasible because the two operations could share common infrastructure." Presumably they could share common ocean intakes of fresh seawater and outfalls of cooling water and salty reject waters.

Although each of these examples proposes somewhat different systems, both show interest in complexing plants to share raw materials and wastes.

Z. Vegetable Pickling Cannery–Inorganic Chemical and Chlorine Plant Complex

Sometimes an industry produces a waste that does not easily lend itself to treatment and "begs" for reuse by the proper industry partner. Such a situation exists in this example where brine waste is the nemesis and an inorganic chemical plant serves as the "willing and able" partner.

The vegetable pickling (brine) wastewater usually is only one of several wastes from this type of cannery (such as a pickle plant). Other wastes would include lime water, alum and tumeric wastes, and syrupy (vinegar and sugar) wastes. These latter three wastes could be impounded and reused to exhaustion before being monitored for proper pH and biologically treated or reused by ancillary industries. However, it is the brine waste that defies reuse more than once or twice before it must be discharged. This waste is mainly salt with some dirt and preliminary fermentation products.

Because of its sodium and chloride content, the brine waste could be used as a raw material for an inorganic chemical and chlorine manufacturer. By using a mercury cell, continuously fed brine can be partly ionized in one compartment (electrolyzed) between a graphite anode and a moving mercury cathode. Chlorine gas liberated at the anode can be sold directly as the gas or converted into bleaching powder by reacting it with the lime water from the cannery.

The sodium (amalgam) hydroxide can be passed through a bed of limestone to produce a Na_2CO_3 (soda ash) product. This product can be sold to the local water plant to soften its drinking water and/or sold to chemical companies for a variety of uses.

This type of complex rids the vegetable pickling plant of one of its most troublesome wastes while creating a new ancillary industry making chlorine or its compounds, as well as soda ash. In Figure 18.33, the cannery and chemical plants are depicted as "partners" in this industrial complex. Presumably other, primarily organic, wastes

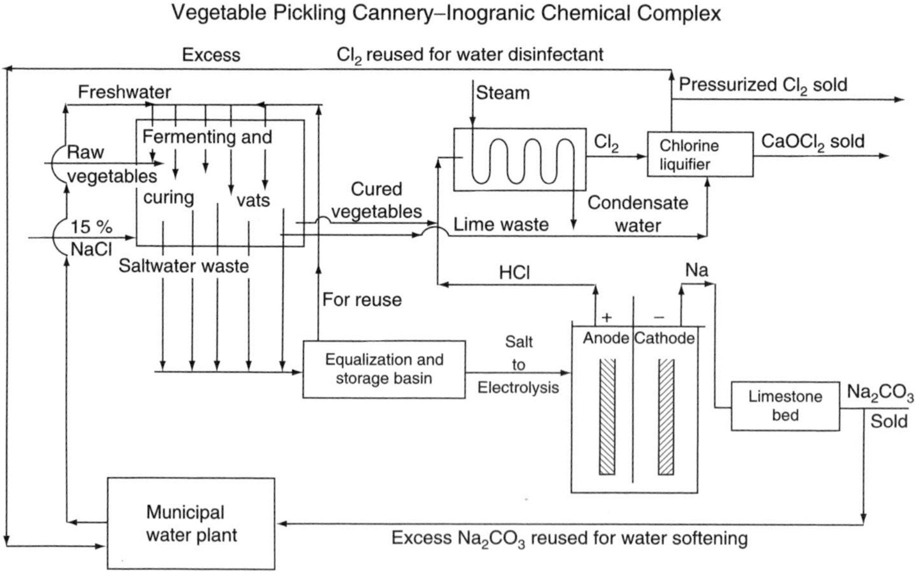

FIGURE 18.33. Vegetable pickling cannery–inorganic chemical complex.

produced by the cannery are treated biologically and/or reused in the cannery or in another ancillary industry. On the other hand, the "difficult" brine waste is equalized, stored, and then sent to the adjacent chemical plant as a raw material. Here, it is essentially electrolyzed to produce chlorine gas at the anode and sodium (amalgam) hydroxide at the cathode. The former is pressurized to a liquid and sold to the water purification plant or to other chemical companies. Or it is reacted with the lime vat water from the cannery to make bleach powder, also for sale to chemical companies. The sodium hydroxide is passed through a limestone bed to produce, on a batch basis, soda ash for sale to the local water plant or other chemical plants.

In this way, a "nasty" waste that may be detrimental to receiving waters if discharged has been reused by an industrial complex partner (the chemical plant). In addition, two or more new products result from the partnership, adding to the economic advantage of the overall combined operation.

AA. Sugar–Ethanol–Gasoline Plants Complex

While we still use gasoline to power most of today's automobiles, we need to reduce our dependency on foreign oil to produce it. One way of doing this is to dilute the gasoline with an additive such as ethanol, which is reputed to have no adverse effects on gas burning or power efficiencies.

The cost of this production can be reduced by manufacturing the ethanol at a site adjacent to the oil refinery. To accomplish this, we can grow an agricultural crop such as sugarcane in the environmental complex, convert it to sugar, ferment it to alcohol, and finally feed it to the oil refinery for dilution of the gasoline.

Traditionally, most of the ethanol produced in the United States is manufactured by fermenting corn in mid-America. The ethanol must then be shipped to the various oil refineries throughout the country for dilution with the gasoline. The EBIC avoids this transportation step, cost and potential danger. Our proposed complex also allows for the use of other crops indigenous to the local area as an ethanol source.

In a typical complex as shown in Figure 18.34, sugarcane is grown in a field within the complex and adjacent to the oil refinery. The cane is harvested and transferred to the sugar refinery within the complex. The sugar product is delivered for sale or sent to the fermenter for conversion by anaerobiosis to ethanol. The sugarcane bagasse is burned in the refinery's boilers for steam heat and/or power. The cachaza from the sugarcane plant is fed to the fermenter as an added source of ethanol. The ethanol is stored for periodic transfer to the adjacent oil refinery where it is diluted (usually 90-10) with refined gasoline for direct sale to the gas stations. The typical distillation gas wastes from the oil refinery are passed through lime filters before being released to the air environment. Periodically, lime (now primary calcium sulfate) filters are removed and distributed on the sugarcane-growing field for extra fertilizer.

In this way, no wastes leave the complex from the sugarcane field, the sugar refinery, fermenter, or oil refinery to contaminate the surrounding environment. California is already considering converting part of its agricultural land to sugarcane for the specific purpose of producing ethanol ("A cash crop?" 2002). "Dozens of

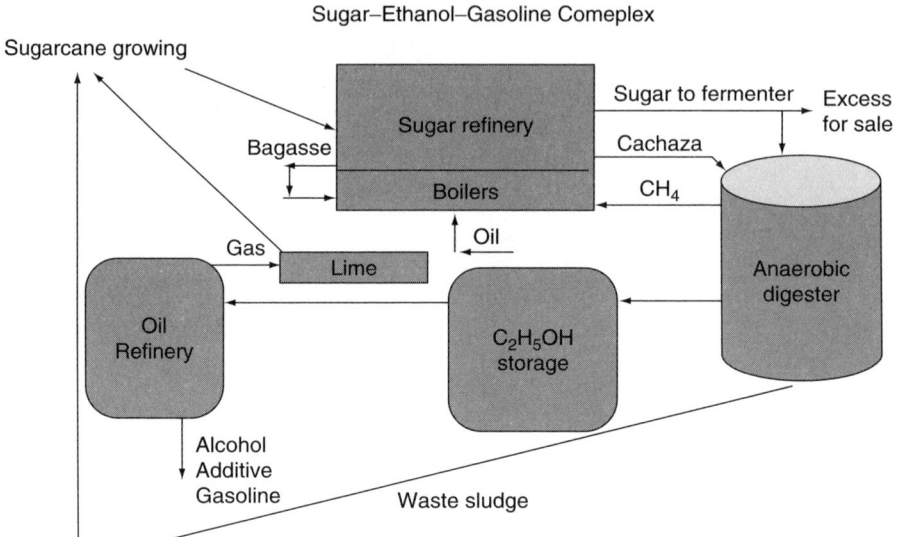

FIGURE 18.34. Sugar–ethanol–gasoline complex.

Imperial Valley alfalfa and cotton farmers are mulling over whether to grow the unlikely tropical grass for conversion into the gasoline additive ethanol." The rationale for this is the "considerable transportation costs" of importing ethanol from the midwest (corn) belt. Previously used methyl tertiary-butyl ether (MTBE) as an additive is being discontinued because "it has been found to contaminate waterways" and is "a suspected carcinogen that is believed to have contaminated more than 10,000 sites in the state."

In a further follow-up ("Ethanol use is headed" 2002), BP, the largest gasoline supplier in California acknowledged that the "federal clean air laws require use of either ethanol or MTBE, as an oxygen-enhancing agent to help gasoline burn cleaner, in a third of the United States gasoline supply" and "there are strong environmental and economic reasons for refiners to go ahead and make that move" (to ethanol).

BB. Reclaimed Cell Phones–Cement Plant–Concrete Products Complex

As a consequence of our "high tech," we are faced with the disposal of used and outdated electronic equipment such as wireless cell phones. These types of materials build up in the environment at a fast rate because of their rapid outdating, as contrasted to computer or VCR-type electronics. They usually cannot be refurbished, sold, and/or reused because of their relatively low production cost that renders them disposable.

If these small cellular phones are wasted into refuse landfills, they will contaminate the ground and groundwater with the heavy toxic metals contained in them. Once again, the challenge to the industrial waste engineer is to recover and reuse the basic ingredients in an economical and practical manner. The basic ingredients are plastics and heavy metals. The suggested EBIC is shown in Figure 18.35.

In this complex, the reclaimed cellular phones are heated in an oil-fired furnace to 275°F, at which point the plastics are melted and separated from the residual metal parts retained on a fine screen.

A cement plant, located within the industrial complex, uses the usual sand, shale, and limestone as raw materials, fed to a kiln (heated by oil), and ground to cement. A portion of the cement is sold on the open market, but most is used to feed the adjacent concrete products plant. The latter also receives recovered cement dust as a supplemental material. The heat from the cement stack gas is circulated through a separate oil heater equipped with hot water tubes.

The concrete products plant is divided into three separated operations. All three import normal cement, stone, and cement dust and heater steam condensate as raw materials. One accepts liquid plastic from the furnace as an added raw material, the second accepts recovered furnace metals as an added raw material, and the third operates normally without other external recovered additives.

The first produces plastic–concrete products for uses such as park benches and so on. The second produces metal–concrete products such as ocean reef materials and so on. The third produces normal concrete products such as pipe and other building materials.

No wastes—liquid, solid, or airborne—leave this complex and four products are made and sold externally to lower production costs.

These wastes have been loosely described as "e-wastes" and have been termed as "the most perplexing refuse issue facing the nation" ("Technology's toxic trash" 2002). "Last year, the state [California] banned televisions and computer screens from landfills, joining only Massachusetts in designating those items as hazardous waste, but diverting e-waste from landfills is only a small step toward solving a burgeoning problem. There's no clear plan—not in California, not anywhere—about what should be done with the abandoned TVs and computers. The discarded computers and TVs are often shipped to developing countries, where the toxic waste leaks into waterways or contaminates landfills." As far as recycling of these wastes is concerned "industry experts say there aren't nearly enough recycling companies to dismantle all the computers and TVs Californians discard, because until recently, there wasn't much demand for the service. Plus, it's expensive because the materials often must be shipped out of state, where companies are doing the recycling."

In an attempt to cover the costs and problems of recycling, the State of California is considering two bills. "One would require manufacturers to pay a fee of up to $30 for every TV or computer they sell, which would be used to subsidize a recycling program. A second proposal requires manufacturers to label all electronics that contain CRTs as

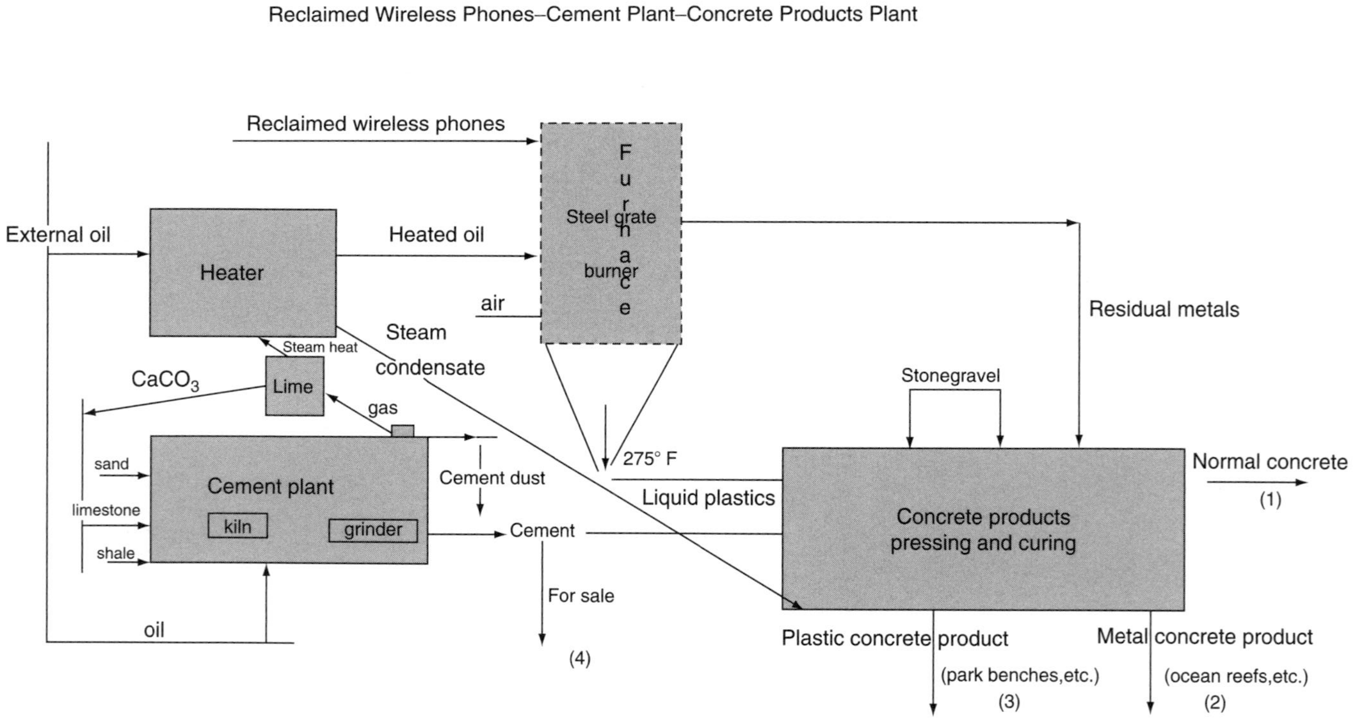

FIGURE 18.35. Reclaimed wireless phones–cement plant–concrete products plant.

hazardous materials, and to include information about how the materials can be recycled when they're obsolete." It's obvious that recycling presents many problems with these wastes, as is the case with other wastes such as glass bottles or aluminum cans. These problems accentuate the potential for using the EBIC method to avoid such situations and costs.

CC. Sugarcane–Fuel Briquet Industrial Complex

The growing and processing of sugarcane results in two major wastes that have an adverse impact on the surrounding environment: bagasse and cachaza (see Chapter 17, Section C, for a detailed description). The sugarcane stalks are chopped into small pieces by rotary knives and the cane juice is extracted from these pieces by crushing them through roller mills. The solid residual material (fibrous) is known as *bagasse* (Nemerow 1995). After the cane juice is extracted from the stalks, the juice is treated with lime in the boiler room. The resulting precipitate is vacuum-filtered to separate the insoluble sugar—the wasted filter cake is termed *cachaza*. Then the clarified juice is further processed by thickeners, with little or no wastage, to produce sugar crystals for sale.

Most of the bagasse is usually burned in the field or in the mill's boilers. This results in local air pollution either directly from the fields or indirectly from the boiler flue gas. The cachaza also represents a solid waste that must be disposed of in some typically costly manner.

If these two troublesome solid wastes can be collected and processed into a useful product, an environmental solution will be attained. The keys here are in the processing, which must be simple and economical, as well as in the usefulness of the product. Such processing is suggested in Figure 18.36.

The bagasse and cachaza are collected mechanically from the cane cutters and filter presses, respectively, and blended together in kettle mixers. The blending and mixing should be facilitated by the moisture content inherent in the filter cake (cachaza). The homogeneous mixture is transported on belt conveyers to the fuel briquette plant. Here, the mixture is pressed and cut into briquettes for ease of air-drying, packaging, and sale for fuel. The utility of the briquettes depends on the moisture content and Btu value of the fuel.

The economics of the processing and the environmental benefits will also be quantified in order to facilitate acceptance of the complexing. If quantification proves positive, this industrial complex will become a reality and the environment will benefit as well.

DD. Hog Production–Animal Feed–Energy Production Environmental Complex

Pig growing and slaughtering compromise a major food industry. Unfortunately, the wastes associated with this production are voluminous and objectionable. They must be disposed of safely and economically even when occurring in rural environments.

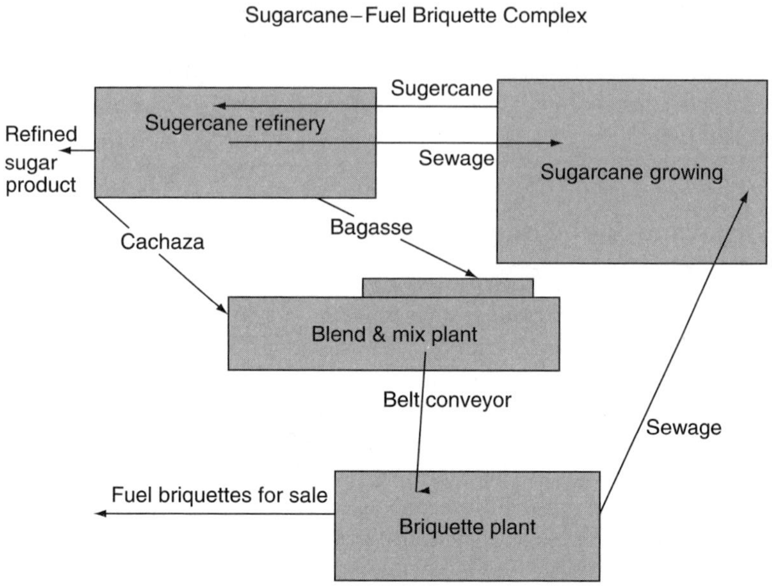

Sugarcane–Fuel Briquette Complex

FIGURE 18.36. Sugarcane–fuel briquette complex.

Piglets are fed, fattened, and grown in pens. These pens are generally elevated above a concrete sluiceway that carries away the urine and detritus from the pigs. The grown hogs are periodically removed, slaughtered, and replaced with new piglets.

Generally, the urine and feces wastes are washed into large adjacent lagoons. Here, the solids settle to the bottom; some wastewater seeps into the underlying ground and eventually the groundwater, and some liquid is evaporated from the lagoon surface and/or sprayed on nearby agricultural land or growing fields.

Problems of environmental pollution occur when dikes of the lagoons break, surface wastes overflow the lagoons, or nearby well waters are contaminated with the pig lagoon wastewaters.

All of these problems can be avoided and societal production costs can be reduced by using an industrial complex system such as that shown in Figure 18.37.

Here, the pig wastewaters and feces are sluiced into an anaerobic digester. The waste material is decomposed anaerobically into methane gas and digested sludge. The gas is either fed to and burned in the plant's heating system or generates steam to drive a turbine and electrical system to produce power for sale to local plants. The "ripe" digested sludge is transported to the nearby corn or bean field for fertilization; these crops are subsequently given to the growing piglets.

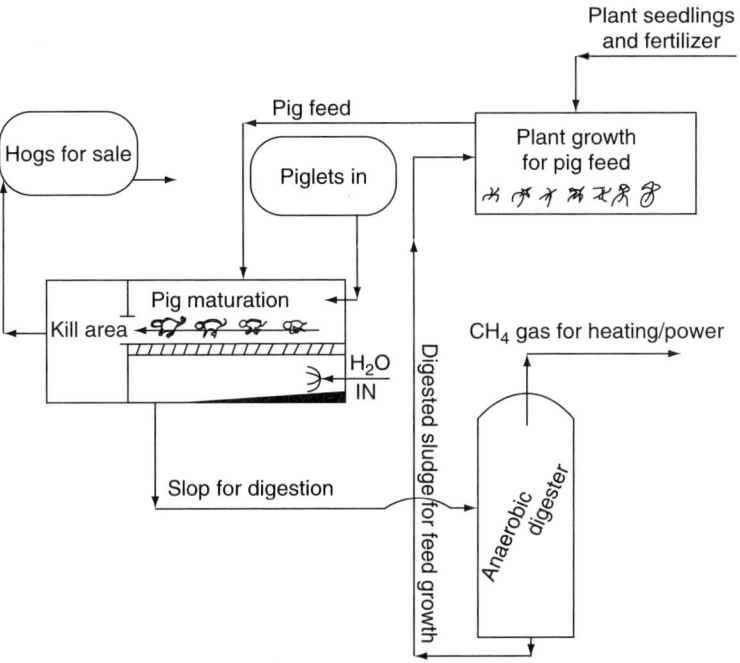

FIGURE 18.37. Hog production–animal feed energy production–environmental complex.

No wastes leave the complex and electrical power and agricultural feed production all enhanced, if not actually increased.

EE. Seawater Desalination Plant–Boric Acid (Borax) Plant Complex

Because of the continuing shortage of freshwater for drinking, the advent of obtaining drinking water from seawater has been of increasing importance. While producing drinking water, salts are extracted from the seawater supply, resulting in a saline waste. This salty waste is usually discharged back to the ocean from where it originated. Often the saline wastewater may interfere, alter, or otherwise harm the near-shore ocean environment.

Because the saline wastewater contains a rich supply of minerals such as borates, it can be used as a raw material for a borax manufacturing plant located in the same complex as the desalination plant. In fact, Chui (2005) reports that "mining companies pump dense brine from the lake [Searles Lake] and use it to produce boric acid, borax, and other products."

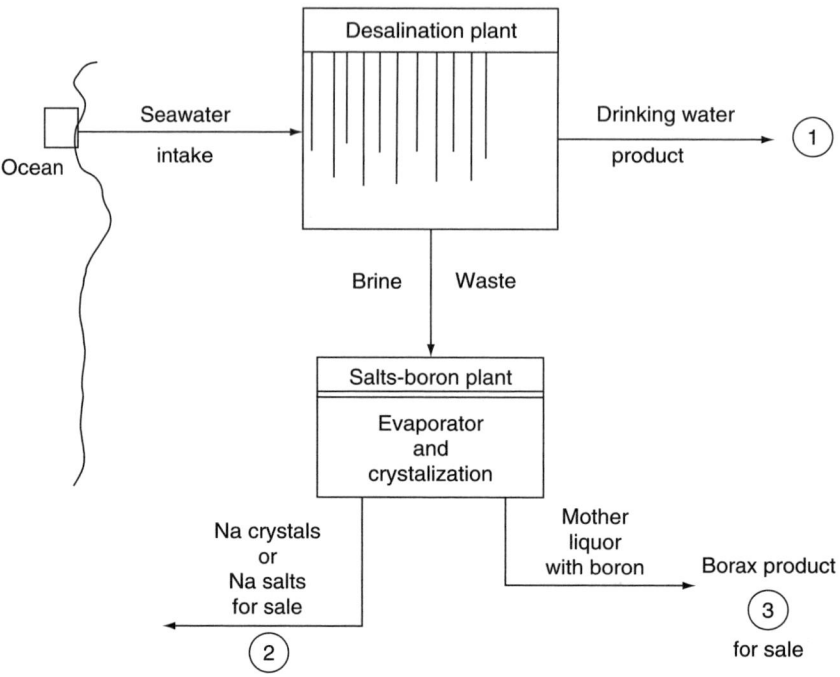

FIGURE 18.38. Seawater desalination plant–Borax plant complex.

Such a complex would eliminate any potential source of saline wastewater and its potential effect on the nearby ocean habitat. An example of such a complex is shown in Figure 18.38.

In this complex, seawater is converted to freshwater by membrane filtration and a brine waste is discharged. Instead of returning it to the sea, it is sent to an evaporation and crystallization plant where sodium salts are removed as crystals and borax recovered from the "mother liquor," and later sold to chemical distributors for sale to industry or the public. No brine is wasted to the environment and three separate product types are obtained.

FF. Cow Feedlot–Power Plant–Fertilizer Complex

As mentioned earlier in this chapter, feces, odors, and organic matter contamination are the important adverse environmental impacts of feedlot production. Animal feedlot wastes contain valuable (but polluting) organic matter. Krueger (2005) reported that each cow on a 500-cow dairy farm produces about 80 lb of waste manure each day. This is being converted into 100 kW of electrical energy by anaerobic digestion of the manure into methane fed to a generator. Because 1 kW of electricity is required (on the average) for a household, this is enough energy to service about 100 homes. Other farm animal wastes such as chicken and pig could also be used in this complex. Hog production wastes can be handled in almost the same manner as shown in Section DD.

Krueger (2005) also verified the expected result that the digested manure is devoid of its objectionable odor but still contains both residual organic matter and minerals. The organics and minerals are valuable constituents for a commercial-grade fertilizer—similar to that of Milorganite produced similarly using Milwaukee, Wisconsin, using digested sewage sludge.

Therefore, we can group all three industries in a single complex, as shown in Figure 18.39, and eliminate all adverse environmental impacts.

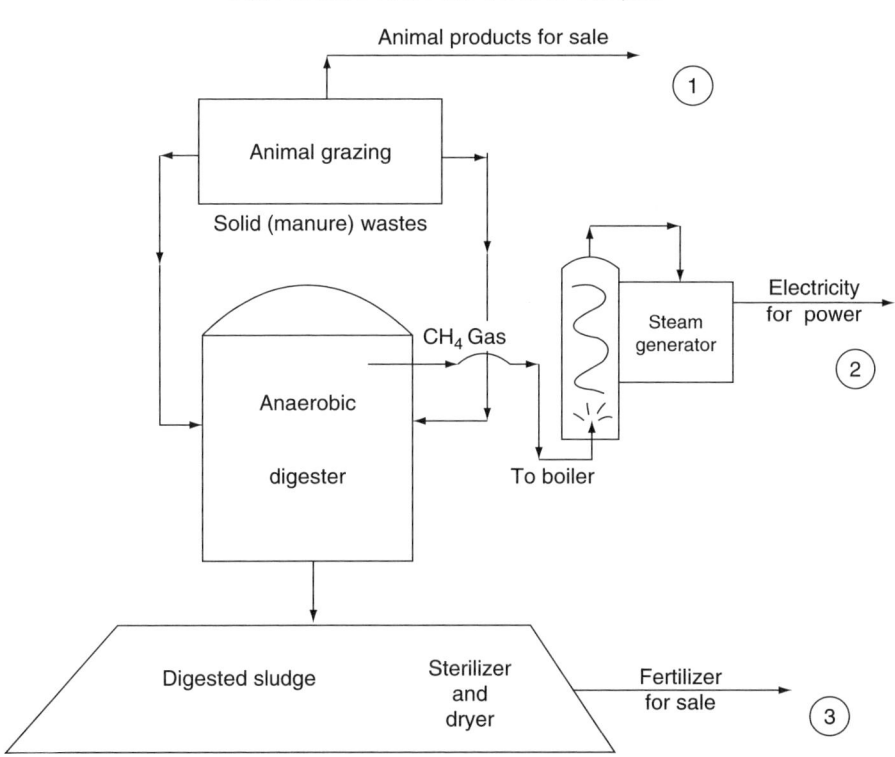

FIGURE 18.39. Cow feed lot–power plant–fertilizer complex.

508 Twenty-First Century

In this complex, animals such as cows, chickens, and pigs, graze normally in a large land area. But, instead of "smelling up" the surroundings with dung, the manures are fed into an anaerobic digester to produce a steady stream of methane gas. The methane is burned in a small boiler yielding steam that feeds a turbine driving a generator to produce electricity for sale to local homeowners or power plant grid systems. An alternative would be to use the methane gas directly to drive the turbine-driven generator. The waste-digested sludge from the digester is dried and sterilized and then bagged and sold commercially for use as a fertilizer.

Three industrial plants produce three useful products and generate no wastes to the outside environment.

GG. Reused Plastic Waste–Consumer Products Complex

Plastic products after useful service to society are usually wasted—one way or another—into the environment. The usual way is to release them into the trash, which generally finds its way into our "sanitary" landfills. Here, they either remain for the life of the fill or disintegrate, yielding toxic organics and sometimes metals, depending on the nature of the plastic and the operating condition of the fill. The result is both a woeful waste of resources and a danger to the environment and those who may come in contact with that portion of it.

The main deterrent to remedying the situation has been the difficulty in collecting the used plastics and relatively low value that it brings to the recoverer.

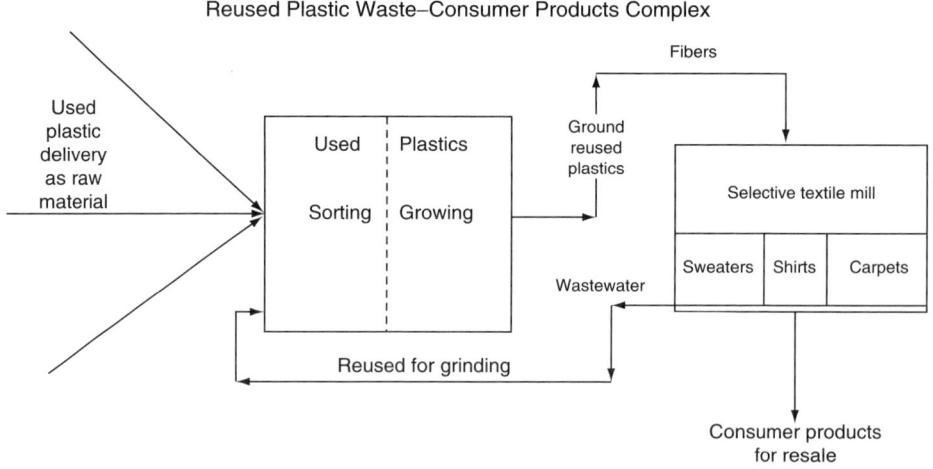

FIGURE 18.40. Reused plastic waste–consumer products complex.

I propose an ideal and logical solution to this conflict between the environment and the reuser. Create an industrial complex in which the used plastic is reused for an economic gain!

In Mexico, the bottling and soda industries have joined forces to collect and recycle the ubiquitous bottles made from polyethylene terephthalate, a thin plastic known as *PET* ("Mexico's PET project" 2004). They formed a company known as *ECOCE*, which will collect the used plastics and transport them to a recycling plant that grinds the plastics into fibers to produce carpets, shirts, sweaters, and other consumer products. One such potential complex is shown in Figure 18.40.

In this complex, used plastics are brought into the complex and sorted and ground at one plant and then the recovered fibers are selectively remade into either sweaters, shirts, or carpets in an ancillary plant. The key to the success of such a complex obviously lies in the ability of the managers to obtain a sufficiently high value for the consumer end-products to justify the expenditure of recovering and reusing the waste plastics. Woven into the economics is the value to a cleaner environment. In such a complex, plastic soda bottles will no longer pollute the environment and useful products will be the result.

HH. Lumber–Textile–Corn Growing–Alcohol Producing Industrial Complex

A potential complex combining Sections P and S of this chapter uses three sources of starch (lumber sawdust, textile de-sizing, and corn) to produce alcohol for an automotive energy alternative to part or all of the gasoline. The three plants all produce useful products and objectionable industrial wastes. The *lumber mill* usually leaves both sawdust, and wood chips, and scraps as solid wastes from the production of finished lumber for commercial and residential customers. These wastes are normally hauled to landfills or burned at a cost to the land or air environments. The *textile mill* must remove the starch sizing from woven goods before dyeing and finishing its consumer goods products. The de-sizing waste is either sent to a municipal treatment plant or biodegraded in its own waste-treatment plant. Both are costly for the textile mill or, if untreated, harmful to the water environment into which it is discharged. Cooking and dyeing wastes must be treated in a biological treatment basin. An *agricultural industry* growing corn is the major plant (among the three) contributing a source of starch for the ethanol plant. But, it also results in the wastage of corn husks, which must be composted and returned to the farmland as a mulch to retain water and fertilizer.

Sawdust and wood scraps, de-sizing wastes, and corn are all fed to the cooker, along with hydrolyzing enzymes for conversion to sugar (Figure 18.41). "Sugar" from the cooker after cooling is led to the fermenter along with a yeast culture to convert it to alcohol. The fermenter evolves a carbon dioxide gas and a liquid ethyl alcohol. The gas is compressed and sold as a product to the beverage industry.

The liquid ethyl alcohol, along with the fermenter mash, are discharged to a distilling vessel, which evolves pure ethyl alcohol at or near the top and residual grain

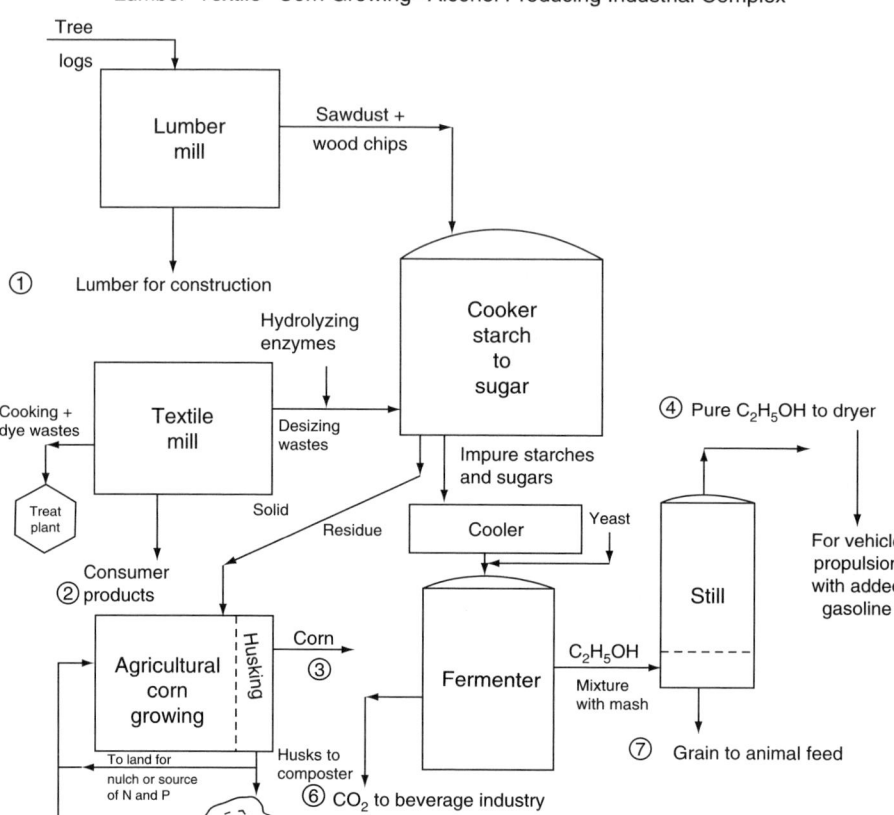

FIGURE 18.41. Lumber–textile–corngrowing–alcohol producing industrial complex.

mash at the bottom. The pure C_2H_5OH (ethyl alcohol) is dried and sold to the transportation industry as a fuel to power vehicles.

It may be difficult to "arrange" for all three raw materials (starch sources) to locate in the same complex with the starch converter. In that case, it may be necessary to omit one or both of the textile and/or lumber mills from the complex. However, from environmental and economic standpoints, it would be preferable to locate all four industrial operations in a single complex.

Products from this four-industry complex are (1) lumber, (2) consumer products, (3) corn, and (4) ethyl alcohol, as well as reusable byproducts (5) compost, (6) carbon dioxide, and (7) distillery grain mash.

Little or no waste from these four industries reach the environment untreated or unused.

References

$20 billion sits in nuclear dump fund, or does it? 2002. *San Diego Union Tribune,* July 14, p. A-1. 2002.

A cash crop? *San Diego Union Tribune,* April 30, Business Section, p. C-1.

A farm for fish. 2000. *San Diego Union Tribune,* Sept. 30, p. C1.

Austin, G. T. 1984. *Shreve's Chemical Process Industries,* 5th ed. New York: McGraw Hill Publishing Co.

Bradsher, K., Revkin, A. 2001. A preemptive strike on global warming. *The New York Times,* March 15, Business, p. 1.

Caffrey, A. 2001. 4800 Megawatts under the sea? Jules Verne-esque idea to import power. *The Wall Street Journal,* May 26, p. 36.

Calbreath, D. 2001. A taste of irradiation. *The San Diego Union Tribune,* Oct. 19, p. C-1.

Chachere, V. 2001. Tapping into the Gulf. *The San Diego Union Tribune,* Aug. 5, p. H2.

Cheremismoff, N. P., P. N. Cheremismoff. 1989. The plastic waste problem report. *Pollution Engineering,* August, pp. 58–68.

Chui, G. 2005. Microbes live in nasty environs and thrive on arsenic. *The San Diego Union Tribune*

Codd, I. 1973. Pollution Control and the Iron and Steel Industry. Presented at the Third Interregional Symposium on the Iron and Steel Industry Brasilia, Brazil, October 14–21.

Costle, D. M. 1980. *Development Document for Proposed Effluent Limitations Guideline, New Source Performance Standards, and Pretreatment Standards for the Steam Electric Point Source Category.* Washington, DC: Environmental Protection Agency. EPA 440/1-80/029b September.

Electric Power Research Institute. 2000. New aquaculture technology could boost fish-farm profits. *Water Environment Technology* 12(4):8.

Ethanol use is headed for state soon, BP says. 2002. *Reuters San Diego Union-Tribune,* May 3, p. C2.

Ex-official ends support for Nevada nuclear waste site. A.P. Report. *The San Diego Union Tribune,* May 31, p. A6.

Hauter, W. 2001. Environmental group likes other solutions. *The San Diego Union Tribune,* Oct. 26, p. B-9.

Herrick, T. 2001. The afterlife of oil rigs. *The Wall Street Journal,* June 27, p. B1.

High costs weigh on Alcoa's plant. 2006. *San Diego Union Tribune,* Jan. 10, p. C5.1.

Kilman, S. 2001a. California gets no-exemption on rules for making cleaner-burning gasoline. *The Wall Street Journal,* June 12, p. A6.

Kilman, S. 2001b. Environmentalists sue Smithfield Foods. *The Wall Street Journal,* March 1, p. B19.

Kollin, J. 1991. Recycling firm puts old plastics to use. *Fort Lauderdale Sun Sentinel,* July 10, News Section, p. 4.

Krueger, A. 2005. Farmers get a charge out of manure. *The San Diego Union Tribune,* July 24, p. B-1.

LaRue, S. 2001. Desalting idea back on table in county. *The San Diego Union Tribune,* May 30, North Coast Section, p. 1.

Lawhome, A. R. 1989. Integrated cement, power and lime facility obtains maximum cost efficiency. *The Florida Specifier,* December, p. 1.
Lindquist, D. 2002. Marathon is planning major Baja energy unit. *San Diego Union-Tribune,* March 1, Business Section, p. C-1.
Loehr, R. C. 1967. Effluent quality from anaerobic lagoons treating feedlot wastes. *J. Water Pollution Control Fed.* 39:384.
Mathews, R. G. 2001. Steelmakers say they are a key component of security. *The Wall Street Journal* Sep. 19, p. B4.
McCormac, P. 2003. Green-waste power plant considered for Fallbrook. *San Diego Union Tribune,* November 19, p. 2, North Coast.
Mexico's PET project. 2004. *Sam Quinones The San Diego Times Union,* February 26, p. Cl.
Monthly Power Plant Report Energy Information. 1986. Administration Form E1A-759. Washington, DC.
Myerly, Nicholson, Katzen, et al. 1981. The forest refinery. *Chemtech* 11(3):186.
Nemerow, N. L. 1976. *Iron and Steel Environmental Management* [prepared for UNIDO, Vienna, Austria], p. 13.
Nemerow, N. L. 1984. *Industrial Solid Wastes.* New York: Ballinger Publishing Company.
Nemerow, N. L. 1995. *Zero Pollution for Industry.* New York: John Wiley & Sons.
Nemerow, N. L., F. J. Agardy. 1998. Strategies of Industrial and Hazardous Waste Management. New York: Van Nostrand Reinhold (John Wiley) Publishing Company.
Nemerow, N. L., A. Dasgupta. 1987. Zero Pollution for Textile Waste. *Proceedings of 7th Alternative Energy Sources Conference Miami, Florida, December 1985,* Vol. 6, p. 499. Hemisphere Publishing Co.
Nemerow, N. L., A. Dasgupta. 1991. *Industrial and Hazardous Waste Treatment.* New York: Van Nostrand Reinhold Publishing Company.
Nemerow, N. L, A. Dasgupta. 1991a. *Industrial and Hazardous Wastes.* New York: Van Nostrand Reinhold Publishing Company.
Newman, B. 2001. The sewer-fat crisis stirs a national stink; sleuths probe flushing. *The Wall Street Journal,* June 4, p. A-1.
Pflum M. 2000. Correspondent web posted 9/1/2000. Available at: ww.cnn.com/2000/nature/09/01/cara.bil.
Process converts wood waste into chemicals. 1991. *Pollution Engineering,* March, p. 45.
Profitability problems plague plastics recycling. 1990. *The Chemical Environment Chemical Business,* March, p. 34.
Radioactive waste cleanup approved. 2000. *The San Diego Union Tribune,* Nov. 1, p. A9.
Reicher. 1999. *Mother Earth News* Oct.-Nov. 1999:24.
Roletti, A. Carlsbad approves $80,000 for study of desalination plant. *San Diego Union Tribune,* February 13, North Coast Section, p. 2.
Sell, N. J. 1992. *Industrial Pollution Control,* p. 17. New York: Van Nostrand Reinhold Publishing Company.
Sherman, S. P. 1989. Trashing of a $150 billion business. *Fortune Magazine,* August, p. 96.
Shreve, R. N., J. A. Brink. 1977. *Chemical Process Industries.* New York: McGraw-Hill Publishing Co.
Steelmakers get Bush warning on tariff relief. 2002. *The Wall Street Journal,* June 26, p. A2.

Technology's toxic trash. 2002. *San Diego Union Tribune,* July 7, p. 1.
Thayer Chemical and Engineering News Bureau (Northeast). 1989.
Toxins and water. 2001. *San Diego Union Tribune,* May 24, p. B10.
U.S. Department of Energy. 1987. *Impacts of Proposed RCRA Regulations and Other Related Federal Environmental Regulations on Fossil Fuel-Fired Facilities.* Engineering Science DOE/ET/13543-2316. Washington, DC: U.S. Department of Energy.
U.S. Department of Interior. 1967. *The Cost of Clean Water,* Vol. III, *Industrial Waste Profile,* No. 10, *Plastic Materials and Resins.* Washington, DC: U.S. Department of Interior.
Web page created by Dr. Alastair Grant University of East Anglia Norwich, NR 47TJ, United Kingdom, 12/5/2000.
Wilkie, D. 2001. Nuclear waste plan lacks core. *San Diego Union Tribune,* May 21, p. 1.
Yucca Mountain. 2002. *San Diego Union Tribune,* July 15, Business, p. 2.

CHAPTER 19

Potential Municipal–Industrial Complexes

Introduction

For many years, municipalities have been cooperating with industries by permitting them to use their public plants to dispose of local factory wastes. Your author first wrote about this "joint treatment" in 1963. These systems allowed participating industries to contract with their municipal sewer service agencies to accept certain amounts and types of wastes into their systems with and without payment provision. Thus, precedent has been established, provisions made, and experiences gained from these prior associations of city and industry. Some of the past arrangements of combined treatment turned out to be a boon to both parties, whereas others resulted in a bane to both. The reader is urged to read mine and other discussions of joint treatment to aid in comprehending the problems of the past and the rationale for optimism for what we now recommend for the future.

For the present and the future, I am recommending the municipal–industrial complex concept similar to what was reviewed in Chapter 18. To accomplish this type of "pollution solution," the industrial plant must be located at the site of the municipal wastewater treatment plant. This should really not be a great burden to the industry because it would find it advantageous to operate at the lower elevation and secluded location usually selected by cities.

We will consider in Chapter 19 only one of many potential municipal–industrial complexes in which municipal solid wastes are recovered and reused within the complex by two industrial plants.

Municipal Solid Wastes–Industrial Complexes

Municipal wastewaters typically contain approximately 5–10% settleable suspended solids. In addition, some of these solids are grease-like in nature and will separate from the denser settleable solids. Usually cities use large settling tanks to remove both types of solids—one by the process of sedimentation and the other by flotation. Both types of solids are eventually usually wasted into the environment or treated extensively before some type of ultimate disposal is used. Both the treatment and discharge into the environment are costly and damaging.

In Figure 19.1, I present a schematic concept of one municipal–industrial complex in which both types of solids are recovered and used by industries in the complex to make additional products. In this complex, the municipal sewage's settled solids are rotary-dried to produce a 5–10% cake, which is conveyed to the agricultural growing area to enhance the growth of selected fruits and vegetables. These food crops are then harvested and sold to outside canners or canned, if possible, on-site of the complex.

The lighter than water solids (grease) are skimmed from the settling tank surface and conveyed directly to the on-site renderer, which concentrates and converts these solids by enclosed heating to an edible animal food additive, natural animal glue, or a soap base for sale as products. Transportation of the solids to distant industrial factories is avoided, as is the importation of raw materials for industrial production. No municipal solids are released to the air, water, or land environments before or after costly treatments.

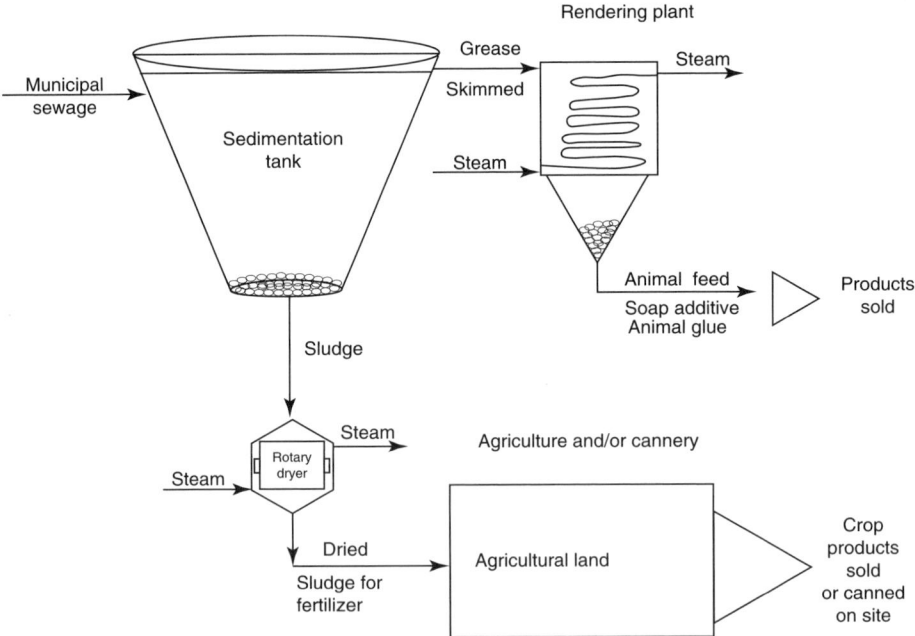

FIGURE 19.1. Schematic diagram of one type of municipal-industrial complex.

In addition, other advantages may exist for such a complex. For example, the wastewater effluent from the municipal system may be reused to irrigate the agricultural area rather than discharging it to a nearby watercourse.

Municipal Wastewater–Industrial Complex

Municipal wastewaters also contain about 1,000 parts per million (ppm) of dissolved and colloidal solids, most of which are organic. These solids are costly to remove by effective treatment (usually biological). Hence, some, if not all, of these solids are discharged into the environment, causing degradation of water courses.

An ideal solution to this problem is to incorporate the municipal treatment plant into a complex with other industries that can use these solids to evolve a product for commercial use. One such complex is shown in Figure 19.2.

In this complex illustration, two industries and the municipality are involved: fishery food, agriculture crop food, and municipal sewage. The sewage is first treated by sedimentation. The supernatant wastewater is directed to an algal production pond. Here, algae grow aided by natural sunlight and minerals and other nutrients remaining in the sewage.

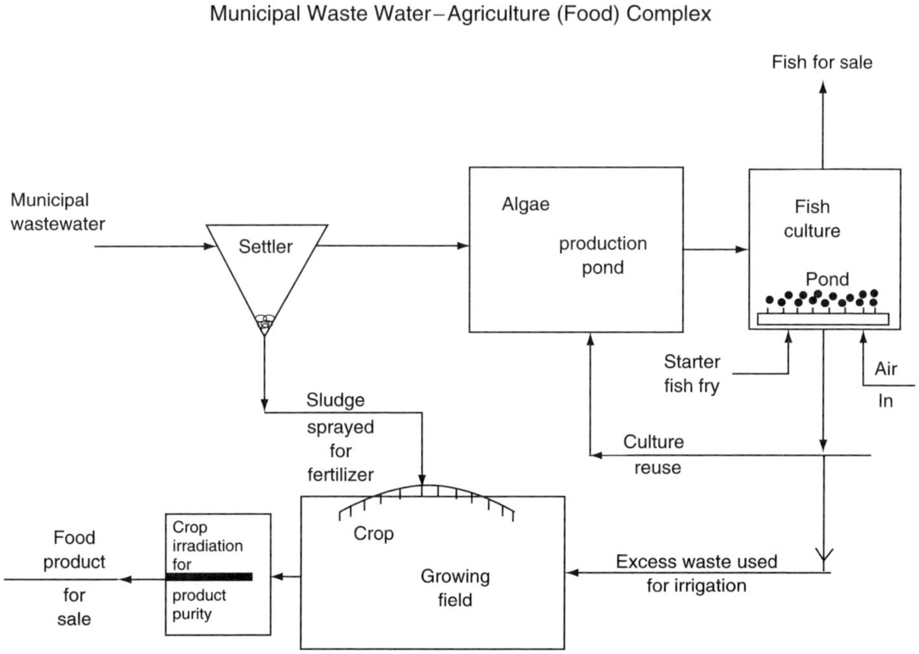

FIGURE 19.2. Municipal waste water–agriculture (food) complex.

After sufficient detention time, the algae-laden overflow is discharged to the fish culture pond. Starter fish fry and air are introduced into this pond to enhance the growth of fish such as Tilappia for sale as animal and human food. The fish growth is enhanced by the amount and nature of algae fed to the culture pond. Some fish pond culture is recirculated to the algae production pond to stimulate the growth of more algae.

Excess culture pond effluent is distributed in a crop-growing area. Crops such as corn, beans, and tomatoes can be grown in this field. The crop growth is also aided by applying (spraying) settled sewage sludge from the sedimentation basin. After harvesting and crop irradiation, the food products can be sold again for animal and human consumption.

By using a complex system such as this one, neither municipal sludge solids nor contaminated liquid effluent reaches the land, air, or water environment outside the complex. This complex represents just one of several that can accomplish the same objective while evolving other industrial products.

Lake Industry–Villagers Complex

Lakes can serve as valuable industries for their users. For example, some lakes can be used by boaters, fishermen, and even bathers and swimmers. If, at the same time, these lakes are used by downstream owners for generating electricity and for irrigation water by farmers, conflicts over the lakes' use and operation may exist.

We are just beginning to amass information about ways and means of using and operating such multipurpose lakes. Your author has proposed economic methods for allocating water usage in lakes as far back as 1970 and repeated in summarized form in this book in Chapter 21. However, until now, administrative decisions largely determine which water use gets priority. Administrative decisions are usually made by political pressures rather than by rational economic values.

For example, downstream water users may sometimes exert enough influence on upstream lake property owners to release (or lower) water levels in the lake, thus interfering with lake-industry uses. I suggest one technical (or nonadministrative and noneconomic) method whereby such situations can be ameliorated.

Usually lakes are surrounded by nearby villages or even cities whose residents use the lake for recreation, as mentioned in the first paragraph. When lake levels are lowered because of release of water for downstream users, village lake users lose some or all of the recreational benefits available to them. One way of overcoming, to some extent, this dilemma is to treat and discharge its domestic wastewater (sewage) at the headwaters of the lake.

This will add about 100 gallons of water per person per day to the lake instead of sending the sewage water immediately downstream as an eventual waste. A village of only 1,000 people, for example, could increase the water volume of the lake by about 6.35 million gallons each year (1,000 × 100 × 365). This may, by itself, be enough to preserve boating and fishing activities in the lake. In a manner of speaking, this is a form of municipal wastewater reuse that may not be as objectionable to the public as other uses, such as irrigating edible crops and street cleaning.

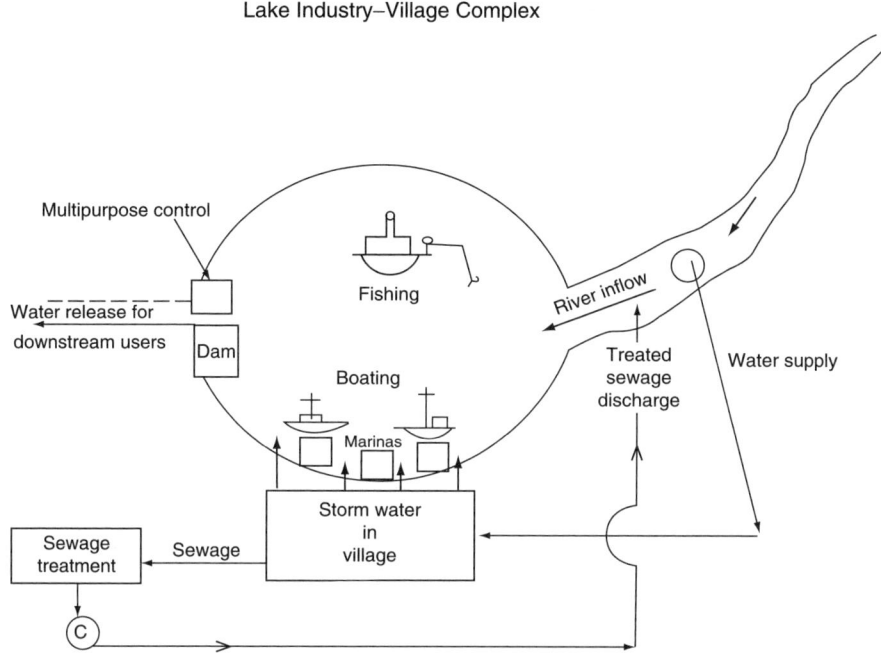

FIGURE 19.3A. Lake industry–village complex.

I have depicted such a complex in Figure 19.3. In this complex, the lake-bordering village treats and pumps its sewage to the upper part of the lake instead of discharging it below the lake dam to downstream river users. This in no way affects the villages' upstream drinking water quality or quantity. The fishing, boating, and recreation industry existing in the lake will be enhanced, especially when downstream water users are lowering dam overflows to serve their interests.

Thus, we have a type of complex in which the lake industry is aided by the wastes of the villagers. An illustration of a situation in which this complex solution might be used is Detroit, Oregon (Gavin 2001). In Detroit Lake, boating businesses disappeared because of competition for water prompted by the Northeast's near-record drought. This caused the Army Corps of Engineers to decide against bringing the water level high enough to support recreational lake activities.

The Ultimate Natural Resource Conservation and Resource Preservation Plant

This section is for those whose imagination has been stimulated sufficiently to envision the ultimate in this concept. Why not use all liquid and solid wastes of a municipality to produce some of the products that municipality needs to survive and grow such as food, electricity, and water? It can be done, and again with no wastage of material to harm the surrounding environment.

If you doubt that it is possible, I refer you to Figure 19.3 as proof of the concept. In this depiction of an ultimate environmentally balanced industrial complex (EBIC), all municipal wastewater and combustible refuse represent the EBIC plant inputs.

The wastewater is settled to produce grease and sludge, which are fed to the fermenter to produce methane gas for burning and steam formation in the power plant for electrical energy.

The burnable refuse is also fed directly to the power plant for burning, steam formation, and power plant electricity.

The settling basin effluent after chlorination is fed to an algae growth pond fortified with tricalcium phosphate fertilizer and carbon dioxide from the power plant stack gas.

The algae are introduced into the fish growth pond to produce an edible fish product for sale after irradiation.

The entire fish pond effluent is used to irrigate a crop-growing field. The field is also supplemented with waste-digested sludge from the fermenter. The agricultural product is periodically harvested, irradiated, and sold to the city folk.

Meanwhile, all the excess wastewater overflow that was fed to the growth field for irrigation is filtered through the underground soil and becomes a source of reusable groundwater for reuse by the municipality. In summary, all municipal wastewater and burnable refuse have been treated and reused as a source of water, electricity, and food for the same municipality.

A current example of an intended ultimate waste treatment facility is located in Fallbrook, California. A proposed joint project by Fallbrook and Camp Pendleton would reuse wastewater that now flows out to the ocean (2003). Following four sequences of treatment—preliminary, primary, secondary, and tertiary—the clean wastewater would then flow through a wetland into the Santa Margarita River, and then finally into various recharge ponds where the water would percolate into an underground aquifer. Later, water from the underground aquifer would be reclaimed with nitrogen compounds removed and stored in an open surface reservoir. From this reservoir, the water would overflow into the Santa Margarita River where it would be partly diverted into recharge ponds. Groundwater from these ponds would be pumped into the water-treatment plant, stored, and used for a domestic water supply for the Camp and Fallbrook.

The developers of this plan recognize that "Fallbrook's proposal to recycle waste water to make it safe for drinking must overcome negative sentiment." They counter with the fact that "treated waste water would spend six months percolating into the ground and mixing with naturally occurring water before it would be treated again and reused." In this instance, I would add that "dilution—with treatment—would provide the solution to pollution."

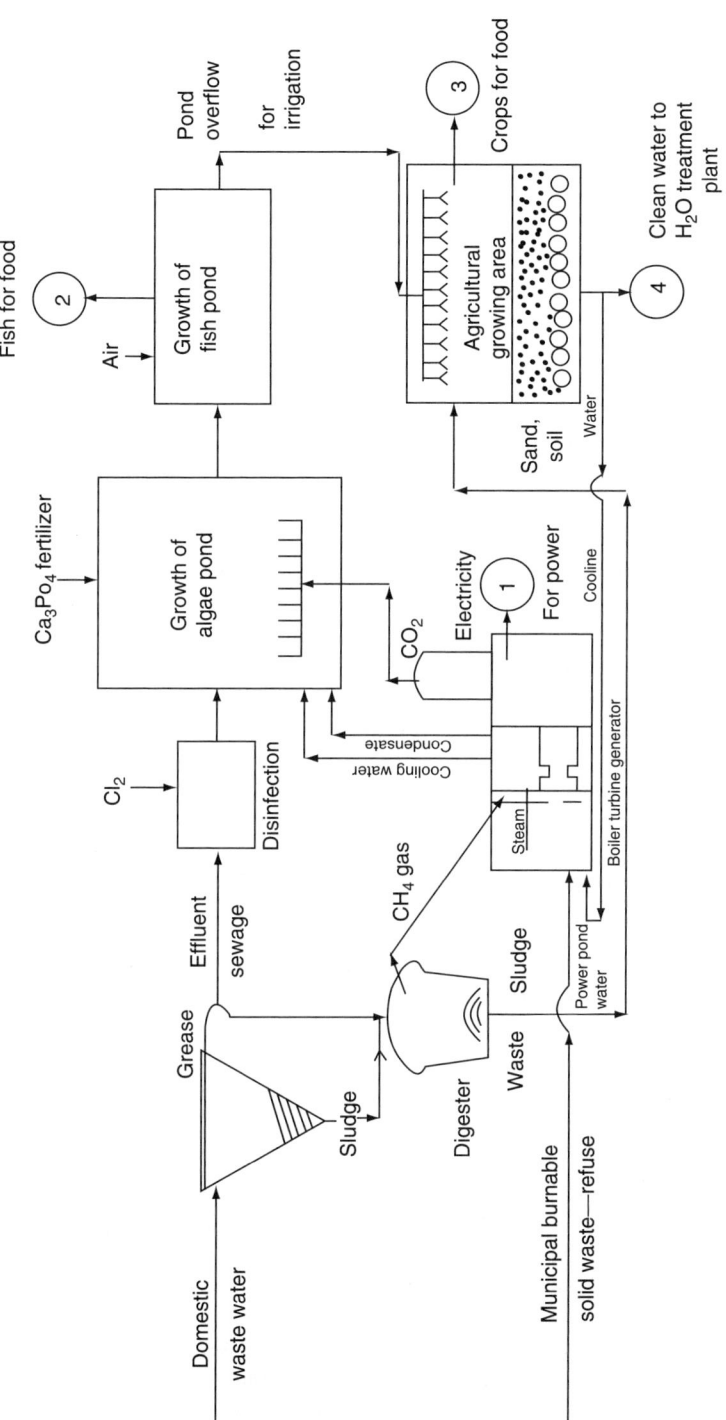

FIGURE 19.3B. Food–electricity and water production plant.

Byproduct Synergy

The U.S. Business Council for Sustainable Development (2003) has proposed the principle of byproduct synergy as "one industry waste stream can be used by another as a primary resource." It is "as a simple idea, but one which has enormous potential for reducing waste volumes and toxic emissions to air and water, as well as cutting operating costs." To facilitate the system, the U.S. Business Council for Sustainable Development states that "businesses need to work together to determine what unwanted by-products exist, and what their potential applications are. The resources can then be exchanged, sold, or passed free of charge between sites, creating a by-product synergy." The Environmental Protection Agency (EPA) and the World Business Council for Sustainable Development have defined this system as "the synergy among diverse industries, agriculture, and communities resulting in profitable conversion of by-products and wastes to resources promoting sustainability." They claim that it is the principle that "underpins the concept of industrial ecology—a holistic view of industry in which organizations exchange energy and material between one another, rather than operating as isolated units. Industrial ecology promotes a shift away from traditional open linear systems towards closed loops and interdependent relationships of the kind found in nature."

The byproduct synergy system is really an advanced stage of marketing and exchanging wastes. Previously, the older waste exchanges did not flourish for several reasons (discussed more fully in Chapter 13 of this book). However, this system is a new attempt to list wastes and target industries' needs much more directly and completely. For these reasons, and for the timeliness of its inception, as well as the advanced education of industries, it has a better chance for success.

The Industrial Ecosystem Development Project (International Institute for Sustainable Development, 2002)

The EPA instigated a 2-year project "to identify potential by-product partnerships in an industrialized area of North Carolina, encompassing Raleigh, Durham and Chapel Hill." It was modeled after or inspired by the Kalundborg, Denmark, system (which is described in Chapter 20) with the exception that the area is much larger with a larger population and dominated by pharmaceutical, computer, and telecommunications equipment manufacturers. A brief description of their findings is in order here to illustrate how the system works.

Of the 343 facilities (industrial), 182 agreed to take part in the study. It is important to realize that this again is a project system "after the fact." That is, all the industrial plants had already been built and were in operation in the area. It had not been any kind of planned complex. They used a geographic information system (GIS) to ascertain the byproducts arising there and the inputs they required. Its goal was to ascertain with "matches" among nearby plants. About half the sites yielded 49 different byproducts, of which 12 were deemed "viable" for short-term partnerships, namely acetone, carbon, desiccant, hydrochloric acid, methanol, packaging, plastic bags, sawdust, sodium hydroxide, wood ash, wood chips, and wood fluff. In addition, 24 byproducts were found for which partnerships could be developed with more effort, including copper, electricity, floppy disks, glass fibers, ink, plastic, and wire.

They give as examples that in "one instance, a company which used vermiculite as a packaging material realized that it could use waste sawdust from a furniture shop directly across the street—waste material that would otherwise have been landfilled." In another case, they found that one plant could save 5,000 truck transportation miles a year by taking leftover acetone to a local business that could use it, rather than haul it to a hazardous waste facility 150 miles away.

The report of this study concludes that "the main obstacle to industrial ecology is the absence of a 'champion' to bring the various industries together." In addition, "what is lacking in most communities is an agent to promote the vision of a web of materials, water and energy flowing between neighbours, and to gather the local information about by-products available or raw material requirements needed to build this web."

The U.S. Business Council for Sustainable Development (2003) has taken on the responsibility of furthering the implementation of this system. It repeats the purpose of byproduct synergy as the "practice of matching under-valued waste or by-product streams with potential users, helping to create new revenues or savings for the organizations involved while simultaneously addressing social and environmental impacts."

The Council lists the following benefits of the process:

1. Reduced operating expense
2. Reduced energy use
3. Reduced emissions
4. Waste transformed into product
5. Surpassed regulatory targets
6. Improved community
7. Improved productivity
8. Improved profitability

It also admits the following barriers:

1. Technical barriers
2. Economic barriers
3. Regulatory barriers and liability
4. Perception and reputation
5. Lack of incentives

Examples of Byproduct Synergies

The Counsil explains that each regional project involves getting 10–20 unlike companies to pay a fee and engage local, state, and federal governments to support the study. It gives as an example one of the earliest companies to adopt byproduct synergy, the Chaparral Steel Company. As one of its first synergies, it discovered the "potential for steel slag to be used as a raw material for the cement manufactured by nearby Texas Industries." The steel slag contained Ca_2SiO_3, formed by the high temperatures of the steel-making process and a building block of Portland cement. By using the steel slag instead of purchased lime, which would then have to be heated to calcination, Texas Industries reduced the energy requirements and related emissions of CO_2, NOx, and SO_2 of the cement-making process. The company found that profits for both companies also increased.

In Table 19.1, the byproduct synergy presents a large opportunity for reducing raw material consumption, energy use, and emissions and waste generation, along with associated cost savings.

The Business Council for Sustainable Development and its Byproduct Synergy Production unit, Applied Sustainability (AS), launched new projects in the United States, Mexico, and Canada starting in 1997. "In each of these projects synergies emerged with potential for measurable financial, social and environmental values." I summarize some of the interesting byproduct synergies here for illustrative purposes.

In Tampico, Mexico
> PVC residuals were converted into shoe soles.
> Excess acetonitrile was substituted for a more expensive solvent.
> CO_2 was recovered and used in a new manufacturing CO_2 plant.
> Plastic polymers were recovered and pulverized with liquid nitrogen to homogenize these scrap wastes for reuse.
> Waste plastic residuals were reused in construction.
> Plastic packaging bags were used in construction of platforms for ship loading.
> Waste hydrocarbons and municipal waste were used in a waste-to-energy project.

Alberta, Canada
> Spent caustic (NaOH with contaminants) from a Kraft paper mill and a refinery was reused in the Kraft process to make up for Na losses.

North Texas
> Wood waste from tree trimming was used for a biomass-fueled electricity generation unit.
> Copper cooling wastewater was used to recover copper.
> Electrostatic dust from a precipitator was recovered for use as a fertilizer because of its high concentration of boron.

Montreal, Canada
> Reprocessing of lead smelter baghouse dust for metal reuse.
> Reuse of a relatively pure supply of hydrogen gas if it can be transported economically.
> Reuse of auto-shredder fluff for energy production.

TABLE 19.1
Annual Cost and Environmental Benefits of Successful Synergies

Implemented Synergies	Ecological/Biological	Energy Savings	Residue Reduction	Cost Savings
CemStar® 130,000 tons of steel slag used in place of lime (single plant operation)	Reduced SO_2 (acid rain) through coal displacement	Displacement of 11,800 tons of coal used to calcine lime (3.5 billion Btu)	130,000 tons of steel slag not land-filled Emission reductions from coal displacement: 65,000 tons CO_2, 800 tons of NOx, 33 tons of hydrocarbons	*Steel producer:* Reduced/eliminated steel slag treatment/disposal costs *Cement producer:* Less costly raw material Calcination is not required; energy consumption and associated emissions for cement production are reduced
Auto shredder residue (ASR) 120,000 tons of ASR mined for metal reclamation, and ASR remaining after metal recovery used for power generation	Reduced SO_2 (acid rain) through coal displacement	18,000 tons of metals (Al, Cu, Mg, Sn) recovered from ASR and not mined 98,000 tons of carbon-based ASR displaces 66,000 tons of coal for power generation (20 billion Btu)	120,000 tons of ASR not landfilled Energy savings associated with metal recovery vs. mining prevent 151,000 tons of CO_2 emissions SO_2 emissions reduced by substitution of ASR for coal	*ASR producer:* Reduction/elimination of ASR disposal fees Increased revenue from recovered metals Revenue from sale of ASR as alternative fuel *ASR consumers:* Lower-cost, less energy-intensive method of obtaining metals Lower-cost fuel
Graphite/copper sludge 37,500 lb of sludge saved from landfills and municipal water systems	Landfill biota not exposed to toxicity of copper waste	18,750 lb of copper recovered and not mined (5.6 million Btu)	37,500 lb of graphite/copper sludge not landfilled 412,500 gallons of graphite/copper-tainted wastewater not released to municipal wastewater treatment	*Sludge producer:* Reduced/eliminated waste disposal fees Revenue from sale of sludge to copper extraction company *Metal recovery company:* Lower-cost source of copper

New Jersey

Dow Chemical Company plants are pursuing the use of three wastes: (1) a latex emulsion stream from paint production for reuse in road construction and in agricultural operations to control dust, (2) an off-grade, or scraps of polyethylene, for possible use in making shoe soles, and (3) reusing rigid polyurethane scraps for potting soils to increase aerability.

Once again, all of these studies are aimed at reducing wastes of industrial plants that are already in operation. Although this is an admirable and important step in the elimination of industrial wastes and in potentially reducing production costs, ultimately we must design new facilities right from the beginning to include so-called byproduct synergy. That task may be even more difficult to implement because of the inertia it must overcome at the beginning.

References

Gavin, R. 2001. Missing: Little Lake in Oregon. *The Wall Street Journal,* October 3, p. B1.

International Institute for Sustainable Development. 2002. www.iisd.org.

International Institute for Sustainable Development. 2006. By-product synergy and industrial ecology. Business and Sustainable Development: A Global Guide. Available at www.BSDglobal.com.

Nemerow, N. L. 1963. *Theories and Practices of Industrial Waste Treatment.* Addison Wesley Publishing Company.

U.S. Business Council for Sustainable Development. 2003. *An Introduction to By-Product Synergy.* Austin, TX: U.S. Business Council for Sustainable Development.

CHAPTER 20

Naturally Evolving Industrial Complexes

In the 1900s, industrial plants located in the general vicinity of one another began to use wastes discharged by one another. The necessity to dispose of wastes and to import raw materials has led each plant to search for both buyers of wastes and suppliers of raw materials. The search sometimes resulted in finding these industrial plants relatively nearby, because for some time manufacturers were located in the same geographical area. These areas were sometimes referred to as "industrial zones" or "industrial parks." The original purpose of these zones or parks was to confine manufacturing to certain physical places in the overall community. This freed the area from transgressing traditional residential needs (quiet, clean air, less street traffic, etc.). Earlier, the purpose of these industrial areas was certainly not to avoid pollution of the environment. In fact, just the opposite resulted, as it allowed more environmental degradation. However, this polluted area was at least partially segregated from the rest of the community that served the people: residences, schools, hospitals, and recreational facilities such as parks. The philosophy was similar to that used early by state stream pollution agencies: to classify streams from A to F according to their best usages. For example, historically an F classification allowed a stream to be used for the discharge of all industrial wastes as its best usage. The stream zone in this case was segregated and allowed to deteriorate to its lowest level of water quality. It wasn't long, however, before these agencies realized that such classification, and indeed stream usage, was detrimental to industry and the whole society. "F" and other lower stream usages were soon abandoned. We discovered that we could not afford to degrade any one area to avoid contaminating another.

This same realization may already be taking hold in the so-called *naturally evolving industrial complexes,* and hence, the need to promote closer working relationships between industrial plants in order to avoid or at least minimize environmental pollution.

Ehrenfeld and Gertler (1997) reported on a naturally evolving industrial ecosystem in Kalundborg, Denmark. The ecosystem slowly evolved over a 25-year period and contains 11 physical linkages (shown in schematic form by Ehrenfeld and Gertler and reproduced here in Figure 20.1). They reported that four main industries comprise the "heart" of the ecosystem: (1) the coal-fired Asnaes Power Station, (2) Statoil oil refinery, (3) NovoNordisk Pharmaceutical and enzyme manufacturer, and (4) Gropoc Plasterboard Manufacturer. They also reported that several users in Kalundborg "trade and make use" of wastes and energy produced, as well as "turn byproducts into products." In addition, firms outside the city also receive byproducts as raw materials. The authors reported that this symbiotic relationship between the ecosystem industries developed slowly and quite naturally.

Over a 25-year period, they claim that the ever-increasing need for energy and the availability and use of byproducts as feed was "fundamental to this approach." They maintain that "entropy-minimizing states in stable biological systems are accompanied by increases in the interdependence among the entities." This infers that old systems of independent industrial location and production results in entropy-increasing processes (a wasteful system). Naturally, as the authors point out, entropy is a measure of disorder in any system and will increase as plants operate independently to generate products and wastes.

The authors describe this slowly evolving industrial ecosystem as an attempt to make "economic use of their byproducts and to minimize the cost of compliance with new ever-stricter environmental regulations." Also, it remains inferred or at least implied that this system will ultimately result in lower industrial production costs.

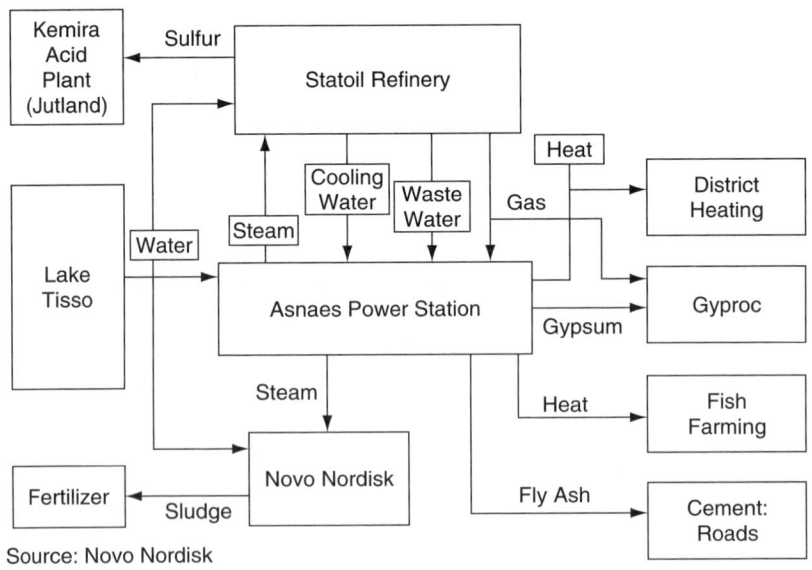

FIGURE 20.1. The Industrial Ecosystem at Kalundborg, Denmark.

In their paper, Ehrenfeld and Gertler (1997) tabulate in chronological order from 1959 to 1993 the evolution of this entire ecosystem. They place the Asnaes Power Station, the largest in Denmark, as the "heart of this system," and they further report that by altering its operation of wasting energy, it improved its energy efficiency to about 90% from about 40% when used as a free-standing plant.

The authors report some of these energy-efficiency improvements as follows:

1. Distributing heat from the power plant through a network of underground pipes.
2. Delivering steam to two of its industries. The 2-mile-long steam pipeline paid for itself in 2 years and eliminated thermal pollution of the nearby fjord at the same time.
3. Providing gypsum-containing byproduct to the wallboard manufacturer for incorporation into its product. The gypsum resulted from the power plants sulfur dioxide scrubber.
4. Selling fly-ash and clinker from the power plant for road building and cement production.
5. Piping gas from the refinery to fire the wallboard manufacturing plants' drying ovens.
6. Trucking desulfurizing plants' sour gas to a conversion plant to make sulfuric acid. The remaining sour gas is reburned in the power plant.
7. Piping the refineries' cooling water to the power plant where it is purified and reused as boiler feed water.
8. Piping the refiners' biologically treated wastewater to the power plant for cleaning uses.
9. Distributing the pharmaceutical waste sludge to nearby farms for fertilizer.

Ehrenfeld and Gertler (1997) also present a table itemizing the "resource savings through interchanges," as well as brief descriptions of some other potential partial ecological complexes that they refer to as *symbiotic ecosystems*. They discuss the subject of "barriers and limits" of symbiotic ecological systems. One significant presentation by the authors is the statement that "it would take some form of public intervention in imposing large disposal costs or subsidies to the recovery firms to create the favorable economics that led to the evolution of Kalundborg." As a practical matter, this is the direction to which the public has been moving. The references in the authors' paper are recommended for reading for the serious pursuer of originating "industrial ecosystems."

Woodard (2002) reported on the operation of this complex. Per the manager of Asnaes, "we're all making money from this" (referring to the city's system of turning waste into raw resources). "We have a bit of difficulty understanding why the rest of the world isn't doing it."

Lowe et al. (2000) wrote a handbook for local development teams entitled *Eco-Industrial Parks*. This comprehensive treatise included topics such as the benefits, costs, risks, challenges, and foundations of eco-industrial parks (EIPs). They also describe how to begin setting up a team to evolve the EIP and to design, plan, construct, and manage it. An important contribution of this report is the section in the Appendix titled

"Cases." These cases include the Burnside Industrial Park in Nova Scotia, the Chattanooga, Tennessee Parks, the Brownsville, Texas Park, the Baltimore Empowerment Zone Eco-Industrial Park, the Port of Cape Charles Sustainable Technologies Industrial Park, the Zero Emissions Research Institute (ZERI) in Tokyo, Japan, the Rotterdam Harbour Industrial Ecosystem Project (ENIS), the Haymouth and Virginia sustainable mixed use developments. The reader of this book is urged to review the entire report to obtain a historical view of EIPs, as well as the authors' views of their utility in today's industrial development.

Lowe (2000) created a paper on industrial ecosystems and byproduct exchanges from a document originally prepared for the Philippine Board of Investments. He expands on the Kalundborg case (described earlier in this chapter) to illustrate potential byproduct exchanges of Philippine industries. This reference provides evidence that the EIP system is moving ahead and receiving increased attention of not only environmental engineers, but also environmentalists in the broadest sense.

References

Ehrenfeld, J., N. Gertler. 1997. Industrial ecology in practice. *J. Industrial Ecol.* 1(1).

Lowe, E. 2000. *Industrial Ecosystems: Methods for Creating By-Product Exchanges among Companies.* Oakland, CA: Indigo Development.

Lowe, E. A., S. R. Moran, D. B. Holmes. 1996. *Co-Industrial Parks.* R.P.P. International Emeryville. E.P.A. Research Triangle CR822666010.

Woodard, C. 2002. Danes' great green machine. *San Francisco Chronicle,* March 3, p. A3.

CHAPTER 21

Benefit-Related Expenditures for Industrial Waste Treatment

At this juncture, some discussion is useful to clarify my position of how much industry should pay to treat its industrial waste. Clearly waste treatment represents a significant cost of doing business. That cost must not only be predetermined by industry, but managers also must decide on the amount they are willing and able to spend. Understanding this expenditure will aid the reader in reaching the conclusion that regardless of the accepted cost of waste treatment, the environmentally balanced industrial complex (EBIC) is less costly and more efficient and leaves no residual environmental pollution.

In 1972, I made the following public statements, which are still valid today (Nemerow 1972):

> Industrial waste treatment is a necessity to preserve our water resources. Economical stability of our society is also equally vital to our well-being. Waste treatment may cost more than an industrial plant is willing or even able to spend. This is true especially in situations where the stream resources are limited and intense competition exists between water users and consumers. Unfortunately for us, these latter situations are becoming more and more prevalent. What then is industry to do in these cases? Move? Cease production? Enter into a legal maneuver in order to delay or prevent excessive costs for waste treatment? None of these is really desirable for industry or society. What should governmental regulatory agencies do in these critical situations? Force industry into one of the above alternatives or ignore the need to protect the stream resources and allow the plant to continue to pollute? Neither of these positions is satisfactory. How then do we solve the problem of apparently conflicting interests of two factions of our society? In this chapter, some answers should become apparent.

Lest any reader question whether treatment costs are justifiable, I have recalled a listing of primary, secondary, and intangible benefits of industrial waste treatment.

Primary Benefits

1. Savings in dollars to the industrial firm by the reuse of treated effluents instead of freshwater.
2. Savings in dollars resulting from compliance with regulatory agencies, that is, avoidance of legal and expert fees and management's time involved in court cases.
3. Savings in dollars from increased production efficiency that are made possible by improved knowledge of waste-producing processes and practices.

Secondary Benefits

1. Saving in dollars to downstream consumers from improved water quality and, hence, lowered operating and damage costs.
2. Increase in employment, higher local payroll, and greater economic purchasing power of the labor force used in construction and operation of waste-treatment facilities.
3. Increased economic growth of the area because of the commitment of industry to waste treatment and potential for expansion at the existing plant.
4. Increased economic growth of area with more clean water available for additional industrial operations, which in turn yield more employment and income for the area.
5. Increased value of adjacent properties as a result of a cleaner, more desirable receiving stream.
6. Increased population potential for the area because cleaner water will be available at a lower cost. The limiting factor of water cost and quantity has been pushed back further into the future.
7. Increased recreational uses, such as fishing, boating, swimming, as a result of increased purity of water; recreational opportunities previously eliminated are available again.

Intangible Benefits

1. Good public relations and an improved industrial image after installation of pollution abatement devices.
2. Improved mental health of citizens in the area confident of having adequate waste treatment and clean water.
3. Improved conservation practices that will eventually yield payoffs in the form of more clean water for more people over more years.
4. Renewal and preservation of scenic beauty and historical sites.
5. Residential development potential for land areas nearby because of the presence of clean recreational waters.
6. Elimination of relocation costs (of persons, groups, and establishments) because of contaminated waters.

7. Removal of potential physical health hazards of using polluted water for recreation.
8. Industrial capital investment ensures permanence of the plant in the area, thus lending confidence to other firms and citizens depending on the output produced by the industry.
9. Technological progress, resulting from the conception, design, construction, and operation of industrial waste-treatment facilities.

The most obvious and prominent observation from the listing of benefits is that one must quantify these in some manner to arrive at a specific level of justifiable expenditure. I have made an attempt to do just that in Chapter 16B. However, at this point I would like to express my opposition to the view expressed by some that all industrial waste-treatment costs are justifiable to protect stream resources. Advocates of this position make light of any attempt to quantify benefits because of their foregone conclusions. These advocates further believe that wastewater resources engineers are "poaching" on other fields in applying economic measures to treatment decisions. What these overexuberant conservationists fail to consider is that our economic ability to ameliorate society's ills is limited by not being able to afford to do everything to improve the environment instantly. Therefore, someone has to establish priorities setting forth the proper amount of waste treatment required. We are obliged to provide government with formulas or at least methods for making more objective decisions in pollution abatement situations.

Quantification of Benefits

We can begin the process of quantification by defining "benefits" as a willingness to pay or the value of avoiding payment of a given number of dollars at the given water quality by actual and potential water users (Nemerow and Faro 1970).

The dollar benefit of a water resource at a given quality may be determined by listing all the uses that are affected by water quality, by valuing each use individually, and by summing the resultant values. The major uses that are affected by water quality may be grouped in the following categories: (1) recreation uses, (2) withdrawal water uses, (3) wastewater disposal uses, (4) bordering land uses, and (5) in-stream water uses. The value of these uses may be estimated by taking surveys of the users to determine the extent of demand for each use and the amount each user is willing to pay for a unit of use or the unit benefit. Annual dollar benefits for a given use are the product of the total demand times unit benefit. Total annual dollar benefit at a given water quality is the sum of these benefits for each use.

Total annual dollar benefit at a higher water quality may be estimated by determining the probable demand for beneficial water uses at the new quality. This demand may be estimated by surveying the present need for comparable uses at a nearby lake or stream with this new water quality, or it may be estimated by questioning potential water users to determine latent demand likely to be present at this new quality for possible beneficial uses that are presently being foregone.

An expanded description of the above five receiving water uses is given in the following subsections.

Recreation Use Benefits

Water-oriented recreation uses will include sightseeing, walking and hiking, swimming, fishing, picnicking, boating, hunting, camping, water skiing, canoeing, sailing, and skin and scuba diving. These recreation uses may be valued by including all the expenditures of the average recreations as a measure of their willingness to pay. These include the costs of equipment, food, travel, and recreation area user fees.

Withdrawal Water Use Benefits

Withdrawal water uses include municipal water supply, industrial water supply, and agricultural and farmstead water supplies. The water-quality benefits reflected in the municipal water supply may be estimated to be at least equal to the cost of water treatment by chemical coagulation, sedimentation, and rapid sand filtration. Water-quality benefits for the industrial water supply may be estimated by using water-treatment costs, not to exceed those for municipal treatment. Industrial costs to produce ultrapure water are not assigned as water-quality benefits, because these costs are more related to overhead costs of particular manufacturing process, in contrast to the cost of a normally supplied public utility. Agricultural and farmstead water use benefits may be estimated as negative values, if damages have occurred to irrigation, poultry and livestock watering, and farmstead family or dairy uses.

Wastewater Disposal Benefits

Wastewater disposal benefits may be estimated to be the total annual costs for waste treatment required to meet existing minimum stream or effluent standards. The difference in annual costs between the existing level of treatment and the level required to meet the minimum standards may be considered a present benefit to the waste discharger. These costs include those for the common waste-treatment plants and the costs of industrial wastewater reduction practices, interceptor sewers, water-quality surveillance, stream low-flow augmentation, and possibly in-stream aeration.

Land Value Benefits

Bordering land value benefits at a given water quality may be estimated for a given land use by comparing the per-acre market value of shoreline property with the value of nearby nonshoreline property. These market values may be estimated using local tax records and the tax equalization rates. The difference in these per-acre values will then reflect unit benefits or damages of the shoreline location. Values at a higher water quality may be estimated by applying this technique to a nearby lake and projecting the ratio of shoreline to nonshoreline per-acre values back to the original lake.

In-Stream Water Use Benefits

In-stream water uses include commercial fishing, barge and ship navigation, flood control, and hydroelectric power generation. The value of commercially caught fish may be taken as a benefit, whereas the other uses involve damages or negative benefits.

Relationship of Treatment Costs to Benefits to Arrive at Unit Charge for Resources

As far back as 1970, I proposed that a regional board be empowered to sell the assimilative capacity of a specific water resource to dischargers using the resource (Nemerow and Karanik 1970). Such a board would require at least the following information from the users (industrial and others):

1. Identity of all discharges
2. Quantity of dischargers
3. Existing and desired pollution index (a measure of water quality)
4. Benefits of waste treatment
5. Assimilative capacity of the water resource

Then, using a method developed and proposed by our group (Nemerow and Sumitomo 1970), one can determine the present value of the pollution index of the water resource. The pollution index is developed for specific water uses when multiple items of water quality are considered. It is specific for one of three classifications of water use: human contact, indirect contact, and remote contact.

An overall pollution index can be developed as a weighted average of the three specific indices, the weight of each being related to the relative type use of the watercourse. Essentially, the formulation is

$$PI_{ij} = \left[\frac{Max(C_i^2/L_{ij}) + Mean(C_i^2/L_{ij})}{2} \right]^{1/2}$$

where i = specific water use,
 C_i = concentration of contaminant, and
 L_{ij} = allowable concentration of contaminant at the specific water use.

After itemizing all the annual benefits of uses of a water resource at its existing quality level, we can obtain two points on the following curve (as illustrative of actual data collected on Onondaga Lake, Onondaga County, New York State [see Figure 21.1]).

Before we examine the figure, I would like to quote some words of wisdom by former Vice President Albert Gore (1992) on our economic system:

> The hard truth is that our economic system is partially blind. It "sees" some things and not others. It carefully measures and keeps track of the value of those things most important to buyers and sellers, such as food, clothing, manufactured goods,

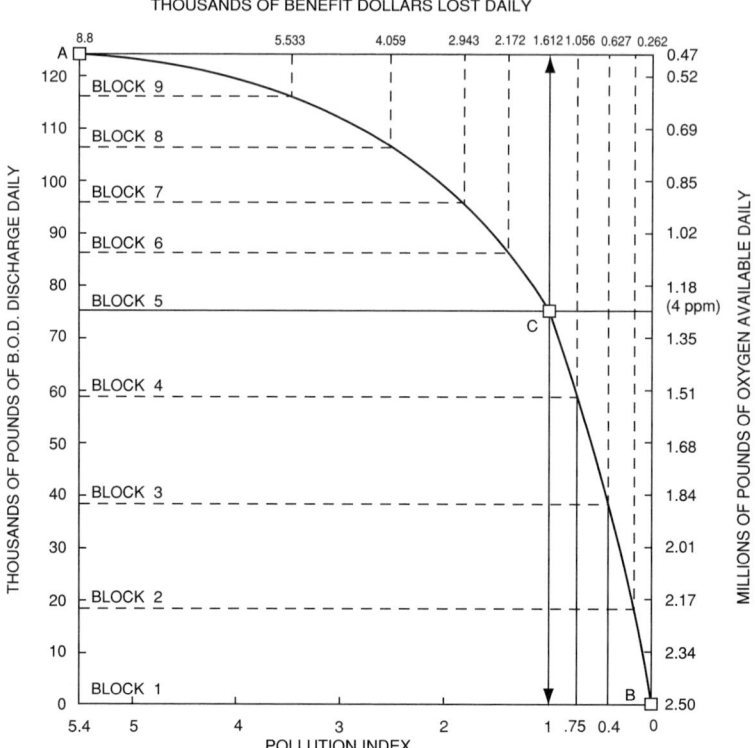

FIGURE 21.1. Unit price calculation.

work, and, indeed, money itself. But its intricate calculations often completely ignore the value of other things that are harder to buy and sell: fresh water, clean air, the beauty of the mountains, the rich diversity of life in the forest, just to name a few. In fact, the partial blindness of our current economic system is the single most powerful force behind what seem to irrational decisions about the global environment.

In Figure 21.1, two points are very significant: A, the coordinates of 5.4, the computed present pollution index of the lake, and 123,000 lb of BOD per day, the present daily BOD inflow into the lake. The other point, B, has coordinates of 0 as a pollution index (representing absolutely no contaminants) and 0 pounds of BOD being discharged into the lake each day. Although no precise data are available between these two factual points, we can make some assumptions that will help define the shape of the curve. First, the curve should descend in BOD towards a resultant pollution index of zero, but probably not precisely in a linear fashion. We also know that point A corresponds to $8,800 of benefits lost daily at the present discharge level and that point

B corresponds to no loss in benefits when there is no discharge of wastes. A third point, C, can be approximated at the intersection of the two water-quality objectives, PI = 1 and dissolved oxygen = to 4 mg/L. The curve can now be drawn with considerable certainty through the three points A, B, and C. The latter point presumes that a direct relationship exists between the daily BOD discharged (76,000 lb) and dissolved oxygen available (1.2 million 16/day at 4 ppm). From this curve, we can proceed to make a firm decision of what water quality to maintain based on constraints of dollar benefits and amount of oxygen with a resultant allowable BOD discharge. In our illustration (Figure 21.1), we decide to improve the PI to 1.0 from its current value of 5.4. This will result in a decrease in dollars of benefits lost from $8,800/day to $1,612/day (a savings of $7,188 at a discharge of 76,000 lb BOD per day) (37% reduction) or sale of BOD capacity resources up to a level of 76,000 lb/day are the two alternatives open to the board. If sale of resources is chosen, four blocks of 19,000 lb BOD each can be sold at costs directly related to the dollar benefits lost for each incremental amount of BOD contamination allowed. For example, the third 19,000 block results in a benefit loss of $1,056–627, or $429/day. Therefore, a potential and reasonable unit charge could be $429/19,000 or 2.3 cents/lb. Customers (water-polluting users) could opt to build their own waste-treatment plants to eliminate the need for purchase of these BOD units. You will note that each block of BOD capacity purchase becomes increasingly more expensive. Presumably, no more than 76,000 lb of BOD would be sold unless a lower water quality was desired and less would be sold if a higher quality was selected by the board. Customers in this case will probably select the least-cost alternative to comply with allowable BOD discharge.

Purpose of Quantified Benefits

Quantified benefits serve two major purposes. First, they allow one to compare the total annual dollar benefit with a total annual expenditure or cost of maintaining or achieving a particular water quality (*pollution index*). In the case of the Onondaga Lake analysis, it was found that the annual benefits of an improved water quality (PI = 1.0) would be about $7.5 million or $3.2 million more than at its existing water level (PI = 5.4). Annual expenditures of capital and interest payments to obtain this improved water quality have been estimated at about the same $3.2 million. Therefore, we can show that the quantifiable benefits alone would equal the annual costs. In addition, intangible benefits would certainly exist and be ample to dictate a "cleanup" policy to the board.

Second, they allow one to compute a unit charge for a pollution-carrying capacity resource. For example, in our illustration the sale of the third block of BOD at $2.3 cents/lb will decrease our water quality from a pollution index of about 0.4 to about 0.75 (see Figure 21.1). All pollution capacity consumers (BOD purchasers) are now in a position to select one of the two alternatives: (1) to purchase BOD capacity at $2.3 cents/lb, or (2) to treat its own wastes at a lower cost. In deciding between the two, the consumer must consider both capital and operating costs. Typical capital costs for many organic industrial wastes ranged from $50 to $150 per pound of BOD per day at the time of this study. Operating costs for industrial waste-treatment plants are relatively high and extremely variable and were not available for public analysis at the time of the study. It was unfortunate that many industrial plants themselves did not assess their waste-treatment operating costs at that time.

How to Determine Whether an Industrial Plant Can Afford Waste Treatment

Neither an industrial plant nor a governmental regulatory agency really knows whether a given industrial plant can afford to build and operate waste-treatment facilities. We can discount the idea that a plant must treat its wastes regardless of the cost as being unrealistic. We must provide an objective and feasible method for ascertaining a treatment cost that a plant can afford and still remain competitive in its industrial category. Vice President Gore observed that "the bad things economists want to ignore while they measure the good things are often said to be too difficult to integrate into their calculations" (Gore 1992).

In a survey conducted during the 1970s, detailed information was obtained from a questionnaire from four of nine preselected plants (Nemerow 1971). These plants were selected because of the my previous knowledge of their cooperative participation in effective waste treatment while remaining highly competitive. The four dependable replies were evaluated and are summarized in Table 21.1.

Study of the table reveals that two relationships of industrial economics have potential for application in water pollution abatement. The first is the ratio of dollar values of waste-treatment cost to the production cost or "value added." Three of the four plants reported a 1% (±0.1%) ratio. The fourth plant in the other industry showed only a 0.42% ratio. Indications are that a plant may allocate a percentage of its production cost to waste treatment. A very preliminary rough approximation of this percentage showed about 1% as a fair value. However, these plants used very economical and effective waste-treatment methods.

The second ratio results are $0.18–0.39 cost of waste treatment for each 1,000 gallons of waste treated. All costs of waste treatment include annual costs of amortized capital expenditures, maintenance, power, and chemicals. A 10-year useful life of equipment and a 7% interest rate were used to compute annual capital costs.

The most important deficiency of the usefulness for both methods is that of determining whether a plant can really afford to spend 1% of its production cost or even $24 cents/1,000 gallons of waste for treatment costs. The financial profitability as a result of business and professional management determines the ability of a plant to compete with others and, thus, be able to provide adequate waste-treatment facilities.

An attempt was then made to compute significant values such as marginal income and profit to sales ratios (Ganotis and Nemerow 1972). Total sales revenue of a plant minus the direct costs of the plant would be classified as "marginal income." Direct costs include raw basic materials, materials used in the process to produce the product, variable labor costs, utilities used in production of the product, and waste-treatment costs. All other costs are fixed costs and include such items as management salaries,

TABLE 21.1
Summary of Cost Data of Four Plants

			Cost Ratios Based on Product Economics				Cost Ratios Based on Waste Flows		Cost Ratios Based on BOD Removals		Cost Ratios Based on Suspended Solids Removal		Ratio Cost Based on Total Plant Assets
Type of Industry	Plant No.	Type of Product	Waste Treatment Cost / Dollar Value Added[a] (%)	Waste Treatment Cost / Dollar Value of Raw Materials	Waste Treatment Cost / Dollar Value Added Plus Raw Material (%)	Waste Treatment Cost / Dollar Value of Selling Price (%)	Waste Treatment Cost / Dollar Value of Profit and Overhead (%)	Waste Treatment Cost / Gallons per Day	Waste Treatment Cost / 1,000 gals. Waste Treated ($/1000 gals)	Waste Treatment Cost / Pound BOD per Day ($/# BOD per Day)	Waste Treatment Cost / Pound BOD ($/# BOD)	Waste Treatment Cost / # Pound SS Removed in a day ($/# SS/day)	Cost / Suspended Solids Removed ($/# SS)	Waste Treatment Cost / $ Plant Assets (%)
Cannery	1	Beans	1.11	2.16	0.795	0.51	1.64	0.0236	0.374	10.30	0.1635	18.50	0.294	0.574
Cannery	2	Tomatoes Peaches	1.08	—	—	—	—	0.007	0.233	1.71	0.057	2.28	0.076	1.50
Poultry processing	3	Chickens	1.096	0.159	0.139	0.122	0.89	0.0484	0.186	1.08	0.00417	1.92	0.0074	0.925
Tannery	4	Upper leather	0.42	0.245	0.155	0.101	0.29	0.1035	0.393	47.00	0.004	0.957	0.000329	1.04

[a] Best potential for meaningful relationship for water pollution abatement.
SS, Suspended solids.

rent, mortgage and interest, loans, taxes depreciation, waste-treatment capital costs, and so on. When fixed and direct costs are added and the sum subtracted from sales revenue, one obtains the true profit income.

The resulting ratios of marginal and profit income values to the sales revenues for each plant of a given industry can be compared as follows:

$$\frac{\text{Marginal Income}}{\text{Sales Revenue}} + \frac{\text{Profit Income}}{\text{Sales Revenue}} = \text{Financial Index}$$

The higher the financial index, the more financially able a plant should be to afford additional waste treatment. When comparing indices of all plants in an industry, one can decide by using the statistical value of the financial index and the industry's agreement of which plant should be able to provide waste treatment and still remains economically competitive.

Unfortunately, although the method is sound, industry either does not know its real direct and fixed costs or does not want to disclose these to "outsiders." Without this information, the method is useless. Perhaps sometime soon, if and when marginal income and profits for each separate plant are known and made public, this method will provide a means for decision making in the environmental area.

Compromise on Practical and Feasible Method of Determining Industrial Firm's Capability to Provide Waste Treatment

A method can be developed and tested to establish a basis for determining both a reasonable amount of expenditure that potentially could be spent for treatment and for compelling the most financially able companies to instigate treatment first. Financial potential is measured by the pollution abatement cost (to meet required environmental quality) to sales ratio for each firm, hereafter referred to as the *sales index*. For a firm with sales revenue of $100 million annually and a treatment cost requirement of $5 million, the sales index would be 5/100, or 0.05. Indices can be calculated in a similar manner for all firms in the industry, placed in an array, and analyzed statistically.

A hypothetical comparison of the indices of five firms in an industry is shown in Figure 21.2 (Nemerow 1971). Indices vary from $0.025 to $0.11 per sales dollar. In our illustration, the firms in the industry arrived at a consensus that an index of more than 10 cents per sales dollar would result in an economically impossible situation. Therefore, firm A must increase its sales dollars, internalize an increased profitability or receive some sort of subsidy in order to be able to pay for pollution abatement. Its other alternative is to cease or alter production. On the other hand, firm D, with the lowest index of 2.5 cents, would be the most financially able to treat its waste first (with a resulting change in its index).

A new redistribution of the indices is then computed (see Stage II in Figure 21.2), with a mean index increasing towards the cutoff point of 10 cents and, firm E, with an index of 4 cents next in line for treatment.

Hypothetical Index Calculations:

Statistical Treatment of Hypothetical Data of Treatment Costs and Sales

Firms	A	B	C	D	E
1. Proposed Treatment Cost (85% removal)	$ 110,000	$120,000	$30,000	$20,000	$18,000
2. Gross Sales:	$1,000,000	$1,3000,000	$600,000	$800,000	$450,000
Index ($^{(1)}/_{(2)}$) $/per sales $	$1.1	$.093	$.05	$.025	$.04

Assume: Prior industrial Agreement has been made that it is feasible to spend up to 10c per sales dollar before financial crisis occurs.

STAGE I Average index: 5.3 c/$sale, Firm D potentially will be most able to treat *first*.

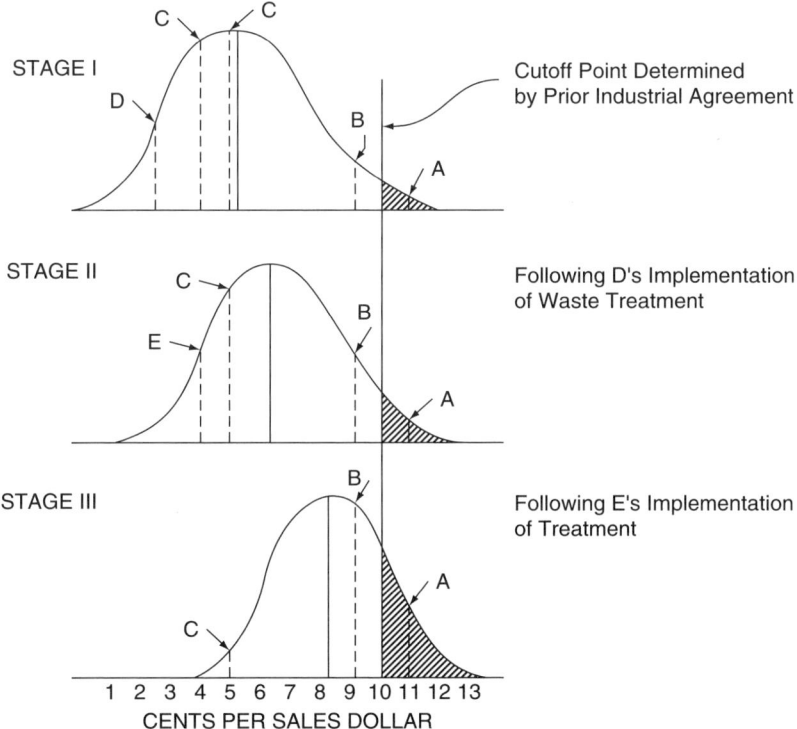

FIGURE 21.2. Sales values, treatment costs, and sales indices of the pulp and paper industry.

The sales index method was applied to the pulp and paper industry, for which data were published in 1971 ("Paper process pollution" 1971). The sales values, treatment costs, and sales indices are shown in Table 21.2.

The industrial average treatment cost of $30.68 million per company would be needed to meet requirements. In addition, it would cost an average of $4.34 cents per

TABLE 21.2
Pulp and Paper Industry Treatment Costs, Sales Revenues (1970), and Sales Indices (Computed)

Corporation	(A) Treatment Cost $ Millions	(B) 1969 Sales in $ Millions	Sales in $\frac{A}{B} \times 10$
1. Amer. Can	12.5	1,723.7	7
2. Boise-Cascade	42.1–50.1	1,726.0	26
3. Consolidated Paper	9.29	127.7	7
4. Cont. Can	16.5	780	9
5. Crown Zeller	58.7	919.3	64
6. Diamond Int.	8.6	498.1	17
7. Fibreboard	4.8	181.8	26
8. Ga. Pac.	23.6	1160.2	20
9. Gt. Northern Nek.	25.2–34.2	340.7	86
10. Hammermill	41.5	353.3	1175
11. Hoerner-Wald.	17.5	237.3	73
12. Int. Paper	101	1777.3	57
13. Kimb-Clark	13.5	834.7	16
14. Marcor	28.3	2500.7	11
15. Mead	26	1038	15
16. Owens Ill.	3.9	1294.4	3
17. Potlach	28–37	337.1	96
18. Riegal	11.5	184.0	62
19. St. Regis	59.8	867.8	69
20. Scott	78.8	731.5	107
21. Union Camp	9.0	449.5	20
22. U. S. Plywood	13.2–21.2	1455.5	11
23. Westvaco	36.5	419.6	87
24. Weyerhaeuser	32.5	1239.2	26

Note: Capital and estimated operating cost to reach desired treatment level as required by federal & state governments (average costs used in computation of sales index).

sales dollar (sales index) to attain this level of waste treatment. Descriptive statistics of sales index data showed widespread differences among firms, as indicated by a range of 0.3–11.75 cents per sales dollar. The relatively high standard deviation implies that pollution abatement activity in this industry varies significantly among firms. A cutoff point (upper limit of sales index) would then have to be determined by a caucus of the 24 companies, and presumably Company 16 (sales index 0.3) is in the best financial position to install necessary waste-treatment facilities while Company 10 (sales *index* 11.75) may already be financially incapable of affording waste treatment.

We only know that Company 16 is potentially capable of additional expenditure for waste treatment and likewise Company 10 is potentially incapable of the same expenditure. The method does not assess the companies' actual financial ability to pay. It gives the companies and regulatory agencies a valid indication, however, that one company should be able to afford treatment costs and the other should not be able to afford the same costs. These conclusions, in themselves, should reveal facts to both the companies and society that heretofore had been undisclosed. The findings should provide both with directives for future action that will lead to elimination of excessive pollution by an industrial sector.

References

Costs of Water Pollution Control, National Symposium North Carolina State University, Raleigh, North Carolina, April 6–7, 1972, pp. 230–245.

Ganotis, C., N. L. Nemerow. 1972. Development of a Rational Procedure to Determine Reasonable Expenditures for Industrial Waste Treatment. In: *Proceedings of the 5th Annual Northeast Regional Antipollution Conference Kingston, Rhode Island, July 18–20, 1972.*

Gore, A. 1992. *Earth in the Balance,* p. 182. New York: A Plume Book.

Nemerow, N. L. 1971. *Economics from the Viewpoint of Clean Streams for Our Future Environment. The National Specialty Conference Los Angeles, California, March 24, 1971,* pp. 279–291.

Nemerow, N. L., R. C. Faro. 1970. Total dollar benefit of water pollution control. *J. San. Eng. Div. Proc. A.S.C.E.* SA3 7323, June.

Nemerow, N. L., J. Karanik. 1970. *Benefits of Water Quality Enhancement.* Water Pollution Control Research Series DAJ 16110 12/70, Part C. Washington, DC: Environmental Protection Agency, Water Quality Office.

Nemerow, N. L., H. Sumitomo. 1970. Part A. In: *Benefits of Water Quality Enhancement.* Water Pollution Control Research Series DAJ 16110 12/70. Washington, DC: Environmental Protection Agency, Water Quality Office.

Paper Process Pollution in the Pulp and Paper Industry, p. 58. New York: Council on Economic Priorities, 1971.

CHAPTER 22

Summary

Nelson Leonard Nemerow

I have attempted to build the case for a change in industrial manufacturing from one consisting of waste treatment to one encompassing waste utilization. This significant change is made possible by using the concept of environmentally balanced industrial complexes (EBICs) instead of free-standing industrial plants. EBICs are designed so that each industrial plant utilizes some wastes from other adjacent plants with the result that no wastes escape into the external environment. The ultimate results are elimination of industrial environmental pollution and lower industrial production costs. Many examples, both realistic and potential, of such EBICs are given throughout the book.

Other useful resource information is given, such as the evolution from the Industrial Revolution era to the present. The academic and scientific development of environmental knowledge in the United States is also presented.

I have also expanded the EBIC concept to include examples of industrial–municipal complexes with the same overall objectives. In addition, byproduct exchanges and eco-industrial park examples of others are described to show a temporary bridge between the old and new EBIC practices. I have referred to these developments as *naturally evolving industrial complexes*.

I hope the case studies and background rationale for EBICs will provide more clarity and proof that the new method of industrial production is not only practical and economical, but also necessary and vital to our future.

Critique of the Subject of Environmentally Balanced Industrial Complexes

Franklin J. Agardy

Agardy, a partner in Forensic Management Associates, Inc., holds a bachelors degree in civil engineering from City College of New York, and master's and Ph.D. degrees in sanitary engineering from the University of California, Berkeley. He is one of the nation's leading experts in the water pollution control field and has extensive experience in hazardous waste management, solid waste, and landfill management, as well as environmental planning and analysis.

As a widely published authority on biological treatment systems and the treatment of industrial wastes, Agardy has served as a consultant to the U.S. Environmental Protection Agency, as well as local and state agencies with similar responsibilities. His experience includes air pollution control and solid waste management projects involving recycling and resource recovery, management of environmental impact studies for oceanographic and transportation systems, oil refineries, offshore terminal facilities, and power plants. His efforts have ranged from water-quality studies of lakes, lagoons, and marine systems to preparation of oil spill contingency and utility emergency plans. Litigation management support and expert witness activities have been a natural extension of these efforts.

Critique

Efforts to reduce and, indeed, eliminate "pollution" have been ongoing for well over 100 years. However, the concept of "industrial groupings" targeted to entirely eliminate pollution has been and continues to be voiced most prominently and vocally by one Dr. Nelson L. Nemerow.

How many times have we heard that the lessons of history point to the events of tomorrow. And, how many times have we ignored these lessons and "assumed" that we still had time to address and correct mistakes, mistakes often made in haste or in the name of "expediency." Even today, we hear that old slogan, The Solution to Pollution is Dilution. And although that slogan long ago wore out its welcome, we certainly still implement dilution as a solution, at least partially. Indeed, if the scientific projections are correct, global warming is being exacerbated by the ever increasing release of carbon dioxide to such a degree that it is exceeding the ecosystem's ability to "assimilate" the releases. A similar picture can be painted for the release of fluorocarbons to the atmosphere and both domestic and industrial liquid wastes to numerous bodies of water around the globe. Once assimilative capacity has been exceeded, the rate of overall degradation increases rapidly and possibly to the point of no return. Indeed, is it possible, for example, to replace rain forests? I do not believe so.

Although there have been papers written and facilities constructed in which residuals from one industry or operation have been employed as feedstock for other industries, the examples did not lead to broad (or even narrow) acceptance of the approach. In a 1971 paper (1), the term "misplaced resources" was coined to describe residuals, so although the concepts described by Dr. Nemerow are not new, they were simply not fully developed and certainly not embraced. However, there is value in reviewing some of these efforts, in time sequence, to shed more light on the "historical perspective" leading to the "environmentally balanced industrial complex."

In early 1969, I had occasion to travel to Amarillo, Texas, and visit a 50,000-head cattle feedlot. The accumulation of feedlot wastes had reached such proportions that on windy days downtown Amarillo played host to a fine dust, with an associated odor that caused considerable grief to the community. The feedlot happened to be owned by a company that also had extensive natural gas holdings, with a gas transmission pipeline running directly through (actually under) the feedlot. So, how to take a waste stream and use it as raw materials such as was the case with the natural gas? The obvious solution was to anaerobically digest the feedlot waste, converting the majority to methane gas, which could (after removing the impurities) be added to the pipeline. Alas, the cost analysis demonstrated that the feedlot methane production exceeded the natural gas production by \$0.1 cent/ft^3. This was a sufficiently large enough differential to kill the project. The feedlot waste continued to pile up, the wind continued to blow, and the next attempt at a solution came years later. The story did have a positive ending in that the firm of CH2M-HILL did indeed design and build a bioconversion plant in the state of Colorado in the mid-1970s, converting feedlot waste to methane.

A similar story, also dealing with an agricultural waste, in this instance rice hulls, can be told. In the early 1970s, the State of California banned the burning of rice hulls, citing air pollution restrictions. One has only to visit the delta area of California to fully appreciate the extensive rice growing operations. So, once again a residual "needed a new home" and horse collars had long since gone out of style. Looking at the composition of rice hulls, one finds essentially a pure mixture of carbon and silica. Reflecting on the fact that the raw materials required in the manufacturing of cement include silica and that heat must be applied in the manufacturing of cement, it does not require a great leap of faith to appreciate that rice hulls contain the silica required for the cement and the carbon required for the fuel source. Again, the devil was in the economics, and California was just the wrong environment for such a combination of cement manufacturing and agribusiness, at least in the 1970s. Again, however, the idea did not die, and in the early 1980s, just such a plant was designed and constructed in Indonesia.

In 1971, the industrial/utility complex concept was described using several examples in which co-location of power-generating facilities (producers of large quantities of hot water) with other neighboring operations were "projected" to have a decided advantage over "going it alone." One of the examples included co-location of power generators with desalination plants, wherein the heat could be used to assist in evaporation/separation. This simple co-mingling is, however, a one-directional dependency wherein the power plant benefits the desalting operation (other than reducing the ultimate heat load released to the environment) but continues to operate independently. This same one-directional dependency can be exhibited by co-location of power-generating facilities with municipal wastewater treatment plants (this concept works equally well with industrial wastewater facilities) where the increased temperature positively affects both chemical and biological processes through increased process efficiency (on a theoretical basis by at least 5%). As interesting as these ideas were, we as authors of the paper concluded by saying, "From a practical standpoint, we are also faced with people and political problems, mostly related to a reluctance to do things differently than they have been done in the past."

Have we finally reached that point of no return? I would hope not, but time is of the essence, and Dr. Nemerow has developed a carefully thought out and documented solution, which if (when) implemented will have a profound effect on our appreciation of waste residues as, indeed, misplaced resources and will represent a significant beginning to an entirely new age in the continuing battle to reduce and finally eliminate pollution—and, I might add, in a profitable manner.

In the final analysis, we no longer have the luxury of using "time" as "assimilative capacity." We no longer can depend on assimilative capacity at all. Finally, we can no longer equivocate, employing the age-old formula of "more studies." The birth of the "environmentally balanced industrial complex" signals the beginning of an entirely new approach to saving our environment. Having said this, I urge every reader to take this message of Dr. Nemerow's to heart and press on to an implementation strategy as soon as possible.

Index

Accidental spills, 31, 32
Acetic acid, 152
Acetone, 290
Acid corrosive waste, 256, 257
Acid waste deep-well injection, 129
Acid waste neutralization, 36, 37, 38, 39
Acid waste utilization, 43
Activated carbon adsorption, 307, 308, 309, 310
Activated-sludge treatment, 109, 110, 111, 112, 136, 137, 138, 219, 224, 226
Acute hazardous industrial waste, 264, 265, 266, 267, 268, 269, 270
Adsorption, 85, 307, 308, 309, 310
Aeration of equalization basins, 47
Aerobic digestion of sludge solids, 149, 150, 151, 152, 153
Agitation, 47, 84
Agriculture, 485, 486, 506, 507, 508, 509, 510
Agriculture–biomass power plant–municipal–forestry environmentally balanced industrial complex, 452, 453
Agriculture–cannery environmentally balanced industrial complexes, 473, 474
Agriculture–municipal wastewater complex, 517, 518
Agriculture–power plant environmentally balanced industrial complexes, 472, 473
Air contaminant limits, 334, 335, 336, 337, 338, 339, 340, 341, 342, 343
Air contaminant removal, 355, 356, 357, 358, 359, 360
Air stripping, 306, 307
Alcohol, *see* Ethanol
Algae in lagoons, 107, 108
Algae mineral removal, 95, 96, 97
Alkaline corrosive waste, 257
Alkaline waste neutralization, 40, 41, 42, 43
Alkylbenzene sulfonate (ABS), 100, 131
Alternative energy sources, 396, 397, 398, 399, 400, 401, 402, 403
Alum, 81, 83, 84, 263, 288
Aluminum–electric power–brick plant environmentally balanced industrial complex, 483, 484, 485
Aluminum oxides, 82
Aluminum recycling, 365
Aluminum sulfate, 81
American Chemical Society (ACS), 253
American Society for Testing and Materials (ASTM), 327, 344
American Toxic Waste Disposal, Inc., 271

549

550 *Index*

Ammonia, 357, 358
Ammonium sulfate–fertilizer–cement environmentally balanced industrial complex, 405, 406, 407, 410, 411, 412
Anaerobic digestion, 125, 149, 150, 151, 152, 153
Animal feedlot–plant food environmentally balanced industrial complex, 477, 478
Animal feed–hog production–energy production environmentally balanced industrial complex, 503, 504, 505
Antibiotics, 266
Applied Sustainability (AS), 524
Asbestos, 32, 269, 270
Asbestos Hazard Emergency Response Act (AHERA), 269
Asnaes Power Station, 528
Atomic Energy Commission (AEC), 475
Atomized suspension, 161, 162
ATW/Calweld, Inc., 308
Automatic flow-control systems, 48
Automotive oil waste composition, 255

Baffling, 46, 47, 59
Bagasse, 421, 422, 423, 424, 425, 426, 503
Barging of sludge, 168
Barrett Sludge Extractor, 302
Basins for equalization, 45, 46, 47
Batch discharge, 19
Bauer Company, 74
Benzol–coke–tar plant environmentally balanced industrial complex, 479, 480
Bio-Disc system, 133, 134, 135, 136, 137, 138, 139, 140
Biochemical oxygen demand (BOD), 17, 26, 28, 29, 45, 54, 68, 71, 79, 108, 109, 111, 113, 125, 126, 134, 135, 138, 213, 214, 218, 394, 536, 537

Biological treatment of hazardous waste, 289, 290
Biological wastes, 266, 267, 268
Biomass energy, 403
Biomass power plant–municipal–forestry–agriculture environmentally balanced industrial complex, 452, 453
Biosolid settling tank, 20, 21
Biosorption, 114, 115, 115
Blackstrap molasses, 31
Blood, 30
Boiler-flue gas, 40, 41
Boric acid plant–desalination environmentally balanced industrial complex, 505, 506
Brick plant–aluminum–electric power environmentally balanced industrial complex, 483, 484, 485
Brine waste, 29, 31
Briquette–sugarcane environmentally balanced industrial complex, 503
Briquette–sugarcane–fertilizer environmentally balanced industrial complex, 420, 421, 422, 423, 424, 425, 426, 427, 429
Bromine, 141
Brownian movement, 79
Brush aeration, 131, 132
Bubbler System, 49
Buoyancy, 67
Business Council for Sustainable Development, 523
Byproduct recovery, 29, 30, 31
Byproduct synergy, 522
Cachaza, 425, 426, 503

Cannery–agriculture environmentally balanced industrial complexes, 473, 474
Cannery–inorganic chemical and chlorine plant environmentally balanced industrial complex, 497, 498, 499

Index 551

Cannery–nuclear power plant waste processing environmentally balanced industrial complex, 493, 494, 495, 496
Canning industry, 108, 124
Carbohydrate digestion, 150
Carbon dioxide, 36, 40, 41, 42, 125, 152, 357, 358, 460, 461
Carbonic acid, 40
Cattle industry, 547
Caustic soda, 36, 38, 39, 40, 300
Cavitation, 125, 126
Cayadutta Creek, 196, 198, 199, 200, 204, 210, 211, 212, 213, 214
Cell phone reclamation–cement plant–concrete product environmentally balanced industrial complex, 500, 501, 502, 503
Cellophane membrane, 92
Cellulose nitrate membrane, 92
Celotex, 31
Cement–ammonium sulfate–fertilizer environmentally balanced industrial complex, 405, 406, 407, 410, 411, 412
Cement-based fixation, 296
Cement–concrete block plant–coal power plant environmentally balanced industrial complex, 454, 455, 456, 457, 458, 459, 460, 461
Cement–fertilizer environmentally balanced industrial complex, 381, 382, 383, 384, 385, 386, 387, 388, 389, 390, 391, 392, 393
Cement–lime–power plant environmentally balanced industrial complexes, 466, 467, 468, 469, 470
Cement plant–concrete product–cell phone reclamation environmentally balanced industrial complex, 500, 501, 502, 503
Cement production costs, 388
Cement raw materials and wastes, 407
Cement–steel mill–fertilizer environmentally balanced industrial complex, 453
Centrifuging, 165, 166, 167, 168
Ceramic membranes, 93
Charge neutralization of colloids, 82, 83, 84
Chemical coagulation, 81, 287, 288
Chemical Conservation Corporation (ChemCon), 347, 349
Chemical fixation of hazardous waste, 296
Chemical oxidation of organic matter, 141, 142
Chemical oxygen demand (COD), 18, 79
Chemical Process Industry (CPI), 345
Chemical treatment of hazardous waste, 287, 289
Chemical Waste Management, 286
Chemical weapons, 286, 287
Chem-Trol Pollution Services, Inc., 141
Chlorella, 95
Chlorine, 141
Chlorine dioxide, 141
Chlorine, water, electric power, and lye plant environmentally balanced industrial complex, 482, 483
Chlorine–pickling cannery–inorganic chemical plant environmentally balanced industrial complex, 497, 498, 499
Chlorofluorocarbon (CFC), 303
Chlorophyceae, 96
Chromates, 99, 306
Chromium, 27, 30, 227
Churchill multiple-regression, 212
Clarifiers, 59, 63, 64, 65, 66
Clarithickener, 58
Coagulation of minerals, 99
Coal mine, 26
Coal power plant–cement–concrete block plant environmentally balanced industrial complex, 454, 455, 456, 457, 458, 459, 460, 461
Coal-derived granular carbon, 86

Coffee
  decaffeination–dry–cleaning–metal plant environmentally balanced industrial complex, 490, 491
Coilfilter, 154, 155
Coke and gas–steel mill–fertilizer environmentally balanced industrial complex, 446, 447
Coke dry quenching, 26
Coke–tar–benzol plant environmentally balanced industrial complex, 479, 480
Collection and Reclaiming, 140, 141
College Retirement Equity Fund (CREF), 379, 380
Colloid characteristics, 79, 80, 81
Colloidal solid removal, 79, 80, 81, 82, 83, 84, 85, 86
Colorado River, 18
Completely mixed system, 117
Composite waste analyses, 215, 218, 219
Composite waste sampling, 2215
Composting sludge, 171
Comprehensive Environmental Response, Compensation, and Liability Act (CERCLA), 269, 270, 322, 327, 344
Concrete block, 457, 458, 459
Contact stabilization, 114, 115, 116
Contaminant concentration reduction, 25, 26, 27, 28, 29, 30, 31, 32, 33
Cooling water, 13, 16
Copper recovery, 30, 299, 300
Copperas, 81, 83
Corn growing and processing–ethanol environmentally balanced industrial complex, 485, 486
Corn growing–lumber–textile–ethanol environmentally balanced industrial complex, 509, 510
Corrosive hazardous waste, 256, 257
Cotton gin, 5
Cow feedlot–power plant–fertilizer environmentally balanced industrial complex, 506, 507, 508

Credits, 2
Cyanide, 26, 27, 30, 99, 257, 258, 306
Cyanophyceae, 96

Dairy industry, 27, 30, 506, 507, 508
De-rusters, 26
Deep-well injection, 126, 127, 128, 129, 130, 276, 291, 292
Dense fixed bed, 356
Desalination–boric acid plant environmentally balanced industrial complex, 505, 506
Desalination–electric power plant environmentally balanced industrial complex, 496, 497
Detention time, 68
Detoxification of hazardous wastes, 305, 306, 307, 308, 309, 310
Dialysis, 91, 92, 93
Dibromochloropropane (DBCP), 266
Dioxin, 266, 267, 286
Direct costs of waste treatment, 385, 387, 389
Dispersed solid system, 357
Dispersed-air flotation, 67
Dispersed-growth aeration, 112, 113, 114
Dissolved-air flotation, 67, 70
Dorr-Oliver, Inc., 468, 470
Dow Chemical Company, 526
Drinking water plant–electric power plant environmentally balanced industrial complex, 496, 497
Dry-cleaning industry, 273
Drying beds, 157, 158, 159
Dry–cleaning–metal plant–coffee decaffeination environmentally balanced industrial complex, 490, 491
DuPont Company, 283
Dye recovery, 304, 305

EbaraInfilco, Ltd., 42
Eco-industrial parks (EIPs), 529, 530
El Paso Products Company, 17

Index 553

Electrical storing–wax manufacturing environmentally balanced industrial complex, 492, 493
Electric Power Research Institute (EPRI), 473
Electrochemical treatment of hazardous waste, 316, 317
Electrodialysis, 92, 93
Electron beam treatment, 268
Elutriation, 154, 156, 157
Emission trading program, 394, 395
Encapsulation fixation, 296
Enhanced high-rate clarification (EHRC), 50
ENRECO, Inc., 297
Environmental knowledge, 6, 7
Environmental Protection Agency (EPA), 245, 246, 247, 249, 250, 254, 259, 275, 280, 286, 305, 308, 309, 310, 317, 321, 322, 323, 325, 326, 344, 345, 455, 522, 523, 524, 525
Environmental site assessment (ESA), 327, 343, 344
Environmentally balanced industrial complex (EBIC), 2, 3, 366, 367, 545
Environmentally balanced industrial complex benefits, 531, 532, 533, 534, 535, 536, 537, 538, 539, 540, 541, 542, 543
Environmentally balanced industrial complex critique, 546, 547, 548
Environmentally balanced industrial complex economic justification, 379, 380, 381, 382, 383, 384, 385, 386, 387, 388, 389, 390, 391, 392, 393, 394, 395, 396, 397, 398, 399, 400, 401, 402, 403
Environmentally balanced industrial complex implementation, 377, 378
Environmentally balanced industrial complex planning, 373, 374, 375, 376, 377
Environmentally balanced industrial complex rationale, 369, 370, 371, 380, 381

Equalization, 19, 28, 29, 45, 46, 47, 50
Equipment modification, 26, 27
Ethanol, 152
Ethanol–corn growing and processing environmentally balanced industrial complex, 485, 486
Ethanol–lumber–textile–corn growing environmentally balanced industrial complex, 509, 510
Ethanol–sugarcane–power environmentally balanced industrial complex, 418, 419, 420
Ethanol–sugar–gasoline plant environmentally balanced industrial complex, 499, 500
Ethanol–wood environmentally balanced industrial complex, 481, 482
Evaporation, 89, 90, 91, 157, 158, 312
Evaporator load, 163
E-wastes, 501
Expanded bed, 356
Explosives manufacturing, 274
Exportation of hazardous wastes, 298
External costs, 8

Fair's f, 212
Fenton's reagent, 142
Fermentation, 150, 151, 152
Ferric chloride, 81, 156
Ferric sulfate, 81, 84, 289
Fertilizer, 261, 262, 263
Fertilizer–ammonium sulfate–cement environmentally balanced industrial complex, 405, 406, 407, 410, 411, 412
Fertilizer–cement environmentally balanced industrial complex, 381, 382, 383, 384, 385, 386, 387, 388, 389, 390, 391, 392, 393
Fertilizer–cow feedlot–power plant environmentally balanced industrial complex, 506, 507, 508
Fertilizer–steel millcoke and gas environmentally balanced industrial complex, 446, 447

Fertilizer–steel mill–cement environmentally balanced industrial complex, 453
Fertilizer–sugarcane–briquette environmentally balanced industrial complex, 420, 421, 422, 423, 424, 425, 426, 427, 429
Financial index, 540
Fischer and Porter Co.49
Fish, 35, 250
Flameless thermal oxidation, 287, 288
Flocculation, 55, 56, 70
Flotation, 66, 67, 68, 69, 70, 86, 289
Fluidized bed, 356
Fluorides, 100
Fluorine, 263
Fluorosilicious acid, 263
Fluosolids system, 468, 470
Fly-ash, 284, 294, 295
Foam phase separation, 130
Food–electricity–water production, 519, 520, 521
Formaldehydes, 272, 273

Galvanizing wastes, 260
Gasoline–sugar–ethanol plant environmentally balanced industrial complex, 499, 500
GDS, Inc., 308
Geothermal energy, 400, 401, 402
Glass block–nuclear power plant environmentally balanced industrial complex, 475, 476, 477
Glass recycling, 366
Gloversville–Johnstown joint treatment plant, 228, 229, 233, 234, 235, 236, 237, 238, 239, 240, 241, 243
Grease trap, 487
Gross bed loading, 158
Gypsum, 405

Halogen organic waste, 274
Halophiles, 261
Hazardous waste, 245, 246, 249

Hazardous waste treatment risk evaluation, 277, 280, 281
Hazardous waste types, 251, 252, 253, 254, 255, 256, 257, 258, 259, 260, 261, 262, 263, 264, 265, 266, 267, 268, 269, 270, 271, 272, 273, 274, 358, 359, 360
Heat transfer rate in evaporation, 90, 91
Henry's Law, 67
High-rate aerobic treatment, 116, 117
Hindred settling, 64
Hoak's nomograph, 39
Hog production–animal feed–energy production environmentally balanced industrial complex, 503, 504, 505
Hospital waste, 266, 267, 268
Hydrasieve, 74
Hydrochloric acid, 141, 165
Hydrochloric acid pickling, 26
Hydroclone, 74
Hydrofluorocarbons (HFCs), 303
Hydrogen fuels, 396, 397
Hydrogen peroxide, 141, 142
Hydrogen sulfide, 41, 357
Hydrolytic bacteria, 150
Hydropower, 4

IIT Research Institute, 286
Illegal dumping, 346
Illite, 85
Incineration of hazardous waste, 267, 284, 285, 286, 287
Incineration of sludge, 162, 163, 164, 165
Indirect costs of environmental damage, 385, 387, 388, 389, 411, 412
Industrial ecology concept, 371
Industrial Revolution, 3, 4, 5, 6
Industrial waste characteristics, 180
Industrial waste treatment overview, 9, 10
Industry types
   agriculture, 27, 30, 452, 453, 472, 473, 474, 477, 478, 503, 504, 505, 506, 507, 508, 517, 518

briquette production, 420, 421, 422, 423, 424, 425, 426, 427, 429, 503
canning, 108, 124473, 474, 493, 494, 495, 496, 497, 498, 499
cement production, 381, 382, 383, 384, 385, 386, 387, 388, 389, 390, 391, 392, 393, 405, 406, 407, 410, 411, 412, 453, 454, 455, 456, 457, 458, 459, 460, 461, 466, 467, 468, 469, 470, 500, 501, 502, 503
chlorine production, 482, 483, 497, 498, 499
coke, 446, 447, 479, 480
dry-cleaning, 273, 490, 491
ethanol production, 152, 418, 419, 420, 481, 482, 485, 486, 499, 500, 509, 510
fertilizer, 261, 262, 263, 381, 382, 383, 384, 385, 386, 387, 388, 389, 390, 391, 392, 393, 405, 406, 407, 410, 411, 412, 420, 421, 422, 423, 424, 425, 426, 427, 429, 446, 447, 453506, 507, 508
metal finishing, 448, 449
metal plating, 26, 30, 90, 260, 299, 300, 301, 302
oil, 256, 258, 488, 489, 490
paint, 273, 486, 487
paper, 14, 16, 30, 79, 432, 434, 436, 437, 438, 439, 440, 444, 445, 446
plastics, 366, 448, 449, 461, 462, 463, 464, 465, 466, 508, 509
power generation, 262, 418, 419, 420, 452, 453, 454, 454, 455, 456, 457, 458, 459, 460, 461, 466, 467, 468, 469, 470, 472, 473, 482, 483, 484, 485, 496, 497, 506, 507, 508
rendering, 30, 413, 414, 417, 418
slaughterhouse, 30, 413, 414, 417, 418
steel mill, 4, 6, 14, 17, 18, 94, 446, 447, 453

sugar, 418, 419, 420, 421, 422, 423, 424, 425, 426, 427, 429, 499, 500, 503
tanning, 50, 74, 198, 199, 226, 227, 228, 229, 233, 234, 235, 236, 237, 238, 239, 240, 241, 243, 413, 414, 417, 418
textiles, 4, 6, 27, 28, 29, 48, 261, 289, 290, 304, 305, 509, 510
timber, 263, 444, 445, 446, 449, 450, 451, 452, 470, 471, 472, 481, 482
wax manufacturing, 492, 493
Inorganic dissolved solid removal, 89, 90, 91, 92, 93, 94, 95, 96, 97, 98, 99, 100, 101
Insecticides, 85, 100
Insurance, 344, 345, 346
Intangible costs of environmental damage, 385, 388, 389, 412
Ion exchange, 93, 94
Iron and steel, 4, 6, 26
Iron ore reduction, 26
Iron oxides, 82
Isopropanol, 290

Kanawha River, 50
Kaolinite, 85
Kayaderosseras Creek, 198
Kennison nozzle, 49
Kepone, 266
Kraus process, 111

Lagooning, 106, 107, 108, 109, 159, 233
Lake industry–villagers complex, 518, 519
Lake Tahoe, 18
Landfill hazardous waste, 275, 276, 293, 294, 295
Landfill leachate treatment, 313, 315
Landfill sludge disposal, 168, 169
Land treatment of hazardous waste, 281, 292, 293

Land value benefits of environmentally balanced industrial complexes, 534
Laundry wastewater recovery, 303, 304
Leachate treatment, 313, 315
Lime, 83, 84, 99
Lime slurry, 36, 37, 38
Lime-based fixation, 296
Lime–cement–power plant environmentally balanced industrial complexes, 466, 467, 468, 469, 470
Limestone, 36, 37, 159, 358
Liquid injection incinerator, 285
Lumber mill environmentally balanced industrial complexes, 470, 471, 472
Lumber–textile–corn growing–ethanol environmentally balanced industrial complex, 509, 510
Lye, water, electric power, and chlorine plant environmentally balanced industrial complex, 482, 483
Lyophilic colloid, 80, 81
Lyophobic colloid, 80, 81

Manufacturing wastes, 13
Marketing unused waste resources from environmentally balanced industrial complex, 393, 394, 395
Material-balance equation, 130
Meat packing, 30, 125
Mechanical aeration, 125, 126
Membrane separation of hazardous waste, 315, 316
Mercury, 259
Mercury recovery, 300, 301
Metal-containing sludge, 259, 260
Metal finishing–plastic plant environmentally balanced industrial complex, 448, 449
Metal plant–dry-cleaning–coffee decaffeination environmentally balanced industrial complex, 490, 491
Metal plating, 26, 30, 90, 260, 299, 300, 301, 302
Methane, 125, 474, 478
Methane bacteria, 150
Methane fermentation, 152

Methyl t-butyl ether (MTBE), 485
Methylisocyanate (MIC), 257, 258
Micro-strainer, 73, 74
Microfiltration (MF), 98, 99
Mineral removal, 89, 90, 91, 92, 93, 94, 95, 96, 97, 98, 99, 100, 101
Mixing wastes, 36, 37
Modified aeration, 112
Moench Tannery, 74
Molten salt process, 282, 283
Monitoring waste streams, 31
Montmorillonite, 85
Municipal effluent reuse, 16, 17
Municipal ordinances, 185, 186, 190
Municipal plant operator, 178, 179
Municipal solid wastes–industrial complexes, 515, 516, 517
Municipal waste treatment, 2, 176, 179, 180, 182, 183, 184, 185, 315
Municipal wastewater–industrial complexes, 517, 518, 519
Mycobacterium tuberculosis, 266

National Conference on Waste Exchange, 311, 312
National Institute for Occupational Health and Safety (NIOSH), 358
Naturally evolving industrial complexes, 527, 528, 529, 530
Navy hazardous waste, 251, 252
Neutralization, 28, 29, 35, 36, 37, 38, 39, 40, 41, 42, 43, 291, 306
Newton's draw law in laminar flow, 166
Nickel recovery, 30
Nitrates, 100
Nitric acid, 38
Nitrifying aeration tank, 111
Nitrogen organic waste, 274
Nitrogen oxides, 287
Nitrogenous fertilizer, 262, 263
North water filter, 72
NovoNordisk, 528
Nuclear power plant waste processing–cannery environmentally balanced industrial complex, 493, 494, 495, 496

Nuclear power plant–glass block environmentally balanced industrial complex, 475, 476, 477
Nuclear Waste Policy Act, 265

Occupational Safety and Health Administration (OSHA), 325, 347
Ocean dumping of hazardous waste, 296
Ocean wave energy, 402, 403
Oil drilling–seashore recreation environmentally balanced industrial complex, 488, 489, 490
Oil Recovery Systems, Inc., 307
Oil recovery, 302, 303
Oil refinery sludge, 256, 258
Ore extraction wastes, 262
Organic acid digestion, 151
Organic chemical–wood processing environmentally balanced industrial complex, 449, 450, 451, 452
Organic dissolved solid removal, 105
Organic polymer fixation, 297
Overflow rate (OFR), 54, 55, 136
Oxidation ditch, 132
Oxygen treatment of organic matter, 143, 144
Ozonation, 142, 143

Paint industry, 273
Paint manufacturing–restaurant environmentally balanced industrial complex, 486, 487
Paper mill, 14, 16, 30, 79
Paper mill and pulp environmentally balanced industrial complex, 432, 434, 436, 437, 438, 439, 440
Paper mill–wood environmentally balanced industrial complex, 444, 445, 446
Paper recycling, 365
Parchment membrane, 92
Patents, 5
PCBX process, 271, 272
Perchloroethylene (PCE), 349
Permanganate, 141
Petrochemical industry, 274

pH neutralization, 28, 29, 35, 36, 37, 38, 39, 40, 41, 42, 43, 291, 306
Phanerochaete chrysosporium, 308
Phenols, 272, 273, 358
Philips process, 465
Phosphate, 261, 262
Phosphochalk, 412
Phosphogypsum, 406, 412
Phosphoric acid pickling, 26
Photographic wastes, 260, 261, 301, 302
Photolysis, 143, 287
Photovoltaic (PV) cells, 399
Physical treatment of hazardous waste, 289
Pickle factory, 26, 29, 31, 261
Pickling cannery–inorganic chemical and chlorine plant environmentally balanced industrial complex, 497, 498, 499
Pickling liquors, 260
Plant food–animal feedlot environmentally balanced industrial complex, 477, 478
Plasma arc detoxification, 306, 307
Plasma technology, 282
Plastic plant environmentally balanced industrial complexes, 461, 462, 463, 464, 465, 466
Plastic plant–metal finishing environmentally balanced industrial complex, 448, 449
Plastic recycling, 366
Plastic waste–consumer products environmentally balanced industrial complex, 508, 509
Political compatibility of industry and municipality, 178
Pollution, 1, 2
Pollution trading credits, 2
Polychlorinated biphenyls (PCBs), 252, 267, 270, 271, 272, 286, 297, 298, 305, 306, 449, 450, 492, 493
Polyethylene, 462, 464, 465
Polystyrene, 462, 463
Poultry plant, 28
Power, 4, 6
Power companies, 395

Power plant–agriculture environmentally balanced industrial complexes, 472, 473
Power plant–aluminum–brick plant environmentally balanced industrial complex, 483, 484, 485
Power plant–cement–lime environmentally balanced industrial complexes, 466, 467, 468, 469, 470
Power plant–cow feedlot–fertilizer environmentally balanced industrial complex, 506, 507, 508
Power plant–drinking water plant environmentally balanced industrial complex, 496, 497
Power plant–sugarcane–alcohol environmentally balanced industrial complex, 418, 419, 420
Power plant wastes, 262
Power, water, chlorine, and lye plant environmentally balanced industrial complex, 482, 483
Precedent, 177
Pressure flotation, 67, 71
Price-Anderson Act, 265
Process changes, 25, 26
Products of incomplete combustion (PICs), 287
Proform, Inc., 349
Proportioning, 19, 31, 47, 48, 49, 50
Protein digestion, 151
Pseudomonas, 144
Public Service Electric and Gas Company, 265
Pulp and paper mill environmentally balanced industrial complex, 432, 434, 436, 437, 438, 439, 440
Pyrolysis, 281, 282

Radiation treatment, 494
Radioactive waste, 264, 265, 266, 476, 477, 493, 494, 495, 496
Rail accidents, 348
Railroad, 5
Rayon, 16

Reactive industrial waste, 257, 258
Receiving-stream water quality, 180, 181
Recirculated water, 14
Recovery and reuse of hazardous wastes, 298, 299, 300, 301, 302, 303, 304, 305
Recreation use benefits of environmentally balanced industrial complexes, 534
Recycling, 365, 366
Recycling Laboratories, 140, 141
Refractory removal, 99, 100
Regional exchange of hazardous waste, 310, 311, 312
Reliable Water Company, 98
Rendering plant, 30
Rendering–tannery–slaughterhouse environmentally balanced industrial complex, 413, 414, 417, 418
Resource Conservation Recovery Act (RCRA), 246, 260, 276, 322, 323, 326, 345, 455
Restaurant–paint manufacturing environmentally balanced industrial complex, 486, 487
Reuse, 2
Reuse, 15, 16, 17, 18
Reverse osmosis, 97, 98, 315
Reynold's number, 55
Rice hulls, 547
Rockwell International, 283
Rotary kiln incinerator, 285
Rotating biological contractor (RBC), 134, 144

Sales index, 540, 541, 542
Salmonella, 268
Salt disposal, 99
Salts, 261
Sanitary landfill, 168, 169
Sanitary wastewater, 14
Scenedesmus, 95
Schulze–Hardy Rule, 83
Scotscraft, Inc., 30
Screening, 72, 73, 74

Seawater demineralization, 95, 96
Sedimentation process unit design, 59, 63, 64, 65, 66
Sedimentation theory, 53
Segregation of waste, 27, 28, 277
Self-cementing fixation, 297
Sequencing batch reactor (SBR), 309
Settling phases, 64
Settling tanks, 54, 55, 56, 57, 58, 59, 60, 61, 62
Settling velocity, 64, 65
Sewage degradation, 106, 107
Sewage effluent, 15, 16, 17, 18, 19
Sewage plant, 30
Sewage–raw industrial waste joint treatment, 175, 363, 364, 365
Sewer service charge, 178
Sewer-rental charges, 190, 192, 193
Ship incineration, 284, 286
Short-circuiting, 57, 58
Silver recovery, 301, 302
Slaughterhouse, 30
Slaughterhouse–tannery–rendering environmentally balanced industrial complex, 413, 414, 417, 418
Sludge solid treatment and disposal, 149, 150, 151, 152, 153, 154, 155, 156, 157, 158, 159, 160, 161, 162, 163, 164, 165, 166, 167, 168, 169, 170, 171, 172, 219, 224, 226, 236
Sludge volume index, 111
Slug discharge, 19
Social relationship of industry and municipal officials, 177
Soda ash, 36, 498, 499
Sodium carbonate, 38
Sodium hydroxide, 39, 40, 91, 92
Solar energy power, 399, 400
Solar evaporation, 312
Solid waste definition, 246
Solvents, 273, 274
Solvents recovery, 303
Spent oil emulsions, 255, 256
Spent solvents, 273, 274

Sphaeroltilus, 111, 114
Spill prevention, 348, 349
Spray irrigation, 123, 124
Spreading, 132, 133
Stannous chloride, 304
Starch, 26
Statoil oil refinery, 528
Steam engine, 4
Steel mill, 4, 6, 14, 17, 18, 94
Steel millcoke and gas–fertilizer environmentally balanced industrial complex, 446, 447
Steel mill–fertilizer–cement environmentally balanced industrial complex, 453
Step aeration, 112
Sterling Drug Company, 161, 236
Stokes' law, 68, 166
Storage of hazardous wastes, 346, 347, 348
Stream survey, 199, 200, 204, 210, 211, 212, 213, 214
Stream water use benefits of environmentally balanced industrial complexes, 535
Streeter–Phelps method, 212
Submerged combustion, 41, 42
Subsurface disposal, 132, 133
Sugarcane–briquette environmentally balanced industrial complex, 503
Sugarcane–briquette–fertilizer environmentally balanced industrial complex, 420, 421, 422, 423, 424, 425, 426, 427, 429
Sugarcane–power–alcohol environmentally balanced industrial complex, 418, 419, 420
Sugar–ethanol–gasoline plant environmentally balanced industrial complex, 499, 500
Sulfide residues, 258
Sulfite recovery, 437, 438
Sulfite waste liquor, 30
Sulfur dioxide, 357, 458
Sulfur organic waste, 274

Sulfuric acid, 36, 38, 39, 40, 42, 43
Sun Oil Toledo refinery, 17
Superfund, 321, 322, 323, 325, 326, 327, 343, 344
Supersorbon process, 273
Suspended solid removal, 53, 54, 55, 56, 57, 68, 69, 60, 61, 62, 63, 64, 65, 66, 67, 68, 69, 70, 71, 72, 73, 74, 75, 76
Sweco separator, 72, 73
Symbiotic ecosystems, 529

Tagamet, 266
Tannery–slaughterhouse–rendering environmentally balanced industrial complex, 413, 414, 417, 418
Tanning industry, 50, 74, 198, 199, 226, 227, 228, 229, 233, 234, 235, 236, 237, 238, 239, 240, 241, 243
Tar–coke–benzol plant environmentally balanced industrial complex, 479, 480
Telegraph, 5
Telephone, 5
Tetrahydrofuran (THF), 290
Textile mill environmentally balanced industrial complexes, 429, 430, 431, 432
Textiles, 4, 6, 27, 28, 29, 48, 261, 289, 290, 304, 305
Textile–lumber–corn growing–ethanol environmentally balanced industrial complex, 509, 510
Thagard Research Corporation, 283
Thermal treatment of hazardous waste, 281
Thermatrix, Inc., 287, 288
Thermoplastic fixation, 297
Thickening sludge, 170
Titanium dioxide, 143
Tokyo Bay, 281
Total organic carbon (TOC), 309
Total oxidation, 116, 117
Total suspended solids (TSS), 63, 64
Toxic industrial wastes, 258, 259, 260, 261, 262, 263, 270, 271, 272, 273, 274

Transport of hazardous waste, 348, 349
Transport of sludge, 169
Trichloroethylene (TCE), 140, 349, 358, 490, 491
Trickling filtration, 117, 120, 121, 122, 123
Trihalomethanes, 258
Trinity River, 50
Turbulence, 55
Turpentine, 273
Tyndall effect, 81
Tyson Foods, 380

Ultrafiltration (UF), 98, 99
Union Carbide Corporation, 143

Vacuum filtration, 153, 154, 233
Vacuum flotation, 67
Vara International, 303
Velocity of particles in viscous fluids, 55
Voltage-converting–wax manufacturing environmentally balanced industrial complex, 492, 493
Volume reduction, 13, 20, 21

Waste boiler-flue gas, 40, 41
Waste classification, 13, 14
Waste-flow proportioning system, 49
Waste Management Company, 284
Waste mixing, 36, 37
Waste oils, 254, 255
Waste segregation, 27, 28
Waste stream monitoring, 31
Wastewater disposal benefits of environmentally balanced industrial complexes, 534
Water conservation, 14
Water effluent reuse, 15, 16, 17, 18
Water, electric power, chlorine, and lye plant Water Pollution Control Federation (WPCF), 185
Water quality limits, 328, 329, 330, 331, 332, 333
Water recovery from laundry wastewater, 303, 304

environmentally balanced industrial complex, 482, 483
Wax manufacturing, 492, 493
Well injection, 126, 127, 128, 129, 130, 276, 291, 292
Wet combustion, 124, 125
Wet combustion process, 159, 160, 161
Whey, 43
Wind power, 397, 398, 399
Withdrawal water benefits of environmentally balanced industrial complexes, 534
Wood–ethanol plant environmentally balanced industrial complex, 481, 482
Wood–paper mill environmentally balanced industrial complex, 444, 445, 446
Wood mill environmentally balanced industrial complexes, 470, 471, 472
Wood processing–organic chemical environmentally balanced industrial complex, 449, 450, 451, 452
Wood-preserving wastes, 263
World Bank, 380, 381

Yeast factory, 30, 125

Zeolites, 93
Zero Emissions Research Institute (ZERI), 530
Zeta potential, 82, 83
Zimpro process, 159, 160, 161, 236
Zinc oxide, 143
Zinc titanate, 143